The Ocular Lens

The Ocular Lens
Structure, Function, and Pathology

Edited by

Harry Maisel
Wayne State University
Detroit, Michigan

MARCEL DEKKER, INC.   New York and Basel

Library of Congress Cataloging in Publication Data

Main entry under title:

The Ocular lens.

Includes bibliographies and index.
1. Cataract. 2. Crystalline lens. I. Maisel,
Harry . [DNLM: 1. Cataract. 2. Lens,
Crystalline. WW 260 021]
RE451.038 1985   617.1'42   85-4408
ISBN 0-8247-7297-0

MARCEL DEKKER, INC.
270 Madison Avenue, New York, New York   10016

Current printing (last digit):
10 9 8 7 6 5 4 3 2 1

PRINTED IN THE UNITED STATES OF AMERICA

Preface

The lens as an organ represents a specialized and important function in the human visual process. To the scientific researcher, the lens represents a challenge to uncover the biological basis of its transparency, of its ability to refract light, of its capacity to remold as demanded by accommodation, and of the derangement in its functions in pathological processes. To the ophthalmologist, the lens represents a pathological source of human blindness, and one which until now has been dealt with the less-than-desirable functional limitations of surgical excision. The highly unique and specialized functions of the lens are evidenced by its unique genetic, developmental, morphological, physiological, metabolic, and photobiological realities. The purpose of this book is to assemble in one composite volume a comprehensive view of the progress in the accumulated body of knowledge of lens structure and function accomplished in the last decade, and which will, hopefully, serve both ophthalmologists and research scientists as a record with which to build an appropriate answer to the problem of cataracts.

Cataracts increasingly represent a leading cause of blindness throughout the world, and their prevention or delay in progression would represent a major achievement for human welfare. Recent progress in lens research has resulted in the institution of clinical trials in an attempt to achieve this goal. It is the editor's wish that this book serve as a further stimulus to both basic and clinical scientists to engage in more intensive efforts in these endeavors, and to both attract and orient new participants in the pursuit of the biological basis of lens structure, function, and pathology.

I wish to express my appreciation to all those who contributed to this book, and especially to my colleague Dr. Jose Alcala for his advice and support. Ms. Clarice Perazza deserves a special thanks for her secretarial support.

HARRY MAISEL

Contributors

Jose Alcala, Department of Anatomy, Wayne State University School of Medicine, Detroit, Michigan

Frederick A. Bettelheim, Chemistry Department, Adelphi University, Garden City, New York

Walter F. Bobrowski, Department of Opthalmology, Kresge Eye Institute, Wayne State University, Detroit, Michigan

Hong-Ming Cheng, Howe Laboratory of Ophthalmology, Harvard Medical School, Boston, Massachusetts

Leo T. Chylack, Jr., Howe Laboratory of Opthalmology, Harvard Medical School, Boston, Massachusetts*

Ruth M. Clayton, Institute of Animal Genetics, University of Edinburgh, Edinburgh, Scotland

James Dillon, Department of Opthalmology, Columbia University, New York, New York

Clifford V. Harding, Jr., Kresge Eye Institute, Wayne State University, Detroit, Michigan

Peter F. Lindley, Department of Crystallography, Birkbeck College, University of London, London, England

---

*Present affiliation: Brigham and Women's Hospital, and Massachusetts Eye and Ear Infirmary, Boston, Massachusetts

Woo-Kuen Lo, Department of Ophthalmology, Kresge Eye Institute, Wayne State University, Detroit, Michigan

Harry Maisel, Department of Anatomy, Wayne State University School of Medicine, Detroit, Michigan

Richard T. Mathias, Department of Physiology, Rush-Presbyterian-St. Luke's Medical Center, Chicago, Illinois

Michael E. Narebor, Department of Crystallography, Birkbeck College, University of London, London, England

James L. Rae, Department of Physiology, Rush-Presbyterian-St. Luke's Medical Center, Chicago, Illinois

Nancy S. Rafferty, Northwestern University, Medical School, Chicago, Illinois

Abraham Spector, College of Physicians and Surgeons, Columbia University, New York, New York

Lesley J. Summers, Department of Crystallography, Birkbeck College, University of London, London, England

Stanley R. Susan, Department of Ophthalmology, Kresge Eye Institute of Wayne State University, Detroit, Michigan

Graeme J. Wistow, Department of Crystallography, Birkbeck College, University of London, London, England*

Seymour Zigman, Ophthalmology Research Laboratory, University of Rochester, School of Medicine and Dentristy, Rochester, New York

---

*Present affiliation: Laboratory of Molecular and Developmental Biology, University of London, London, England

Contents

The Ocular Lens

1
Lens Morphology

NANCY S. RAFFERTY / Northwestern University, Chicago, Illinois

The lens of the eye has been referred to variously as an organ, a suborgan, or a tissue. Its sole function is to transmit and refract light reaching the eye on the retina. Its structure, which at first glance appears remarkably simple, has been shown by studies over a century and a half to be, instead, remarkably complex.

As an organ, the lens is unique in its derivation from one cell type; it is unique in its retention throughout the life of the animal of all the cells that are ever produced; it is unique in having no blood or nerve supply; it is unique in its clarity; it is unique in synthesizing unique proteins; and it is unique in that its function of focusing an image on the retina is a finely controlled process that is still incompletely understood.

This chapter considers changes in shape of the anterior face of the lens during accommodation and aging, and the relation of these changes to the development and normal structure of the lens. The last part of the chapter undertakes a description in more detail of the plasma membranes and the cytoskeleton of lens epithelial and fiber cells. These two structural systems are currently eliciting burgeoning interest among lens cell biologists in relation to specialized lens functions, such as cell-to-cell transport of metabolites, maintenance of lens clarity, and accommodation.

Regarding the electron micrographs presented in this chapter, considerable attention was given to selection of pictures of well-fixed normal human and animal lenses. Good fixation of lens material is important to be able to differentiate changes in ultrastructure resulting from aging or development of cataract from those due to improper fixation. This has classically been a difficult task. A fixative/buffer solution that produces artifact-free ultrastructure in one species of a given age, may not be optimal for the lenses at different ages or of different species. Furthermore, various fixatives may enhance the visualization of one organelle while destroying others. The lens tissue used to illustrate this chapter was fixed with the protocols indicated below and noted in the legends by numbers in brackets; most of these protocols have been derived empirically.

1. 1% OsO_4 in 0.1 M s-collidine buffer, pH 7.2, 4°C, 2 hr
2. 2% glutaraldehyde in 0.1 M PO_4 buffer, pH 7.2, room temperature, 1 hr, followed by 1% OsO_4 in 0.1 M PO_4, pH 7.2, 0°C, 2 hr
3. 2% glutaraldehyde in 0.1 M s-collidine buffer, pH 7.2, room temperature, 1 hr, followed by [1].
4. 2% glutaraldehyde, 0.05 M PO_4 buffer, 0.2% tannic acid, 0.002% $CaCl_2$, pH 7.2, 30 min to 1 hr, followed by 1% OsO_4 in 0.05 M PO_4 buffer, pH 7.2, 0°C, 2 hr
5. 4% glutaraldehyde, 0.1 M PO_4 buffer, pH 7.2, 4°C, 4 hr

GENERAL MORPHOLOGY

Lens Shape

Dimensions of the Human Lens

The human lens is usually described as a biconvex structure, with the anterior surface facing the cornea having a less spherical form than the posterior surface facing the vitreous. The human lens does not have a static shape, its shape changing rapidly, depending on the visual focus, and slowly with aging.

Considerable attention has been directed toward measuring the changes in lens dimensions with growth and with accommodation. The most valid measurements are those made in the living emmetropic eye using optical and ultrasonographic techniques.

Fincham (1) utilized the rare opportunity of measuring optically the lens dimensions in a 22-year-old patient whose iris had been lost as a result of a penetrating ocular injury at the nasal limbus, but whose eye function was otherwise normal. The lens equator was visible. The data obtained by Fincham (1) in this case of aniridia and in other young patients are shown in Table 1.

Brown (2–4) developed a slit-lamp/camera apparatus through which nearly the entire lens profile can be visualized; and he developed an artificial eye by which corneal refraction can be corrected in the lens measurements. His data are presented in Table 1.

Ultrasonography has been used to measure lens sagittal thickness in the living eye (Table 1), but this technique cannot be used to measure equatorial dimensions (5–7). As is seen in Table 1, ultrasonography generally gives smaller dimensions for lenses of the same age groups than those obtained optically.

After removal from the eye and from the forces exerted by the zonula fibers of the ciliary body and by the vitreous, the isolated human lens acquires slightly

Table 1 Data on Dimensions of the Human Lens Derived from the References Listed

Age (years)	Number of eyes	Equatorial diameter (mm)		Sagittal diameter (mm)		Method of measurement	Reference
		Unaccommodated	Accommodated	Unaccommodated	Accommodated		
12[a]	2	10	—	—	—		
18	2	—	—	3.7–3.8	—	Optic	1
22[a]	1	10.2	9.75–9.8	3.79	4.34		
11	1	—	—	3.95	4.31		
19	1	—	—	3.67	4.13	Optic	9
29	1	—	—	3.71	4.13		
45	1	—	—	4.66	4.83		
Male							
7–12	11	—	—	3.53 ± 0.15	—	Ultrasonography	5
13–19	45	—	—	3.57 ± 0.18	—		
20–29	23	—	—	3.63 ± 0.23	—		
34–47	3	—	—	4.17 ± 0.15	—		
Female							
7–12	17	—	—	3.64 ± 0.16	—		
13–19	31	—	—	3.60 ± 0.19	—	Ultrasonography	5
20–29	9	—	—	3.80 ± 0.11	—		
20–29	44	—	—	3.91 ± 0.31	—		
30–39	34	—	—	4.15 ± 0.28	—	Ultrasonography	11
40–49	37	—	—	4.33 ± 0.35	—		
50–60	44	—	—	4.66 ± 0.36	—		
20–29		—	8.67	—	—	Mechanically on	
30–39		—	8.96	—	—	extracted lens	
40–49		—	9.09	—	—	*in vitro*	8
50–59		—	9.44	—	—	1 hr. after	
60–69		—	9.49	—	—	death	
70–79		—	9.64	—	—		

[a]In eyes suffering aniridia.

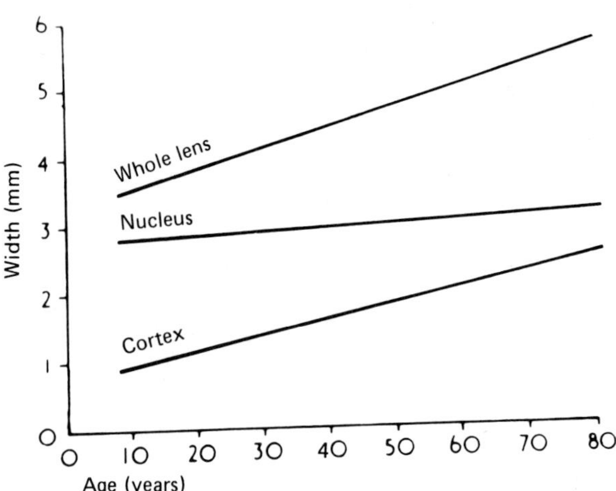

Figure 1 Growth of the normal human lens with age and of the cortex and nucleus separately. [Reprinted with permission from N. Brown (13) and the Ophthalmological Society of the United Kingdom.]

different dimensions from those measured in situ. These dimensions are shown in Table 1, as measured by Smith (8). The isolated lenses appear to have a smaller equatorial diameter than the accommodated lens in the living eye, measured by Fincham (1), which may result from the complete release of ciliary traction.

From the data shown in Table 1 and Fig. 1 (9), it is readily apparent that the human lens increases in size with each decade of life without measurable slowing. Most of the growth is accounted for by an increase in the cortex; the nucleus is nearly static (Fig. 1). This observation of continued lens growth throughout life has been confirmed in animal studies of the proliferating population in the pre-equatorial epithelium (10).

Several investigators have agreed that the human lens increases in thickness at a rate of about 0.02 mm/year (11–13). From Smith's data it appears that equatorial growth is also about 0.02 mm/year. The study of Weekers et al. (11), which includes a large number of cases in each age group, demonstrates the large variation in lens thickness between individuals of the same approximate age (Table 1). The data of Sorsby et al. (5) (Table 1), which are separated by sex, indicate that female lenses may be slightly thicker than male lenses, although Harding et al. (14) have shown that male lenses are nearly 8% heavier than female lenses.

Lens growth is also illustrated by the gradual increase in weight and in volume. In the first 3 months of life, the lens has a weight of about 93 mg and a volume

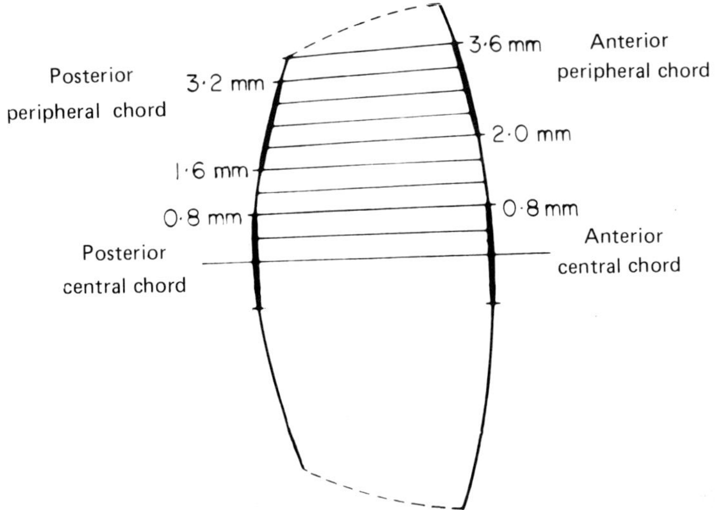

Figure 2 Scheme for measuring radii of curvature of central and peripheral chords. [Reprinted with permission from N. Brown (4) and Academic Press.]

of 90 mm^3; by age 90 the lens has a weight of 258 mg and a volume of nearly 239 mm^3 (8). The large increase in total lens weight is a reflection of an increase in wet weight and less so of an increase in dry weight (15).

Changes in Lens Shape with Aging

Changes in lens shape can be accurately determined from optical measurements of the radii of curvature of the anterior and posterior surfaces (1). Helmholtz (16) measured the anterior radius of curvature of the unaccommodated emmetropic human lens to be about 10 mm, but Fincham (1) found a variable value ranging from 8.4 to 13.8 mm with 11 mm about average. The posterior surface was recorded by Fincham as having a radius of curvature of between 4.6 and 7.5 mm. These values are, of course, age related.

Brown (4) standardized measurements of the radii of curvature of chords taken through the lens at precise distances from the central chord (Fig. 2). Using individuals over a wide range of ages, he showed that, in unaccommodated emmetropic lenses, a linear reduction in the radius of curvature occurs with age (Fig. 3). In the 8-year-old child the radius of curvature of the anterior central chord is 15.98 mm, and in the 82-year-old adult it is only 8.26 mm. The radii of curvature of the anterior peripheral chords show a similar decline, whereas those of the posterior central and peripheral chords decrease much slower. These data were utilized by Brown (4) to construct an outline of the shape of the lens and

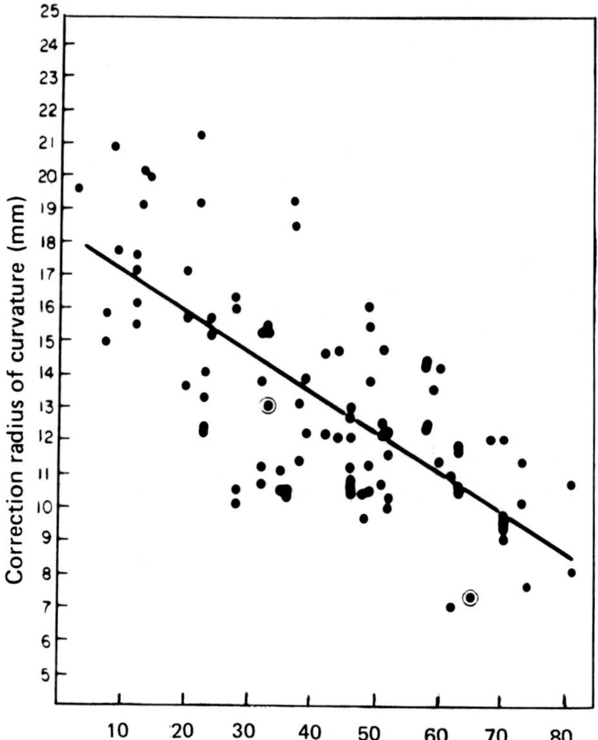

Figure 3 The change in radius of curvature of the central chord of the anterior lens surface with age. [Reprinted with permission from N. Brown (4) and Academic Press.]

of the nucleus of the young and old lenses (Fig. 4). Thus, while the cortex of the lens increases in sagittal thickness with age, giving the lens a more spherical shape, the aging nucleus remains constant in sagittal width and becomes reduced in equatorial diameter.

Changes in Lens Shape with Accommodation

To increase the refractive power of the lens for focus of objects closer than 20 ft, either the anterior surface of the lens must take on a more spherical shape or else the whole lens must translate forward away from the retina. Human and other primate eyes have adopted the former means of near-point focus (1,16,17).

With accommodation of the young adult human lens, the anterior surface moves forward toward the cornea and its central area takes on a more conoidal form than exists in the unaccommodated state (1,9,18). Fisher (19) describes

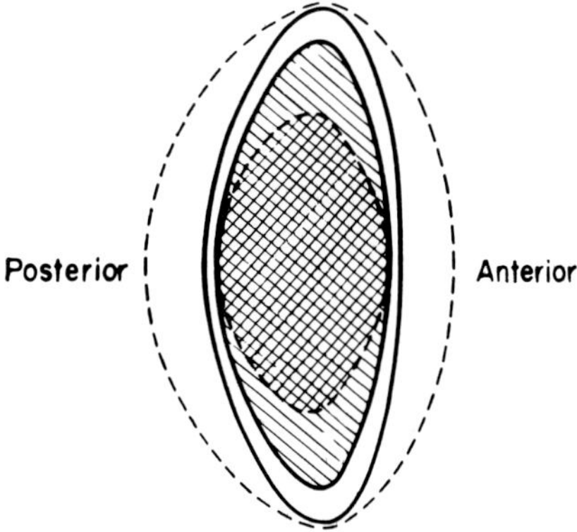

Figure 4 Comparison between the shape of a 10-year-old lens (continuous line) and nucleus (single-hatching), and that of an 82-year-old lens (broken line) and nucleus (cross-hatching). [Reprinted with permission from N. Brown (4) and Academic Press.]

the accommodated form of the anterior surface as ellipsoid, and calculated that the molding force of the lens capsule and ciliary muscle is not sufficient to produce a conoid shape. The posterior surface of the lens moves backward, but to a much smaller extent than the anterior surface. Thus, the central part of the lens thickens (Table 1). The change in shape of the unaccommodated and accommodated lenses of the young adult are shown in Fig. 5 (9).

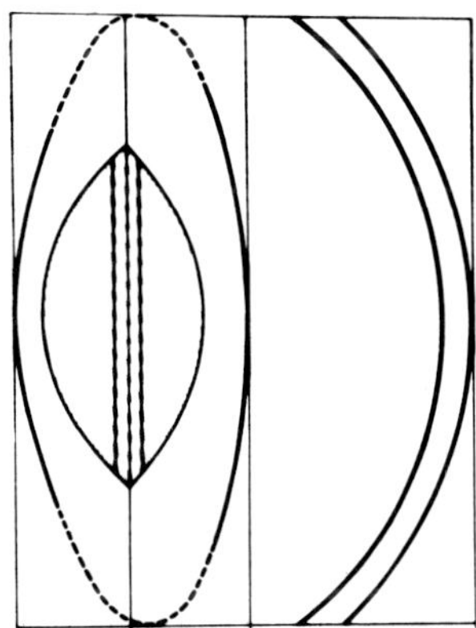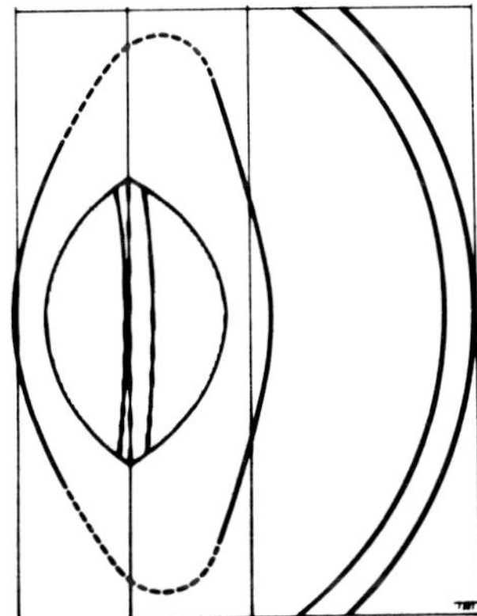

Figure 5 Change in shape of the whole lens (by extrapolation) in a 29-year-old at 0 D (left) and at 10 D (right) of accommodation. [Reprinted with permission from N. Brown (9) and Academic Press.]

The changes in shape of the lens with accommodation are reflected by changes in the radii of curvature, which become reduced the greater the amplitude of accommodation, for example, from 12 to 5 mm anteriorly over a 9 diopter (D) range in a 19 year old, and from 5.74 to 4.87 mm posteriorly (1). Brown (9) has reported similar figures, and, also, data on the decline in accommodative amplitude with age, when the radius of curvature of the anterior central chord changes only from 9.7 to 7.6 mm over the maximum 4 D of accommodation in a 45-year-old individual. Loss of accommodation with age of the emmetropic eye was plotted by Duane (20) and by Walls (17) and is seen to decline to a value approaching zero at age 70. However, using a cleverly designed apparatus to radially stress lenses of various ages, Fisher (21) found that the maximum amplitude of accommodation was reduced to near zero by age 57.

Lens Shape and Accommodation in Other Species

The shape of the lens of a wide variety of animals was studied by Rabl (22,23) and Kahmann (24), and the subject has been reviewed more recently in the books by Walls (17) and Duke-Elder (25) and in review articles by Hughes (26) and Sivak (27). These authors correlate the large lens relative to the volume of the eye of lower mammals with reduced or absent accommodative amplitude. The dimensions of the lenses of species frequently used in experimental studies on lens are given in Table 2, along with their amplitude of accommodation. The greater sagittal thickness of these animal lenses in relation to eye size and consequent low stored capsular energy (19) is supposed to preclude accommodative changes in lens shape, but in fact, the lenses with a flattened anterior surface have been shown to have 1 or 2 D of accommodation, for example, the gray squirrel (24,25) and the ground squirrel (Fig. 6) (28). By contrast, the nearly spherical lens of the mouse is agreed to have negligible accommodative amplitude and that of the laboratory rat and rabbit, minimal (<1 D). Nonhuman primates have lenses similar in shape and accommodative amplitude to that of the human. The monkey, *Cynomolgus* or *Rhesus*, has been shown to have at least 10 D of accommodation (29).

Birds have been shown to have extremely large accommodative amplitudes (25), especially birds that must use their vision under water to obtain food, such as diving ducks (30,31). In diving ducks, the accommodated lens has an anterior lenticonus, which has never been observed in a mammalian accommodated lens (Fig. 7). It has been suggested that this extreme anterior bulging, assisted by the tightening iris sphincter, compensates for the loss of corneal refraction in a water environment.

Development

Development of the human lens begins early in week 5 of gestation when an outgrowth of the forebrain, the optic vessicle, contacts the surface ectoderm on

Table 2 Dimension of Animal Lenses

Animal	Diameters (mm)		Equatorial/sagittal Index	Accommodation	References
	Equatorial	Sagittal			
Horse	20.14	12.28	1.64	—	22
Pig	9.50–10.00	7.37–7.62	1.28–1.31	—	22
Sheep	15.53–15.73	12.04–12.44	1.26–1.28	—	22
Cow	18.65–19.55	14.38–14.76	1.29–1.33	—	22
Rabbit	11.89	8.79	1.35	0–4	24, 26
Guinea pig	5.53	3.84	1.44	0–2	24
Rat	4.2	3.5	1.20	0.0	24
Mouse	2.49	2.00	1.24	0.0	17
Squirrel	4.87–5.53	3.40–4.23	1.29–1.43	1–2	17, 24
Dog	10.60–10.92	7.13–7.69	1.42–1.48	2–3.5	24, 26
Cat	10.56–13.25	7.50–8.93	1.40–1.48	1.5–3.0	24
Monkey (*Macacus*)	7.72	4.67	1.65	10–12	17, 24, 26
Chicken	5.69	3.63	1.56		23

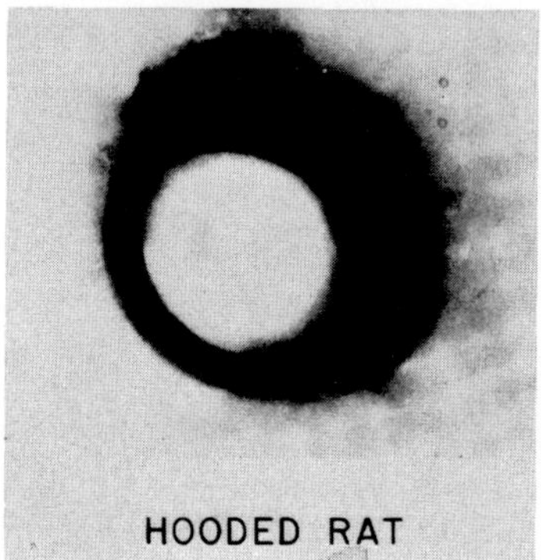

Figure 6 Sagittal sections of the eye of a hooded rat and of a ground squirrel (not to scale) emphasizing the difference in relative size, shape, and position of the lens. [Reprinted with permission from J. Sivak (28) and the William and Wilkins Company.]

each side of the head fold. In week 6, the optic vesicle induces the adjacent ectoderm cells to elongate and to undergo a series of shape and biochemical modifications that result, first, in the formation of a thickened lens placode and then in a lens vesicle pinched off from the invaginating surface (Fig. 8). The inductive interactions are exceedingly complicated. This fascinating area has been reviewed recently by McAvoy (32) and is covered in Chap. 2.

As the lens vesicle breaks away from the adjacent ectoderm, cells in the deep (posterior) part of the vesicle elongate and obliterate the cavity of the vesicle by the end of week 6 (33). The elongating cells form the primary lens fibers, whose

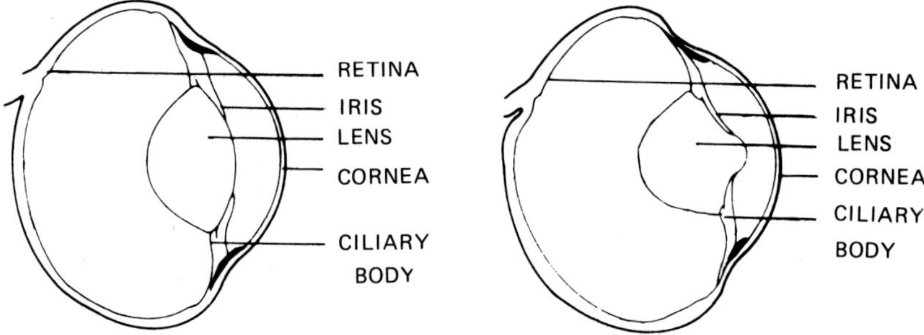

Figure 7 Tracing of sagittal sections of the eyes of hooded mergansers (diving ducks) with accommodation relaxed (top) and with accommodation induced with nicotine sulfate (bottom). [Reprinted with permission from J. Sivak (30) and Elsevier Biomedical Press.]

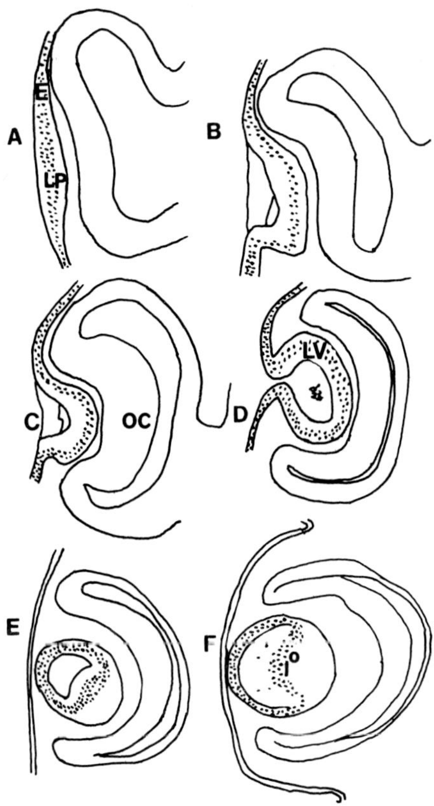

Figure 8 Sketches drawn from many sources of sagittal sections through the developing lens placode (LP), lens vesicle (LV), and lens with 1° lens fibers (1°); head ectoderm (E); optic cup (OC).

bases lie within the invaginating optic vesicle or cup. Although the anterior cells of the lens vesicle continue to divide, the elongating posterior fibers gradually cease cell division. McAvoy (32) divides the lens into two functionally distinct compartments, a proliferation and an elongation compartment, which are related to their position in the optic cup. Thus, the optic cup is believed to provide an elongation environment. The anterior cells, outside the margins of the optic cup, do not elongate and retain the ability to divide; those within the optic cup elongate and lose their ability to divide.

It should be mentioned at this point that, beginning in week 6, the developing lens becomes surrounded by a network of fine vessels, the tunica vasculosa lentis, which is connected proximally to the hyaloid (central) artery of the retina

(34–37). The function of these vessels may be to provide the lens with a mechanism for nutritional exchange before the anterior chamber and aqueous humor are operational. Whether these arteries contribute to development of the lens capsule and zonules is discussed later. The tunica vasculosa lentis begins to disappear in gestational month 4 and is normally absent at birth.

Throughout life, the anterior cells that form the single cell layer of lens epithelium retain their division potential. However, cell division in the anterior central area is infrequent in most adult lenses under normal conditions, but division can be activated by chemical (38) or mechanical injury (39–41). With time in development, the proliferating cells of the lens epithelium become largely confined to the pre-equatorial area called the proliferative or germinative zone (42–44). Cells produced in this region move posteriorly toward the lens equator, where morphologic and biochemical differentiation occur and secondary fibers are formed. Movement has been confirmed by autoradiography following injection of radioactive thymidine, which labels cells in the mitotic cycle in the proliferative zone (42–45). There are some data to suggest that the majority of cells in the proliferative zone usually undergo one and no more than two division cycles before moving into the differentiating zone at the equator (46). It is not known whether one or both progeny of a cell division move together or in alternate cycles, but it is believed that a terminal division is necessary for differentiation to take place.

As the epithelial cells begin their movement toward the equator, their axes rotate $90°$. During elongation, the apical end of the cell is insinuated anteriorly between the overlying epithelial cells and the underlying cortical lens fibers. However, it has recently been shown by Kuszak and Rae (47) that the anterior end of each elongating fiber is flared and makes contact with several epithelial cells, and thus there is no single initial fiber layer beneath the epithelium (Fig. 9A). The basal end of the cell moves posteriorly between the overlying capsule and the underlying cortical lens fibers. As new cells elongate, they are added superficially to the next most recent fibers, assuming their characteristic hexagonal shape from the molding of adjacent fibers. The lens fibers thus are pushed inward by the continual laying down of new fibers. The nuclei of the young cortical fibers describe what is called "the bow" in histologic sections (Fig. 9B), until the nuclei eventually disintegrate. The ends of the fibers terminate in a suture, which can take the form of a single linear junction between the ends of lens fibers arriving from different locations in the lens or the form of a complex branched junction. Kuszak and Rae (47) have shown that, in the frog lens, the elongating bow cells grow initially faster in the posterior direction but anterior-directed growth soon begins and proceeds more rapidly. This point is illustrated by the fact that the first fiber to reach the anterior suture has not reached the posterior suture but is still attached at its posterior end to the capsule and that the first fiber to reach both anterior and

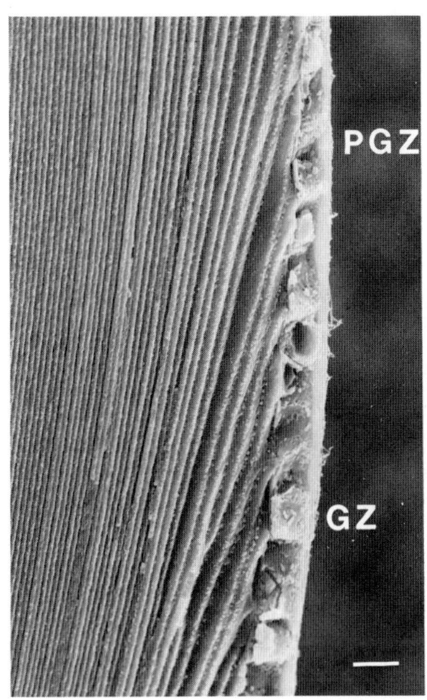

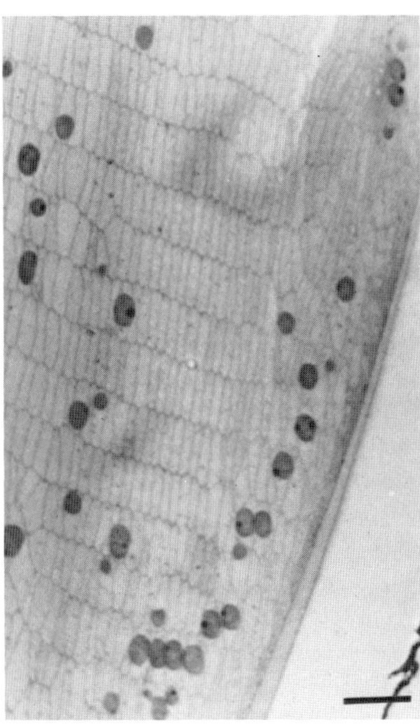

A   B

Figure 9 (A) Scanning electron micrograph of a broad view of an adult frog lens periphery showing pregerminative (PGZ) and germinative zone (GZ) cells overlying the fiber mass (×600); bar = 10 μm. (Courtesy of J. Kuszak.) (B) Light micrograph of a frontal section through the equatorial bow region of a human 10-day-old lens [3]. Methylene blue Azure II stain (×512); bar = 10 μm.

posterior sutures is the thirtieth fiber beneath the anterior epithelium. Since lens fibers are not discarded but are continually pressed inward by ever-younger additions, the entire life history of the lens is conserved in the organ.

Lens development from a sheet of surface ectoderm involves numerous morphogenetic movements and cellular shape changes. The mechanisms are not understood, although much experimental work has been and continues to be performed. Regarding the elongation of the ectodermal cells of the lens placode, Byers and Porter (48) and Porte et al. (49) observed in chicks the marked development of long straight microtubules in the cortical region of the presumptive lens cells just prior to elongation. With ensuing cell elongation, the microtubules are aligned with the long axis of the cell, especially in the apical end. Other cell organelles move away from the apical end, and this end becomes smaller in diameter than the basal end. Invagination would seem to occur as a natural outcome of the change in shape of the cells (50). A second surge of microtubule formation was observed to accompany elongation of the primary lens fibers in the posterior lens vesicle; these disappear when the primary fibers obliterate the lens vesicle cavity (48). However, microtubules persist in the elongating secondary lens fibers differentiating at the equator (48,51).

Although microtubules, perhaps as part of a cytoskeletal framework, may play a role in embryonic lens cell elongation, Pearce and Zwaan (52) showed that chick lens placode cells maintain their elongated shape after the microtubules are destroyed. Furthermore, Piatigorsky (53) found no increase in rate of tubulin (subunit of microtubules) synthesis during chick lens cell elongation in vitro. More recently, Beebe et al. (54) showed that elongation of secondary lens fibers in the chick can proceed without the presence of microtubules. Beebe et al. (55) propose that cell elongation results from an increase in cell volume in conjunction with the elasticity of the lens capsule.

Other mechanisms involved in lens morphogenesis have been proposed. Wrenn and Wessells (56) showed the presence of bundles of microfilaments in the apical ends of presumptive lens cells aligned such that contraction of these filaments would narrow the cell apex, as would a purse string, allowing for invagination of the sheet of cells. However, Maloney and Wakely (56a) tested this idea by incubating embryonic chick heads in an ATP-contraction medium and noted that, although the treatment had some effect on optic vesicle invagination, it had no effect on either the presumptive lens ectoderm or on the lens placode. Another mechanism proposed by Zwann and Hendrix (57) involves the elongation of cells resulting from population pressure of the rapidly proliferating placode cells. Rapid proliferation, coupled to the limiting of spreading of the organ laterally, which results from increased adhesion of the lens to the optic vesicle, would cause the invagination of the lens placode.

Morphology of the Lens Capsule

A noncellular capsule surrounds the vertebrate lens. The capsule of the young adult human lens measures about 13 μm in thickness anteriorly (58) and about 4 μm posteriorly (59). In histologic section the capsule appears homogeneous. It stains positively with periodic acid-Schiff, indicating the presence of a sulfated mucopolysaccharide component (60), and it dissolves in collagenase, indicating a collagen component. The capsule is considered a replicated basal lamina, but it is called a basement membrane because it is visible with the light microscope.

With the electron microscope the lens capsule is relatively uniformly fibrillar throughout its major part but has been described as having a denser outer layer, the zonular lamella (Fig. 10), which has different staining properties from the rest of the capsule in histologic section (60). It is now generally believed that this outermost layer is composed of a mixture of capsular collagen filaments and

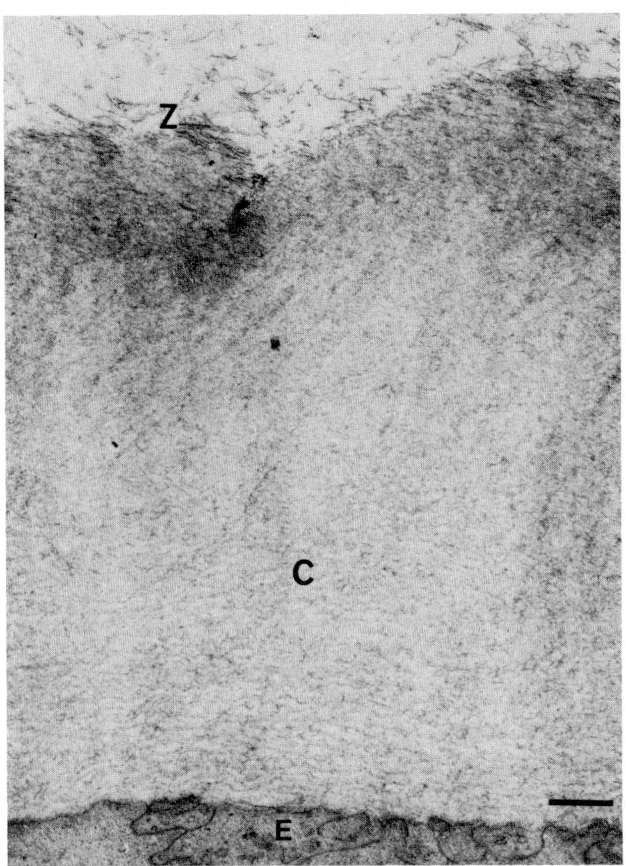

Figure 10 Transmission electron micrograph of equatorial lens capsule (C) of a juvenile squirrel showing collagen lamellae near the epithelium (E) and a denser outer layer, the zonular lamella. Zonules (Z) can be observed entering the capsular surface obliquely [1] (×8075); bar = 1 μm.

zonular elastic microfibrils (61,61a). The latter run tangentially to the lens surface at their attachment into the capsule (Figs. 10 and 11) (59,62–66). Although most of the zonula fibrils penetrate the human lens capsule for only a short distance, 0.5–1 μm deep (59,64), bundles of fibrils occasionally can be observed in pockets at the epithelial surface (62,67). The innermost layer of the capsule adjacent to the epithelium is clear and is continuous with the clear intercellular space of subjacent epithelial cells (59).

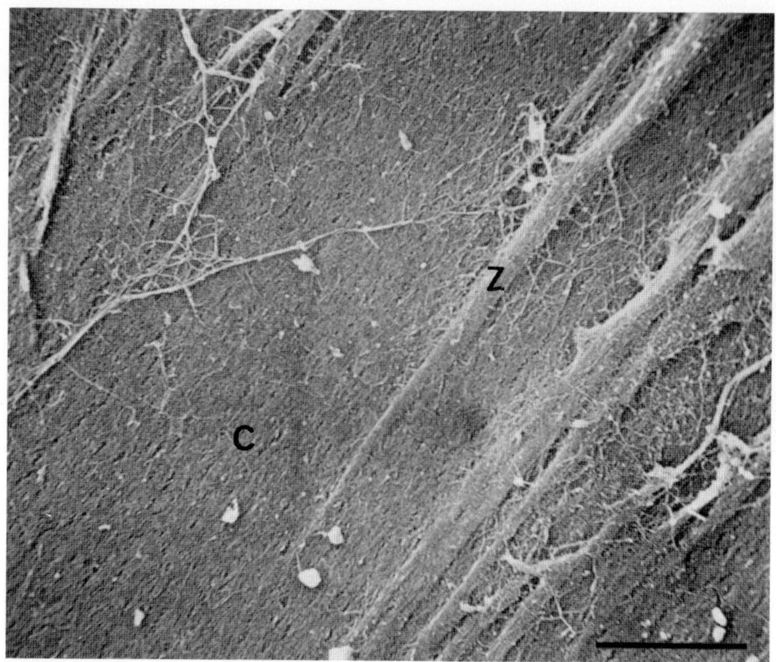

Figure 11 Scanning electron micrograph of human lens capsule (C) showing insertion of zonules (Z) into anterior surface [3] (× 2000); bar = 10 μm.

A recent experiment showed that, although the outer surface of the capsule is interwoven with zonula fibers, thereby introducing a different biochemical substrate, detached epithelial cells from bovine lenses reattach and spread equally readily to either the inner or the outer surface of the anterior or posterior capsule (68).

Studies in both mouse (34,69) and human embryos (70) have shown that the capsule forms from the original basal lamina of the epithelial cells of the lens placode and vesicle. Although Cohen (34,69) points out that the basal lamina of the mouse lens vesicle appears to fuse with the basal lamina of the surrounding vessels of the tunica vasculosa lentis, the lens basal lamina is formed about 4 days earlier than that of the capillaries adjoining the lens; therefore, the lens basal lamina cannot be derived from the capillary endothelium. However, the endothelium of these vessels has been implicated in the production of zonula fibers (71).

Wulle and Lerche (70,72,73) noted that the basal lamina in the human embryo lens early in gestational week 6 is a single thin lamella anteriorly, but at

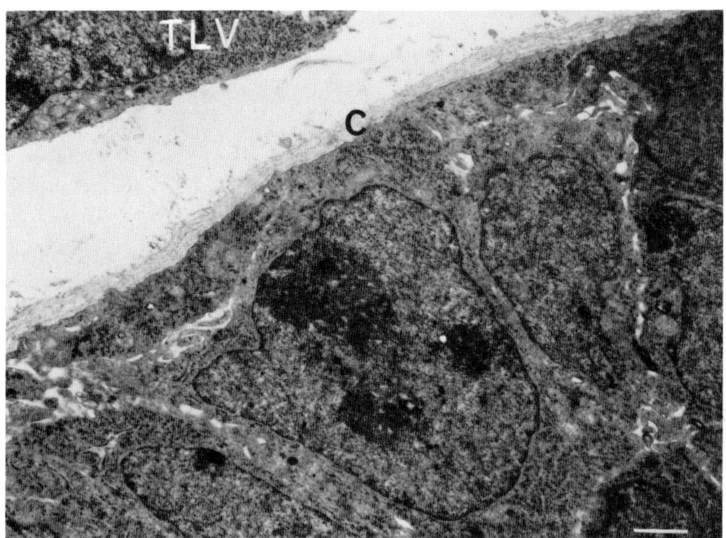

Figure 12 Transmission electron micrograph of the anterior epithelium of a day 7 embryonic mouse lens showing the developing lens capsule (C), which at this stage has about six lamellae. Tunica vasculosa lentis (TVL) is adjacent [2] (×6750); bar = 1 μm.

the equator and posteriorly, it is double or triple. By the end of week 6, the anterior capsule consists of two to three lamellae, each 6 nm in thickness, with the outermost discontinuous. Early in week 8 the anterior capsule has four to five lamellae, each 15 nm thick and separated by electron-lucent clefts. By the end of week 8, the number of lamellae cannot be readily separated from one another. The mouse lens capsule has been shown to develop in the same manner (Fig. 12) (69).

Both Cohen (69) and Wulle and Lerche (70) concluded that the capsule in the embryo increases in thickness by deposition of new lamellae on the outside of older lamellae. However, Young and Ocumpaugh (74) and Rafferty and Goossens (75), using young and old rats, and young mice, respectively, showed autoradiographically that radioactively labeled precursors incorporated into new capsule move from the epithelium and through the capsule; they are added at the inner capsular surface and are pushed through the capsule by subsequently synthesized unlabeled precursors (Fig. 13A). After a double application of [^{3}H]proline 4 weeks apart, and followed for intervals up to 9 months, a double band of label was seen to progress through the capsule (Fig. 13B) but stopped short of the capsular surface (75). These findings suggest that the capsule is

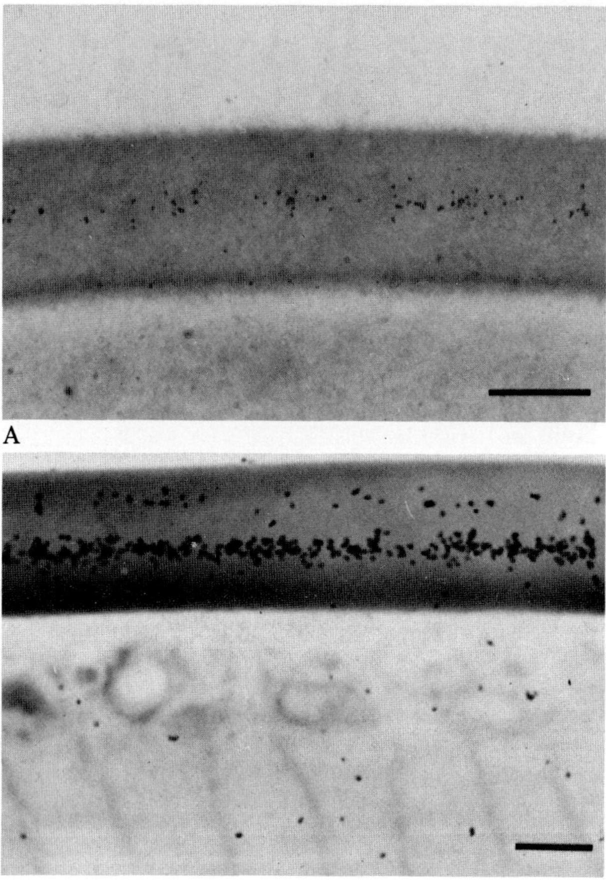

B

Figure 13 (A) Autoradiograph of anterior lens capsule of a mouse that had received a single injection of [³H]proline 20 weeks earlier. A band of blackened grains of the photographic emulsion indicates that the radioactive capsular precursor has moved about halfway through the capsule in this time. The animal was 8 weeks old when the precursor was given. Methylene blue Azure II stain (× 833); bar = 10 μm). (B) Autoradiograph of anterior lens capsule of a mouse that received two injections of [³H]proline, the first (outer band) 12 weeks previously and the second (inner dense band) 8 weeks previously. The animal was 4 weeks old when the first precursor was given (× 833); bar = 10 μm).

deposited in lamellar fashion at the inner surface and that mixing of the newly formed collagen with collagen formed earlier does not occur. Furthermore, these data indicate there is neglible turnover of capsular collagen.

The lens capsule continues to grow throughout most of life, growing to accommodate the increasing volume of the lens and growing in thickness anteriorly. It has been shown that the maximum capsular thickness is at the equator until later in life, when the capsule underlying the anterior zonule attachment assumes the greatest thickness (76,77). Fisher and Pettet (58) and Salzmann (76) have also shown that the capsule under the posterior attachment of the zonule and at the posterior pole does not increase in thickness with age. Seland (59) confirmed lack of posterior capsular growth, with the thickness 3.5 μm at birth and 4.0 μm at 25 years. In the autoradiographic study with [^{3}H]proline given to young mice (75), the posterior capsule was found to cease growing postnatally at 4 months of age, but the anterior capsule grew in thickness for 8 months. Since the postequatorial surface of the lens does not have an epithelium, the mechanism by which capsule is laid down posteriorly is not known. It has been suggested (59) that synthesis of the capsule may be accomplished by young posterior lens fibers, but this idea has not been tested experimentally. However, Fisher and Wakely (77) found that the posterior capsule of 2-month-old rabbits, subjected to capsulotomy, still retained a patent hole after 3 years, as large as at the time of injury. On the other hand, anterior capsulotomy results in rapid repair of the capsule in experimental animals (77–80).

There is some evidence that the capsular material is assembled extracellularly from precursors in the epithelial cells, which reach the cell surface by diffusion or are transported in small membrane-bound vesicles. Von Sallmann et al. (81) noted granules with the staining characteristics of capsule first within the cytoplasm of cultured lens epithelial cells and, a week later, within strands of material extracellularly.

It is known that basement membranes are composed of a collagen protein and one or more noncollagenous glycoproteins (82–86). The collagen component has been shown by Grant et al. (84) to contain larger quantities of hydroxyproline and hydroxylysine than interstitial collagens. Furthermore, these workers (85) found that the precursor form of collagen synthesized by whole lens or lens cells in culture consists of polypeptide chains with a considerably higher molecular weight (140 kd) than that isolated from interstitial connective tissue. These authors also found that the precursor form is the molecule that is secreted into the medium in vitro. Its conversion into the α_1 chain form and its assembly into type IV collagen microfibrils occurs extracellularly.

Fisher and Wakely (77), using high-magnification electron microscopy of undegraded fresh capsule flakes or sonicated fragments, resolved the material into sheets of parallel filaments with a periodicity of 4.1 nm along their length and little tendency to cross over and form nets. Optical transform, furthermore,

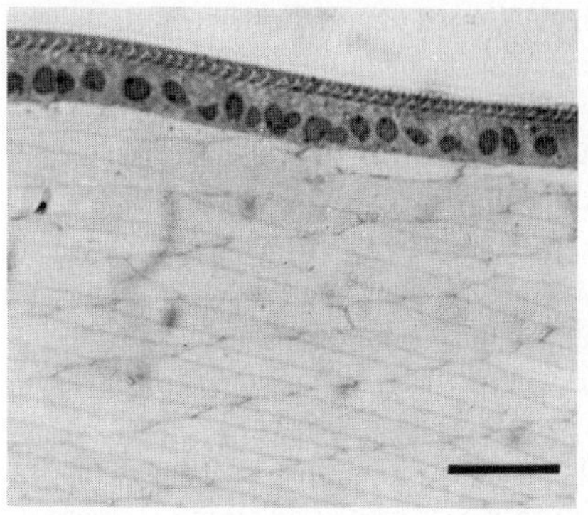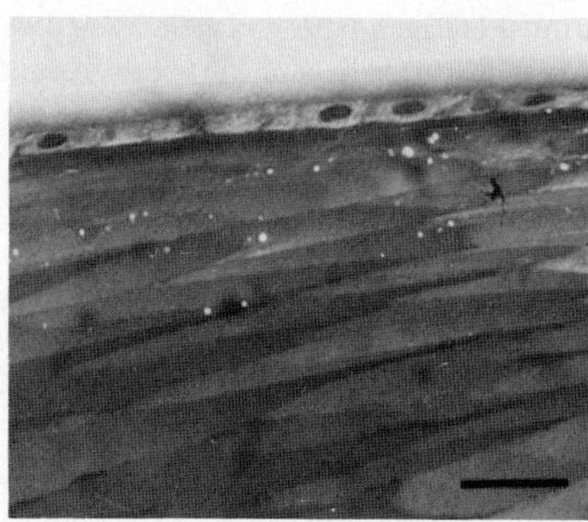

A   B

Figure 14 (A) Light micrograph of the anterior epithelium of a 10-day-old
human lens showing the high cuboidal epithelial cells. The striated appearance of
the lens capsule is a sectioning artifact [3]. Methylene blue Azure II stain
($\times$1031; bar = 10 μm. (B) Light micrograph of the peripheral epithelium of a
72-year-old human lens showing flatter cuboidal epithelial cells [3] ($\times$1031);
bar = 10 μm.

revealed that the filaments are helical with a tilt angle of 53°. These observations, in conjunction with those on the elastic constants of the lens capsule, lead
these authors to the conclusion that the basement membrane elasticity depends
on the extension of collagen filament coiled superhelices and not on the extension of a mesh of chains.

Schwartz and Veis (87,88) have extracted bovine lens capsule collagen by
mild pepsin digestion and dialysis against ATP. With electron microscopy the
solubilized molecules have a rodlike triple helical cross-banded portion measuring about 300 nm and a large globular noncollagenous region at each end. These
segment long-spacing crystallites tend to aggregate in end-interacted strings or
nets. Whether the native capsular collagen molecules are organized in networks
without preferred orientation remains to be seen but is contradicted by the work
of Fisher and Wakely (77). Schwartz and Veis (87,88) suggest that the globular
noncollagenous ends may be either procollagen extensions or covalently associated glycoproteins. Noncollagenous glycoproteins exist in lens capsule in a
complex and poorly understood relation to capsular collagen. One predominant
polypeptide (along with three of lower molecular weight) separated by Dische
and Murty (86) has a molecular weight of 35-40 kd. The posterior capsule has
twice the ratio of the concentration of glycoproteins to the concentration of
collagen than the anterior capsule, leading these authors to suggest that the
glycoprotein is related to the lens epithelial surface and not to the mass of the
lens capsule. In other words, the glycoproteins may lie anteriorly between the

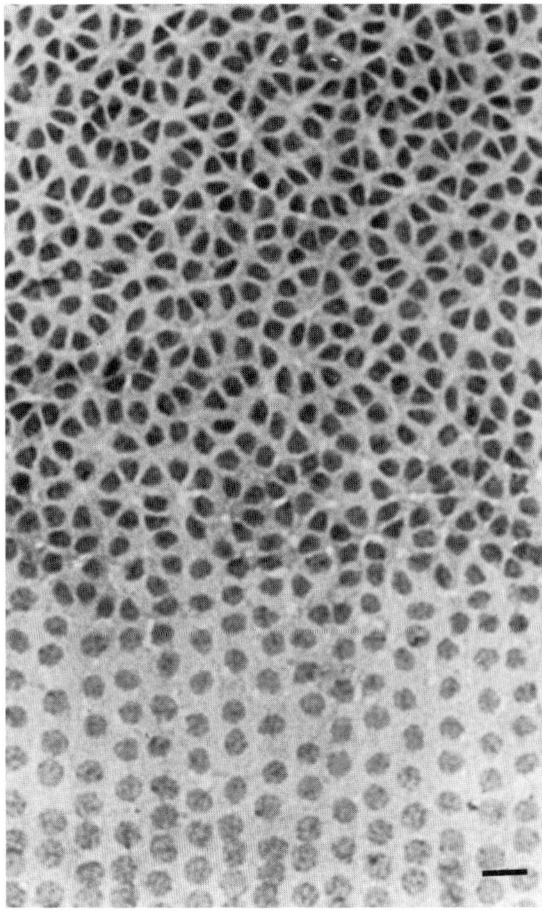

Figure 15 Flat-mount preparation of a portion of the epithelium in the proliferative and meridional row region of an adult squirrel lens. The cells appear to be arranged in rosettes, but each cell is polygonal. This arrangement has not been seen before in the lens epithelium of other species. Delafield's hematoxylin stain ($\times$ 550); bar = 10 μm.

capsule and epithelium and posteriorly between the capsule and lens fibers. This suggestion fits the observation of a nonfibrillar clear area in these locations (see above).

Morphology of the Lens Epithelium

A single layer of epithelial cells underlies the lens capsule on the anterior surface of the lens. Except in the equatorial region, where the epithelial cells undergo

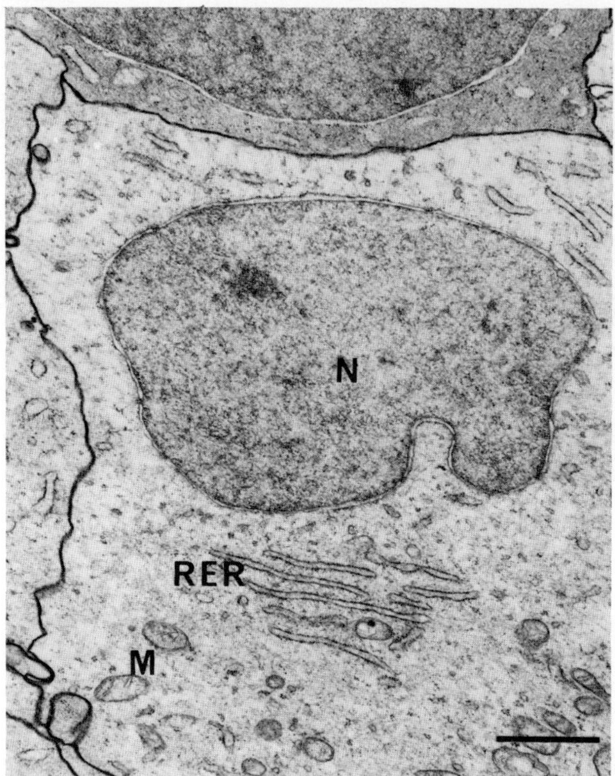

Figure 16 Peripheral epithelial cells of a day 17 chick embryo showing complex
lateral interdigitations of plasma membranes [4] ; N, nucleus; RER, rough endoplasmic reticulum; M, mitochondria ($\times$ 13,440, TEM); bar = 1 μm.

differentiation and elongation, all other epithelial cells in the pre-equatorial and
central areas have a roughly cuboidal shape, as seen in section with light microscopy (Fig. 14A). However, in older human lenses the epithelial cells are more
flattened than in younger lenses (Fig. 14B). In flat mount preparations of the
whole epithelium, the cells appear roughly hexagonal (Fig. 15).

With electron microscopy the so-called cuboidal cells are shown, in fact, to
have complex interdigitations with each other laterally (Fig. 16). The interdigitations increase in number between cells in the pre-equatorial region (47,89). The
basal cell surface underlying the capsule and the apical cell surface adjacent to
the lens fibers are smoother; that is, they do not interdigitate with the capsule
and only in a minor way with the lens fiber surfaces. The complex lateral interdigitations are present in human lens (61,78,90–92) and all animal lenses studied
(34,35,79,93-96).

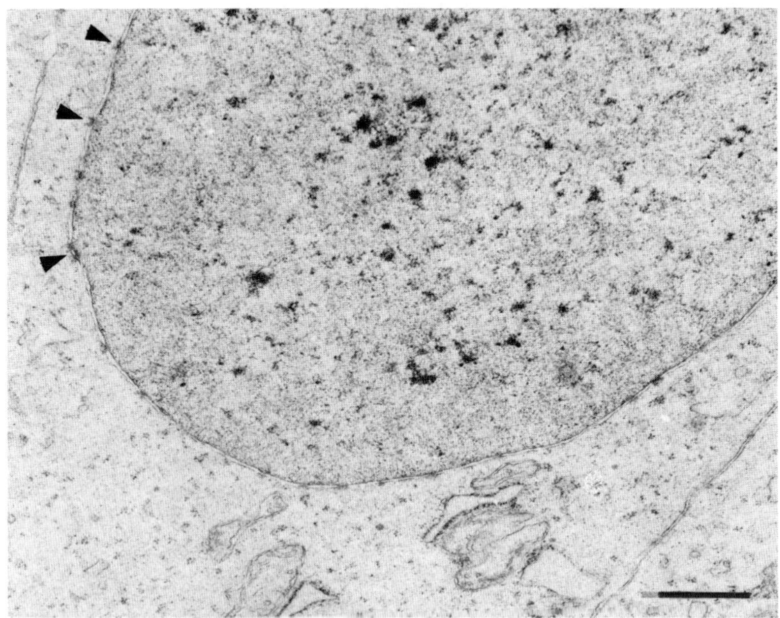

Figure 17 Elongating epithelial cell at the equator of a neonatal squirrel lens. Nuclear chromatin is dispersed, and nuclear pores (arrowheads) are numerous [1] (× 14,000, TEM); bar = 1 μm.

The epithelial cells in all these species contain the organelles for the energy-requiring functions that have been attributed to them, including cell division, synthesis of α- and β-crystallins, active cation transport, and secretion of capsular precursors. These organelles are similar in all the species studied, but changes are apparent with aging.

Nucleus: In the young lens the cell nucleus is large, spherical, or elliptic and often indented or lobed (Fig. 16). Chromatin is usually dispersed, not clumped at the nuclear envelope. Each of the two nucleoli is small and consists of a central fibrillar and a surrounding granular part. During differentiation of the epithelial cell at the equator, the nucleoli undergo a series of morphologic changes until they are dispersed into the nucleoplasm at the time when the nucleus is undergoing disintegration (97–99). Nuclear pores are prominent in lens epithelial cell nuclei (Fig. 17) (100).

In the older lens, epithelial cell nuclei are flatter, probably reflecting the overall flattening of the cells. Also, the chromatin appears more condensed at the periphery (Fig. 18).

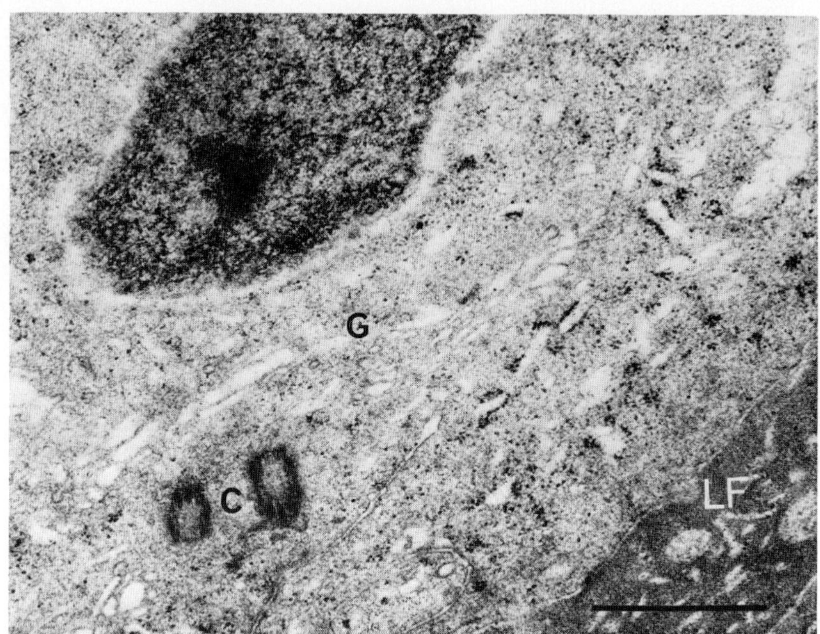

Figure 18   Anterior epithelial cell of a 57-year-old human lens. Nuclear chromatin is condensed at the nuclear envelope [3]; C, centrioles; G, Golgi figures; LF, lens fiber; other organelles as in Fig. 16 (× 23,250, TEM); bar = 1 μm.

Cytoplasm: The cytoplasm of the epithelial cell contains free ribosomes, polyribosomes, and smooth and rough endoplasmic reticulum in modest numbers. Golgi figures, small mitochondria with irregular cristae, centrioles, and small vesicles are also present (Fig. 16 to 18). There is nothing morphologically remarkable about any of these organelles that distinguishes the lens epithelial cell, with the possible exception of the mitochondria. Thus their small size, indistinct cristae, and often twisted shape may distinguish the lens cell from a nearby epithelial cell of a ciliary process or an unpigmented iris epithelial cell in an albino animal.

The organelles occur throughout the cell, although they are less concentrated in the interdigitating processes and more concentrated in the apical part between the nucleus and the cell membrane adjoining the cortical fibers (Figs. 16, 18, and 19). The Golgi figure is also located on the apical side of the nucleus. According to Wanko and Gavin (93), mitochondria increase in number from central to pre-equatorial zones. With increasing age, epithelial cell organelles become more difficult to distinguish because of the increasing density of the cytoplasmic matrix; they have been reported to be reduced in number in older lenses (36,92).

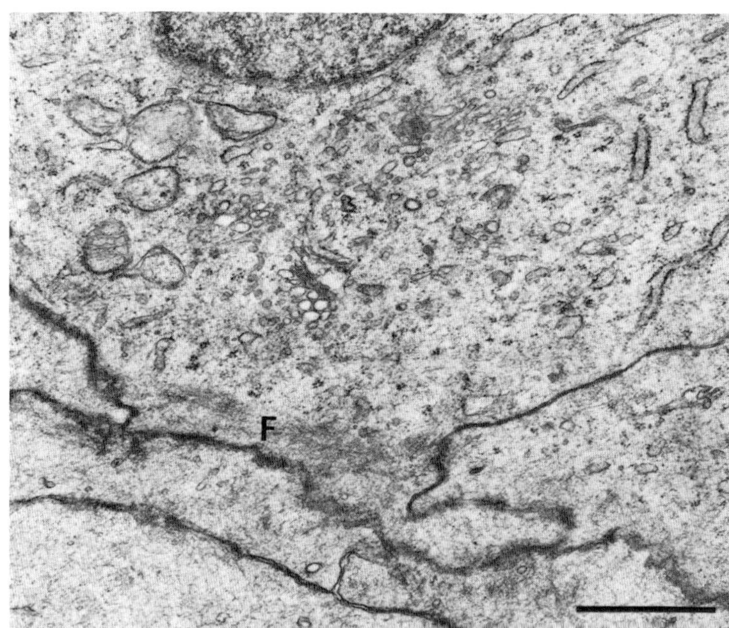

Figure 19 Apical end of a peripheral epithelial cell of a day 17 chick embyro [4]. Organelles as in Figs. 16 and 18; F, filaments (×18,200, TEM); bar = 1 μm.

Organelles and inclusions that may not be present in the lens epithelium of all species are lysosomes, glycogen, and lipid droplets. Dense bodies and lysosomes have been identified in rat lens using acid phosphatase histochemistry by Gorthy et al. (101), and they have been provisionally identified by morphologic criteria in human (91) and other animals (92,94,102) but were not observed in mouse (95) lenses. Glycogen particles, likewise, have been described in human epithelial cells (36,92,95,103) and primate lens (104) but have not been observed in mouse lens (95) based on morphologic criteria (105). Cilia have also been noted in human and rat lens epithelial cells at the equator (35,36,70). These occur in the apical end of the cell.

The plasma membranes and cytoskeletal elements of lens epithelial and fiber cells are eliciting intense interest and attention currently, and these are described in depth in Cellular Morphology.

Morphology of Lens Fibers

The great mass of the lens is composed of lens fibers. Because of optical and structural differences between fibers lying at different depths, the lens is usually

Figure 20   Transverse section of day 17 chick embryo anterior cortical lens fibers. The shapes are less hexagonal and the arrangement less ordered at this stage than in the adult. Compare with the transverse section of cortical fibers in Fig. 27 [4] (×5610, TEM); bar = 1 μm.

spoken of as having a cortical and a nuclear area. The nucleus or core contains the original embryonic and fetal lens fibers in the center (fetal nucleus), and these are overlain by the adult nuclear fibers (adult nucleus or deep cortical fibers). These are overlain in turn by the superficial cortical lens fibers, the most superficial or youngest ones lying subjacent to the lens epithelium anteriorly and to the lens capsule posteriorly.

Cortical lens fibers are generally flattened hexagons in transverse section, and elongated beltlike cells (103). In transverse section, the cortical fibers have two parallel long sides and four short sides forming the boundaries of the hexagon

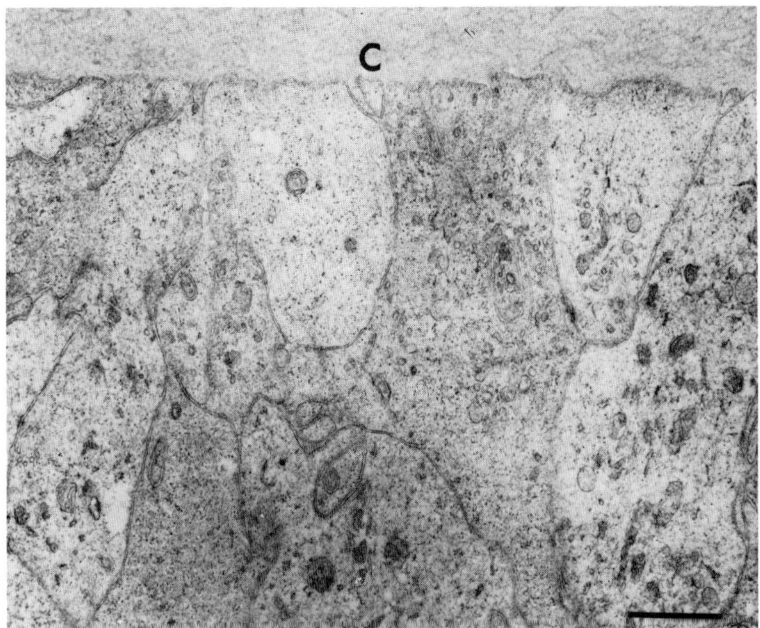

Figure 21 Elongating equatorial epithelial cells of a juvenile squirrel lens. The cells have many of the organelles present in more anterior epithelial cells [1] ; C, capsule (× 12,200, TEM); bar = 1 μm.

(Fig. 20; see also Fig. 27). Hence they fit together in the closest packing order. The fibers are oriented with the long sides parallel to the lens surface. Adajcent fibers form rows apposed at their long sides; adjacent rows are staggered at their apices.

The length of lens fibers varies considerably with the species, age of the lens, and the depth in the lens. In the human, the length of cortical fibers in 7–10 mm. They are 8–12 μm wide, apex to apex, and 2 μm thick except in the region of the cell nucleus, where the width is 5 μm (76). Fibers have been described as being thickest in the middle of their length and thinnest at their anterior and posterior ends (47).

The substructure of lens fibers is similar in all species studied (106). The cytoplasm of elongating cells contains the same organelles as the epithelial cells from which they are derived (Fig. 21). The youngest fibers are nucleated, but as described above, the nuclei are lost with time and the deep cortical and nuclear fibers are anuclcate. In the mouse, cortical lens fibers are nucleated to a depth of about 50 fibers (Rafferty, unpublished data). Cytoplasmic organelles gradually

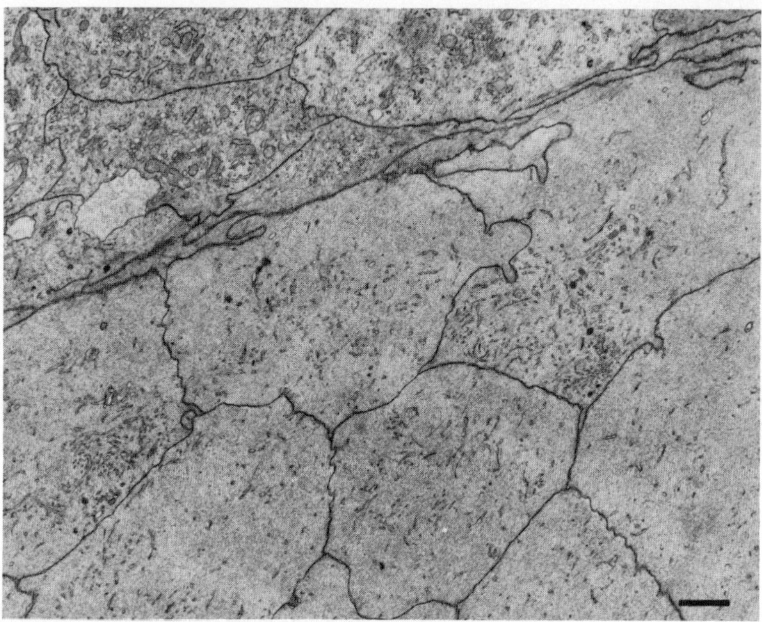

Figure 22 Superficial cortical lens fibers in a day 17 chick embryo showing loss of most of the organelles that are present in the overlying peripheral epithelial cells. Smooth endoplasmic reticulum is retained by diminishes in deeper fibers [4] (×6580, TEM); bar = 1 μm.

disappear the deeper the fibers lie from the surface. Free ribosomes and some mitochondria persist in fibers throughout the cortex, but rough and smooth endoplasmic reticulum, Golgi figures, and vesicles are not seen in fibers more than a dozen deep (in the mouse) (Fig. 22). This observation is compatible with the continued production in the superficial cortex of the crystallins, which are retained within the elongating lens fibers, and with the loss of production of capsular precursors and their export from the cell. Crystallin synthesis is lost in the deeper fibers, and without organelles the fibers appear empty.

At the sutural (polar) ends of the fibers, a bispheroid accumulation of ribonucleoprotein has been described (107,108). These "terminal bodies" are partially digested by ribonuclease and have been found to result from aggregations of ribosomes and polyribosomes with numbers of vesicles in the elongating fibers. Their significance is not at present understood.

The relatively dense matrix of lens fibers is rich in granular and filamentous elements corresponding in part to the crystallins (109) and in part to filaments sectioned transversely and lengthwise. The filamentous cytoskeleton will be described in detail in the next section, along with the membranes of lens fibers.

CELLULAR MORPHOLOGY

Lens Membranes

Epithelial Cells

The elaborate infoldings of the lateral plasma membranes of the central epithelial cells have been shown to be the site of important enzymatic activity. Sodium/potassium-ATPase (Na^+ pump) has been located cytochemically on the extracellular surface of these membranes (104,110) and also along the apical epithelial and adjacent cortical fiber membranes (111). The same membranes are also the sites of localization of extracellular acid phosphatase (112,113). Acid phosphatase increases with age and aging cataract, and it is thought to be related to tissue breakdown in these conditions (112). The basal epithelial membrane of the epithelial cell does not have these enzyme activities.

The lateral and apical plasma membranes also have junctional specializations, gap junctions and desmosomes, which function in ionic and metabolic communication, and in cell adhesion, respectively. Occluding junctions have been identified to date only in the lens epithelium of the chick (114,115) and possibly human (116). Thus, in the intact lens, the passage of water, ions, and certain other materials through the intercellular space can take place if passage through the lens capsule is possible. The passage of a 40 kd protein, horseradish peroxidase, through the capsule allows this electron-dense material to infiltrate the intercellular spaces of the epithelium and cortical lens fibers (94,114). Procion yellow dye and lanthanum salts also penetrate the capsule and fill the intercellular spaces (114,117). Such observations negate the presence of zonulae occludentes around the epithelial cells, which would seal off the intercellular spaces deep to the occluding band, and provide the lens with a barrier function. However, maculae of the occluding type would allow passage of material around them, as would "leaky" occluding junctions that possess incomplete anastomosing ridges.

Desmosomes have been described in the human lens epithelium and in that of most species (91,92,94,101,103,115). Their presence in mouse lens is questionable (95). Desmosomes consist of a round dense attachment plaque on the cytoplasmic side of the membrane of two apposed cells into which filaments of the intermediate size insert. The intercellular space at the desmosome is filled with a medium-density extracellular material often bisected by a central electron-opaque line (Figs. 23 and 24). Desmosomes in intact lens epithelium occur alone and not in conjunction with other membrane junction specializations, and without such association the lens epithelium cannot be said to possess a

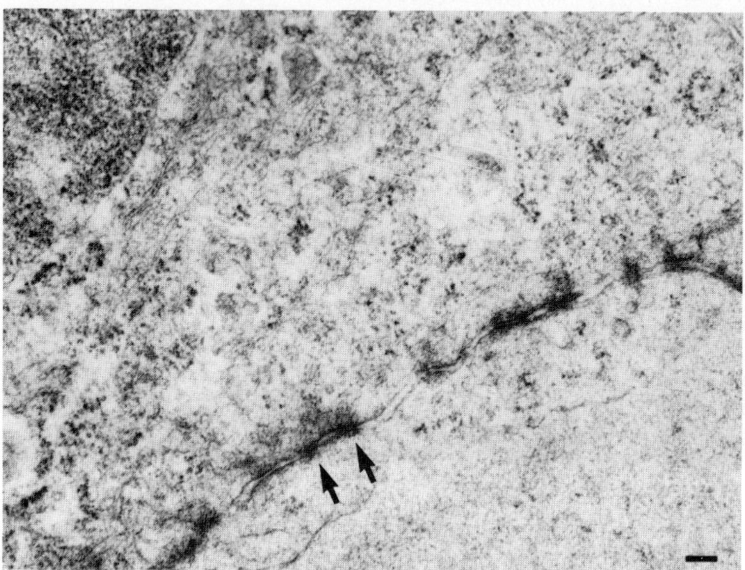

Figure 23 Desmosomes (arrows) at the epitheliofiber junction of equatorial cells of a mature squirrel lens [1] (×37,000, TEM); bar = 0.1 μm.

junctional complex in the strictest sense. Desmosomes are the strongest points of attachment between cells and may keep cells of the epithelium in close apposition during accommodative shape changes.

Gap junctions are present along the membranes of lens epithelial cells, and their presence is consistent with the need for cell-to-cell communication and passage of intercellular materials in an organ lacking innervation and a blood supply. Intercellular communication across the low-resistance gap junction has been shown in several epithelia (118–120). Metabolic (114) and electrotonic coupling (121) have been demonstrated between cells in the lens (but see below).

With thin-section electron microscopy, the gap junction appears as a narrowing of the intercellular space between adjacent cells from the usual 15–20 nm to 2–4 nm (Fig. 24). Gap junctions characteristically fill with intercellular tracers, such as lanthanum nitrate or ruthenium red (122–125).

After freeze-fracture along the hydrophilic plane of the membrane, gap junctions are further characterized by the arrangement of particles and pits on the two hydrophobic fracture faces of the plasma membrane. Whereas nonjunctional membranes have diffusely scattered 9 nm particles on the P face (face of cleaved membrane closest to the cytoplasm) and complimentary pits on the E face

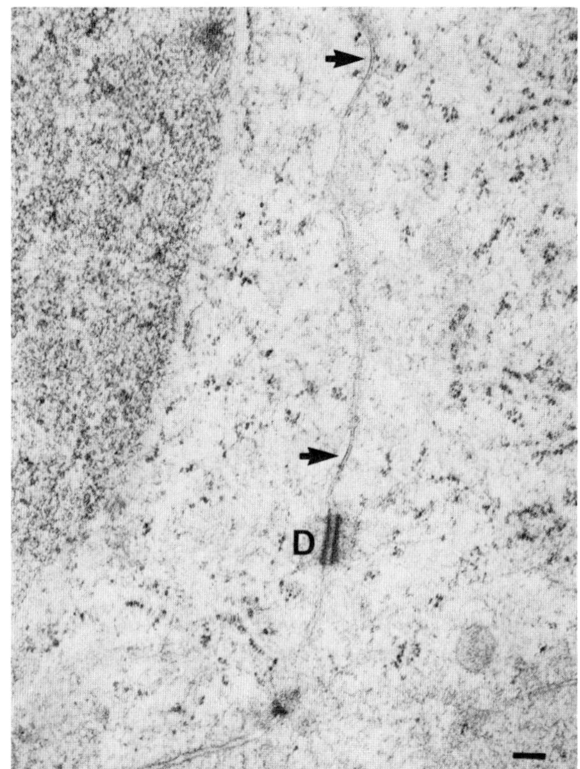

Figure 24 Gap junctions (arrows) and a desmosome (D) on the lateral plasma membranes of anterior epithelial cells of a mature frog lens. Note the narrowed intercellular space at the gap junction [1] (× 33,750, TEM); bar = 0.1 μm.

(cleaved face closest to the extracellular space) (126,127), the particles and pits are more densely packed at the gap junctions.

The particles in the junctional membrane are units called connexons (128), which contain channels 1.5–2 nm in diameter that traverse the thickness of the membrane. The connexons of the junctions of apposed cells are believed to extend into the intercellular gap and merge (128). Thus, ions and metabolites up to 600–700 d can pass from cell to cell through these channels. When the membrane is fractured, the connexons pull out of the E face, creating the pits, and the particles are left on the P face.

The connexons may be randomly packed, the state in which the cells are believed to be coupled, or they may be packed in a crystalline array, the uncoupled state. In most tissues, uncoupling occurs as a result of some kind of physiologic insult, such as anoxia, increased divalent cations, injury, and lowered pH (128a). Goodenough et al. (114) have found that, in mouse anterior lens epithelium, gap junction particles are both randomly packed and crystalline packed at the junction with the cortical lens fibers and packed in hexagonal arrays between epithelial cells. The latter has been substantiated by Peracchia (129) in calf lens. However, in chick lens the epitheliofiber membranes contain only noncrystallizing gap junctions (114), which suggests a species difference may exist.

Lens Fibers

Scanning electron microscopy has been used to study the surface architecture of lens fiber cells. The patterns observed appear to be similar but not identical in the species studied (130).

In the bow region the elongating immature lens fibers, still connected basally to the lens capsule, have numerous fingerlike protrusions that interdigitate at their angles with indentations on adjacent fibers (Fig. 25). More deeply and throughout the superficial cortex, the protrusions acquire a more spherical shape, as do the complimentary receptacles: these have been called "jigsaw interlocking processes" (131), "ball-and-socket interdigitating processes" (130, 132–136), or "knobs and sockets" (103). In the cortical fibers of some rodents, the apical border is scalloped or undulating (Fig. 26). Lens fibers in the deeper superficial cortex, which have lost contact with the posterior capsule, in addition to ball-and-socket interdigitations, acquire arrays of fine elongated ridges or reticulations on the long and short sides of the fibers (Fig. 27) (103,132,134, 136,137). The surface of deep adult cortical fibers displays prominent interlocking ridges and clefts oriented in every direction (Fig. 28), and these give way in the fetal nucleus to small mounds and pits (Fig. 29) (103,132,134,136,137). There is no consensus about differences in surface architecture of anterior and posterior parts of the same fibers at different depths, and a variety of claims has been made. However, one of the earliest groups of workers (89) noted that interdigitating processes are most numerous in those areas of the lens believed to undergo the greatest shape changes during accommodation, that is, the anterior and equatorial cortex. All the surface modifications are believed to function to

---

Figure 25   (A) Superficial cortical fibers of adult frog lens showing fingerlike and ball-and-socket protrusions from the apices of the two short sides of each fiber. The concavities between the apical protrusions (arrows) receive the protrusions from the angles between long and short sides of the next layer of fibers (removed) [5] (×6000, SEM); bar = 1 μm. (B) Thin section along the long axis of superficial cortical fibers of adult squirrel lens showing fingerlike and ball-and-socket interdigitations with adjacent cells [1] (×12,420, TEM); bar = 1 μm.

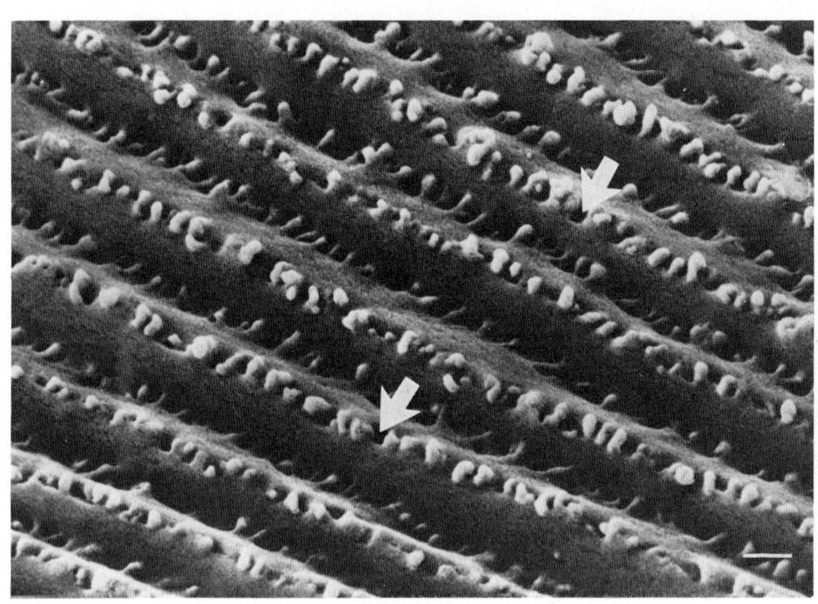

A

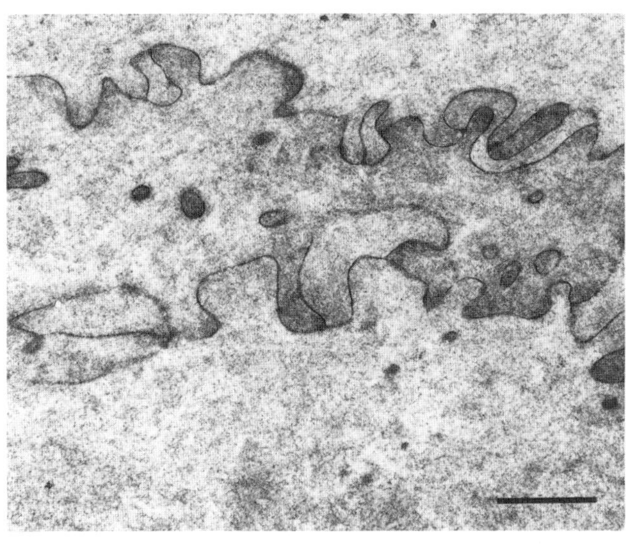

B

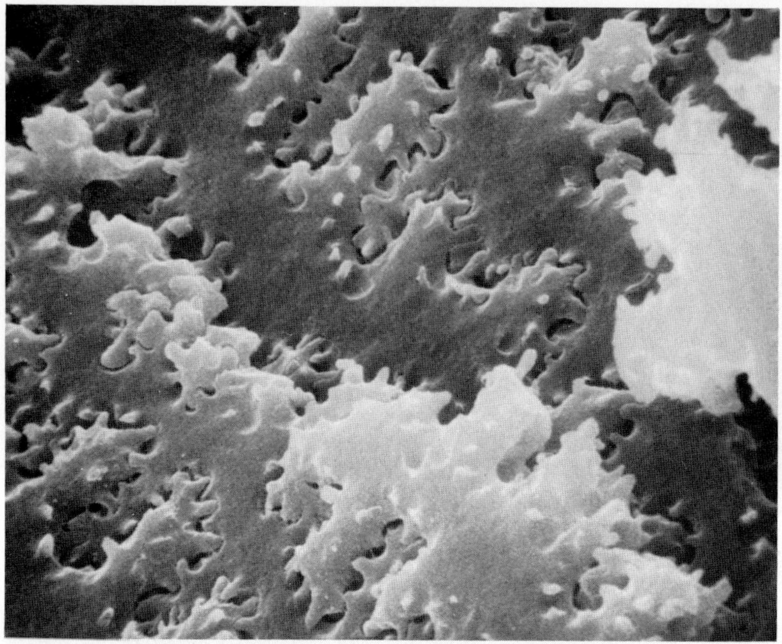

Figure 26 Superficial cortical fibers of mouse lens have scalloped borders with a row of processes delineating the long and short sides of the fiber [5] (SEM).

keep the lens fibers in apposition and to prevent slipping during lens shape changes.

Occluding junctions have not been seen between lens fibers. Desmosomes occur along embryonic lens fibers (Fig. 30) but are lost in the adult.

Abundant gap junctions have been identified along the plasma membrane of lens fibers in a wide variety of species (123,124,130,138–141). The junctions are often associated with the protrusions of the surface membrane. Gap junctions have been reported to comprise 50–60% of mature chick lens fiber membranes (142) but appear to be fewer in frogs (Fig. 31). Since a cortical lens fiber has a 1000-fold increase in plasma membrane over that of the lens epithelial cell, a staggering number of gap junctions are present in lens fibers. By comparison, Benedetti et al. (143) have calculated that gap junctions comprise only 3% of the epithelial cell surface.

Benedetti's group (123,139,143) has studied the development of gap junctions from the epithelial to the deep cortical layers of calf lens. They describe the interepithelial cell junctions as typical gap junctions consisting of small arrays of geometrically packed particles (see also Refs. 114 and 129). As the

Figure 27 Deeper superficial cortical fibers of frog lens separated to show the fine reticulations on the long sides (upper) and cut transversely to show the flattened hexagons aligned in rows like stacks of lumber (lower) [5] (×2000, SEM); bar = 10 μm.

cells differentiate and begin to elongate, junctional areas increase in number and generally consist of linear arrays of particles, often embracing small polygonal clusters of 9 nm particles. In the fully differentiated cortical fibers, large polygonal aggregates of particles are found. These results have been confirmed in developing chick lens by Kuszak et al. (144), who correlate the uncoupled crystalline junction with fiber elongation and the coupled state with differentiated fiber. Furthermore, there is some evidence to suggest that the onset of transparency of the chick embryo lens on day 5 is correlated with the development of coupled (noncrystalline) gap junctions (130,144).

There is currently some disagreement about the form (and, therefore, the functional state) of the gap junctions between lens fibers. The junctions between mature lens fibers have been described as pleomorphic, that is, noncrystalline, arrays (Figs. 31 and 32A) (123,141,144-146), in contrast to the hexagonal, crystalline lattices usually found between lens epithelial cells (see above). The occasional small crystalline arrays observed in superficial cortical fiber

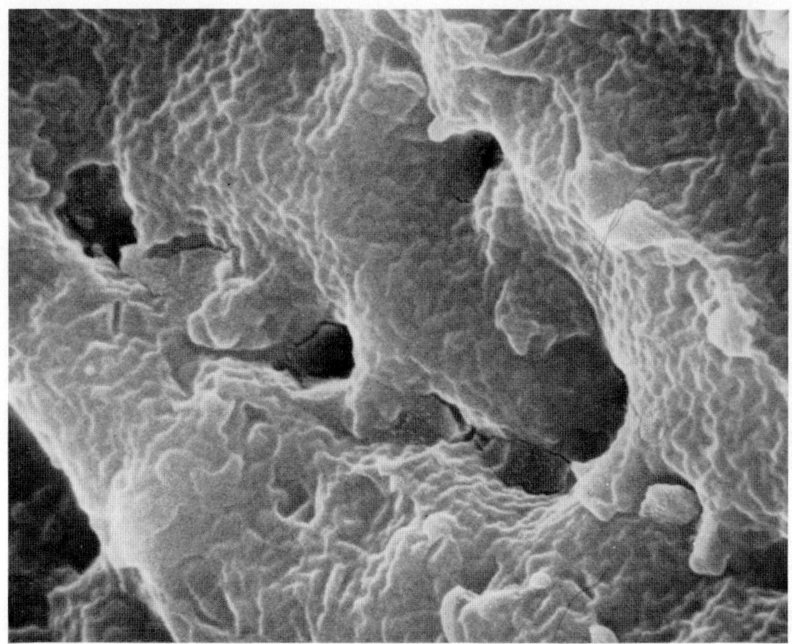

Figure 28   Deep cortical fibers of mouse lens showing interlocking ridges and clefts on surface [5] (SEM).

membranes have been interpreted as remnants of these junctions present before differentiation into the elongating cell took place (Fig. 32B) (143,147). For the most part, the junctions between fibers appear to be coupled, which is supported by the work of Rae (148), who noted diffusion from cell to cell of Procion dyes iontophoresed into lens fibers through glass capillary microelectrodes. The junctions can be uncoupled by exposure to 5×10^{-7} M Ca^{2+}, by freezing with liquid nitrogen (129,149,150), or by trypsinization at pH 6–6.5 (151). However, Kistler and Bullivant (151) believe the orthogonal arrays produced by the low pH are unrelated to lens gap junctions because the latter are left intact by this treatment. Zampighi et al. (152) rarely observe typical gap junctions, either random or crystalline, in freeze-fracture replicas of isolated calf lens fiber junctions. Instead they find extensive square (tetragonal) arrays covering the exterior hydrophilic surfaces, and it appears that the fracture plane has bisected the gap of the isolated junctions, not the center of the lipid bilayer. Although this kind of junction may result from the isolation procedures by this group, it should be pointed out that square arrays are occasionally observed in the replicas of other workers (e.g., Ref. 140).

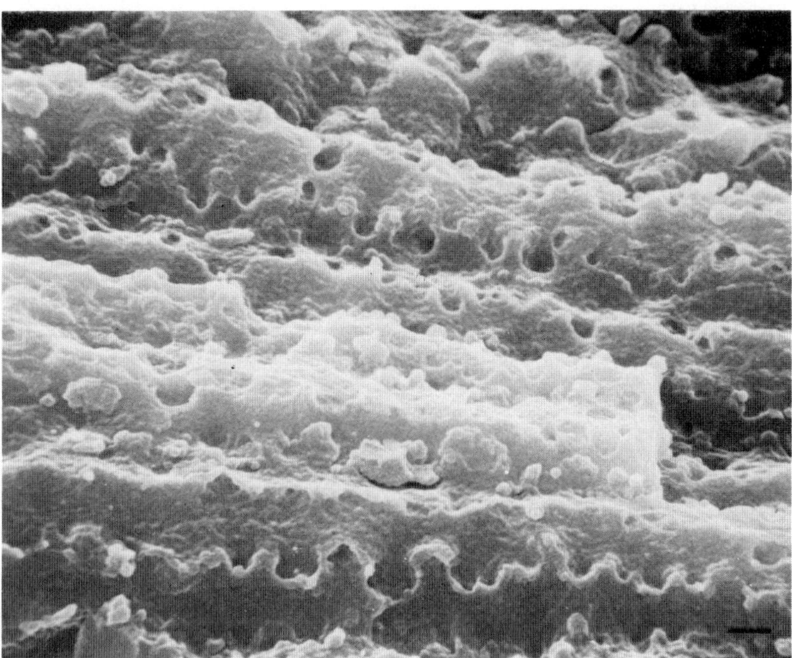

Figure 29 Surface of nuclear fibers of mouse lens showing complex ridges, mounds, and pits [5] (× 5000, SEM); bar = 1 μm.

The diversity of opinions concerning the kind of junctional types between lens fibers is further complicated by differences in the structure of the connexon. Four subunits have been proposed (153) and six subunits also (Figs. 32B and 33) (147,154). In addition, using rotary-shadowed freeze-etch replicas, Kuszak et al. (147) identified two types of connexons. One type, seen predominantly in the uncoupled crystalline junction, is characterized by a central electron-lucent zone surrounded by six electron-dense subunits; the second type, seen predominantly in the coupled, noncrystalline gap junction, has a central electron dense core 1.5–2 nm in the middle of the central electron-lucent zone surrounded by the six subunits. These observations appear to correlate well with the model of the structure of the connexon proposed by Unwin and Zampighi (154).

Cytoskeleton

The larger organelles share the volume of the cell with the elements of the cyto-skeleton, a term that implies an integration of these filamentous structures into

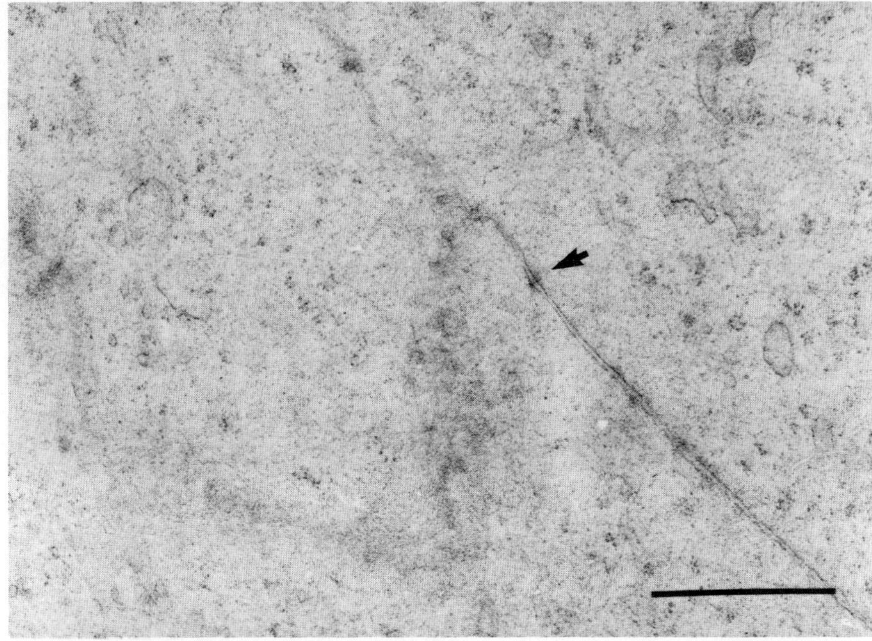

Figure 30 Desmosomes (arrow) between lens fibers of neonatal squirrel lens
[1] (× 16,000, TEM); bar = 0.1 μm.

a structural and functional unit. Indeed, stereoviews of the matrix of a variety of
cultured cells with the high-voltage electron microscope by Wolosewick and
Porter (155) and by Schliwa and Van Blerkom (156) suggest that the filaments
interact and are probably physically attached to one another, possibly through
a microtrabecular lattice. Elements of the cytoskeleton include actin filaments,
intermediate filaments, microtubules, and, in most cell types, myosin filaments.
These elements, along with their regulatory proteins, have been implicated in cell
motility (both internal movements, such as cytokinesis and exocytosis, and cell
locomotion) and in maintenance of cell shape.

   Although filaments had been noted in lens cells in some of the earlier studies
with the electron microscope (61,94,106), they were not singled out for special

Figure 31 Low magnification of unidirectionally shadowed freeze-etch replicas
of cortical fiber cell membrane from adult chick (A) and frog (B) lens. Compare
the size and density of gap junction (GJ) in these two specimens (× 40,000);
bar = 0.1 μm. (Courtesy of J. Kuszak.)

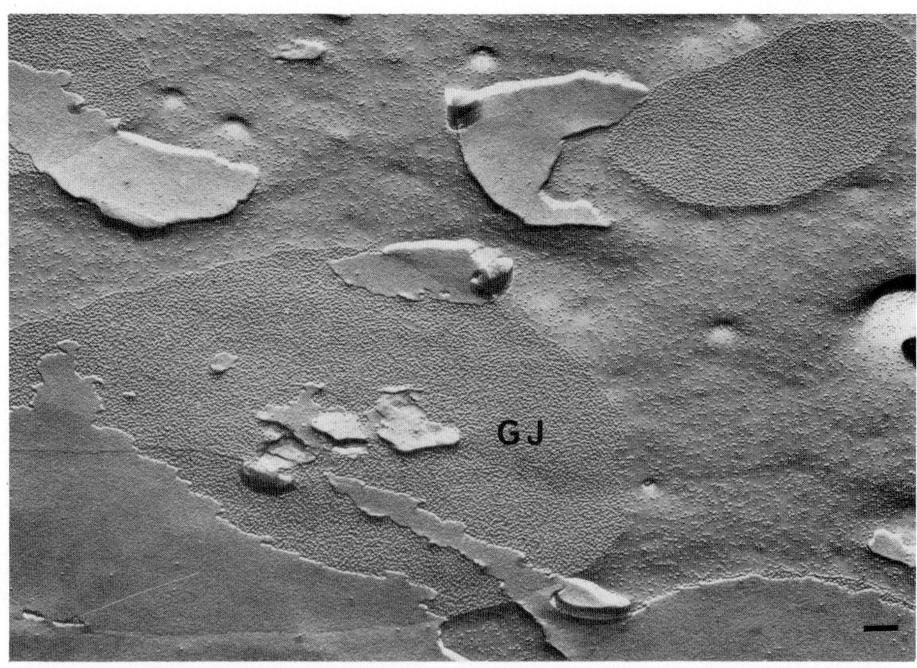
G J
A

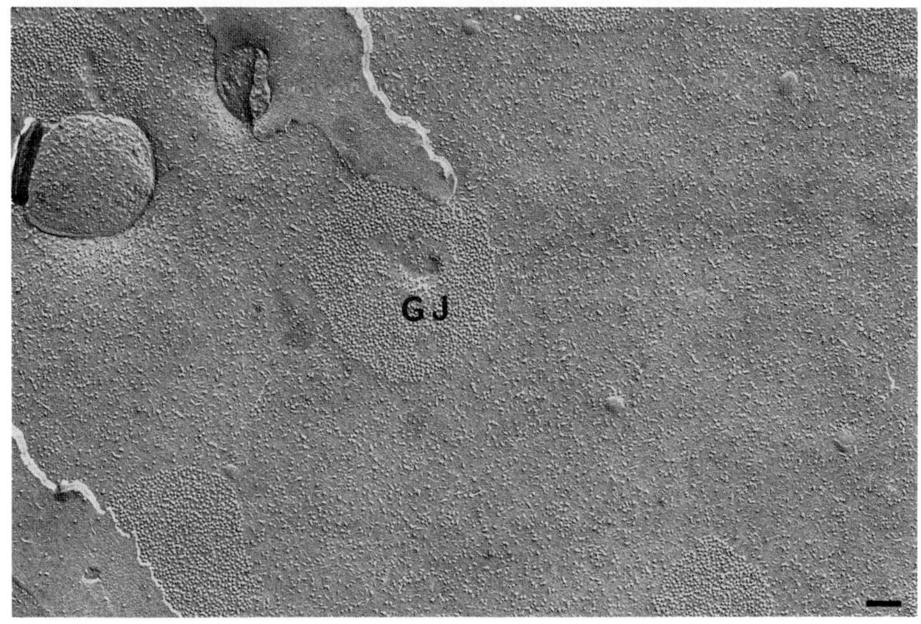
G J
B

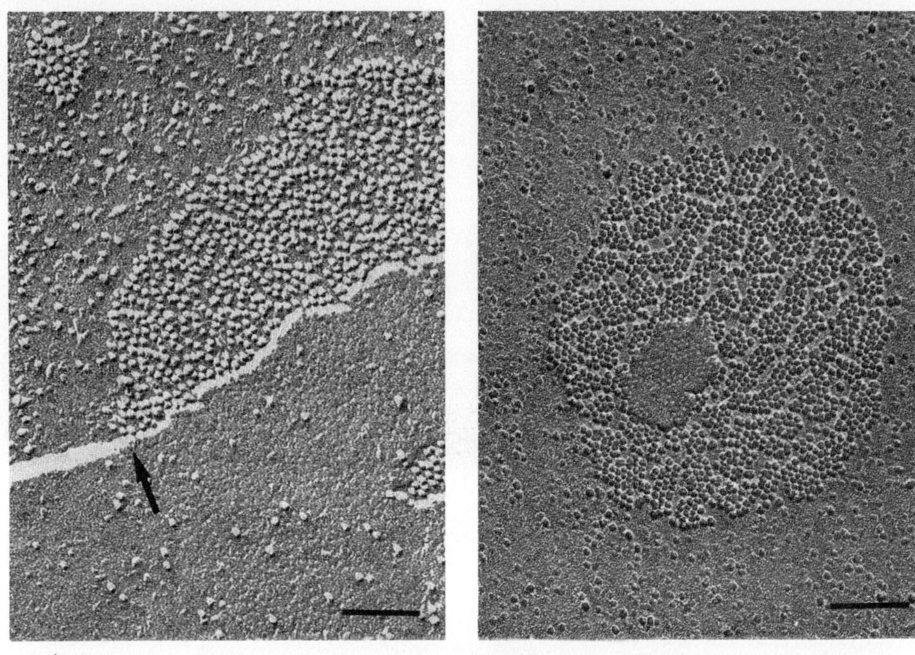

A   B

Figure 32 (A) High magnification of unidirectional shadowed freeze-etch
replica of adult frog lens cortical fiber cell gap junction. Note the narrowing of
the intercellular space at the site of the gap junction (arrow) ($\times$ 100,000); bar =
0.1 μm. (Courtesy of J. Kuszak.) (B) High magnification of rotary shadowed
freeze-etch replicas of an adult chick elongating fiber cell gap junction. Note the
hexagonal packing and the six subunit structure of the connexons ($\times$ 100,000);
bar = 0.1 μm. (Courtesy of J. Kuszak.)

study until considerably later when their functional significance in other cell
types was being explored.

Microtubules

Microtubules are long, generally straight cylinders measuring 24 nm in diameter
and with a 15 nm diameter hollow core. Their length if determined in part by
the cell type and is as much as 10–25 μm long in the nerves of some inverte-
brates. Microtubules are built of globular dimer subunits, α- and β-tubulins
(molecular weight, MW 55 kd each), which are aligned usually into 13 protofilament strands, sides touching, forming the wall of the tubule. In cross section,
microtubules appear as circles, and with high magnification the 13 subunits can
be differentiated. Cilia, flagella, centrioles, and basal bodies represent special
aggregates of microtubules.

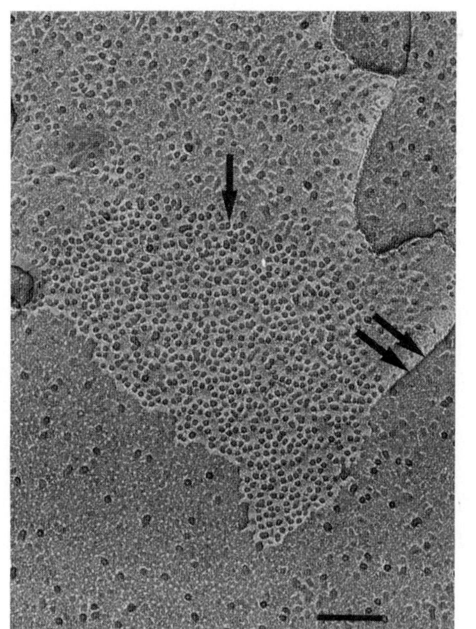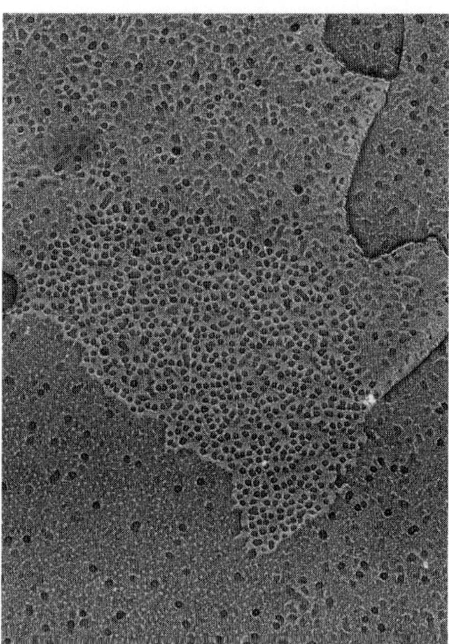

Figure 33 Stereopair micrographs (±12°) of rotary shadowed freeze-etch replica of an adult frog lens cortical fiber gap junction. Note the junction is a raised plaque (arrow) and the intercellular space is narrowed at the site of the gap junction (double arrow). The six subunits of the connexons can be seen (× 80,000); bar = 0.1 μm. (Courtesy of J. Kuszak.)

Although microtubules are good candidates for providing the skeleton of a cell, in actual fact, except those composing cilia and flagella, they can be rapidly disassembled and reassembled. Several large proteins called microtubule-associated proteins (MAP), facilitate the reassembly process (157). MAP project from the surfaces of the microtubules and are believed to assist in the sliding or ratcheting of one on the other during cell movement or in maintaining cell shape. Disassembly occurs in high Ca^{2+} concentrations, low temperature, and in the presence of colchicine and vinblastine. In vitro microtubules exhibit steady-state opposite end assembly and disassembly, called treadmilling (158), and although treadmilling could be a mechanism by which cellular organelles are moved, it has not been proved to operate in the intact cell (158).

Microtubules are the least studied element of the lens cytoskeleton. The classic description was made by Kuwabara (90) in human lenses of various ages. He observed that microtubules are sparse and randomly oriented in the central lens epithelium. In the elongating equatorial cells, they increase in number and become oriented with the long axis of the cell, and in cortical fibers they are found to be oriented lengthwise in the peripheral cytoplasm. They are not found in nuclear fibers. These observations have been confirmed by Bradley et al. (159) in chick lens and by Ireland et al. (160) in rabbit lens. The ultrastructural appearance of microtubules in cultured bovine lens epithelial cells has been confirmed by light microscope immunofluorescence staining with tubulin-specific antiserum (161).

The function of microtubules in lens cells is unknown. Kuwabara (90) suggested they may help maintain the long curved shape of the cortical fibers. As discussed in the section on lens development, the role of microtubules in elongation of lens placode and primary and secondary fibers is questionable. At the present time it is probably safe to suggest that microtubules play a structural role as a part of the cytoskeleton.

Actin

Actin is ubiquitous in plant and animal cells (e.g., see Refs. 162 and 163). Actin is present in either or both a filamentous (F or polymerized) form and a soluble (G or monomeric or unpolymerized) form. The two forms are interconvertible. Actin filaments, also called thin filaments, polymerize in high ionic strength buffers and depolymerize in low ionic strength buffers and at low temperatures (163,164). Actin filaments consist of 5 nm globular monomers arranged in a double helix with a half-pitch of 36–38 nm (163). With transmission electron microscopy, they measure 5–8 nm in diameter and they take the form of straight or slightly bent rods of indeterminate length (164).

The actin polypeptide has a molecular weight of 42 kd as determined from mobility in SDS-PAG electrophoresis (165,166) and by amino acid sequencing of muscle actin (167). In tissues, actin exists as three different molecular types with different isoelectric points (pI): α_1, the most acidic (pI 5.4) is characteristic of striated muscle, and β and γ, with slightly higher pI, are characteristic of actins found in nonmuscle cells and in smooth muscle (168,169).

Actin is a highly conserved protein evolutionarily, and most actins are similar in their properties, including immunologic cross-reactivity between tissues of widely divergent species. Using highly purified antibodies against actin indirectly coupled with an immunofluorescent substance, the overall distribution of actin can be visualized in the cell. The technique is especially useful on cells that are spread out in culture. Actin filaments are generally localized in the cell cortex and in networks beneath the plasma membrane, or concentrated in processes and on the side of the cell attached to the substrate as bundles of stress fibers (170).

Since the diameters of actin and intermediate filaments (see below) overlap, the two types are difficult to distinguish by standard transmission electron microscopy. A more dependable method is to label actin filaments with the heavy meromyosin (HMM) fragment or the subfragment 1 (S1) of skeletal muscle myosin obtained by limited tryptic or chymotryptic digestion of myosin (162,171). The label appears as a characteristic arrowhead pattern along the actin filament. This method has revealed that the actin filament has polarity. All the arrowheads point in one direction along the filament, and furthermore, assembly of the filament occurs preferentially at the barbed end and not at the pointed end (172). Those filaments perpendicularly associated with the membrane have the arrowheads pointing away from the membrane; those filaments comprising stress fibers running parallel to the membrane have arrowheads pointing in either direction (173). HMM and S1 do not bind to intermediate filaments or microtubules. The major disadvantage of the method is that the cells must be treated with glycerol or a detergent to permeabilize the plasma membrane to the large muscle fragments (HMM, 120 kd; S1, 90 kd), and the cytoarchitecture is disrupted. The advantage of the method is that the cytoskeleton is preserved (166) while much of the cytoplasmic matrix is lost, resulting in less background interference in distinguishing the filaments.

Although suspected to be actin filaments on the basis of their diameter in electron micrographs (Fig. 34) (80,95,96,103), the identity of the thin filaments in lens epithelial cells should be attributed to Lonchampt et al. (161), who detected the protein in cultured bovine lens using anti-actin antibody immunofluorescence and correlated the pattern with that seen ultrastructurally. The finding was later confirmed by others (174–178). From isoelectric focusing gels of lens homogenates, Mousa and Trevithick (177) determined that actin comprises 10.7% of the total protein of the newborn rat epithelial cell, but only 1.1% of the total protein of the newborn or adult cortical fiber cell. They identified the molecular species of rat lens actin as consisting of β- and γ-actins, which is consistent with the actin species occurring in other nonmuscle cells. Furthermore, they showed that the ratio of β to γ shifts from 2.2 in the newborn rat epithelium to 1.2 in the lens fibers.

Using the cortex of calf lens, Kibbelaar et al. (178) isolated and purified actin by DNAase I affinity chromatography, which binds the water-soluble G-actin. These workers also extracted actin from the water-insoluble fraction, showing that lens actin exists in both forms, as it does in most other tissues. They also determined by amino acid analysis and peptide mapping that lens actins are similar to actins in most other tissues. Ramaekers et al. (178a) determined the proportion of F/G actin in different regions of bovine lenses. G-actin predominates in the epithelium and cortex, and F-actin increases from superficial to deep cortical zones.

The presence of actin filaments in lens has been further verified by labeling the filaments with HMM or S1. The technique was first accomplished by Ireland et al. (160) and by Bradley et al. (159) on lens epithelial cells stripped with the capsule from rabbit and chick lens, respectively. Actin filaments were also

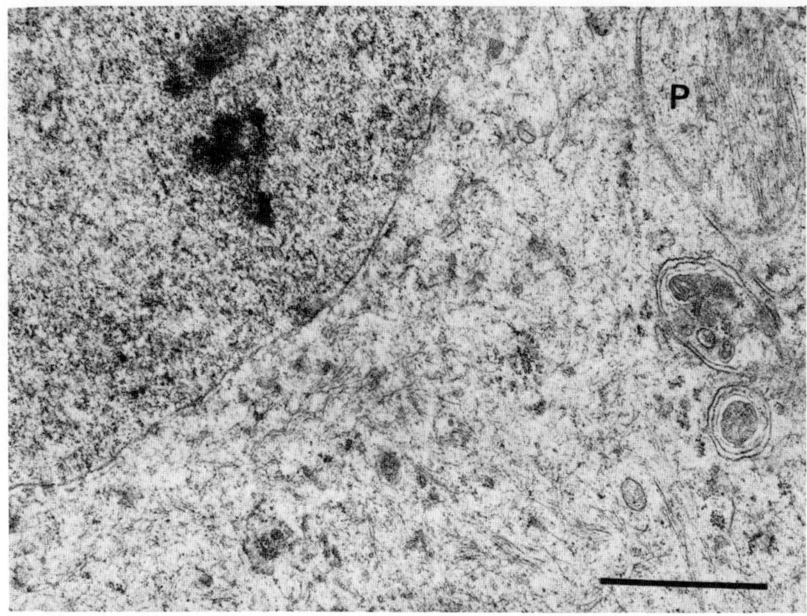

Figure 34   Filaments in an anterior epithelial cell of a 4-month-old rat lens. The cytoplasm is dense with filaments of both the thin and intermediate size. The interdigitating process (P) has an especially dense bundle of filaments [1] ($\times$ 22,600); bar = 1 μm.

labeled with HMM in human lens epithelium from newborn, 24-year-old, and 80-year-old lenses (179). Typical actin filaments extracted from chick lens fibers have also been labeled with S1 (180). More recently, Ireland and Maisel (180a) have labeled actin filaments in glycerinated chick lens fibers, thereby showing that actin filaments exist as separate entities from the beaded chains (see below).

Linking lens actin with accommodative lens shape changes has become a popular suggestion, but the concept has proved to be difficult to test. As a first attempt, the filament pattern (including both actin and intermediate filaments) in large spherical nonaccommodating lenses was compared with that in anteriorly flattened accommodating lenses (96). It was shown that poorly accommodating lenses (mouse and rat) possess epithelial cells densely packed with filaments, whereas accommodating lenses (human neonate and squirrel) have dispersed filaments throughout the cell (compare Figs. 34 and 35) and an accumulation of filaments only at the epitheliofiber junction (Figs. 36 and 37). Also, with aging, the filaments are seen to increase in rodent lenses (181), and in human lenses (36,96) when accommodation is decreased to a near-zero value.

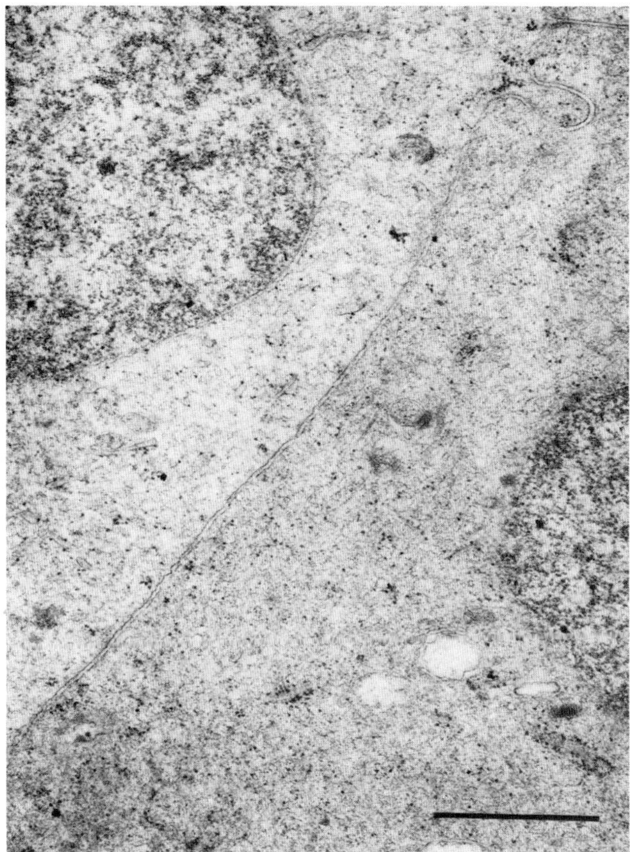

Figure 35 Peripheral epithelial cell of a day 10 human lens. Filaments are not present in bundles, only scattered [3] (× 22,000); bar = 1 μm.

Kibbelaar et al. (182) investigated the immunofluorescent labeling pattern in cortical fibers of lenses with different accommodative amplitudes using anti-actin antibodies. The most intense labeling occurs along the cortical cell membranes of the poorly accommodating lenses of calf and rat, and in the cytoplasm of the nuclear fibers of these animals. In highly accommodating lenses of the pigeon, there is far less fluorescence in cortical lens membranes and no fluorescence in nuclear fibers. The authors concluded that actin filaments have nothing to do with accommodative lens shape changes and that they serve to preserve the spherical shape of the nonaccommodating lens.

Actin filaments appear to insert on the plasma membrane. Although it has not been shown in other cell types that the actin filaments insert directly into

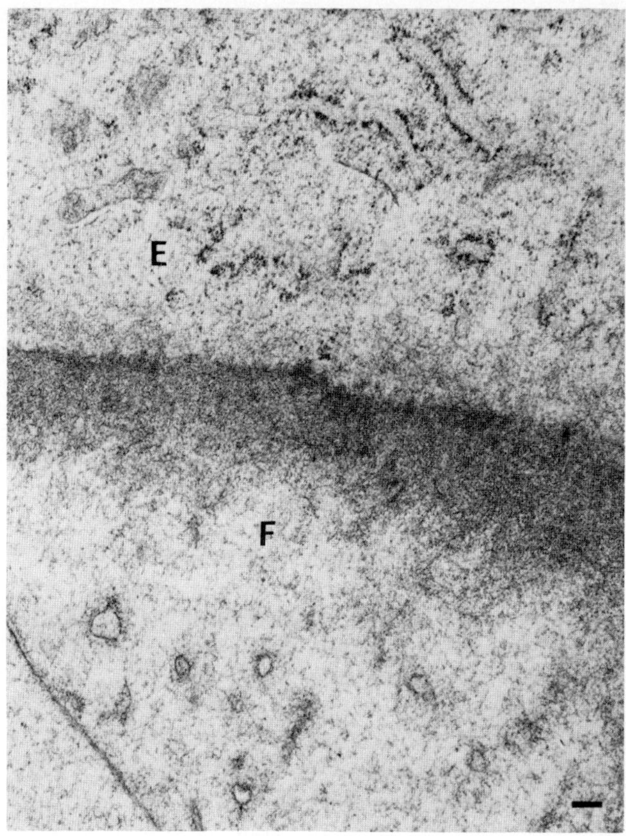

Figure 36 Peripheral region of a juvenile squirrel lens showing the dense mat of thin filaments on both sides of the epitheliofiber junction [1] ; E, epithelial cell; F, fiber cell ($\times$ 31,000); bar = 0.1 μm.

the membrane, it is clear in the case of the red blood cell that anchorage is through an intermediary protein, spectrin, and its attachment to ankyrin and band 3 protein in the membrane (183). Recently, intermediary attachment proteins have been found in neuronal and other cells (184,185). The situation is unclear in lens. However, plasma membranes isolated by a cytoskeleton-stabilizing protocol carry along a certain amount of actin (and intermediate filaments) (186). Furthermore, both immunofluorescence staining (178,187,187a) and studies with actin-binding fluorescent phallacidin (188) have shown the close association of actin with lens fiber and epithelial plasma membranes.

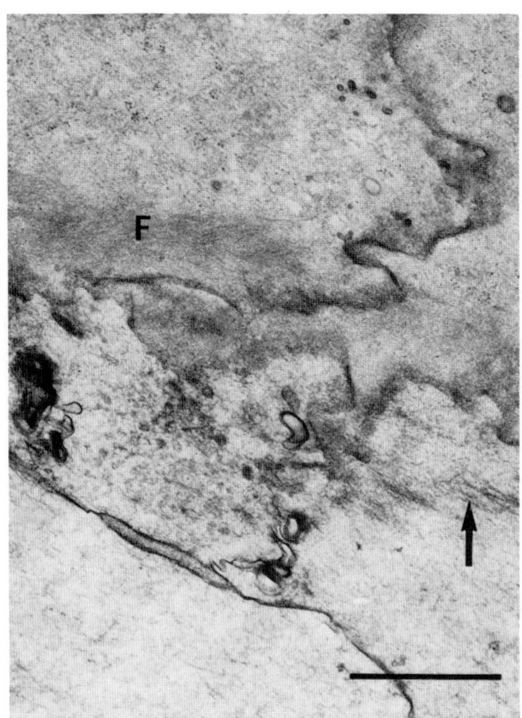

Figure 37 Epitheliofiber junction of anterior lens of a day 5 chick. Note the dense mat of thin filaments (F) on the epithelial side of the junction. Some thicker filaments are present on the fiber side (arrow) [4] (×20,000); bar = 1 μm.

The mat of filaments at the epitheliofiber junction is of special interest to the author of this chapter. As mentioned above, it is prominent in accommodating lenses and nearly nonexistent in lenses that undergo no accommodative shape changes. In S1-labeling experiments on whole chick lens, the filaments in this mat are nearly entirely actin with only a few intermediate filaments interspersed (Fig. 38) (Roth et al., in preparation). In another approach, the filaments at the epitheliofiber junction of accommodating lenses (squirrel, chipmunk, and chick) are being studied after drug-induced accommodation versus nonaccommodation. Preliminary results indicate that the flattened state is associated with loss of the epitheliofiber junction filaments, whereas the spherical, accommodated, state is associated with a substantial increase in filaments at this site (Rafferty, unpublished). It has been shown that when the anterior surface of the chick lens is injured, a band of actin filaments (identified by S1 labeling) is mobilized in the

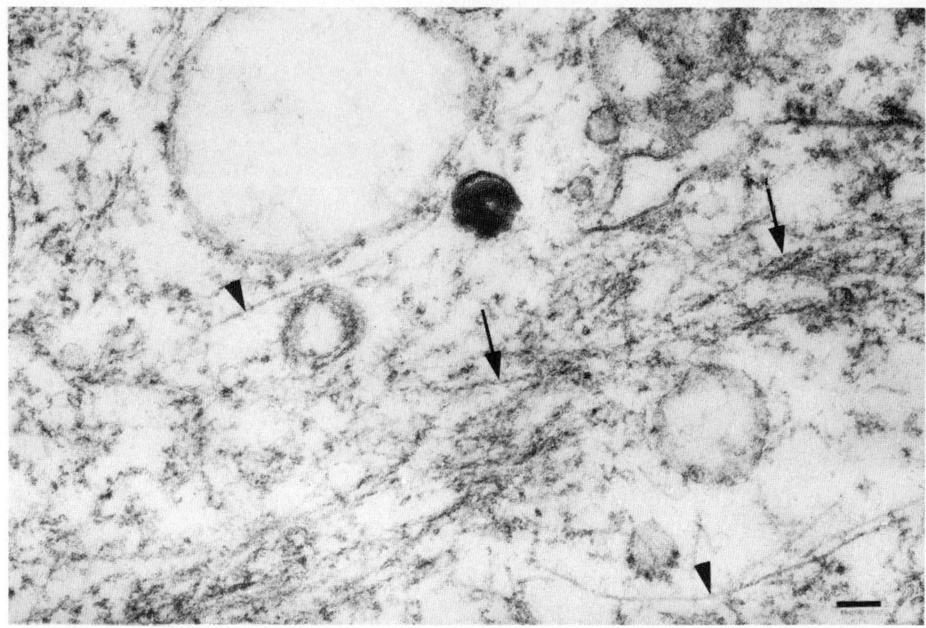

Figure 38   S$_1$-labeled actin filaments in the anterior epithelial cell of a 3-day-old chick lens (arrows). An occasional unlabeled intermediate filament (arrowheads) is present. The epitheliofiber junction is just off the bottom of the picture [4] ($\times$ 55,500); bar = 0.1 μm.

basal end of the epithelial cells beneath the capsule, in addition to the band at the epitheliofiber junction (Fig. 39) (Roth et al., in preparation).

It may be hypothesized that a pool of actin monomers exists in lens epithelial cells and cortical fibers, which can rapidly associate with actin-nucleating centers to form filaments when the tension on the lens is relaxed. Polymerization of filaments at the epitheliofiber junction is especially notable, and the filaments may act in concert to cause the anterior lens surface to rebound into the more spherical shape. Polymerization is an energy-dependent process, and thus the rebound process could be described as an active process without involving a contractile role of actin filaments. Contraction of actin filaments is less easily envisioned because of the arrangement at the epitheliofiber junction. It is known in cell motility that actin filaments "pull" the cell forward and never "push" (163). Another mechanism could be passive deformation of the actin (and intermediate) filaments by the flattening of the anterior lens surface by the tension of the zonules and passive rebound when the tension is relieved. There is as yet

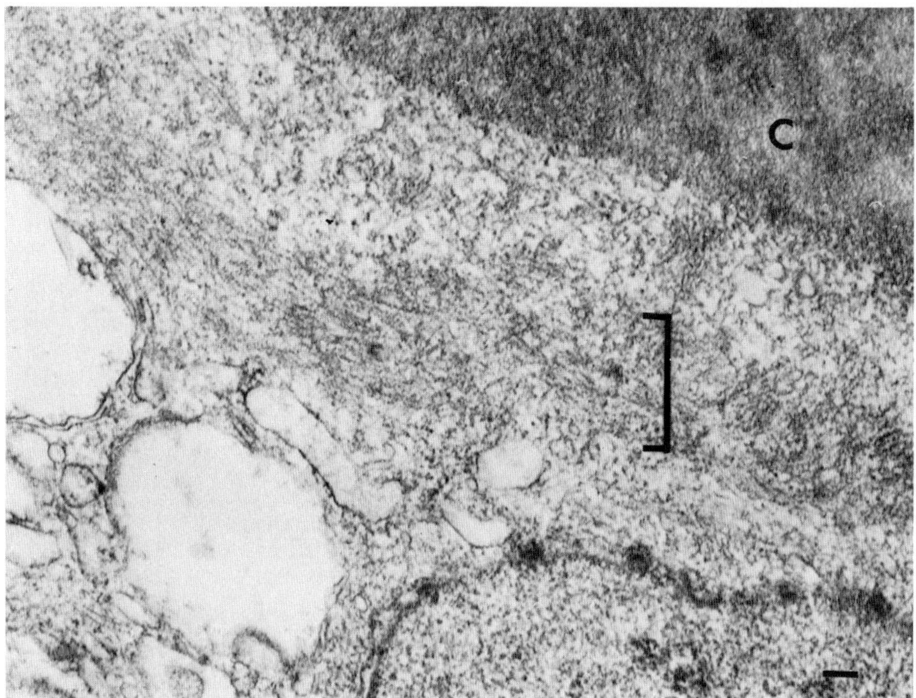

Figure 39 Subcapsular band (bracket) of S_1-labeled actin filaments in the anterior epithelial cell induced by mechanical injury 24 hr earlier; 8-day-old chick [4] ; C, capsule ($\times$ 31,500); bar = 0.1 μm.

no evidence to support any hypothesis except that actin filaments at membranes appear to be associated with the more spherical accommodated state of the lens.

Intermediate Filaments

Intermediate filaments, also called 10 nm filaments, desmin, vimentin, neurofilaments, glial filaments, and keratin filaments according to cell type of origin, are believed to consist of four protofilaments assembled into a coiled-coil α-helical configuration (189). The diameter of intermediate filaments is 8–12 nm and their length varies with the cell type (Fig. 40).

Intermediate filaments comprise a heterogeneous group of filaments, all of which appear similar in the electron microscope, but which have different molecular weights, charge, and immunologic specificity. Intermediate filament proteins of BHK-21 cells, for example, resolve into a 54 kd and a 55 kd doublet on SDS-PAGE (190); the purified 10 nm filamentous protein from skeletal and

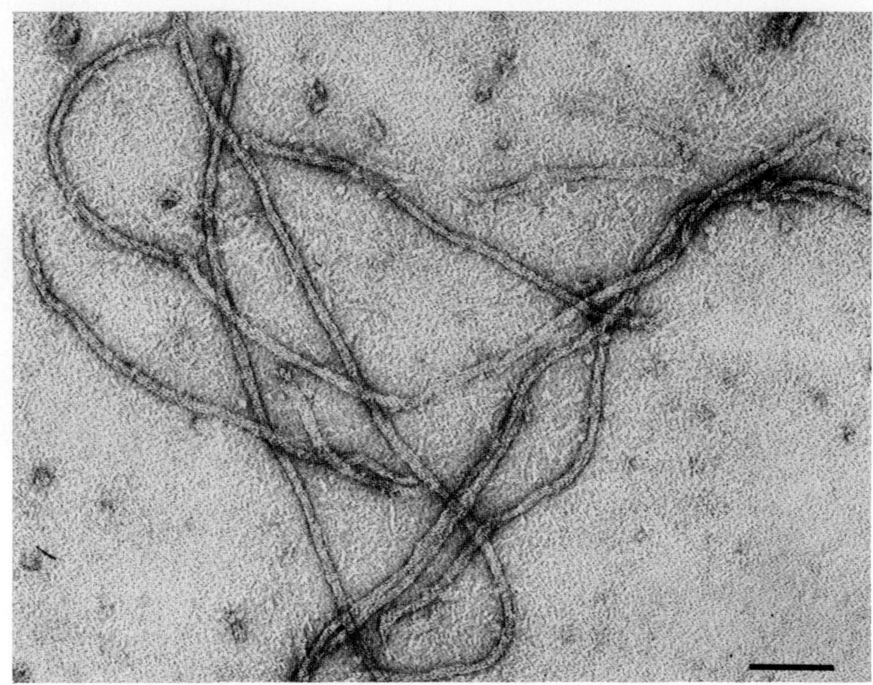

Figure 40 Negative stained preparation of intermediate filaments regenerated from purified α and β subunits. The frayed end in the upper right reveals strands measuring about 5 nm in diameter. Each strand is believed to consist of two proto-filaments ($\times$ 112,500); bar = 0.1 μm. (Courtesy of N. Lieska.)

cardiac muscle, desmin, is a single subunit with a molecular weight of 52 kd (191). As is actin, intermediate filaments are resistant to treatment with glycerol and nonionic detergents, and they can be visualized in the "cleared" cells treated in this way.

Some types of intermediate filaments appear to have a 2.5–3.5 nm electron-lucent core (192,193) when viewed in cross section. This property, their slightly larger size, and the fact that they are not labeled with HMM or S1 distinguish intermediate filaments from actin filaments in electron micrographs. With anti-bodies against a particular intermediate filament protein and immunofluorescence staining, intermediate filaments have been located in the cortex, but especially form a perinuclear net or basket in some cells (194,195).

Lens intermediate filaments were first isolated by Maisel and Perry (196) from chick lens fiber homogenates using high-speed centrifugation and later by Perry (197) by lens fiber rupture at an air-buffer interface and by Maisel et al. (198) by selective urea solubilization. The filaments were mixed with beaded chains (see below). Intermediate filaments had been seen in situ with the electron microscope in the lenses of a wide variety of species, but serious attention was not given to defining their distribution until later (80,95,96,160,199, 200).

In adult rabbit (160), chick (159), and human lenses (179), intermediate filaments are present in epithelial cells and in cortical fibers but are absent in nuclear fiber cells. In the human neonate, however, intermediate filaments are present in lens nuclear fibers (179,201). This interesting observation suggests intermediate filaments are shorter lived than actin filaments, which do persist in the adult nucleus.

In lens epithelial cells, intermediate filaments occur throughout the cytoplasm but are concentrated in a perinuclear net or basket. The filaments appear to be more numerous in the epithelial cells of older individuals (36,96) and of nonaccommodating lenses (96).

Using antibodies directed against three types of intermediate filament protein, vimentin, desmin, and prekeratin, Ramaekers et al. (201), found that immunofluorescent staining occurred in cultured bovine lens epithelial cells only with anti-vimentin, not with the other two. These findings were confirmed by SDS-PAGE and two-dimensional electrophoresis (201,202). Since vimentin is the intermediate filament type associated with fibroblasts, the significance of this type in the ectoderm-derived epithelial cell of the lens is being sought. However, more recently it has been suggested that the process of culturing epithelial cells in itself brings about expression of vimentin rather than of the native cyto-keratin pattern (202a).

Intermediate filaments appear to be associated with the plasma membrane, frequently showing end-on attachment (143,159,160,203). In a translation system in vitro, when purified lens fiber membranes are added to a reticulocyte lysate programmed with lens polyribosomes, newly synthesized vimentin is associated with the plasma membranes (204), along with actin and crystallin polypeptides. These authors also showed, in freeze-fracture replicas, fragments of filaments measuring about 12 nm in diameter associated with the P face. On the other hand, Ramaekers et al. (187a) demonstrated with double indirect immunofluorescence after incubation of whole calf lens frozen sections with anti-vimentin and anti-actin antibodies that the intermediate filament protein is more often associated with the general cytoplasm of the lens fibers than with the plasma membranes, the latter "lighting up" with the anti-actin antibodies.

The function of intermediate filaments in lens cells (or any cell) is speculative. The consensus is that they serve a structural role in maintaining cell shape, especially in conjunction with other elements of the cytoskeleton. Recently, based on measurements of filament thickness and angles of pitch of the strands composing the filaments (Fig. 40) which vary along the length of individual filaments, Lieska et al. (205) have proposed that the filaments are elastic. The wider diameter and angle would represent the relaxed shortened form, and the thinner diameter and angle would represent the stretched-out high-potential energy state. When the forces stretching out the filaments were released, the filaments, attached to the cell membrane, would recoil into their shortened state. In this scheme, recoil is used in the opposite direction from the scheme utilizing actin; in the latter, recoil would be into their extended state.

Beaded Chains

When lens fiber cells are studied by electron microscopy in situ, or when their contents are separated by various means, another cytoskeletal component in the matrix becomes obvious (159,160,179,196,197,199). These are beaded chains consisting of a 7–9 nm backbone filament studded with particles 12–20 nm in diameter. Beaded chains are not present in lens epithelial cells. They appear as short scattered elements in elongating cells and are present throughout the cortex and nucleus of the lens as increasingly more complex and aggregated structures. Both the filamentous backbone and the associated particles have a controversial nature. An excellent discussion of this subject can be found in the review by Benedetti et al. (143). There is some evidence that the backbone filament is actin (143,186) and that the particles may be a combination of α-crystallins and of polyribosomes whose organization in the lens fiber cell is through their attachment to the actin filament backbone. However, as noted above, native actin filaments without "beads" also are present in lens fiber cells (180a).

Myosin and Tropomyosin

Myosin, the thick filament in muscle, is known to occur in nonmuscle cells as well. It has been localized with immunofluorescent antibody techniques, and it has been extracted and purified from a variety of cell types (see Ref. 206 for a review). However, myosin filaments are usually not seen in cells with transmission electron microscopy because of the low molar ratio of myosin to actin. They have been photographed by Pollard (163) in their reaction with actin in human platelets and by Heuser and Kirschner (207) and Mooseker and Tilney (208) in the microvilli of intestinal epithelium. Each myosin molecule consists of a tail (160 nm long) and two globular heads, each 7 nm in diameter, attached to one end of the tail (209). Cytoplasmic myosin molecules self-assemble into 10–11 nm wide (at the naked zone) and 325 nm long bipolar filament arrays containing 25–30 units (163). The heads of the myosin units project laterally at each end of the filament. Interaction with actin, and ATPase activity, are associated with the heads.

Tropomyosin is an α-helical molecule attached in the groove of the actin helix along its length. Tropomyosin is believed to regulate actin filaments as a "negative modulator" by imparting structural rigidity to an otherwise plastic filament (210). Thus, tropomyosin has been located by immunofluorescent antibody staining in those parts of cultured fibroblasts where structural rigidity is necessary to keep the cell adherent to the substratum, but it is absent in membrane ruffles where dynamic motile activity occurs (210).

Other cytoskeletal-associated proteins occur in nonmuscle cells that regulate the gelation and cross-linking of actin and the self-assembly of microtubules. For a review of the extensive literature on regulatory proteins, consult Korn (210). To date, factors regulating intermediate filament polymerization and depolymerization, if these processes occur at all in living cells, are not known.

Studies of myosin, tropomyosin, and α-actinin in lens cells have been slow to develop. Quantitation has not been reported. Ramaekers and Bloemendal (187) have stained cultured bovine lens cells with antibodies against these three structural proteins. The fluorescent patterns are similar to those obtained with anti-actin antibodies, which points to the association of the proteins with the actin filaments.

ACKNOWLEDGMENTS

The author is indebted to Diane L. Scholz, William Goossens, and Amy R. Roth for their contributions to this work. Especial thanks are given to Diane Scholz for her expert photography.

The previously unpublished photographs have accrued from research supported by National Eye Institute Grant EY 00698.

REFERENCES

1. E. F. Fincham. Br. J. Ophthalmol., *21* (Monograph Suppl., 8): 7 (1937).
2. N. Brown. Br. J. Ophthalmol., *56*: 624 (1972).
3. N. Brown. Trans. Ophthal. Soc. U.K., *92*: 303 (1972).
4. N. Brown. Exp. Eye Res., *19*: 175 (1974).
5. A. Sorsby, G. A. Leary, M. J. Richards, and J. Chaston. Vision Res., *3*: 499 (1963).
6. J. Luyckx-Bacus and R. Weekers. Bull. Soc. Belge Ophthalmol., *143*: 552 (1966).
7. D. J. Coleman and B. Carlin. Arch. Ophthalmol., *77*: 124 (1967).
8. P. Smith. Trans. Ophthalmol. Soc. U.K., *3*: 79 (1883).
9. N. Brown. Exp. Eye Res., *15*: 441 (1973).
10. C. Hanna and J. E. O'Brien. AMA Arch. Ophthal., *66*: 103 (1961).
11. R. Weekers, Y. Delmarcelle, J. Luyckx-Bacus, and J. Collignon. In *The Human Lens in Relation to Cataract*, Ciba Foundation Symposium 19, Elsevier, New York, 1973, p. 25.

12. J. Luyckx-Bacus and Y. Delmarcelle. Bull. Soc. Belge Ophthalmol., *152*: 507 (1969).
13. N. Brown. Trans. Ophthalmol. Soc. U.K., *96*: 18 (1976).
14. J. J. Harding, K. C. Rixon, and F. H. C. Marriott. Exp. Eye Res., *25*: 651 (1977).
15. J. Nordmann, H. Fink, and O. Hockwin. Graefes Arch. Klin. Exp. Ophthalmol., *191*: 165 (1974).
16. H. Helmholtz. Graefes Arch. Ophthalmol., *1*: 1 (1855).
17. G. L. Walls. *The Vertebrate Eye*, Cranbrook Institute of Science, Bloomfield Hills. Michigan, 1942.
18. B. Patnaik. Invest. Ophthalmol., *6*: 601 (1967).
19. R. F. Fisher. J. Physiol. (Lond.), *201*: 21 (1969).
20. A. Duane. Am. J. Ophthalmol., *5*: 865 (1922).
21. R. F. Fisher. J. Physiol. (Lond.), *270*: 51 (1977).
22. C. Rabl. Z. Wissensch. Zool., *67*: 1 (1900).
23. C. Rabl. Z. Wissensch. Zool., *65*: 304 (1888).
24. H. Kahmann, Zool. Jahrb., *48*: 509 (1930).
25. S. Duke-Elder. *System of Ophthalmology*, Vol. 1, Henry Kimpton, London, 1958.
26. A. Hughes. In *The Visual System in Vertebrates* (F. Crescitelli, ed.), Springer-Verlag, New York, 1977.
27. J. G. Sivak. In *Current Topics in Eye Research* (J. A. Zadunaisky and H. Davson, eds.), Vol. 3, Academic Press, New York, 1980.
28. M. Gur and J. G. Sivak. Am. J. Optom. Physiol. Opt., *56*: 689 (1979).
29. L. Z. Bito, C. J. De Rousseau, P. L. Kaufman, and J. W. Bito. Invest. Ophthalmol. Vis. Sci., *23*: 23 (1982).
30. J. G. Sivak, Trends Neurosci., December: 314 (1980).
31. B. Levy and J. G. Sivak. J. Comp. Physiol., *137*: 267 (1980).
32. J. W. McAvoy. In *Mechanisms of Cataract Formation in the Human Lens* (J. W. McAvoy, ed.), Academic Press, New York, 1980.
33. W. Lerche and K. G. Wulle. Ophthalmologica, *158*: 296 (1969).
34. A. I. Cohen. Am. J. Anat., *103*: 219 (1958).
35. N. R. Willis, M. J. Hollenberg, and C. R. Braekevelt. Can. J. Ophthalmol., *4*: 307 (1969).
36. M. M. Perry, J. Tassin, and Y. Courtois. Exp. Eye Res., *28*: 327 (1979).
37. I. C. Mann. *The Development of the Human Eye*, 3rd ed., British Medical Association, London, 1964.
38. A. Weinsieder, J. Reddan, and D. Wilson. Exp. Eye Res., *23*: 355 (1976).
39. C. V. Harding and B. D. Srinivasan. Exp. Cell Res., *25*: 326 (1961).
40. H. Rothstein, J. Reddan, and A. Weinsieder. Exp. Cell Res., *37*: 440 (1965).
41. N. S. Rafferty. J. Morphol., *121*: 295 (1967).
42. S. E. Brolin, H. Diderholm, and H. Hammar. Acta Soc. Med. Upsalien, *66*: 43 (1961).
43. C. Hanna and J. E. O'Brien. Arch. Ophthalmol., *66*: 129 (1961).
44. A. G. Mikulicich and R. W. Young. Invest. Ophthalmol., *2*: 344 (1963).

45. D. S. Thomson, A. Pirie, and M. Overall. Arch. Ophthalmol., *67*: 104 (1962).
46. N. S. Rafferty and K. A. Rafferty. J. Cell. Physiol., *107*: 309 (1981).
47. J. R. Kuszak and J. L. Rae. Exp. Eye Res., *35*: 499 (1982).
48. B. Byers and K. R. Porter. Proc. Natl. Acad. Sci. USA, *52*: 1091 (1964).
49. A. Porte, M.-E. Stoeckel, and A. Brini. Arch. Opht. (Paris), *28*: 681 (1968).
50. J. Wakely. Exp. Eye Res., *22*: 647 (1976).
51. J. Piatigorsky, H. DeF. Webster, and S. P. Craig. Dev. Biol., *27*: 176 (1972).
52. T. L. Pearce and J. Zwaan. J. Embryol. Exp. Morphol., *23*: 491 (1970).
53. J. Piatigorsky. Ann. N.Y. Acad. Sci., *253*: 333 (1975).
54. D. C. Beebe, D. E. Feagans, E. J. Blanchette-Mackie, and M. E. Nau. Science, *206*: 836 (1979).
55. D. C. Beebe, P. J. Compart, M. C. Johnson, D. E. Feagans, and R. N. Feinberg. Dev. Biol., *92*: 54 (1982).
56. J. T. Wrenn and N. K. Wessells. J. Exp. Zool., *171*: 359 (1969).
56a. C. Maloney and J. Wakely. Exp. Eye Res., *34*: 877 (1982).
57. J. Zwaan and R. W. Hendrix. Am. Zool., *13*: 1039 (1973).
58. R. F. Fisher and B. E. Pettet. J. Anat., *112*: 207 (1972).
59. J. H. Seland. Acta Ophthalmol. (Copenh.), *52*: 688 (1974).
60. G. B. Wislocki. Am. J. Anat., *91*: 233 (1952).
61. A. I. Cohen. Invest. Ophthalmol., *4*: 433 (1965).
61a. B. W. Streeten, P. A. Licari, A. A. Marucci, and R. M. Dougherty. Invest. Ophthalmol. Vis. Sci., *21*: 130 (1981).
62. A. Porte, M. E. Stoeckel, and A. Brini. Arch. Opht. (Paris), *31*: 445 (1971).
63. N. Bornfeld, M. Spitznas, W. Breipohl, and G. J. Bijvank. Graefes Arch. Klin. Exp. Ophthalmol., *192*: 117 (1974).
64. P. N. Farnsworth, J. A. Mauriello, P. Burke-Gadomski, T. Kulyk, and A. A. Cinotti. Invest. Ophthalmol., *15*: 36 (1976).
65. P. N. Farnsworth, and P. Burke. Exp. Eye Res., *25*: 563 (1977).
66. B. W. Streeten. Invest. Ophthalmol. Vis. Sci., *16*: 364 (1977).
67. G. Raviola. Invest. Ophthalmol. Vis. Sci., *10*: 851 (1971).
68. P. R. Cammarata and R. G. Spiro. J. Cell. Physiol., *113*: 273 (1982).
69. A. I. Cohen. Dev. Biol., *3*: 297 (1961).
70. K. G. Wulle and W. Lerche. Graefes Arch. Klin. Exp. Ophthalmol., *173*: 141 (1967).
71. Y. Takei and K. Mizuno. Graefes Arch. Klin. Exp. Ophthalmol., *202*: 237 (1977).
72. K. G. Wulle and W. Lerche. Graefes Arch. Klin. Exp. Ophthalmol., *176*: 126 (1968).
73. W. Lerche and K. G. Wulle. Ophthalmologica, *158*: 296 (1969).
74. R. W. Young and D. E. Ocumpaugh. Invest. Ophthalmol., *5*: 583 (1966).
75. N. S. Rafferty and W. Goossens. Growth, *42*: 375 (1978).
76. M. Salzmann. *Anatomie und Histologie des Menschlichen Augapfels im Normalzustande*, Franz Deuticke, Leipzig, 1912.

77. R. F. Fisher and J. Wakely. Trans. Ophthalmol. Soc. U.K., *96*: 278 (1976).
78. A. Brini, A. Porte, and M. E. Stoeckel. Bull. Mem. Soc. Fr. Ophtalmol.,
    76: 193 (1963).
79. N. J. Unakar, C. V. Harding, J. R. Reddan, and R. A. Shapiro. J. Microsc.,
    16: 309 (1973).
80. N. S. Rafferty and W. Goossens. Am. J. Anat., *142*: 177 (1975).
81. L. Von Sallmann, P. A. Grimes, and D. M. Albert. Am. J. Ophthalmol., *68*:
    435 (1969).
82. D. St. C. Roberts. Br. J. Ophthalmol., *41*: 338 (1957).
83. A. Pirie. Biochem. J., *48*: 368 (1951).
84. M. E. Grant, N. A. Kefalides, and D. J. Prockop. J. Biol. Chem., *247*: 3539
    (1971).
85. M. E. Grant, N. A. Kefalides, and D. J. Prockop. J. Biol. Chem., *247*: 3545
    (1971).
86. Z. Dische and V. L. N. Murty. Exp. Eye Res., *16*: 151 (1973).
87. D. Schwartz and A. Veis. Biopolymers, *18*: 2363 (1979).
88. D. Schwartz and A. Veis. Eur. J. Biochem., *103*: 29 (1980).
89. T. Wanko and M. A. Gavin. In *The Structure of the Eye* (G. K. Smelser,
    ed.), Academic Press, New York, 1961.
90. T. Kuwabara. Arch. Ophthalmol., *79*: 189 (1968).
91. N. S. Rafferty, W. Goossens, and W. F. March. Am. J. Ophthalmol., *78*:
    985 (1974).
92. A. Porte, A. Brini, and M. E. Stoeckel. Ann. Ophthalmol., 7: 623 (1975).
93. T. Wanko and M. A. Gavin. Arch. Ophthalmol., *60*: 868 (1958).
94. W. C. Gorthy. Exp. Eye Res., 7: 394 (1968).
95. N. S. Rafferty and E. A. Esson. J. Ultrastruct. Res., *46*: 239 (1974).
96. N. S. Rafferty and W. Goossens. Exp. Eye Res., *26*: 177 (1978).
97. S. P. Modak and S. W. Perdue. Exp. Cell Res., *59*: 43 (1970).
98. B. D. Srinivasan and T. Iwamoto. Exp. Eye Res., *16*: 9 (1973).
99. T. Kuwabara and M. Imaizumi. Invest. Ophthalmol., *13*: 973 (1974).
100. C. V. Harding and S. R. Susan. Invest. Ophthalmol., *15*: 433 (1976).
101. W. C. Gorthy, M. R. Snavely, and N. D. Berrong. Exp. Eye Res., *12*: 112
    (1971).
102. C. V. Harding, J. R. Reddan, N. J. Unakar, and M. Bagchi. In *International
    Review of Cytology*, Vol. 31 (G. H. Bourne and J. F. Danielli, eds.),
    Academic Press, New York, 1971.
103. T. Kuwabara. Exp. Eye Res., *20*: 427 (1975).
104. M. Palva and A. Palkama. Exp. Eye Res., *19*: 117 (1974).
105. J. P. Revel. J. Histochem. Cytochem., *12*: 104 (1964).
106. T. Wanko and M. A. Gavin. J. Biophys. Biochem. Cytol., *6*: 97 (1959).
107. B. V. Worgul, T. Iwamoto, and G. R. Merriam. Ophthalmic Res., *9*: 388
    (1977).
108. B. V. Worgul and S. Mandel. J. Electron Microsc. (Tokyo), *29*: 151
    (1980).
109. H. Bloemendal, A. Zweers, E. L. Benedetti, and H. Walters. Exp. Eye Res.,
    20: 463 (1975).

110. M. Palva and A. Palkama. Exp. Eye Res., *22*: 229 (1976).
111. N. J. Unakar and J. Y. Tsui. Invest. Ophthalmol. Vis. Sci., *19*: 630 (1980).
112. W. C. Gorthy. Exp. Eye Res., *27*: 301 (1978).
113. N. J. Unakar, G. Price, and J. Tsui. Ophthalmic Res., *14*: 83 (1982).
114. D. A. Goodenough, J. S. B. Dick, and J. E. Lyons. J. Cell Biol., *86*: 576 (1980).
115. H. Maisel, C. V. Harding, J. R. Alcala, J. Kuszak, and R. Bradley. In *Molecular and Cellular Biology of the Eye Lens* (H. Bloemendal, ed.), John Wiley and Sons, New York, 1981.
116. W. K. Lo and C. V. Harding. Invest. Ophthalmol., *20*: 129 (1981).
117. J. L. Rae and T. Stacey. Exp. Eye Res., *28*: 1 (1979).
118. Y. Kanno and W. R. Loewenstein. Nature, *212*: 629 (1966).
119. N. B. Gilula, O. R. Reeves, and A. Steinbach. Nature, *235*: 262 (1972).
120. W. R. Loewenstein. Cold Spring Harbor Symp. Quant. Biol., *40*: 49 (1976).
121. J. L. Rae. Exp. Eye Res., *15*: 485 (1973).
122. J. Wakely. Exp. Eye Res., *18*: 571 (1974).
123. E. L. Benedetti, I. Dunia, C. J. Bentzel, A. J. M. Vermorken, M. Kibbelaar, and H. Bloemendal. Biochim. Biophys. Acta, *457*: 353 (1976).
124. B. T. Philipson, L. Hanninen, and E. A. Balazs. Exp. Eye Res., *21*: 205 (1975).
125. J. P. Revel and M. J. Karnovsky. J. Cell Biol., *33*: 7 (1967).
126. D. Branton. Proc. Natl. Acad. Sci. USA, *55*: 1048 (1966).
127. D. Branton, S. Bullivant, N. B. Gilula, M. J. Karnovsky, H. Moor, K. Muhlethaler, D. H. Northcote, L. Packer, B. Satir, P. Satir, V. Speth, L. A. Staehlin, R. L. Steere, and R. S. Weinstein. Science, *190*: 54 (1975).
128. D. A. Goodenough. Cold Spring Harbor Symp. Quant. Biol., *40*: 37 (1976).
128a. E. Raviola, D. A. Goodenough, and G. Raviola. J. Cell Biol., *87*: 273 (1980).
129. C. Peracchia. Nature, *271*: 669 (1978).
130. J. Kuszak, J. Alcala, and H. Maisel. Am. J. Anat., *159*: 395 (1980).
131. T. S. Leeson. Exp. Eye Res., *11*: 78 (1971).
132. D. H. Dickson and G. W. Crock. Invest. Ophthalmol., *11*: 809 (1972).
133. P. N. Farnsworth, S. C. J. Fu, P. A. Burke, and I. Bahia. Invest. Ophthalmol., *13*: 274 (1974).
134. P. N. Farnsworth, P. A. Burke, B. J. Wagner, S. C. J. Fu, and T. J. Regan. Metab. Pediatr. Ophthalmol., *4*: 31 (1980).
135. M. J. Hollenberg, J. P. H. Wyse, and B. J. Lewis. Cell Tissue Res., *167*: 425 (1976).
136. K. J. Nelson and N. S. Rafferty. Exp. Eye Res., *22*: 335 (1976).
137. M. Sakuragawa, T. Kuwabara, J. H. Kinoshita, and H. N. Fukui. Exp. Eye Res., *21*: 381 (1975).
138. H. Bloemendal, A. Zweers, A. J. M. Vermorken, I. Dunia, and E. L. Benedetti. Cell Differ., *1*: 91 (1972).
139. E. L. Benedetti, I. Dunia, and H. Bloemendal. Proc. Natl. Acad. Sci. USA, *71*: 5073 (1974).

140. S. Okinami. Graefes Arch. Klin. Exp. Ophthalmol., *209*: 51 (1978).
141. D. A. Goodenough. Invest. Ophthalmol. Vis. Sci., *18*: 1104, (1979).
142. J. Kuszak, H. Maisel, and C. V. Harding. Exp. Eye Res., *27*: 495 (1978).
143. E. L. Benedetti, I. Dunia, F. C. S. Ramaekers, and M. A. Kibbelaar. In *Molecular and Cellular Biology of the Eye Lens* (H. Bloemendal, ed.), John Wiley and Sons, New York, 1981.
144. J. R. Kuszak, J. Alcala, and H. Maisel. Exp. Eye Res., *34*: 455 (1982).
145. R. M. Broekhuyse, E. D. Kuhlmann, J. Bijvelt, A. J. Verkleij, and P. H. J. T. Ververgaert. Exp. Eye Res., *26*: 147 (1978).
146. J. R. Kuszak, J. Alcala, and H. Maisel. Exp. Eye Res., *33*: 157 (1981).
147. J. R. Kuszak, J. L. Rae, B. U. Pauli, and R. S. Weinstein. J. Ultrastruct. Res., *81*: 249 (1982).
148. J. L. Rae, Invest. Ophthalmol., *13*: 147 (1974).
149. C. Peracchia and L. L. Peracchia. J. Cell Biol., *87*: 708 (1980).
150. G. Bernardini and C. Peracchia. Invest. Ophthalmol. Vis. Sci., *21*: 291 (1981).
151. J. Kistler and S. Bullivant. FEBS Lett., *111*: 73 (1980).
152. G. Zampighi, S. A. Simon, J. D. Robertson, T. J. McIntosh, and M. J. Costello. J. Cell Biol., *93*: 175 (1982).
153. C. Peracchia and L. L. Peracchia. J. Cell Biol., *87*: 719 (1980).
154. P. N. T. Unwin and G. Zampighi. Nature, *283*: 545 (1980).
155. J. J. Wolosewick and K. R. Porter. J. Cell Biol., *82*: 114 (1979).
156. M. Schliwa and J. Van Blerkom. J. Cell Biol., *90*: 222 (1981).
157. R. D. Sloboda, S. A. Rudolph, J. L. Rosenbaum, and P. Greengard. Proc. Natl. Acad. Sci. USA, *72*: 177 (1975).
158. R. L. Margolis and L. Wilson. Nature, *293*: 705 (1981).
159. R. H. Bradley, M. Ireland, and H. Maisel. Exp. Eye Res., *28*: 441 (1979).
160. M. Ireland, H. Maisel, and R. H. Bradley. Ophthalmic Res., *10*: 231 (1978).
161. M. O. Lonchampt, M. Laurent, Y. Courtois, P. Trenchev, and R. C. Hughes. Exp. Eye Res., *23*: 505 (1976).
162. H. Ishikawa, R. Bischoff, and H. Holtzer. J. Cell Biol., *43*: 312 (1969).
163. T. D. Pollarrd. In *Molecules and Cell Movement* (S. Inoué and R. E. Stephens, eds.), Raven Press, New York, 1975.
164. T. D. Pollard and R. R. Weihing. CRC Crit. Rev. Biochem., *2*: 1 (1974).
165. S. Brown, W. Levinson, and J. A. Spudich. J. Supramol. Struct., *5*: 119 (1976).
166. M. Osborn and K. Weber. Exp. Cell Res., *106*: 339 (1977).
167. M. Elzinga, J. H. Collins, W. M. Kuehl, and R. S. Adelstein. Proc. Natl. Acad. Sci. USA, *70*: 2687 (1973).
168. P. A. Rubenstein and J. A. Spudich. Proc. Natl. Acad. Sci. USA, *74*: 120 (1977).
169. K. Zechel and K. Weber. Eur. J. Biochem., *89*: 105 (1978).
170. R. D. Goldman, E. Lazarides, R. Pollack, and K. Weber. Exp. Cell Res., *90*: 333 (1975).
171. H. E. Huxley. J. Mol. Biol., *7*: 281 (1963).

172. F. S. Southwick and J. H. Hartwig. Nature, *297*: 303 (1982).
173. D. A. Begg, R. Rodewald, and L. I. Rebhun. J. Cell Biol., *79*: 846 (1978).
174. G. Y. Mousa and J. R. Trevithick. Dev. Biol., *60*: 14 (1977).
175. D. Drenckhahn and U. Groschel-Stewart. Cell Tissue Res., *181*: 493 (1977).
176. A. Rahi and N. Ashton. Br. J. Ophthalmol., *62*: 627 (1978).
177. G. Y. Mousa and J. R. Trevithick. Exp. Eye Res., *29*: 71 (1979).
178. M. A. Kibbelaar, A.-M. E. Selten-Versteegen, I. Dunia, E. L. Benedetti, and H. Bloemendal. Eur. J. Biochem., *95*: 543 (1979).
178a. F. C. S. Ramaekers, T. R. Boomkens, and H. Bloemendal. Exp. Cell Res., *135*: 454 (1981).
180. M. Ireland and H. Maisel. Ophthalmic Res., *14*: 428 (1982).
180a. M. Ireland and H. Maisel. Exp. Eye Res., *36*: 531 (1983).
181. N. Rafferty, W. Goossens, and A. Roth. Ophthalmic Res., *11*: 276 (1979).
182. M. A. Kibbelaar, F. C. S. Ramaekers, P. J. Ringens, A. M. E. Selten-Versteegen, L. G. Poels, P. H. K. Jap, A. L. van Rossum, T. E. W. Feltkamp, and H. Bloemendal. Nature, *285*: 506 (1980).
183. D. Branton, C. M. Cohen, and J. Tyler. Cell, *24*: 24 (1981).
184. V. Bennett, J. Davis, and W. E. Fowler. Nature, *299*: 126 (1982).
185. C. M. Cohen, S. F. Foley, and C. Korsgren. Nature. *299*: 648 (1982).
186. H. Bloemendal, E. L. Benedetti, F. C. S. Ramaekers, I. Dunia, M. A. Kibbelaar, and A. J. M. Vermorken, Mol. Biol. Rep., *5*: 99 (1979).
187. F. C. S. Ramaekers and H. Bloemendal. In *Molecular and Cellular Biology of the Eye Lens* (H. Bloemendal, ed.), John Wiley and Sons, New York, 1981.
187a. F. C. S. Ramaekers, L. G. Poels, P. H. K. Jap, and H. Bloemendal. Exp. Eye Res., *35*: 363 (1982).
188. S. R. Gordon, E. Essner, and H. Rothstein. Cell Motility., *4*: 343 (1982).
189. R. D. Goldman, A. Milsted, J. A. Schloss, J. Starger, and M.-J. Yerna. Ann. Rev. Physiol., *41*: 703 (1979).
190. J. M. Starger, W. E. Brown, A. E. Goldman, and R. D. Goldman. J. Cell Biol., *78*: 93 (1978).
191. E. Lazarides and D. R. Balzer. Cell, *14*: 429 (1978).
192. R. B. Wuerker. Tissue Cell, *2*: 1 (1970).
193. P. Steinert, R. Zackroff, M. Aynardi-Whitman, and R. D. Goldman. Methods Cell Biol., *24*: 399 (1982).
194. R. D. Goldman and E. A. C. Follett. Science, *169*: 286 (1970).
195. V.-P. Lehto, I. Virtanen, and P. Kurki. Nature, *272*: 175 (1978).
196. H. Maisel and M. M. Perry. Exp. Eye Res., *14*: 7 (1972).
197. M. M. Perry. Exp. Eye Res., *22*: 125 (1976).
198. H. Maisel, N. Lieska, and R. Bradley. Experientia, *34*: 352 (1978).
199. A. Lasser and E. A. Balazs. Exp. Eye Res., *13*: 292 (1972).
200. H. Maisel. Experientia, *33*: 525 (1977).
201. F. C. S. Ramaekers, M. Osborn, E. Schmid, K. Weber, H. Bloemendal, and W. W. Franke. Exp. Cell Res., *127*: 309 (1980).

202. D. Barritault, Y. Courtois, and D. Paulin. Biol. Cell., *39*: 335 (1980).
202a. B. Anderton. Nature, *302*: 211 (1983).
203. H. Bloemendal, E. L. Benedetti, F. Ramaekers, and I. Dunia. Mol. Biol. Rep., *7*: 167 (1981).
204. F. C. S. Ramaekers, I. Dunia, H. J. Dodemont, E. L. Benedetti, and H. Bloemendal. Proc. Natl. Acad. Sci. USA, *79*: 3208 (1982).
205. N. Lieska, H. Maisel, and A. E. Romero-Herrera. Curr. Eye Res., *1*: 339 (1981).
206. M. Clarke and J. A. Spudich. Annu. Rev. Biochem., *46*: 797 (1977).
207. J. E. Heuser and M. W. Kirschner. J. Cell Biol., *86*: 212 (1980).
208. M. S. Mooseker and L. G. Tilney. J. Cell Biol., *67*: 725 (1975).
209. A. Elliott, G. Offer, and K. Burridge. Proc. R. Soc. Lond. [Biol.], *193*: 45 (1976).
210. E. Lazarides. J. Supramol. Struct., *5*: 531 (1976).
210. E. D. Korn. Proc. Natl. Acad. Sci. USA, *75*: 588 (1978).

2
Developmental Genetics of the Lens

RUTH M. CLAYTON / Institute of Animal Genetics, University of Edinburgh, Scotland

A large number of genetic conditions are known leading to anomalies of the lens in humans and other mammals, of which the most common is cataract (opacity of the lens), but aphakia (absence of the lens) and ectopia lentis (extrusion of the lens from the capsule) generally have a genetic basis. However, aphakia in the great majority of individuals is not due to a genetic defect but follows surgery for cataract.

Lenticular anomalies may follow abnormalities of early development or defects of lens fiber cell differentiation in the embryo and fetus or may develop postnatally in an apparently normal structure. Some postnatal cataracts progress rapidly after birth; other forms appear after considerable delay. A further distinction may be made between lenticular conditions that appear alone and those that form part of a syndrome in which extralenticular or extraocular tissues are also involved.

Kratochvilova and Ehling (1) have estimated that, for dominant genetic conditions alone, 3.1% of those listed in the 1975 edition of McKusic's *Mendelian Inheritance in Man* involve cataract, and a rough estimate of published data suggests at least 50 mutations in the mouse associated with cataract. There are also reports of cataracts in other species, such as rats, dogs, and poultry (2-4).

Many metabolic diseases in humans, including some that cause cataract, such as galactosemia or galactokinase deficiency (5,6), or predispose to its development, such as some of the glucose-6-phosphate dehydrogenase (G6PD) or Ah variants (7-10) are well studied. However, an understanding of most congenital cataracts in humans must lean on investigations of other mammals. How closely might an animal model stand in for a human condition? The heterogeneity of genetic cataracts is likely to be much greater than might be deduced from morphopathology and mode of genetic transmission alone. Linkage data, or, in the mouse, the assignment of many cataract mutants to chromosomal locations, are additional discriminants (11–13), and eventual biochemical and molecular studies may be expected to increase even further the likely number of genes affecting the lens. Given such heterogeneity, it would be unsafe to assume that a specific mouse mutant might serve as an experimental model for any human cataract judged to be similar by its morphology or time of onset. We may, however, regard animal studies as providing us with the elements of a map, which, as it is gradually filled in, describes the course of events in the development, differentiation, and functional maintenance of the eye and lens.

The importance of studying genetic anomalies of the lens is twofold. Virtually any anomaly of the lens will affect vision drastically. The development of reliable techniques of differential diagnosis and a search for the underlying nature of the lesion are requirements for genetic counseling and for the evaluation of possible therapeutic procedures. In the case of cataracts that form part of a syndrome, the extraocular features may affect a range of tissues more accessible to an assessment of physiologic response or molecular investigations. An understanding of the biochemical, molecular, or cellular defect may then make it possible to identify heterozygotes or fetuses at risk.

In those cases where a genetic condition is known to be associated with a degree of liability to cataract, an understanding of the mechanism might help to reduce the risk. However, a study of lens mutants must also contribute significantly to our understanding of those fundamental developmental processes that may not be readily accessible to experimentation in mammals and not at all in humans.

TYPES OF GENETIC MODIFICATIONS OF THE LENS

Developmental Processes

Genetic conditions affecting the lens may be divided into four main categories. Genes that are essential for the development of the lens may regulate processes affecting lens placode induction, formation of the lens vesicle, fiber induction, and the development and regression of the embryonic ocular vascularization. Studies of mutants in this category have support and extended the findings on developmental relation based on morphologic records of development and the techniques of experimental embryology (14,15).

Differentiation of Fiber Cells

Qualitative or quantitative changes in the expression of apparently normal crystallins often involve several or all the crystallins. They are probably secondary consequences, either of changes in the signals available to the lens or changes in cellular functions. Crystallin synthesis is affected by genetic failures of fiber differentiation associated with anomalies of cell orientation or nuclear structure (16,17) or by genetic differences in cell membrane composition and rate of mitosis (18,19). The evidence for the response of crystallin synthesis to growth conditions and external signals is not only genetic. Synthesis is also affected by proximity to the retina (20,21), by the use of agents disrupting cytoarchitectural elements (22), and by conditions affecting the rate of mitosis (18).

Postnatal Maintenance of the Lens

Mutants with a progressive postnatal cataract are likely to be associated with metabolic or osmotic anomalies (23–25). Typically, progressive changes in the water content and ionic balance and other lens components are paralleled by opacity (26–28).

Crystallin Mutants

Finally, there might be genetic cataracts due to changes in the amino acid sequence of a crystallin.

The crystallins have evolved slowly (29,30), and a comparison of 19 mammalian and 2 marsupial αA_2 chains found overall only 35 variable positions, and most species differences involve only about three positions. Selective pressures against changes in crystallin structure are evidently considerable and are presumably related to the requirements for close and orderly molecular associations (29–33). It is not altogether surprising that there are, so far, few cases on record of electrophoretic changes found in individuals with normal lenses within a species (34–37), since it might be anticipated both on general grounds (31) and on a structural basis (32,33) that possible substitutions leading to an unacceptable change in conformation might be expected for an even greater proportion of sites than is the case for hemoglobins. Quantitative differences in subunit composition between species are, however, the rule (29,31), and on a priori grounds (29,31,38), in view of the range of possible crystallin heteropolymers found, these would not be expected to be disruptive of crystallin assembly although they might affect the refractive index.

Crystallins presumably differ in their susceptibility to such modifications of chemical structure as are known to be associated with cataract (39), for example, their susceptibility to oxidative cross-linkage, by virtue of having different numbers of accessible sulfydril groups (32). Quantitative changes in the representation of, for example, members of the γ-crystallin gene family, may therefore affect the likelihood of damage to the lens, or its extent.

It is also plausible that some hitherto undetected polymorphisms might include variants that are more susceptible to the various forms of chemical modification. In humans, the bearer of such variants might at present be classed as liable to senile-type cataracts with early onset. A possible example of polymorphism for a crystallin mutant is afforded by the presence of equal amounts of Ala and Thr at position 55 in the hyrax αA_2 chain (40), although duplicate loci was considered a less likely possibility. However, at least some of the possible structural changes in the charge or structure of a crystallin polypeptide would be expected to affect sites involved in heteropolymer assembly or in the association of crystallins with cytoarchitectural elements. Such effects might lead to local light scattering and therefore to cataract.

Cataracts characterized by a stationary opacity, in a specific location with dense, granular, or hazy opacities but no other anomaly, may well turn out to provide a source of mutant crystallins. One such candidate may be the dominant Coppock cataract in humans, which is a stationary annular opacity in the periphery of the nucleus. The protein profiles of two Coppock cataracts were found to be distinct from those of all other cataracts examined, with an anomaly in the γ-crystallin region (41). A dominant cataract of similar appearance, the *Nzc* mutant, has been described in the mouse (42).

THE PROBLEM OF PLEIOTROPY

Pleiotropic Patterns Including the Lens

The lens may be affected alone, together with other ocular structures or in association with extraocular structures. A pleiotropic pattern, or syndrome, is the association together of several anomalies in one mutant. A pleiotropic pattern may have some features invariably present and others that are only occasionally present or that develop later than other features of the syndrome. Both occasional and delayed features are likely to be secondary effects contingent upon an effect that is regularly present.

A consideration of a sample of syndromes in the two best studied species, human and mouse, suggests that the most common association is of cataract or aphakia with retinal anomalies, or microphthalmia, but effects on the cornea, including fusions of lens and cornea and other ocular defects, are also found (Tables 1 and 2). Postnatal retinal degeneration is also frequently associated with progressive cataract.

Cataract, aphakia, or microphthalmia may also occur in association with effects on extraocular systems (Tables 1 and 2). These include

1. Defects of the vestibular apparatus, such as deafness or choreiform behavior
2. Involvement of the central nervous system (CNS), including mental defect
3. Skeletal defects

4. Defects of kidney structure and function
5. Effects of pigmentation
6. Effects on skin, hair, and teeth
7. Effects on reproduction

These categories are not mutually exclusive, and syndromes with other, less common patterns are also known. There are some problems in attempting comparisons between mouse and human. For example, mental defect can be obvious in a human but not in a caged mouse.

Mechanisms of Pleiotropy

Several organs may be affected together in a syndrome by any one of the following mechanisms.

1. The tissues are the several derivatives of one tissue that is affected in early development, or they are dependent upon inductive relations involving the defective tissue (43).
2. Modifications of the structure or biosynthesis of molecules important for general physiologic functions, such as hemoglobins or hormones, must be expected to have many secondary effects in the organism (44,45).
3. A molecular species, such as an enzyme, a structural protein, or cell surface component, may be superabundant in one tissue but may also be present at low but significant levels in cells of different origin, type, and location. This applies to proteins formerly regarded as organ specific, such as myosin and crystallins, and the evidence is discussed elsewhere (46).
4. Anomalies in a molecular feature of organelles, such as lysosomes, cilia, or mitochondria, or factors affecting mitosis or chromosomal behavior may be expected to affect many structures. The *Wh* gene in the Syrian hamster (47) or the Zellweger syndrome in the human (48) appear to exemplify this mechanism.
5. The systemic accumulation of an intermediate metabolite, as in the aminoacidopathies, or a deficiency of a normal product, as in metabolic blocks, may be expected to affect many systems (44,45).
6. The accumulation and deposition of products that are normally metabolized, as in the lipidoses and mucopolysaccharidoses, may also be expected to affect numerous systems (44,45).

Involvement of the Eye and Ear

Anomalies of the eye and ear occur together in several species (Tables 1 and 2). In the early embryo the eye and ear fields overlap. Both areas are able to give rise to either structure, under appropriate experimental conditions (49), and there are some general similarities in their early development (50).

Table 1 Some Syndromes in Rodents that Manifest Cataract

Mutant	Reference	Anophthalmia	Microphthalmia	Eye-cup Retina	Lens	Capsule	Cornea
fidget (fi)	58, 59		+	+	+		c-l
anophthalmic white (Wh)	47		+	+	+		
eye ear (ie)	11				+		
vacuolated lens (Vl)	42				+		
microphthalmia (mi)			+		+		
Iris dysplasia (Idc)	42		+		+		c-l
Anterior lenticolis (Alm)	42				+		c-l
Iris Anomaly Cataract (Iac)	41				+		
Rat cataract	110				+	+	
Ocular retardation (or)	116			+	+		
Sey	18, 102		+	+	+	+	c-l
Anophthalmia (ey)	73	+	+	+	a		
Aphakia (ak)	77	+		+	a		c-l
Rat cataract	120			+	+	+	
Dickie's Small Eye (Dey)	84				+		c-l
Nuclear Cataract (Nuc)	42				+		
Blind (Bld)				+	+		+
Gorthy's Cataract	86		+	+	+		
Cataract	104				+	+	
Anterior polar cataract (Apyc)	42				+		c-l
Eye blebs (eb)	11	+	+		+		
Dysgenic lens (dyl)	83				+		+

c-l : cornea-lens fusions
hl : homozygous lethal
sl : semilethal

The development of the eye involves a sequence of inductive relations between neural and ectodermal tissue, neural and mesodermal tissue, and mesodermal and ectodermal tissues (51–54). The lens is induced in competent head ectoderm by contact with the eye cups, which grow out laterally from the diencephalon, but local mesoderm also plays a role in lens and ocular morphogenesis (55). The ear is induced in competent head ectoderm in association with a local region of hindbrain, but local mesoderm also has a specific inductive role (56). Neuroectoderm and derivative structures, such as the retina, the pineal, and the posterior lobe of the pituitary, are all ciliated in the embryo (47,57). The

Other	Ear	CNS	Pigment	Skin and Appendages	Body size	Skeleton	Viability	Reproduction	Kidney
	+	+			+	+			
	+	pituitary	+		+			♂ sterile	
	+								
		+	+						
			+			+			
iris			+		+		hl		
			+				sl	sterile	
iris				teeth		+	hl		
						+			
+		?							
		?							
		?							
+									
							hl		
								sterile	
−		+					hl		
+									
							hl		
+					small	+			

invaginating lens, otic vesicle, and anterior pituitary are also derived from ciliated ectoderm. Defects in the inductive processes may, in principle, involve the neural tissue, the competent ectoderm, or the mesoderm.

Two syndromes are described here, both involving eye and ear together with other anomalies, which illustrate these relations. The fidget mutant (*fi*) in the mouse (58,59) is highly pleiotropic (Fig. 1). The eye is always reduced in size, and there may be anophthalmia or aphakia. The lens is small and vacuolated; there is some degree of fusion between lens and cornea and occasional corneal perforation and corneal opacity. The lacrimal gland is absent and the Harderian gland enlarged.

Table 2 Some Syndromes in the Human that Manifest Cataract[a]

Mutant	Reference	Microphthalmia	Retina	Lens	Cornea	Other (ocular)
Lowe's oculocerebrorenal syndrome	182, 183	+		+	+	Glaucoma
Alport's syndrome	184			+		
Cochleosaccular degeneration	64			+		
Norrie's disease	186		+	+	+	Pseudo-tumour
Usher's syndrome	195		+	+		
Ear and eye syndrome	198	+		+	+	
Fabry's disease	185			+	+	Glaucoma iris
Crome's syndrome	188			+		
Zellweger syndrome	189			+	+	
Conradi disease	190			+		
Rothmund-Thomson syndrome	191			+	+	
Hallerman-Strieff syndrome	192	+		+		
Myotonic dystrophy	193			+		
Marinesco-Sjögren syndrome	194			+		+
Cataract with icthyosis	197			+		

[a]Although heterozygous carriers may in some cases be detected by the biochemical lesion, for example, α-galactosidase in Fabry's disease, or by lens opacities, as in Lowe's syndrome, Zellweger syndrome, and myotonic dystrophy (48,193,198,199), in most cases the problem still awaits solution.

[b]MR, mental retardation.

The semicircular canals always fail to differentiate. The animals are therefore head shakers and circlers, but the cochlea is present and they can hear. The optic capsule that develops from mesenchyme condensing around the membranous labyrinth is abnormal in shape, affecting skull morphology and causing displacement of the parafloccular lobes of the cerebellum. There are fusions between the skull bones and other variable skeletal manifestations, including defects of the pelvic girdle, limbs, and digits. The interpretation is not unambiguous (58,59). Reduced induction of eye and ear might be due to a general abnormality of the mesoderm, which lies between the head ectoderm and neural tube, and from which skeletal elements are later derived (43). However, the reduced mitotic rate

Ear	CNS	Body size	Skeleton	Skin, appendages	Kidney	Other
+	MR	Small	+		+	Hypotonia
+					+	
+						
+	MR					
Deaf + Balance	Some MR					
+	MR			Teeth		+
	+			+		Cardiovascular, gut
	+				+	
	+	Face			+	Liver, hypotonia
	MR	Small	+	+		Gut
		Small	Phalanges	+		
				Skin, hair		
		Small	Face	Teeth, hair, skin		
						Muscle
	Ataxia, MR					
				+		

in both brain and optic vesicle (59) suggests that the primary defect might lie in the neural tube, which is involved in many mutants affecting the inner ear (60) and which also induces regional specificities in mesoderm.

The dominant white mutant *Wh* in the Syrian hamster is similar to mutants in the mouse and other species affecting the eye, the vestibular apparatus, pigment, and fertility (61,62). Cilia in the retina and pituitary persist into adult life in the mutant (47,62). There is a higher degree than normal of ciliation of cells of the lens and inner ear, both of which are normally ciliated in the adult; lens cells have a single cilium (63), but all the vestibular sense organs are ciliated structures. The defects of eye and ear may therefore be related to abnormal morphogenesis

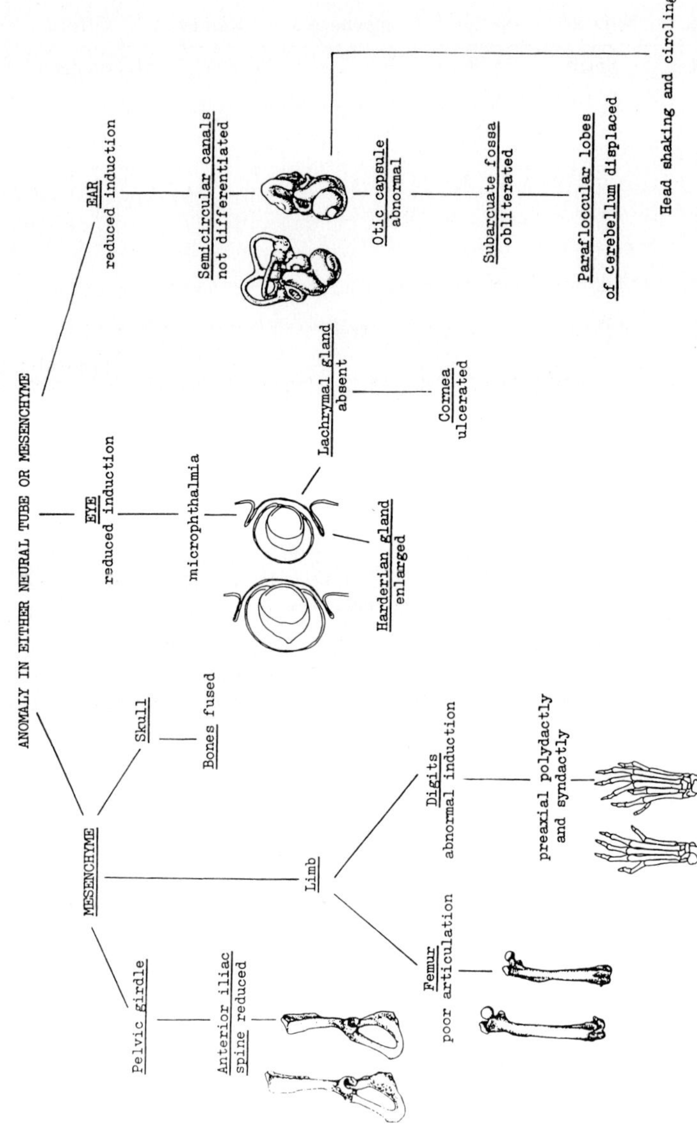

Figure 1 The pleiotropic pattern of manifestation in the fidget mouse, after Truslove (58) and Gruneberg (43), where an abnormal structure is shown; a normal one is shown to the right for purposes of comparison.

associated with the persistence of cilia, the sterility to similar abnormalities of pituitary and pineal, and the lack of pigment to the primary anomaly in the neural tube and neural crest (47,62).

Possible Involvement of a Ubiquitous Organelle

The pleiotropic effects in the Zellweger syndrome, which affects the lens and numerous other tissues and organs (Table 2) (48), may provide an example of an anomaly in a cellular organelle. Abnormal mitochondrial morphology and an absence of peroxisomes occurs in several of the affected tissues.

Degenerative Syndromes with Progressive Cataract: Relation of Lens and Retina

Postnatal cataracts that affect the lens alone, or that occur together with other abnormalities, all of which are secondary consequences of a specific metabolic defect, are described in Postnatal Cataracts, in the section on late-onset cataracts. A shared cellular and molecular characteristic is, however, the simplest explanation of degenerative syndromes affecting the lens and inner ear, such as cochleosaccular degeneration with progressive cataract (64), and may also be the explanation for some degenerative conditions affecting lens and retina. Posterior subcapsular cataract frequently develops in individuals with progressive retinal degeneration, and cataract development may also accompany progressive retinal degeneration in animals (65). Heckenlively (66) examined patients with autosomal dominant, autosomal recessive, and X-linked forms of retinitis pigmentosa (RP), Usher's syndrome, cone-rod degeneration, and "sporadic" cases. Cataract did not develop in all members of a group, and he concluded that it is a secondary consequence of the retinal degeneration.

However, the mode of inheritance of a condition and the overall similarity of the degenerated state is not a guarantee of genetic homogeneity. There are likely to be subgroups that cannot be distinguished from each other by their pathology but that may differ in their liability to specific or nonspecific cataract. There is considerable molecular overlap between retina and lens (67,68), and it remains open whether some forms of retinal degeneration may involve molecules expressed in both tissues. The possibility of biochemically specific cataracts in some forms of retinitis pigmentosa is suggested by particular RP genotypes, in which all individuals have a cataract (69), and also by the finding of three different types of protein profiles in RP cataract lenses. None of the protein profiles resembled those obtained from posterior polar cataracts from non-RP conditions, or cataracts from individuals with other nongenetic retinal diseases (70), but examination of a cataract from a patient with RP showed a number of unusual histologic anomalies (71). Arden and Fox (72) found abnormal nasal cilia in some RP patients. This finding may indicate a subset of retinal degenerative diseases involving a cellular modification in lens, retina, and some other tissues.

GENES AFFECTING ONLY EYE DEVELOPMENT AND LENS DIFFERENTIATION

Some of the processes involved in the development of the lens are outlined in Table 3, and the role of the genotype in regulating those processes may be illustrated by a few brief examples, some of them illustrated in Figs. 2 and 3.

The Lens Placode and Lens Vesicle

Chase and Chase (73) described recessive anophthalmia (*ey*1 and *ey*2) in the mouse. A small proportion of animals were microphthalmic. A small irregular eye cup generally failed to contact the ectoderm and induce a lens, but partial contact was occasionally observed. Subsequent investigations found that some reduced degree of induction was general but was followed by regression. Silver and Hughes (74) suggested an anomaly of embryonic periocular mesenchyme, but Harch et al. (75) postulated that regression followed an irregular location of the lens placode in the eye cup. However, Salaün (76) explanted mutant and normal eye rudiments and found no difference between them in vitro, which implies an abnormality extrinsic to the lens and eye cup.

In the recessive mouse mutant aphakia (*ak*) (77), the eye cup appears to be normal and the placode is induced and forms a vesicle, which generally fails to separate from the ectoderm (16). Mitotic spindles are irregularly oriented in the epithelium, eye cup, and lens vesicle (78). Cells migrate into the vesicle cavity, no primary fibers are formed, and the vesicle regresses. The eye cup of the *ak/ak* embryo develops, sensory cells differentiate, and the optic nerve forms, but since the eye is small, the retina is folded (16). The small size of the eye in *ey*, *fi*, and *ak* is in agreement with the finding by Coulombre and Coulombre (79) that growth of the globe requires the presence of a lens and intraocular pressure. Since α-crystallin was detected in the placodes of *ey*1 and *ak*, in the head ectoderm after regression of the *ey*1 placode, and in the retinal folds after regression of the *ak* vesicle (80–82), it appears that induction by the eye cup activates its synthesis and that β- and γ-crystallins appear in the rodent only in association with fiber formation (20).

Detachment of Lens Vesicle from Cornea

Several mutants apparently fail to completely separate lens vesicle and head ectoderm but differ from *ak* in permitting fiber formation. The lens and cornea remain in contact in some cases with a considerable area of contact between the tissues, both of which are abnormal. In others, the connection is represented by a persistent lens stalk, which leads to a central corneal opacity and anterior polar cataract. An example is the recessive dysgenic lens (*dyl*) in the mouse characterized by a small eye, corneal opacity, iris adhesion, and cataract (83). The lens vesicle fails to separate from the head ectoderm, the fibers are vacuolated and disorganized, and part of the lens is extruded through the region of cornea-lens fusion. The dominant mouse mutant Dickies small eye (*Dey*) is similar (84), and there are four dominant mutants described by Kratchovilova (42) and three by Ehling et al. (85) that suffer from anterior polar cataracts and cornea-lens adhesions. They may be distinguished severally by their other features, such as adhesion to iris, anophthalmia in the homozygote, sterility, white belly spot, or an absence of extraocular manifestation. Adhesion between lens and cornea may be secondary. Gorthy (86) describes a cataractous mouse in which the retina folds and displaces the lens forward. Peter's anomaly in humans (87), with keratolenticular fusion, cataract and corneal opacity, and iris adhesions, presumably forms by one of these mechanisms.

Fiber Differentiation and Capsule Synthesis

Primary fiber formation in the lens vesicle of amphibians, birds, and mammals (79,88,89) requires proximity to the embryonic retina. Degeneration of these fibers may, however, reflect a defect intrinsic to the fiber cell. Fiber degeneration occurs in a number of mutants, but there are differences in morphology, the extent of degeneration, and the sequelae.

A dominant gene in the mouse, eye lens obsolescence (*Elo*), affects primary fiber differentiation (17). The lens vesicle is formed normally, the epithelium and the equatorial fibers are normal, but the central primary fibers have abnormal nuclei and a high lysosomal and low mitochondrial content. The degeneration and loss of the central primary fibers leads to a gross distortion in morphology, with disorganized secondary fiber arrangement and a persistent lens cavity. Organ culture shows that the defect is intrinsic to the eye (90). Although γ-crystallins, which characterize fiber cells, were detected by immunofluorescence, this technique would not show whether any particular members of the γ-crystallin family were absent.

The semidominant gene *cts* (91) leads to swelling and degeneration mainly of the cortical fibers around the lens nucleus, so that *Elo* and *cts* appear to discriminate between primary and secondary fibers. Other mutants, such as Cat^{Fr} and Sey, affect both. Generally, fiber breakdown affects the epithelium and capsule synthesis.

The dominant mouse mutant Cat^{Fr} is characterized by a small lens, with a breakdown of nuclear fibers and vacuolated cortical fibers. The capsule is irregularly thickened, and the lens epithelium is hyperplastic, forming localized irregular whorls embedded in excess capsular material. Zwaan and Williams (92) observed nuclear pyknosis in the primary fibers, which become swollen and vacuolated and then degenerate. The secondary fibers also swell, and the epithelium becomes hypertrophic and capsule synthesis increases. The latter is

Table 3 Effects of Mutants on the Sequence of Events in the Development of the Lens

Process affected in the mutant	Some examples of mutants	Developmental effects
Abnormal properties of neural tube and head ectoderm; different cilia remain on cells	*Wh* (47) (perhaps also similar mutants affecting eye, ear, pigment, and fertility) (61)	Derivatives of neural tube include eye cups (retina), posterior pituitary, pineal, pigment cells; derivatives of head ectoderm include lens, otic vesicle, anterior pituitary
Abnormal growth of eye cup	*ey* (73), *fi* (58,59)	Poor induction of lens placode; no contact or minimal induction—anophthalmia; reduced contact—microphthalmia
Irregular formation of vesicle	*ak* (77), *ey* (73)	May resorb or fail to develop primary fibers; α-crystallin may be synthesized
Vesicle fails to separate or does not completely separate from ectoderm	*ak* (77), *fi* (58,59), *dyl* (83), *Sey* (18,102), *Dey* (84), *vlm* (41), *ldc* (41), *Apyc* (41), *Alm* (41)	Lens corneal fusion or persistent lens stalk; corneal opacity, lenticular material may be extruded to the exterior
Primary fibers differentiate abnormally, break down	*Elo* (19)	Collapse of center of lens, irregular shape affects orientation of secondary fibers
Secondary fibers differentiate abnormally	*Cts* (91)	Abnormal fibers form around a normal lens center
Abnormal retina, possibly fails to regulate fiber differentiation correctly	Gorthy's anterior polar cataract (86)	Congenital defects of both lens and retina

Abnormalities of primary and secondary fibers	Cat^{Fr} (92), Sey (20,102). (Smith et al.) rat mutant (110)	Fiber breakdown may lead to epithelial hyperplasia and capsular thickening
Lens epithelium	Mouse cataract (104), Hy1 chick lens (4)	Multilayered epithelium

Capsule synthesis
  Deficient
  Excessive
	ec (112),
	Cat^{Fr} (92), Sey (20,102,103)
		Thin capsule, later ruptures at posterior

Vascularization of fetal eye	or (116), Browman's rat mutant (118)	Cessation of lens growth, vacuolated fibers
Failure of vascularization to regress	Paget's mouse cataract (119), Bourne et al., rat cataract (120)	Persistent hyaloid artery abnormalities in posterior region of lens
Postpartum metabolism and osmotic maintenance (1) factors intrinsic to lens (e.g., regulation of cation pump) (2) systemic factors (e.g., galactose levels)	Nakano and Philly mice, some aminoacidurias in humans, galactosemia, galactokinase deficiency in humans (5,6,127, 135)	Water content of lens increases; ionic balance disturbed; opacity
Continuing relation between lens and retina	Some types of retinitis pigmentosa in humans	Retinal degeneration may lead to cataract
Factors conducive to liability to cataract at any age	Galactose intolerance, some G6PD variants in humans	Interaction with other factors may lead to cataract
Late-onset cataract	Emory mouse (38)	Late-onset possibly develops because age-related factor or factors interact with genetic liability to cataract

a nonspecific response to trauma or loss of fibers (93-95). The nuclear envelope
of the fiber cells is ultrastructurally abnormal. There is also an increase in endoplasmic reticulum, a decrease in polysomes, and a failure to develop the normal
junctional complexes and interdigitations between cells (96-98). The nuclear
abnormalities are already present in the invaginating lens eye cup (96). Both α
and γ crystallins were detected in the lenses by immunofluorescence (99), but
this technique cannot show changes in subunit composition. There is both a loss
and quantitative distortion in the crystallin profiles in homozygous and heterozygous adult lenses (100,205).

The dominant mutant *Sey* in the mouse is a homozygous lethal. Depending
on the genetic background, the heterozygous eye ranges from almost normal,
with a small anterior polar cataract, to a small structure sunk in the orbit in
which the lens is represented by a grossly thickened and folded capsule lined
with epithelium but devoid of fibers and the retina by massed pigmented cells or
by groups of pigmented cells and groups of unpigmented cells (18,101,102).
Variable effects include dilation of the optic cup with reduced lens induction,
fusions between lens, cornea, and iris, and a persistent lens stalk. Lens fibers may
become vacuolated and distorted. The degree of epithelial hyperplasia and
capsular thickening is related to the amount of fiber cells remaining. Lens
material trapped in the optic stalk may develop into fibers or into epithelial
masses synthesizing capsular material (18,102), and such cell masses may also
adhere to the retina. Capsular glycoproteins are abnormal (103), and Descemet's
and Reichert's membranes are thickened. Capsular ultrastructure shows normal
and abnormal areas intermingled (18). This variable expression at the cellular
level may account for the range of features seen in the heterozygous eye, all of
which may be secondary to a change in extracellular membranes.

An interesting dominant ocular syndrome was reported by Davidorf and
Eglitis in the mouse (104), with degeneration of primary fibers and vacuolization
of the secondary fibers. The anterior capsule is thinner than normal, and the
epithelium becomes multilayered. Blood vessels from the anterior iris then
ramify into the lens underneath the anterior capsule. Vascularization of the lens
epithelium has not been reported in any other mouse cataracts in which primary
fiber breakdown occurs, and an explanation of this peculiarity is required.
Retina and other ocular tissues contains a growth factor EDGF (see Ref. 105),
which affects differentiation and stimulates replication of lens cells and other
ocular and nonocular tissues (106). It is also angiogenic (107). Hypertrophy of
lens epithelium, a failure to maintain fiber differentiation, and angiogenesis are
the types of effect that might be expected in this mutant if the levels of EDGF
were elevated or the number of receptors increased. However, there is no
evidence to this point, and other mechanisms are doubtless possible.

Modification of production of an ocular factor, lentropin, which causes fiber
cell differentiation in vitro (108), or of the putative receptors might also be

susceptible of genetic modification. It might be interesting to examine its effect in vitro on embryonic vesicles that fail to produce fiber cells. Capsular thickening associated with irregular epithelial hyperplasia is found in a number of mutants, both in rodents and in humans (109–111), but is not associated with stratified or folded epithelium (4,112). a thin capsule, leading in some cases to rupture and extrusion of the lens, is found in several rodent mutants (104,112–114). Ectopia lentis is also found in genetic conditions in humans, including Marfans syndrome, Weill-Marchanesi syndrome, homocystinuria, hyperlysinemia, and sulfite oxidase deficiency, reviewed by Jensen (115).

Embryonic Perilenticular Vascularization

Cataract is associated with defective development of embryonic ocular vascularization. Blood vessels, which normally invest the lens for a period before disappearing completely, enter through the choroid fissure. In the recessive mouse mutant ocular retardation (*or*), the choroid fissure closes completely and the blood vessels do not enter the eye (116). Lens development is arrested, and the secondary fibers fail to differentiate and become vacuolated. Retinal differentiation is halted, and no optic nerve is formed. Konyukhov and Ugol'kova (117) reported a reduced mitotic index in the invaginating optic cup and suggested that all subsequent events are secondary. However, they also reported reduced mitosis in *ey* and *fi*, but neither resemble *or* in their subsequent development. A similar failure of the central artery and vein to enter the eye was found in a cataract mutant in the rat (118). A persistent hyaloid artery is also found in association with cataract in some mutants (119,120), and hyaloid remnants have been observed in some cataracts in humans.

POSTNATAL CATARACTS

Early-Onset Cataracts

Postnatal cataracts develop rapidly in galactosemia and galactokinase deficiency (5,6), and their development may be prevented by a galactose-free diet. These are osmotic cataracts of a kind that can be produced by high-galactose diets or other sugars (125,126). The Nakano mouse (127) also develops a progressive osmotic cataract, first seen as a pinhead nuclear opacity at about 30 days after birth, in association with Na^+ influx (128). Denucleation and the loss of organelles are delayed, and the fiber cells are swollen. These changes are detectable at 6 days. Dry weight, protein synthesis, potassium levels, and reduced glutathione content all decrease; water and sodium increase, and there is increased proteolysis (129,130). Polysomes from the Nakano lens synthesize the normal complement of crystallins in a cell-free system (131). The abnormal ionic balance in vivo apparently affects the translatability of the mRNA in vitro.

Similar changes can be brought about by ouabain treatment of control lenses. The changes are due to the presence of an inhibitor of Na^+,K^+-activated ATPase (126) and protein synthesis can be restored in organ culture by correcting the Na^+/K^+ ratio (132). These genetic anomalies are retained in long-term cell culture (133), and the cells also have a greater division potential than normal (134).

The dominant Philly mouse cataract is also an osmotic cataract. It is first observed as a faint anterior opacity at 15 days postpartum; a nuclear opacity develops by day 30, and a total cataract by day 45 (135), but ultrastructural changes including subcapsular particles in the bow cortex cells were seen at 10 days (136). An increase in sodium and water and a decrease in potassium are detectable at 20 days. Protein synthesis, reduced glutathione, and ATP levels fall thereafter as the cataract develops. Unlike the Nakano mouse, there is no Na^+,K^+-ATPase inhibitor present, and the permeabilities of the two mutants to Rb^+ differ. The translatable 27 kdβ^- crystallin mRNA, which is normally detected at about a week after birth, was not found in the Philly homozygote lens and was present at reduced levels in the heterozygote (137). The authors suggest that there might be a mutation or deletion of the coding or regulatory sequences or that nonfunctional mRNA may be present, yet it does not seem likely that the failure of a member of the β-crystallin gene family to appear should affect the permeability of cells that were osmotically competent beforehand. On the other hand, ionic changes can affect crystallin synthesis differentially (131). It is possible, therefore, that mRNA processing as well as translatability may be vulnerable to cell conditions.

Liability to Cataract

Some genetic conditions leading to liability to cataract were described in the first section, and other examples are discussed in Koch et al. (25). It is interesting that, while galactosemia or galactokinase deficiency leads to infant cataract, asymptomatic galactose intolerance leads to a risk of cataract (138).

Late-Onset Cataracts in Association with Extralenticular Degeneration

Cataracts associated with genetically determined degenerations of the retina have been discussed. However, many of the cataracts under this heading, which develop after the onset of retinal degeneration, probably reflect a continuing dependence of the lens on the retina.

"Senile" Cataracts

The autosomal dominant Emory cataract is a late-onset cataract in the mouse (139). Some of the animals also show retinal degeneration. Although the hope was expressed that this mutant may be a model for human senile cataract, the heterogeneity of senile cataracts in terms of protein changes and associated risk factors suggest that any one animal cataract might not be expected to provide a model for all late-onset human cataracts (140–142).

QUANTITATIVE CHANGES IN CRYSTALLINS

Two lines of evidence concerning the evolution of the crystallins and their antigenic sequence, when taken together, point to the importance of quantitative regulation of their expression and suggest that cellular factors may be significant in regulating the relative amounts of crystallin expressed.

Crystallins comprise 80–90% of the lens protein and fall into four families (see Chaps. 4 and 11). Both α- and β-crystallins are found in all vertebrates, but γ-crystallins are replaced by δ-crystallins in birds and reptiles. Each class is distinguished by its antigenic specificities and contains a number of polypeptide chains. In α- and β-crystallins these form polymeric proteins, except for βs and the γ-crystallins, which exist as monomers (29–31). Crystallin heterogeneity is amplified by several forms of posttranslational modification (Chap. 11) (39,102,150).

The crystallins synthesized change during development and differentiation (31,102,143) α-crystallin is the first detected in rat embryos (20,21) δ-crystallin in chicks (19,143,146) and β-crystallins in *Xenopus* (145). Nevertheless, the sequence of gene activation may be strictly conserved. A class specific antiserum will not react equally sensitively to all the polypeptides of that class, which may therefore seem to be relatively delayed in development, if the polypeptide synthesized first is also one that is poorly recognized. Selective processing or differences in translatability may obscure the sequence of transcriptional events. The order of appearance during γ-crystallin ontogeny may be due to post transcriptional events.

The pattern of crystallin synthesis changes steadily throughout development; successively formed fiber cells come to have a composition related to their position in the lens (31,102,144). γ-crystallins, or δ-crystallin in birds, are abundant in the earlier formed fibers, whereas β-crystallins become increasingly prominent in the successively later appearing cortical cells (144,146). α-Crystallin may also increase to a lesser degree (29–31). The polypeptide composition within a class also depends on development and differentiation. Thus, anodal β-crystallins in the chick increase significantly, relative to the cathodal β-crystallins with increasing age (143,148). Various possible signals for the control of crystallin synthesis have been suggested, including proximity to the iris (20,21,149).

The crystallin genes are being intensively studied (150,151). Recombinant or cDNA genomic probes, made to mRNA or the DNA sequences, respectively, have been constructed for crystallin or RNAs, or their genes, in each of the four

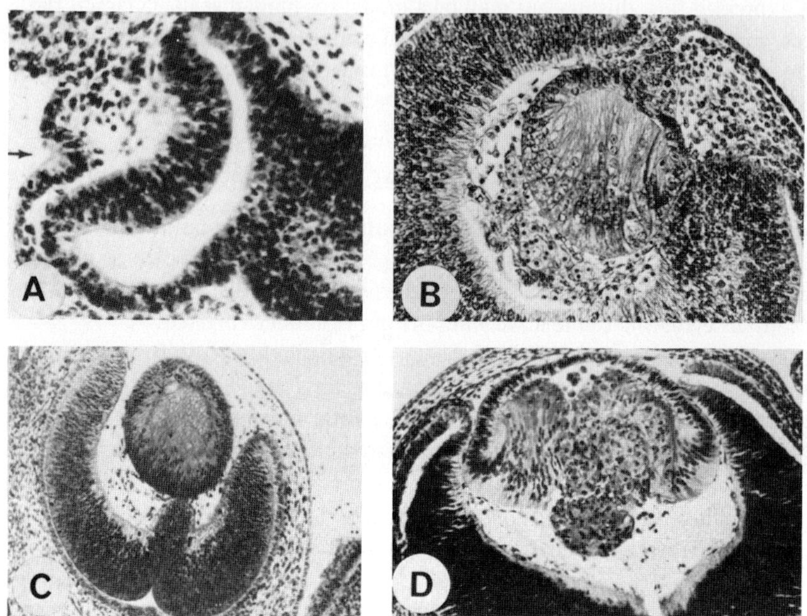

Figure 2   Some abnormalities of development and differentiation of the rodent lens, compiled from several sources. (A) Incomplete induction of the lens placode (75). (B) Primary fiber degeneration (19). (C) Folded retina, followed later by cataract and cornea-lens fusion (86). (D) Cataract following absence of embryonic ocular circulation (116). Anterior vacuoles.

crystallin classes and are being used to examine gene structure and evolution and the regulation of gene expression (150–156). The two δ-crystallin genes are contiguous and very similar. The sequences of the two α-crystallin genes are related to each other with evidence for an internal duplication (29,30). The β-crystallins also appear to be products of a family of at least five or more genes (150,156–158). The γ-crystallins are the best studied, and their relation as members of a gene family is borne out by the similarity of their antigenic structure, amino acid sequences (29,159), tertiary structures (160,161) and by cross hybridization between γ-crystallin clones (162,163). Finally, amino acid nucleotide sequence data all point out to a close relation between β- and γ-crystallins (164–166). These data together with evidence for colinearity of exons and protein structural domains (154,155) indicate that the γ- and β-crystallins have evolved from a smaller ancestral sequence by repeated gene duplication.

   All crystallin genes examined have introns: noncoding nucleotide sequences. A number of mechanisms appear to have been used to generate diversity among the members of a crystallin gene family, including point mutations, deletions, insertions, and duplications. Occasional failure to splice out the first intron in

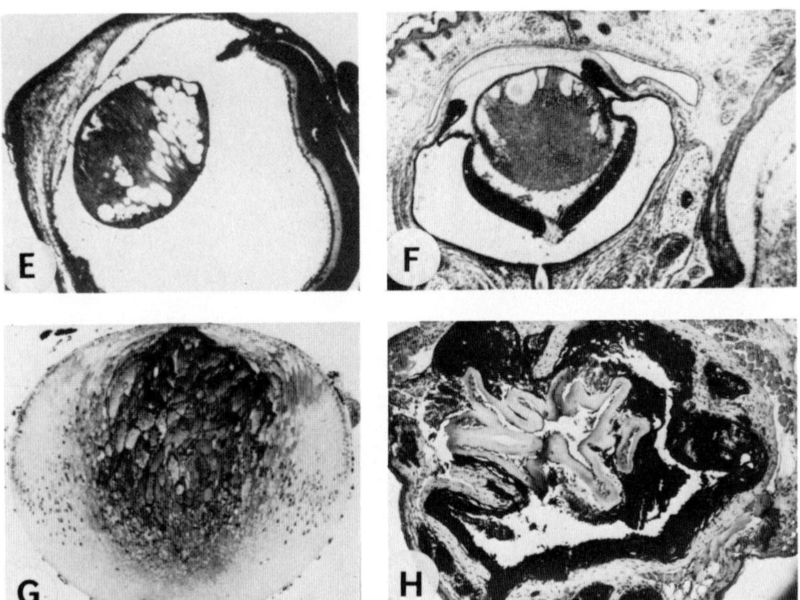

Figure 2   (E) Primary and secondary fiber degeneration (92). (F) Cataract with posterior vacuolization associated with retinal folding (86). (G) Cataract with posterior vacuoles (166). (H) Total fiber loss with folded, thickened capsule and retinal degeneration (202).

rodent α A2 mRNA gives rise to an additional polypeptide α A^{ins}, and it has been suggested (150) that two possible initiation sites in δ-crystallin mRNA might permit the alternative translation of two polypeptides of different length. Quantitative changes appear to evolve faster than the protein sequences (31) and are important during development (31,102,148). The untranslated 5' end of genes carry regulatory sequences and are quite variable among β-crystallins (150).

It has been suggested that the characteristics of the crystallin subunits have evolved to make possible the associated fall in refractive index and rise in water content to permit the lens to function correctly as ocular growth continues (102).

During development the crystallins are not abruptly replaced one by another: each of the γ-crystallins, for example, is most abundant over a particular period of lens development (for example, Ref. 167), but there appears to be an overlapping series of quantitative changes in the representation of each crystallin. These data and comparisons between closely related species that show that quantitative differences in crystallins occur more readily than changes in charge imply that there are genetically determined regulatory factors. The developmental changes in the differential responses of crystallin synthesis to a range of growth

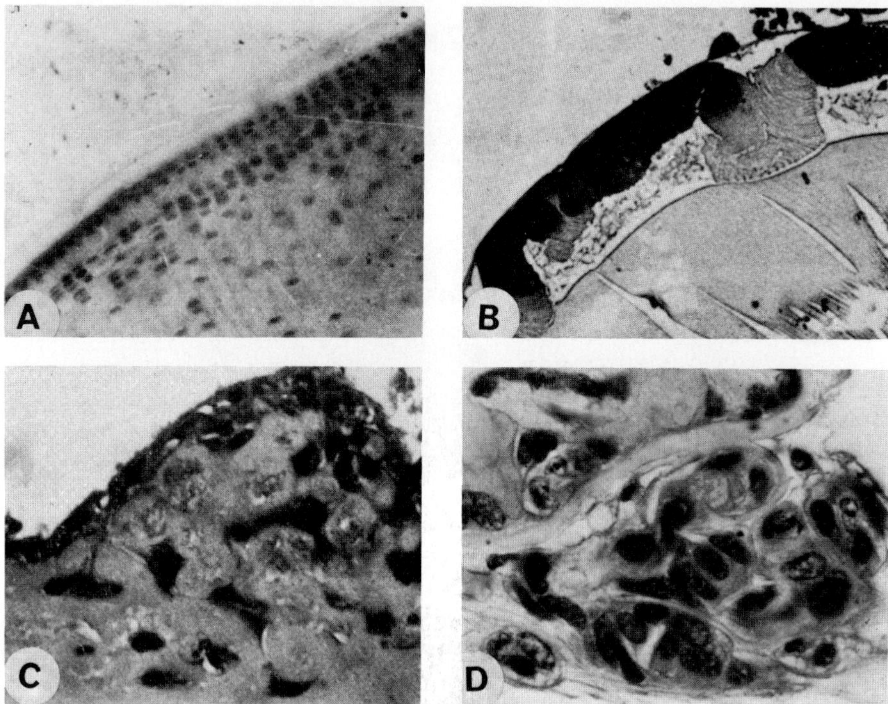

Figure 3 The lens epithelium of different mutants affecting the lens. (A) Multi-layered epithelium, capsule normal or thin (104). (B) Folded epithelium, capsule normal: autoradiograph showing that [^{3}H]proline is not incorporated (and capsular material therefore not synthesized) in epithelial folds pushed inward to the lens body] (200). (C) A nodule of disorganized hyperplastic epithelium and excessive capsular material (92). (D) Widespread disorganized hyperplastic epithelium, excessive capsular material in Sey (Clayton, unpublished).

conditions imply cell-mediated signals able to discriminate between very similar gene products (31,168). Among these factors affecting the quantity, presence, or time of appearance of a crystallin we may classify some of the mutants studied.

A γ-crystallin is apparently missing in *Nop* and *Nzc*, and a γ-crystallin with three allelic variants is at, or close to, the *Elo* locus. The absence of β- and γ-crystallins in the *ak* and *ey* mutants was referred to previously. The temporal sequence of crystallin species is affected in several genetic conditions: crystallin synthesis is delayed, and the location is irregular in *fi* and *or* (59) and in the periodic albinism mutant in *Xenopus* (121). In the *Hyl* chick (4), the epithelium is hyperplastic and folded, the mitotic rate is higher than normal (122), the cell membranes have an abnormal composition and pattern of lectin binding, and there are fewer gap junctions (123). δ-Crystallin appears earlier in both fiber cells and epithelium; β-crystallins are delayed in their appearance (19). At all stages of development δ-crystallin synthesis is higher in *Hyl* than in normal cells. Since agencies increasing the rate of mitosis have the same effect, it is suggested that the accelerated mitotic rate of the mutant affects the ratio of δ- and β-crystallin synthesis (18,124).

EXTRALENTICULAR CRYSTALLINS

If there were a crystallin mutant that grossly distorted the molecular structure, would the consequence be confined wholly to the lens? What follows is frankly wholly speculative.

Myosin, once thought of as organ specific, is now known to be found at low levels in many cell types. It is no longer acceptable to think of myosin in these tissues as a misplaced muscle protein, rather that evolution has, in muscle, produced a tissue that has specialized for a general and widely distributed function.

In the case of the crystallins, these too have been thought of as lens specific, but they are detectable at low levels in certain other tissues, particularly at embryonic stages (reviewed in Ref. 66). Their role, if any, in these tissues (such as retina, adenohypophysis, and midbrain) is unknown, but it is notable that crystallins bind to plasma membranes and ultrastructural elements in the cell (202–204). A partial sequence homology was found recently by Ingolia and Craig (171) between αB-crystallin and *Drosophila* heat shock proteins, but if all the α-crystallin sequences are compared, the degree of homology for this sequence is even higher than their comparison shows: of the order of 74% (Fig. 4). Heat shock proteins are widespread and may be found associated with the nucleoskeleton, chromosomes, and cytoskeletal elements, including microtubules, intermediate filaments, and myofibrils (172–176). α-Crystallins are also able to bind to noncrystallin proteins, such as plasma membranes and cytoskeletal elements (168,188). Should it transpire that extralenticular crystallins have a significant role by forming associations with other cellular elements, it might be that a mutation leading to a sufficient structural distortion might have a serious effect on differentiation or viability.

CONCLUSIONS

This chapter has made no attempt to be comprehensive, but merely to illustrate the types of process elucidated by developmental studies of mutants. The understanding of human metabolic disorders is considerable and increasing rapidly (196), and even if the basis for a particular aspect of a metabolic syndrome is not yet known, it is accessible to study since the nature of the primary lesion is permanent. A defect in an early developmental process is out of reach at present,

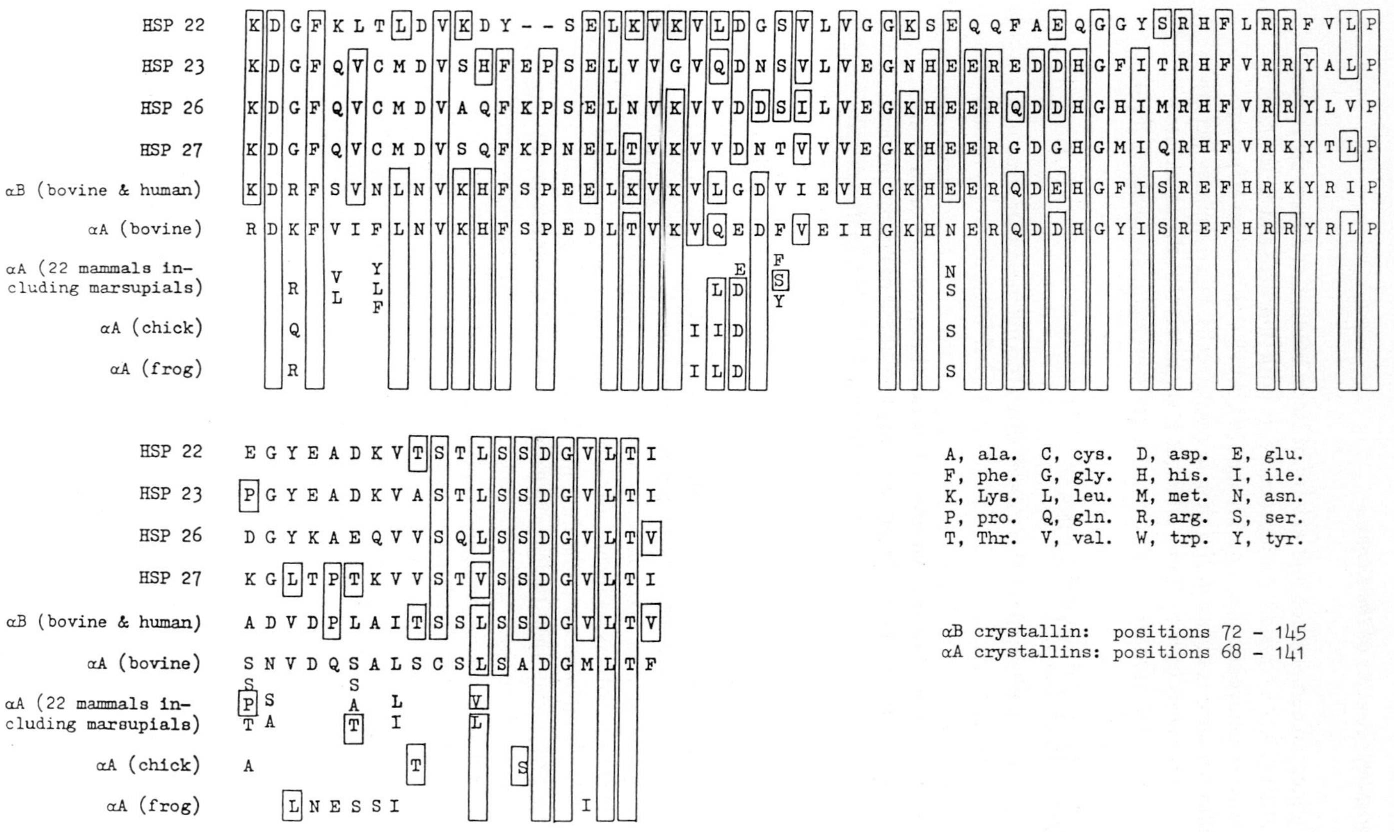

Figure 4 Homologies between *Drosophila* heat shock protein sequences and mammalian α-crystallin sequences. [Compiled from Ingolia and Craig (171), de Jong (29,30), de Jong et al. (40), and Kramps et al. (201).]

so that we are obliged to study other mammals, and some homologies do indeed appear, yet so far the occurrence together of defects in eye and ear in human, mouse, and other species is no more than highly suggestive. The occurrence together of cataract and kidney defects suggests the cataractogenic risk apparently associated with slight but chronic elevations of urea levels (38,141), but other possible linkages can be postulated: inductive defects that as in the fi mouse, affect lens and mesodermal structures; or shared cell membrane components; or enzymes regulating the passage of ions and water. Progressive deafness, retinal degeneration and cataract each present a fairly similar end point for a series of different initial processes (67). Differential diagnosis of apparently similar conditions is the first stage on the long road to molecular understanding. However, the appearance of opacities in heterozygotes for some of the conditions described suggests that the lens is a sensitive indicator and that analysis of a cataractous lens might in some cases provide potentially helpful information.

ACKNOWLEDGMENTS

I am greatly indebted to M. Alexander and L. Dobbie, without whose patient help this would not have been put together, and to Professor H. Maisel for his editorial patience.

REFERENCES

1. J. Kratochvilova and U. H. Ehling. Mutant. Res., *63*: 221 (1979).
2. K. C. Barnett. Trans. Ophthalmol. Soc. U.K., *102*: 346 (1982).
3. J. D. Krehbiel. J. Am. Vet. Med. Assoc., *161*: 634 (1972).
4. R. M. Clayton. Genet. Res. Camb., *25*: 79 (1975).
5. K. J. Isselbacher. In *The Metabolic Basis of Inherited Disease* (J. B. Stanbury, J. B. Wyngaarden, and D. S. Frederichson, eds.), McGraw-Hill, New York, 1966, p. 178.
6. R. Gitzelmann. Lancet, *2*: 670 (1965).
7. W. H. Zinkham. Bull. Johns Hopkins Hosp., *109*: 206 (1961).
8. J. D. Harley, N. S. Agar, M. A. Gruca, M. E. McDermid, and R. L. Kirk. Br. Med. J., *2*: 86 (1976).
9. N. Orzalesi, R. Sorcinelli, and G. Guiso. Arch. Ophthalmol., *99*: 69 (1981).
10. H. Schichi, D. E. Gaasterland, N. M. Jensen, and D. W. Nebert. Science *200*: 539 (1978).
11. M. C. Green. In *Genetic Variants and Strains of the Laboratory Mouse* (M. C. Green, ed.), Gustav Fischer, Verlag: Stuttgart, 1981, p. 8.
12. M. F. Lyon, S. E. Darvis, I. Sayers, and R. S. Holmes. Genet. Res. Camb., *38*: 337 (1981).

13. P. M. Conneally, A. F. Wilson, A. D. Merritt, E. M. Helveston, C. O. Palmer, and L. Y. Wang. Cytogenet. Cell Genet., *22*: 295 (1978).
14. A. J. Coulombre. In *Organogenesis* (R. L. de Haan and H. Ursprung, eds.), Holt Rinehart, New York, 1965, p. 219.
15. G. V. Lopashov. *Developmental Mechanisms of Vertebrate Eye Rudiments*, Pergamon Press, Oxford, 1963.
16. J. Zwaan. Dev. Biol., *44*: 306 (1975).
17. S.-I. Oda, K. Watanabe, H. Fujisawa, and Y. Kameya. Exp. Eye Res., *31*: 673 (1980).
18. R. M. Clayton. In *Genetic Regulation in the Vertebrate Lens Cell* (J. Ebert and T. Okada, eds.), John Wiley and Sons, New York, 1979, p. 129.
19. D. S. McDevitt and R. M. Clayton. J. Embryol. Exp. Morphol., *50*: 31 (1979).
20. J. W. McAvoy. J. Embryol. Exp. Morphol., *45*: 271 (1978).
21. J. W. McAvoy. Differentiation, *17*: 85 (1980).
22. G. Y. Mousa and J. R. Trevithick. Dev. Biol., *60*: 14 (1977).
23. S. S. Vainisi and M. F. Goldberg. In *Genetic and Metabolic Eye Disease* (M. F. Goldberg, ed.), Little, Brown, Boston, 1975, p. 215.
24. M. Landers, J. B. Sidbury, and H. R. Frank. In *Genetic and Metabolic Eye Diseases* (M. F. Goldberg, ed.), Little, Brown, Boston, 1974, p. 357.
25. H.-R. Koch, O. Hockwin, and C. Ohrloff. Metab. Ophthalmol., *1*: 55 (1976).
26. J. Piatigorsky. Biochemistry, *20*: 6427 (1981).
27. G. Duncan and J. M. Marcantonio. Trans. Ophthalmol. Soc. U.K., *102*: 314 (1982).
28. J. J. Harding. Trans. Ophthalmol. Soc. U.K., *102*: 310 (1982).
29. W. W. de Jong. In *Molecular and Cellular Biology of the Eye Lens* (H. Bloemendal, ed.), John Wiley and Sons, New York, 1981, p. 221.
30. W. W. de Jong. In *Macromolecular Sequences in Systematic and Evolutionary Biology* (M. Goodmand, ed.), Plenum Press, New York, 1982, p. 75.
31. R. M. Clayton. In *The Eye* (H. Davson, ed.), Academic Press, London, 1974, p. 399.
32. C. Slingsby. Trans. Ophthalmol. Soc. U.K., *102*: 337 (1982).
33. M. E. Narebor, this publication.
34. L. R. Eckroat. Trans. Am. Fish Soc., *100*: 527 (1971).
35. G. C. Moser and S. Gluecksohn-Waelsch. Exp. Eye Res., *6*: 297 (1967).
36. S. L. Beck and H. V. Bennet. Exp. Eye Res., *29*: 323 (1979).
37. L. C. Skow. Exp. Eye Res., *34*: 509 (1982).
38. R. M. Clayton. In *Stem Cells and Tissue Homeostasis* (B. I. Lord, C. S. Potter and R. J. Cole, eds.), Cambridge University Press, Cambridge, 1978, p. 115.
39. J. J. Harding. In *The Eye Lens as a Model for Molecular and Cellular Studies* (H. Bloemendal, ed.), John Wiley and Sons, New York, 1981.
40. W. W. de Jong, C. C. Nuy-Terwinat, and M. Versteeg. Biochim. Biophys. Acta, *491*: 573 (1977).

41. J. Cuthbert, R. M. Clayton, D. E. S. Truman, C. I. Phillips, and R. S. Bartholomew. In *Interdisciplinary Topics in Gerontology*, Vol. 13 (H. P. von Hahn, ed.), Karger, Basel, 1978, p. 183.

42. J. Kratochvilova. J. Hered., *72*: 302 (1981).

43. H. Grüneberg, *The Pathology of Development*, John Wiley and Sons, New York, 1963.

44. J. B. Stanbury, J. B. Wyngaarden, and D. S. Frederichson. *The Metabolic Basis of Inherited Disease*, McGraw-Hill, New York, 1960.

45. H. Harris. *The Principles of Human Biochemical Genetics*, Elsevier, Amsterdam, 1980.

46. R. M. Clayton. In *Differentiation In Vitro* (M. Yeoman and D. E. S. Truman, eds.), Cambridge University Press, Cambridge, 1982, p. 83.

47. J. M. Asher and S. C. James. Proc. Natl. Acad. Sci. USA, *79*: 4371 (1982).

48. H. M. Mittner, F. L. Kretzer, and R. S. Mehta. Arch. Ophthalmol., *99*, 1977 (1981).

49. A. G. Jacobson. Science, *152*, 25 (1966).

50. T. Horder. In *The Developmental Biology of Plants and Animals* (C. F. Graham and P. F. Wareing, eds.), Blackwell, Oxford, 1976.

51. A. J. Coulombre. In *Organogenesis* (R. L. de Hahn and H. Ursprung, eds.), Holt, Reinhart, Winston, New York, 1966, p. 219.

52. G. V. Lopashov. *Developmental Mechanisms of Vertebrate Eye Rudiments*, Pergamon Press, Oxford, 1963.

53. L. Saxén and J. Wartiovaara. In *The Developmental Biology of Plants and Animals* (C. F. Graham and P. F. Wareing, eds.), Blackwell, Oxford, 1976.

54. N. K. Wessels. *Tissue Interactions in Development*, Benjamin Cummings, Reading, England, 1977.

55. A. G. Jacobson. J. Exp. Zool., *139*, 525 (1958).

56. C. L. Yntema. J. Exp. Zool., *113*, 211 (1950).

57. L. V. Salvini-Plawen and E. Mayr. In *Evolutionary Biology* (M. K. Hecht, W. C. Steene, and B. Wallace, eds.), Plenum Press, New York, 1977, p. 207.

58. G. M. Truslove. J. Genet., *54*: 64 (1956).

59. B. V. Konyukhov and M. P. Vakhrusheva. Teratology, *2*: 147 (1969).

60. M. S. Deol. Nature, *209*: 219 (1966).

61. M. S. Deol. Birth Defects *16*: 243 (1980).

62. J. H. Asher. J. Exp. Zool., *217*: 159 (1981).

63. M. M. Perry, J. Tassin, and Y. Courtois. Exp. Eye Res., *28*: 327 (1979).

64. J. B. Nadol and B. Burgess. Laryngoscope, *92*: 1028 (1982).

65. J. G. Bellows and R. T. Bellows. In *Cataract and Abnormalities of the Lens* (J. G. Bellows, ed.), Grune and Stratton, New York, 1975, p. 272.

66. J. Heckenlively. Am. J. Ophthalmol., *93*: 733 (1982).

67. R. M. Clayton. In *Problems of Normal and Genetically Abnormal Retinas* (R. M. Clayton, J. Heywood, H. W. Reading, and A. Wright, eds.), Academic Press, London, 1982, p. 333.

68. R. M. Clayton. In *Problems of Normal and Genetically Abnormal Retinas* (R. M. Clayton, J. Heywood, H. W. Reading, and A. Wright, eds.), Academic Press, London, 1982, p. 29.

69. O. Valle, H. Erkkila, and C. Raitta. Acta Ophthalmol. (Copenh.), *59*: 695 (1981).
70. J. Cuthbert and R. M. Clayton. In *Problems of Normal and Genetically Abnormal Retinas* (R. M. Clayton, J. Heywood, H. W. Reading, and A. Wright, eds.), Academic Press, London, 1982, p. 369.
71. J. Eshagian, N. S. Rafferty, and W. Goosens. Arch. Ophthalmol., *98*: 2227 (1980).
72. G. B. Arden and B. Fox. Nature, *279*: 545 (1979).
73. H. B. Chase and E. B. Chase. J. Morphol., *68*: 279 (1941).
74. J. Silver and A. F. W. Hughes. J. Comp. Neurol., *157*: 281 (1974).
75. C. Harch, H. B. Chase, and N. I. Gonsalves. Dev. Biol., *63*: 352 (1978).
76. J. Salaün, J. Embryol. Exp. Morphol., *67*: 71 (1982).
77. D. S. Varnum and L. C. Stevens. J. Hered., *59*: 147 (1968).
78. J. Zwaan and B. M. Kirkland. Anat. Rec., *182*: 345 (1976).
79. J. L. Coulombre and A. J. Coulombre. Science, *142*: 1489 (1963).
80. B. V. Konyukhov, N. A. Malinina, E. S. Platonov, and M. I. Yakovlev, Biol. Bull. Acad. Sci. USSR, *5*: 397 (1978).
81. M. I. Yakovlev, E. S. Platenov, and B. V. Konyukhov. Ontogonez, *8*: 115 (1977).
82. N. A. Malinina and B. V. Konyukhov. Ontogonez, *12*: 589 (1981).
83. S. Sanyal and R. K. Hawkins. Invest. Ophthalmol. Vis. Sci., *18*: 642 (1979).
84. K. Theiler, D. S. Varnum, and L. C. Stevens, Anat. Embryol. *155*: 81 (1978).
85. O. H. Ehling, J. Favor, J. Kratochvilova, and A. Neuhauser-Klaus. Mutat. Res. *92*: 181 (1982).
86. W. C. Gorthy. Invest. Ophthalmol. Vis. Sci., *18*: 839 (1979).
87. G. O. Waring, M. M. Rodrigues, and P. R. Laibson. Sun. Ophthalmol., *20*: 3 (1975).
88. Y. Mikami, Jpn. J. Zool., *9*: 269 (1941).
89. Y. Yamamoto. Dev. Growth Differ., *18*: 273 (1976).
90. K. Watanabe, H. Fujisawa, S.-I. Oda, and Y. Kameyana. Exp. Eye Res., *31*: 683 (1980).
91. A. Ikeda, Y. Seki, I. Yoshi, and N. Mishima. Arch. Histol. Jap. *44*: 237 (1981).
92. J. Zwaan and R. M. Williams. Exp. Eye Res., *8*: 161 (1969).
93. S. Duke-Elder. *System of Ophthalmology*, Vol. 9, Kimpton, London, 1969, p. 19.
94. N. S. Rafferty. Anat. Rec., *153*, 111 (1965).
95. B. D. Srinivasan and C. V. Harding. Invest. Ophthalmol., *4*: 452 (1965).
96. Y. Hamai and T. Kuwabara. Invest. Ophthalmol., *14*: 517 (1975).
97. Y. Hamai, Acta Soc. Ophthalmol. Japan, *80*: 717 (1976).
98. B. V. Konyukhov and W. A. Kolesova. Ontogonez, *7*: 271 (1976).
99. E. S. Platonov, M. I. Yakovlev, and B. V. Konyukhov. Ontogonez, *7*: 484 (1976).
100. T. H. Day and R. M. Clayton. Genet. Res. Camb., *19*: 241 (1972).
101. R. M. Clayton and J. C. Campbell. J. Physiol. (Lond.), *198*: 74 (1968).

102. R. M. Clayton. Curr. Top. Dev. Biol., *5*: 115 (1970).
103. D. J. Pritchard, R. M. Clayton, and W. L. Cunningham. Exp. Eye Res., *19*: 335 (1974).
104. F. Davidorf and I. Eglitis. J. Morphol., *119*: 89 (1966).
105. C. Arruti and V. Courtois. Exp. Cell Res., *117*: 283 (1978).
106. J. Tassin, M. Olivier, J. Plouet, M. Laurent, and M. Perry, Adv. Exp. Med. Biol., *158*: 289 (1982).
107. P. Thomson, C. Arruti, D. Maurice, J. Plouet, D. Barritault, and Y. Courtois. In *Problems of Normal and Genetically Abnormal Retinas* (R. M. Clayton, J. Haywood, H. W. Reading, and A. Wright, eds.), Academic Press, London, 1982, p. 63.
108. D. C. Beebe, P. J. Campart, M. C. Johnson, D. E. Feagans, and R. M. Feinberg. Dev. Biol., *92*: 54 (1982).
109. L. J. Pierro and J. Spiggle, J. Exp. Zool., *173*: 101 (1970).
110. R. S. Smith, H. Hoffman, and C. Cisar, Arch. Ophthalmol., *81*: 259 (1969).
111. L. E. Zimmerman and R. L. Font. JAMA, *196*: 684 (1966).
112. A. B. Beasley. J. Morphol., *112*: 1 (1963).
113. F. C. Fraser and G. Shabtach. Genet. Res. Camb., *3*: 383 (1962).
114. G. K. Smelser and L. von Sallmann. Am. J. Ophthalmol., *32*: 1703 (1949).
115. A. D. Jensen. In *Genetic and Metabolic Eye Disease* (M. F. Goldberg, ed.), Little, Brown, Boston, 1974, p. 325.
116. G. M. Truslove. J. Embryol. Exp. Morphol., *10*: 652 (1962).
117. B. V. Konyukhov and T. S. Ugol'kova. Ontogonez, *9*: 475 (1978).
118. L. G. Browman. J. Morphol., *109*: 37 (1961).
119. O. Paget and M. Baumgartner-Gamauf. Osten. Zool. S., *85*: 238 (1953).
120. M. C. Bourne, D. A. Campbell, and M. Pyke. Br. J. Ophthalmol., *22*: 608 (1938).
121. D. S. McDevitt and S. K. Brahma. Differentiation, *14*: 107 (1979).
122. F. E. Randall, D. E. S. Truman, and R. M. Clayton. Genet. Res. Camb., *34*: 203 (1979).
123. P. G. C. Odeigah, R. M. Clayton, and D. E. S. Truman. Exp. Eye Res., *28*: 311 (1979).
124. D. I. de Pomerai, R. M. Clayton, and D. J. Pritchard. Exp. Eye Res., *27*: 365 (1978).
125. R. van Heyningen. Invest. Ophthalmol., *15*: 685 (1976).
126. J. H. Kinoshita. Invest. Ophthalmol., *13*, 713 (1974).
127. T. Nakano, S. Yamamoto, G. Kutsukade, H. Ogawa, A. Nakajima, and E. Takano. Rinsho Ganka, *14*: 196 (1960).
128. S. Iwata and J. H. Kinoshita. Invest. Ophthalmol., *10*: 504 (1971).
129. H. N. Fukui, H. Obazawa, and J. H. Kinoshita. Invest. Ophthalmol., *15*: 422 (1976).
130. D. Roy, M. H. Garner, and A. Spector. Exp. Eye Res., *34*: 909 (1982).
131. T. Shinohara and J. Piatigorsky. Science, *210*: 914 (1980).
132. J. Piatigorsky, H. N. Fukui, and J. H. Kinoshita. Nature, *274*: 558 (1978).
133. Y. Tsunematsu, H. N. Fukui, and J. H. Kinoshita. Exp. Eye Res., *26*: 671 (1978).

134. A. L. Muggleton-Harris, R. D. Lipman, and J. Kearns. Exp. Eye Res., *32*: 563 (1981).
135. P. F. Kador, H. N. Fukui, S. Fukushi, H. M. Jernigan, and J. H. Kinoshita. Exp. Eye Res., *30*: 59 (1980).
136. P. F. Kador, S. Uga, and J. Piatigorsky. In *Ageing of the Lens* (F. Regnault, O. Hochwin, and Y. Courtois, eds.), Elsevier, Amsterdam, 1980, p. 157.
137. D. Carper, T. Shinohara, J. Piatigorsky, and J. H. Kinoshita. Science, *217*: 463 (1982).
138. A. F. Winder, P. Fells, and R. B. Jones. Br. J. Ophthalmol., *66*: 438 (1982).
139. J. F. R. Kuck, T. Kuwabara, and K. D. Kuck. Curr. Eye Res., *1*: 643 (1981/2).
140. R. S. Bartholomew, R. M. Clayton, J. Cuthbert, J. Duffy, C. I. Phillips, J. McK. Reid, J. Seth, D. E. S. Truman, C. Wilson and S.-H. Yim. In *Ageing of the Lens* (F. Regnault, O. Hockwin, and Y. Courtois, eds.), Elsevier, Amsterdam, 1980, p. 241.
141. R. M. Clayton, J. Cuthbert, C. I. Phillips, R. S. Bartholomew, N. L. Stokoe, T. Flytche, J. McK. Reid, J. Duffy, J. Seth, and M. Alexander. Exp. Eye Res., *31*: 553 (1980).
142. R. M. Clayton, J. Cuthbert, J. Duffy, J. Seth, C. I. Phillips, R. S. Bartholomew, and J. McK. Reid. Trans. Ophthalmol. Soc. U.K., *102*: 331 (1983).
143. M. Rabary, Exp. Eye Res., *1*: 310 (1982).
144. J. Genis-Galvez and H. Maisel. Life Sci., *6*: 197 (1967).
145. J. W. McAvoy. J. Embryol. Exp. Morphol., *45*: 271 (1978).
146. J. Zwaan and A. Ikeda. Exp. Eye Res., *7*: 301 (1968).
147. D. S. McDevitt and S. K. Brahma. Dev. Biol., *84*: 449 (1981).
148. P. R. Waggoner, N. Lieska, J. Alcala, and H. Maisel. Ophthalmol. Res., *8*: 292 (1976).
149. R. M. Clayton, P. G. Odeigah, D. I. DePomerai, D. J. Pritchard, I. Thomson, and D. E. S. Truman. In *Biology of the Epithelial Lens Cells* (Y. Courtois and F. Regnault, eds.), INSERM, *60*: 123 (1976).
150. J. Piatigorsky. In *Human Cataract Formation*, Ciba Foundation Symposium, 106, Pitman, London, 1984.
151. H. J. Dodemont, P. M. Andreoli, R. J. M. Moorman, F. C. S. Ramaekers, J. G. G. Schoenmakers, and H. Bloemendal. Proc. Natl. Acad. Sci. USA, *78*: 5320 (1981).
152. T. S. Okada. Cell Differ., *13*: 77 (1983).
153. R. M. Clayton. In *Differentiation In Vitro* (M. M. Yeoman and D. E. S. Truman, eds.), 1982, p. 83.
154. G. Inana, J. Piatigorsky, B. Norman, C. Slingsby, and T. Blundell. Nature, *302*: 310 (1983).
155. R. J. M. Moorman, J. T. Den Dunnen, L. Mulleners, P. Andreoli, H. Bloemendal, and J. F. F. Schoenmakers. J. Mol. Biol., *171*: 353 (1983).
156. J. G. G. Schoenmakers, J. T. Den Dunnen, R. J. M. Moorman, R. Jongbloed, et al. In *Human Cataract Formation*, Ciba Foundation Symposium, 106, Pitman, London, 1984, p. 208.

157. R. M. Clayton and D. E. S. Truman. Exp. Eye Res., *18*: 495 (1974).
158. P. Herbrink, H. van Westreenen, and H. Bloemendal. Exp. Eye Res., *20*: 541 (1975).
159. H. Bloemendal. In *Molecular and Cellular Biology of the Eye Lens* (H. Bloemendal, ed.), John Wiley and Sons, New York, 1981.
160. T. Blundell, P. Lindley, L. Miller, D. Moss, C. Slingsby, I. Tickle, B. Turnell, and B. Wistow. Nature, *289*: (1981).
161. Y. N. Chirgadze. Dokl. Akad. Nauk. SSSR, *259*: 1502 (1981).
162. T. Shinohara, J. Piatigorsky, D. A. Carper, and J. H. Kinoshita. Exp. Eye Res., *34*: 39 (1982).
163. R. J. M. Moorman, J. T. den Dunnen, H. Bloemendal, and J. G. G. Schoenmakers. Proc. Natl. Acad. Sci. USA, *79*: 6876 (1982).
164. G. Wistow, C. Slingsby, T. Blundell, H. Driessen, W. de Jong, and H. Bloemendal. FEBS Lett., *133*: 9 (1981).
165. G. Inana, T. Shinohara, J. V. Maizel, and J. Piatigorsky. J. Biol. Chem. *257*: 9064 (1982).
166. G. Inana, J. Piatigorsky, B. Norman, C. Slingsby, and T. Blundell. Nature, *302*: 310 (1983).
167. C. Slingsby and L. R. Croft. Exp. Eye Res., *17*: 369 (1973).
168. R. M. Clayton, P. G. Odeigah, D. I. de Pomerai, D. J. Pritchard, I. Thomson, and D. E. S. Truman. INSERM, *60*: 123 (1976).
169. E. L. Benedetti, I. Dunia, F. C. S. Ramaekers, and M. A. Kibbelaar. In *Molecular and Cellular Biology of the Eye Lens* (H. Bloemendal, ed.), John Wiley and Sons, New York, 1981, p. 137.
170. R. H. Bradley, M. Ireland, and H. Maisel. Exp. Eye Res., *28*: 441 (1979).
171. T. D. Ingolia and E. A. Craig. Proc. Natl. Acad. Sci. USA., *79*: 2360 (1982).
172. A.-P. Arrigo, S. Fakan, and A. Tissieres. Dev. Biol., *78*: 86 (1980).
173. P. M. Kelley and M. J. Schlesinger. Cell, *15*: 1277 (1978).
174. J. M. Velazques, B. J. DiDomenico, and S. Lindquist. Cell, *20*: 679 (1980).
175. R. M. Sinibaldi and P. W. Morris. J. Biol. Chem., *256*: 10735 (1981).
176. C. Wang, R. M. Gower, and E. Lazarides. Proc. Natl. Acad. Sci. USA, *78*: 3531 (1980).
177. R. M. Clayton. In *Problems of Normal and Genetically Abnormal Retinas* (R. M. Clayton, J. Haywood, H. W. Reading, and A. Wright, eds.), Academic Press, London, 1982, p. 333.
178. J. Zwaan and R. M. Williams. Exp. Eye Res., *8*: 161 (1969).
179. G. Muller. Z. Mikrosk. Anat. Forsch., *56*: 520 (1950).
180. M. L. Watson. J. Hered., *59*: 60 (1968).
181. K. P. Hummel and D. B. Chapman. Mouse Newslett., *28*: 32 (1963).
182. V. T. Curtin, E. E. Joyce, and W. Ballin, Am. J. Ophthalmol., *64*, 533 (1967).
183. L. B. Holmes, B. L. McGowan, and M. L. Efron. Pediatrics, *44*: 358 (1969).
184. E. J. Arnott, M. D. A. Crawford, and P. J. Toghill. Br. J. Ophthalmol., *50*: 390 (1966).

185. W. G. Nance, M. Warburg, D. Bixler, and E. M. Helverston, Birth Defects, *10*: 285 (1974).

186. M. Warburg. In *Genetic and Metabolic Eye Disease* (M. F. Goldberg, ed.), Little, Brown, Boston, 1974, p. 447.

187. D. B. van Dorp and J. W. Delleman. J. Pediatr. Ophthalmol. Strabismus, *16*: 166 (1979).

188. L. Crome, S. Duckett, and A. W. Franklin. Arch. Dis. Child, *38*: 505 (1963).

189. J. M. Opitz, G. M. Zu Rhein, L. Vitale, N. T. Shahidi, J. J. Howe, S. M. Chou, D. R. Shanklin, H. D. Sybers, H. R. Dood, and T. Gerritsen. Birth Defects, Orig. Art. Series, *5*: 144 (1969).

190. B. M. Josephson and M. D. Oriati. Pediatrics, *28*: 425 (1961).

191. H. W. Siemens. In *Genetics and Ophthalmology* (P. J. Waardenburg, A. Franceschetti, and D. Klein, eds.), Charles C Thomas, Springfield, Illinois, 1963, p. 896.

192. H. F. Falls and W. J. Schull. Arch. Ophthalmol., *63*: 409 (1960).

193. G. Pescia and A. E. H. Emery. J. Genet. Hum., *24*: 227 (1976).

194. M. Alter, O. R. Talbert, and G. Croffead. Neurology (N.Y.), *12*: 836 (1962).

195. B. Halgren. Acta Psychiatr. Neurol. Scand. [Suppl] *34*(138): 9 (1959).

196. V. A. McKusick, R. G. Weilbacher, and G. W. Gragg. JAMA, *204*: 113 (1968).

197. O. D. Pinkerton. Arch. Ophthalmol., *60*: 393 (1958).

198. J. W. Delleman, E. M. Bleeker-Wagemakers, and A. W. C. van Veelen. J. Pediatr. Ophthalmol., *14*: 205 (1977).

199. R. J. M. Gardner and N. Brown. J. Med. Genet., *13*: 449 (1976).

200. R. M. Clayton, G. Eguchi, D. E. S. Truman, M. M. Perry, J. Jacob, and O. P. Flint. J. Embryol. Exp. Morphol., *35*: 1 (1976).

201. J. A. Kramps, B. M. de Man, and W. W. de Jong. FEBS Lett., *74*: 84 (1977).

202. F. C. S. Ramaekers, A. E. Selten-Versteegen, and H. Bloemendal. Biochim. Biophys. Acta, *596*: 57 (1980).

203. J. Alcalá, H. Maisel, and N. Lieska. Exp. Cell Res., *109*: 63.

204. J. Alcalá, H. Maisel, M. Katar, and M. Eis. Exp. Eye Res., *35*: 379 (1982).

205. A. T. Garber, L. Stirk, and R. Gold. Exp. Eye Res., *36*: 165 (1983).

206. J. Graw, J. Kratochvilova, and K.-H. Summer. Exp. Eye Res., *39*: 37 (1986).

3

The Physiology of the Lens

JAMES L. RAE AND RICHARD T. MATHIAS / Rush-Presbyterian-St. Luke's Medical Center, Chicago, Illinois

This chapter is presented with the hope that it will be useful for the beginning student of ophthalmic sciences as well as for the seasoned investigator with an interest in lens research. To support these different levels of background, we have organized the chapter into two distinct sections. The first is an overview of lens function as we see it. In this overview we present a general description of lens function, considerable speculation, and no literature references. In the second section we present arguments and references that have led us to our view. When possible, we suggest areas of investigation where work is needed.

At present, the physiology of the lens is poorly understood for at least two reasons. First, it has not been as extensively studied as other tissues, such as nerve or muscle, and second, most past electrophysiologic studies have treated the lens as if it were one giant cell, thus ignoring the structural complexity known to be present. Such studies have described overall average properties but have not dealt with either structural or functional inhomogeneities in the tissue.

It is the purpose of this chapter to provide a modern view of lens functions rather than to provide a complete literature review of lens physiology. Since much of the past work has been descriptions of overall or input properties, these studies are not directly relevant to the modern view of a multicellular tissue with specialized membrane processes within the lens. Such studies are not reviewed here except when they are required for historical perspective or for an understanding of overall properties.

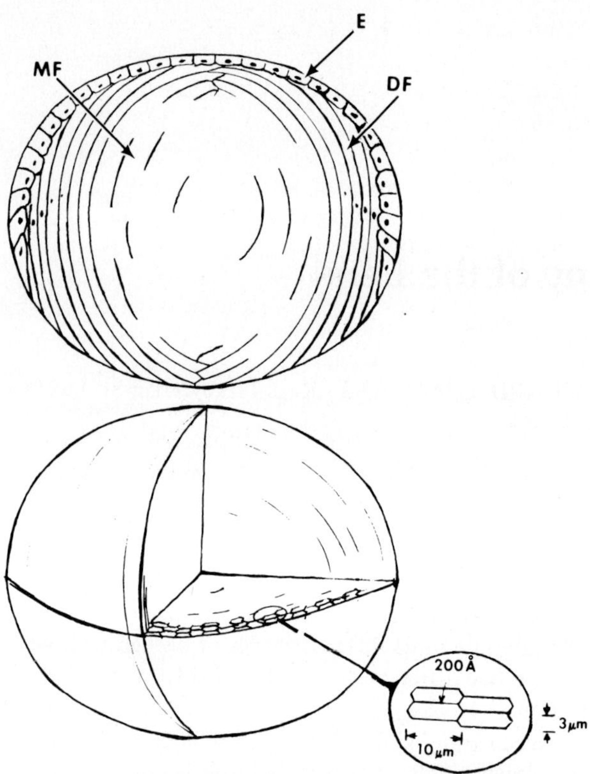

Figure 1 Schematic representation of the structure of the lens. The upper panel depicts the anterior epithelial layer (E), the elongating fiber region (DF) near the equator, and the mature fiber (MF) layer that occurs deeper in the lens than the differentiating zone. The lower panel shows the columnar arrangement of fibers in the lens and gives typical fiber and extracellular cleft dimensions.

OVERVIEW

Structure and Function

The lens, along with the cornea, is responsible for bringing light rays into precise focus on the retina. To do this, it must be transparent and maintain an appropriate composite refractive index difference between aqueous and vitreous humors. Lens physiologic and biochemical processes are thus aimed at the maintenance of transparency and refractive index.

The lens is composed of at least three different populations of cells (Fig. 1). On the anterior surface only, there is a single layer of epithelial cells. These cells, which contain all the organelles characteristic of most cells, are responsible for the majority of active transport of ions, amino acids and lipid precursors and facilitated diffusion of glucose. These processes may also be located to a lesser extent in fiber membranes, but this is at present an area of uncertainty.

At the equator of the lens is a "cap" of differentiating cells (Fig. 1). These are formed by elongation of equatorial epithelial cells. These cells abut epithelial cells on the anterior side and lens capsule on the posterior. Their length is initially very short, but eventually they lengthen to extend essentially from anterior to posterior poles. Hence, this differentiating zone has very short cells in layers near the lens surface and longer cells in each deeper layer. In this zone, the cells lose their nuclei, mitochondria, and other organelles that might give rise to light scattering. The organelles that are present are mostly near the equator where they are "hidden" behind the iris and not usually in the optical path.

The third group of cells are the mature fiber cells (Fig. 1), which by definition extend from pole to pole (suture to suture). Fiber cells exist in "spherical" shells and are mostly layered in a tapering columnar structure. These cells contain a cytoskeleton but lack most other cellular organelles.

All cells of the lens appear to be electrically coupled. This is true for the epithelial cells and the differentiating fibers they contact as well as for all fiber cells. Thus the lens is a highly specialized syncytium where adjacent cells communicate through gap junctions and allow low-molecular-weight substances to move easily from one cell to another.

The fibers of the lens are very closely packed so that the distance separating adjacent cells is only 100–200 Å. This tight packing, which is presumably necessary to limit light scattering, results in a small extracellular space composed of long, tortuous, clefts and that communicate with the aqueous and vitreous. Such restricted clefts offer a high-resistance pathway for diffusion and current flow.

In such a tissue, it is reasonable to locate transport processes near the *surface*, where pumping need not occur into and out of these restricted clefts. Moreover, since transport is a process that can require considerable energy and since production of this energy can occur more efficiently near the lens surface, a *surface* location of transport is even more reasonable. An *anterior* location of pumping processes is also advantageous since pumping would occur into and out of flowing aqueous humor rather than into and out of the more stagnant vitreous.

Furthermore, different cells in the lens have very different conductance or permeability properties. It appears that the epithelial cells have passive ion permeabilities like those of "average" cells, whereas fiber cells have extremely nonleaky membranes. The properties of the differentiating fibers are not really known, but it is reasonable to expect that they lie somewhere between those of the epithelial cells and the mature fibers. Since these differentiating cells are the cells that make up the posterior lens surfacemost layer and since there is good evidence to suggest that the permeability of the posterior surface of the lens is less than that of the anterior, we presume that these differentiating fibers have quite high resistance membranes.

These morphologic and functional features allow the following simple but speculative model of how the lens works. Because the lens must be transparent, its cells are tightly packed to limit the size of extracellular clefts that might cause light diffraction and whose refractive index might be different than that of fiber cytoplasm. Cellular organelles are removed from most cells to reduce light scattering. Active transport processes are located near the surface on the anterior side where they can obtain energy efficiently. Also, with this anterior surface location they can pump into and out of flowing aqueous, thereby ensuring minimal accumulation or depletion of transported solutes. Active transport processes are not located in membrane structures deep in the lens for several reasons: (1) such processes would require energy that cannot be efficiently supplied due to nutrient gradients from surface to center; (2) pumping would have to occur into and out of restricted spaces; and (3) such pumping processes would require extra protein synthesis to support turnover of pump molecules and membrane proteins in a location not well suited to protein synthesis.

The fiber cells are extensively interconnected to each other and to the surface cells through gap junctions. Such interconnections assure that the active transport by the surface cells will be expressed throughout the fiber mass. In other words, the epithelial cells can pump for the fibers. This limited number of epithelial cells is able to maintain the ion gradients for a much larger volume of fiber cells because the fiber cell membranes have a very high ionic resistance and are thus very nonleaky to ions. Consequently, even though there is a very large surface for ion leaks associated with total fiber surface, the actual leak per unit surface is small enough that the surface pump can keep up with it. This arrangement allows both transparency and control of the intracellular milieu of lens fiber cells.

Methodology

The problem of dealing with spatially nonuniform permeabilities or conductances is a well-known problem in physiology. Epitheliologists have long sought to separate adequately the properties of apical and basolateral membranes of transporting epithelia.

When information about permeabilities is sought, physiologists often use electrical methods in place of or in addition to chemical flux methods. Since current in biologic tissues is carried by ions, the measurement of the resistance to ion flow offered by a tissue is also a direct measure of the permeability of the tissue to the ions. Electrical methods are useful because they are precise and

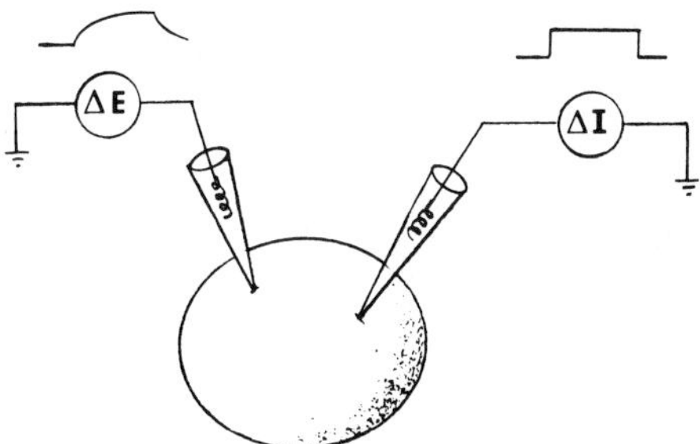

Figure 2 Two microelectrode methods for the measurement of lens resistance or conductance. Current (ΔI) is passed through one microelectrode while the voltage change (ΔE) produced by the flow of current across resistive elements in the lens is measured with a second microelectrode.

sensitive and generally allow the investigator better control over physiologic interventions than is possible with simple isotope uptake or washout experiments. These methods are particularly useful in multicellular preparations, like the lens, where the geometry of cells and clefts can be quite complex. The experimental paradigm for such studies is straightforward. A quantitative model of the tissue is produced that is structurally based on all possible pathways for current flow. Current is injected into the cells of the tissue while the voltage induced by this current is measured. The current and voltage data are compared to the model to assess the relative magnitude of current that flows along each of the possible pathways described in the model. It is very important that the model accurately describe the structure of the tissue, since only then can one hope that the final "permeabilities" determined will be ascribed to the proper structures producing them. For example, the giant cell model of the lens, which does not include the extracellular space, would consider the resistive properties of the extracellular space to have arisen in a membrane and would thus improperly assess the properties of the membrane, in addition to neglecting the role of the extracellular space.

In the lens, the electrical properties are usually determined by placing two glass microelectrodes within different cells in the interior of the lens (Fig. 2). One electrode passes current while the other measures both the lens resting voltage and the change in resting voltage caused by the current flow. Often these data are used to describe simple input properties of the lens. For example, the input resistance is $R_{in} = \Delta E/\Delta I$, where ΔE = the steady-state–induced change in resting voltage due to a step of applied current ΔI. Alternatively, the data are expressed as input conductance: $G_{in} = \Delta I/\Delta E$, where G = conductance. Such an expression of data assumes no "circuit" model for the lens and thus provides

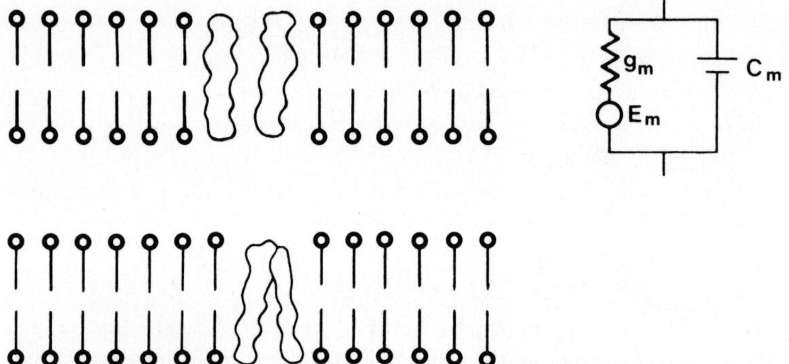

Figure 3   Schematic representation of a single-channel protein embedded in the lipid bilayer matrix of the membrane. In the upper panel the channel is in an open configuration, whereas in the bottom panel the channel is closed. The equivalent circuit for this simple membrane model is shown as a parallel combination of a resistor and a capacitor with a battery in series with the resistor. Channel opening and closing is stochastic.

only a description of average properties from the entire tissue. That is, it lumps the properties of all cell membranes and extracellular space into a single number. An alternative choice of input is to apply sinusoidal currents and study the input impedance of the lens as a function of the sinusoidal frequency. This method has the sensitivity to allow investigation of complex circuit models based on lens structure.

Intracellular Resting Potential

It has been known for more than 25 years that the lens maintains a hyper-polarized resting voltage of about –75 mV, although the resting voltage varies some from species to species. Moreover, the lens maintains a high K^+ and low Na^+ concentration in the cytoplasm of its fiber cells. It is instructive to consider the origin of a resting potential in a single cell and then address the complexities introduced by a syncytial geometry, such as that of the lens.

   Electrically measurable ion permeation through cell membranes is now known to occur through structures called channels. Each channel is thought to be formed from a single large protein molecule embedded in the membrane (Fig. 3). These channel-forming proteins can assume at least two different conformations: in one conformation, a pore is created through the membrane, and in the other conformation this pore is closed. It is through this pore that ions move. In general, membrane channels are capable of a high degree of selectivity in the ions they let through, so different ions may differ significantly in the ease with which they are allowed to pass through a given channel. For example, a channel may allow K^+ ions to move freely but not allow Na^+ or Cl^- to pass at all. For a single channel that is perfectly selective for a single kind of ion, it is easy to determine the driving force for ion movement across the channel. If the channel had a concentration gradient across it for the ion it allows to pass, for example, potassium ions, the K^+ ion would begin to diffuse down its concentration gradient through the open channel. However, as soon as a single K^+ ion crossed it would cause an increase in positive charge on one side of the membrane and a decrease in positive charge on the other side. This would create a voltage difference across the channel. The net ion movement would stop when the voltage across the channel was

$$E_K = \frac{RT}{ZF} \ln \frac{[K]_i}{[K]_o}$$

where R = universal gas constant

Z = valence of the ion

T = absolute temperature

F = Faraday's constant

$[K]_i$, $[K]_o$ = concentrations of K^+ on the inside and outside of the membrane, respectively.

E_K as defined above is called the Nernst potential or the equilibrium potential for potassium. A similar equilibrium potential exists for each channel. It is the potential that occurs when the ion flux due to diffusion exactly balances the ionic current flux due to voltage. Even for the simplest case where a membrane might have perfectly selective channels for Na^+, K^+, and Cl^-, an immediate complexity can be seen. We know from cell ion measurements that the concentration gradients across the membrane for Na^+, K^+, and Cl^- are not the same. Therefore, if a membrane had Na^+, K^+, and Cl^- channels, each channel must have a different driving force across it. The membrane can have only one potential difference across it; thus the resting membrane potential is forced to assume some weighted average of the equilibrium potentials across the various channel types. The weighting depends on the relative conductance contributed by each channel type; that is, the channel type contributing the greatest average conductance is the most important determinant of the resting potential. If the average conductance due to this channel were very much greater than the average conductance of other channels in the membrane, the resting potential would have a value essentially equal to the Nernst potential of this dominant channel. This does not mean that a single channel of the dominant channel type need have a large conductance. Rather, the total average conductance depends on the total number of channels, the conductance of a single channel when it is open, and the average amount of time the channels are open.

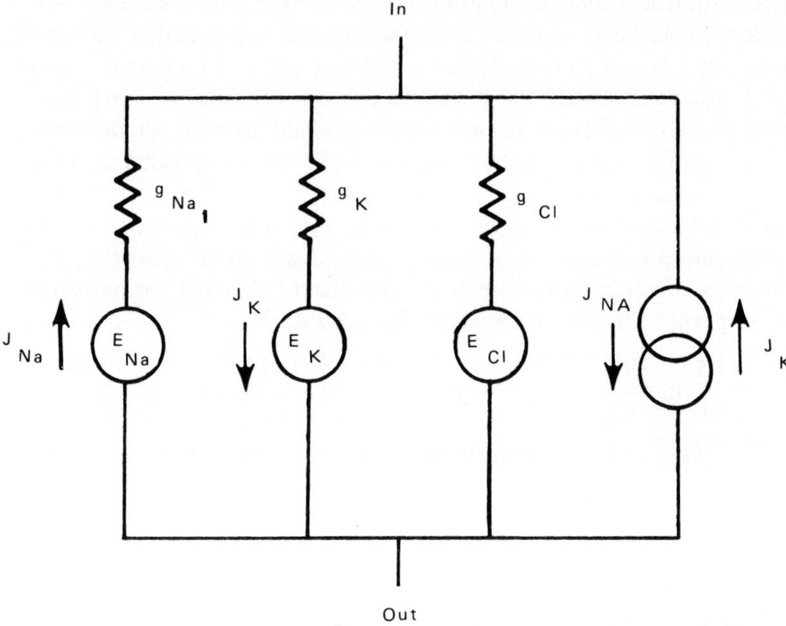

Figure 4 Schematic representation of a membrane containing a sodium, potassium, and chloride channel and a sodium-potassium pump. Each channel is shown as a battery in series, with a conductance with the battery, in general, different for each channel. The sodium-potassium pump is in parallel with the ion channels. The arrows indicate the direction of the net flux.

Actually, it is quite an artificial situation when a single channel type is the dominant generator of the resting membrane potential. Usually Na^+, K^+, and Cl^- channels are all involved (Fig. 4). One formalism that describes the contribution of multiple channel types and electrogenic transport to the resting potential is the following:

$$E_m = \frac{g_K E_K + g_{Na} E_{Na} + g_{Cl} E_{Cl} + J_p}{g_K + g_{Na} + g_{Cl}}$$

Where g_x stands for the membrane conductance for the ion in question, E_x is the Nernst potential for the ion, and J_p is the net current density due to active transport of ions.

The difference between the resting membrane potential and the Nernst potential for a specific ion channel is a driving force for current flow through the channel, and such a driving force is always in the direction to dissipate the ionic gradient. Pumps are required to produce a current equal and opposite to this dissipating current, to maintain the normal ion gradients across the channel and thus across the membrane.

The membrane potential is capable of rapid changes since it will change as the relative contribution of the various channel types changes. The contribution of a particular channel type could be increased by opening closed channels, by synthesizing new channels, by keeping open channels open longer, or by modifying single channels so that the flux through a single open channel changes. Although there is evidence for all these mechanisms, it would appear that opening closed channels and keeping open channels open longer are the most common mechanisms. The classic example of a rapid membrane potential change is, of course, the action potential, where many Na^+ channels are suddenly opened so the membrane potential, which had been dominated by K^+ channels, becomes dominated by Na^+ channels. Similar membrane potential changes probably occur in epithelial cells but on a slower time scale and to a different extent.

Extracellular Resting Voltage and Standing Current

Using membrane potential changes and/or flux measurements from a single cell to determine Na^+, K^+, and Cl^- permeabilities is useful but not very powerful since only average values for the whole cell can be obtained. Epitheliologists now know that transporting cells often have very different ion transport properties associated with different surfaces of the same cell. The apical membranes have one set of permeabilities and pumps and the basolateral membranes another. The resting membrane potential results from a contribution from each of these membranes. Whole-cell permeabilities are due to a spatial average of these two membranes and thus would not in general be the permeabilities of either membrane.

This problem can be considerably worse when resting potential measurements are made from electrically coupled cells. Consider two cells with different relative Na^+, K^+, and Cl^- conductances (Fig. 5). Each cell would have a resting potential dictated by these relative conductances and so each cell would have a different resting membrane potential. If these two cells were to become electrically coupled, their resting potentials would have to become essentially equal since their cytoplasms would be in lower resistance continuity. This is like connecting two batteries of different voltage: a standing current would be constrained to flow. This current would leave through the membrane of one cell, flow in the extracellular fluid, and reenter through the membrane of the second cell. The current would cause a drop in voltage across the extracellular solution, but in this artificial case of only two coupled cells, the extracellular resistance would be low and thus the extracellular voltage drop would be small.

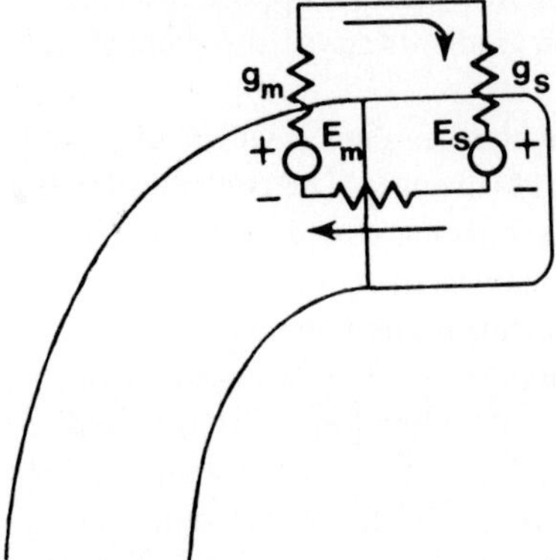

Figure 5 A schematic view of an epithelial cell electrically coupled to a fiber cell. The relative Na^+, K^+, and Cl^- conductances are assumed to be different in the two cells, so that $E_m \neq E_S$. The arrow indicates the current that will flow in response to the difference in electromotive force of these two batteries.

In a tissue like the lens, the situation is very much more complex. Fiber membranes deep in the lens apparently have a different selectivity than surface cell membranes, yet all cells are electrically coupled. Thus one expects a standing extracellular current flow. The extracellular space between lens fiber cells is isolated from the surface by a long, tortuous, restricted extracellular path. The resistance associated with this space is appreciable; hence standing current flow of the kind described above should produce an appreciable drop in voltage along this extracellular space. The voltage at the mouth of the extracellular cleft, where the cleft joins the bathing solution, should be essentially zero, whereas the extracellular voltage near the lens center could be very negative, perhaps even approaching the voltage in the lens fiber cytoplasm. The resting membrane potential of any cell is the difference between the intracellular potential and the extracellular potential. Since the intracellular potential of all lens fibers is expected to be very similar due to the extensive electrical coupling between cells, we are constrained to conclude that the transmembrane potential of lens fibers varies with their depth in the lens. The cells near the nucleus should have the smallest transmembrane voltage, those at the lens surface the largest. At present, reasonable evidence exists that the lens intracellular voltage is relatively uniform spatially and that there is an appreciable voltage drop along the extracellular clefts. Failure to have such an extracellular voltage drop could only occur if fiber cells and epithelial cells have nearly identical selectivity for Na^+, K^+, and Cl^-. There is no evidence to support equal selectivity, whereas considerable evidence suggests that epithelial cells and the majority of fiber cells differ in their conductance characteristics and probably in their selectivity as well. Therefore, the concept of a single resting membrane voltage for the lens is not sensible. Models that use the intracellular resting voltage of the lens to assess Na^+, K^+, and Cl^- permeabilities can only give some overall permeabilities, which should not in general be the permeabilities of any particular membrane in the lens.

Cellular Volume Regulation

Volume regulation in single cells or in syncytial tissues is achieved by membrane proteins controlling the solute content of the cytoplasm. Water moves passively into osmotic equilibrium, whereas the solutes are actively transported. Because the solutes are mostly ionized, the maintenance of cellular volume is intimately related to the electrical properties of the tissue or cell.

The problem faced by all biologic cells is that they contain large, impermeant proteins and other polyelectrolytes, in their interior. These trapped impermeant proteins invariably possess a net negative charge, so if the interior of the cells is to maintain macroscopic electroneutrality, there must be fewer Cl^- ions within the cell than there are Na^+ and K^+ ions. However, water will always move until osmotic balance is achieved between the intracellular and extracellular solutes; thus the sum of intracellular solute concentrations will equal the sum of extra-cellular solute concentrations. Essentially, one-half the extracellular solute concentration is Cl^- (once again because of macroscopic electroneutrality), whereas the intracellular Cl^- concentration must be considerably less than one-half the total solute, owing to the presence of negatively charged impermeant proteins. If there were no resting voltage within the cell, the Cl^- would diffuse down its concentration gradient and into the cell, but water, Na^+, and K^+ would have to follow to keep osmotic balance and electroneutrality. The consequence would be a continuous influx until the cell ruptured. This situation can be avoided if the cell membrane has a resting potential that is negative on the inside, so the tendency for Cl^- to diffuse into the cell is blocked by the electro-motive force.

Interestingly, the diffusion of Cl^- into the cells is blocked by the active transport of Na^+ and K^+, such that large transmembrane gradients in Na^+ and K^+ concentration set up the necessary transmembrane resting voltage to block Cl^- influx. Frequently, in single cells, the resting transmembrane voltage equals the Nernst potential for Cl^-, so there is no net driving force to move the Cl^- into the cell. In the lens, however, there is no single transmembrane resting voltage; hence Cl^- cannot be everywhere in electrochemical equilibrium. Rather, the Cl^- flux is outward across the membranes of cortical fiber cells or epithelial cells an and inward across the membranes of more nuclear fiber cells. The intracellular and extracellular resting voltages within the lens come to steady state when the cortical efflux of Cl⁻ exactly equals the nuclear influx, whereupon there is no net accumulation of Cl⁻ or water, and cellular volume also comes to equilibrium.

One can easily appreciate that the maintenance of fiber cell volume in the lens is a complex process that depends on many factors that are not usually considered relevant to cellular volume regulation. For example, the nuclear to cortical circulating Cl⁻ current depends on the total resistance of the clefts between cells to current flow, and this resistance depends on cleft width as well as lens radius. Moreover, the continuity between fiber cells provided by gap junctions is an essential component of the circulating Cl⁻ current as well as an essential path for expressing the surface membrane transport of cell-to-cell communication, the dimensions of the clefts between cells, the overall dimensions of the lens, the permeability of the various membranes for the major ions, and the viability of active transport processes in the epithelial membranes are all important components of the volume regulation scheme in the lens.

The overview of lens electrophysiology presented thus far has addressed the normal properties of many of the factors mentioned above. An interesting and important question is, what is the role of abnormal electrophysiologic factors in the formation of cataracts? It is not possible to answer that question at this time, but certain speculations are clear candidates as contributing factors to unclear lenses. One has to suspect that, whatever the initiating factor, a step in the formation of any cataract includes disruption of the ability of the lens to regulate fiber cell volume. Moreover, most cataracts begin in a localized region of the lens, so the integrity of cell-to-cell communication, through gap junctions, must somehow be disrupted; otherwise, the lens would become uniformly opaque. Because the regulation of fiber cell volume depends on global properties of the lens, it is possible that, for example, a reduction of active transport in the anterior surface epithelium might have primary effects on volume regulation in nuclear fiber cells or in cortical fiber cells.

Although the role of cellular volume regulation in the formation of cataracts is speculative, the close relation between lens electrophysiology and lens cellular volume regulation cannot be denied. Albeit the relations and experiments are complicated, one must simply learn and do what is necessary to address important problems. We hope this overview focuses more attention on the important relations among volume regulation, electrophysiology, and cataract formation, and we further hope that more investigators will look into this area of research. The field is at present at the pioneer level, and experimental innovations may result in truly significant increments in understanding.

SUPPORT FOR THE OVERVIEW

Structure of the Lens

There can be little doubt today that the lens is composed of many thousands of cells whose cytoplasms are separated from extracellular solution by high-resistance membranes. Evidence for this statement comes from many sources. First, there are numerous studies using scanning electron microscopy that show lens fibers to be *clearly* separated from their neighbors, presumably because each cell is delimited by a cell membrane that surrounds it (1-5). Such discreteness of fibers is true even for fibers near the lens center. In fact, after fixation, single fibers can be teased free even from the lens nucleus. Second, thick sections of whole lenses stained with select Procion dyes show each lens fiber to be surrounded by a structure that almost certainly is a cell membrane (6). Third, recent studies using transmission electron microscopy to study lenses incubated in glutathione suggest the membranes around lens fibers are continuous (7). Earlier studies often suggested breaks in the fiber membranes of the lens nucleus (1), breaks that would now seem to be artifacts of the preparative methods. Of course, these three lines of morphologic evidence tell us nothing of the function of fiber cell membranes. They cannot tell us if the membranes are of high electrical resistance or if they are leaky to ions. The appearance of intact cell membranes is a necessary condition to the conclusion that fiber membranes limit diffusion of substances between cytoplasm and extracellular solution but it is not by itself a sufficient condition. Further functional data are necessary.

A fourth line of evidence for membrane intactness comes from studies in which lenses are incubated in fluorescent dyes, such as Procion yellow (8,9). In these studies, the dye can always be found in the extracellular solution around the fibers but only rarely in the cytoplasm of the fibers. These experiments show that fiber membranes can exclude substances at least as large as the fluorescent dyes. The ability of the nuclear fiber membranes to exclude dye is less certain since the extracellularly placed dyes do not significantly penetrate into the extracellular space of the nucleus (9).

When dye is injected into the cytoplasm of individual fibers through an intracellular microelectrode, it stays within the confines of fiber cells; it does not appear to leak into the extracellular space (10,11). This is then a fifth line of evidence supporting the existence of high-resistance membranes surrounding fiber cells.

Two additional lines of evidence come from electrical measurements. When a microelectrode is carefully aimed between two fiber cell columns and is driven into the frog lens in 1-2 μm increments, the tip of the electrode will alternately fall into two different voltage compartments, one much more hyperpolarized than the other 12-14). When the electrode tip is in the more hyperpolarized compartment, dye iontophoresed from the tip stays within the confines of fiber cell boundaries and the electrical impedance measured simultaneously is compatible with an intracellular location of the electrode (15,16). When the tip is in the more depolarized compartment, dye iontophoresed from the tip diffuses in all directions and quickly disappears. Impedance measurements made with the electrode in this compartment are compatible with an extracellular location of the electrode tip (17). Thus one concludes that there is a voltage difference between the cytoplasm of the cell and its extracellular solution, a voltage difference only possible if fiber membranes can limit ion movements.

When a second microelectrode is placed in the lens for passing current, a step of current produces a voltage change in a remotely placed intracellular microelectrode (18–21). The time constant of this voltage change, defined as $\tau_m = R_m C_m$, where R_m = input resistance and C_m = input capacitance, is about 1 sec or so and depends on the size of the lens. Assuming that the specific capacitance of the membrane is 1 $\mu F/cm^2$, as has been found for most other biologic membranes, the amount of membrane charged in these experiments is 100–300 cm^2 in small frog lenses, depending on the lens size. This is about all the membrane surface expected to be present in these lenses and suggests that current flow occurs across the majority of fiber membrane. Evidence that this long time constant of charging is not due to accumulation or depletion of ions is presented in Rae et al. (21).

Therefore, from structural, dye injection, and electrical evidence we are led to the conclusion that at least the majority of lens fiber cells are surrounded by functional membranes that limit the movement of dyes and ions and presumably most other substances. This evidence does not allow one the simplifying luxury of considering the lens as a giant cell, since the lens is clearly multicellular and has a functional extracellular space in communication with the lens bathing medium.

It is also instructive to consider the distribution of the surface area of lens fibers as a function of depth into the lens. Consider the lens to be constructed of spherical shells whose thickness is equal to the average thickness of lens fibers measured in the equatorial plane. Consider also that each shell is composed of hexagonal fibers with the width measured in equatorial sections but tapering to zero width at both poles (perfect lunar shape). From this, one can sum the areas of adjacent shells to produce the graph shown in Fig. 6. This figure shows that 80% of the surface area available for permeation is located within 40% of the lens surface and that each adjacent shell contributes less and less to the total lens fiber surface area. This model is, of course, somewhat unrealistic since the differentiating fibers do not exist in symmetrical spherical shells. Rather, the new fibers at the equator are very short. Fibers become increasingly longer in each deeper layer until they finally extend essentially from pole to pole (5,22). The surfacemost fiber that abuts an epithelial cell at the anterior pole is some 10–15% of the radius into the lens at the equator (5). This morphologic arrangement ensures that some 25–30% of lens cytoplasm has an almost direct diffusional pathway to the lens surface, since a molecule that gets into one of these surface fibers can diffuse to the interior of the lens along the cross-sectional area of the fiber without crossing through gap junctions.

Even with the inclusion of these differentiating cells, this discussion of surface area distribution is idealized since fibers at different depths have a

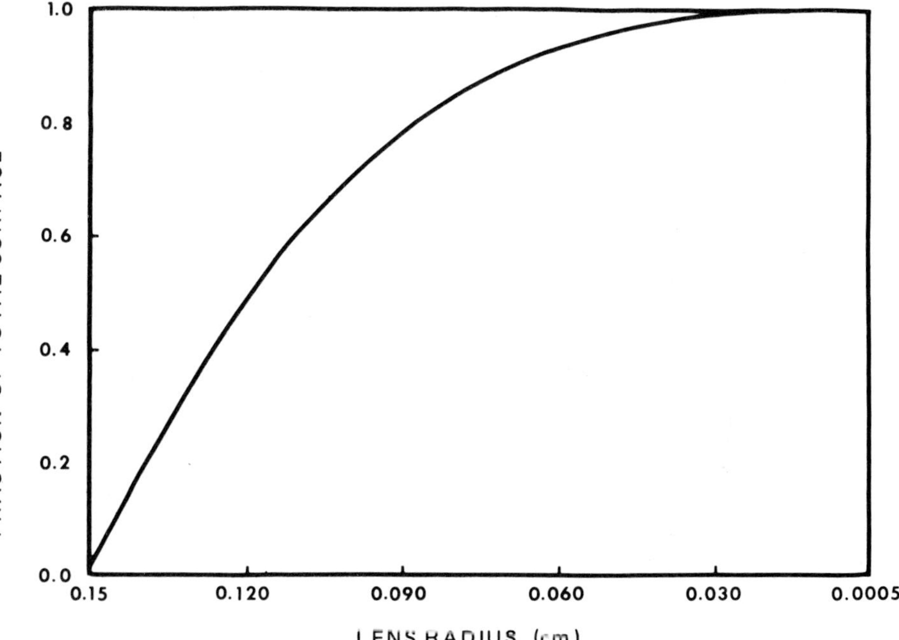

Figure 6 A plot of the distribution of the membrane surface area as a function of lens radius. The relation is calculated from a model in which the lens is assumed to be composed of spherical shells containing fibers of typical dimensions. The X axis represents the radius of spherical shells, with zero at the lens center and the maximum radius at the lens surface. The summation of shell area begins at the lens surface and concludes in the lens center.

different degree of membrane interdigitations and these differences have not been taken into account here. One would welcome quantitative estimates of this radially varying tortuosity, but no such estimates exist at present. However, even with the idealized models presented, one can conclude that fibers deep in the lens do not contribute much surface for permeation and that the differentiating fiber arrangement offers a lower resistance path for internalizing metabolites than would a perfect spherical shell packing with diffusion through layers of gap junctions.

Extracellular Space

By either morphologic or physiologic estimates, the extracellular space of the lens is small. The extracellular clefts seen in transmission electron microscopy are 100-200 Å in width (23-25), from which Paterson (26) has calculated the fraction of extracellular space at about 1%. Yorio and Bentley (27) have recently found the extracellular space of decapsulated amphibian lenses to be between 1 and 2%. These estimates are somewhat less than the commonly quoted value of 5% for other species (26). However, both morphologic and tracer methods might be expected to overestimate the extracellular space. Morphologic measurements to date have not included the effect of cleft narrowing to 20–30 Å in regions of gap junctions. Since in some lenses up to 50% of membrane surface is associated with gap junctions (28), these regions of extracellular narrowing should appreciably reduce overall extracellular volume. Of course, the harsh methods required for lens preservation and embedding and sectioning could introduce appreciable errors in estimates of extracellular cleft sizes, and thus we must be uncertain at present of the numbers obtained with these methods. Tracers used to determine extracellular volume must not cross cell membranes if they are to give accurate estimates of extracellular volume. To our knowledge, no radioautographic evidence exists to rule out some tracer penetration into the cells. Also, some tracer is found in the lens capsule, the actual amount depending on capsule volume, which varies from species to species (26,27).

Numerous studies have suggested that the extracellular space is not uniformly distributed: the nucleus has a lower volume fraction than the cortex (26,29).

From electrical measurements, Mathias et al. (15,30) and Rae et al. (16,21) have found the *effective* extracellular resistivity to be about 50 kΩ-cm and to scale linearly with the conductivity of the lens bathing solution. Since the resistivity of normal Ringer solution in these studies was about 80 Ω-cm, the 50,000 Ω-cm represents a large increase in resistance due to the small cleft width and extensive tortuosity of the clefts. Using the value of tortuosity calculated in the Appendix to Mathias et al. (15), this value of effective resistivity predicts a volume fraction of extracellular space of about 1%.

Therefore, by all existing estimates the lens extracellular space is small and restricted. Several implications can be drawn from this (1) In isotope washout experiments, isotope that fluxes into this high-resistance space will tend to reenter the cell, rather than wash out, so such experiments will not accurately measure the permeability of any lens membrane. (2) Whenever current is passed inside a lens cell, a fraction of it will cross fiber membranes and exit via the extracellular space. This current flowing in a highly resistive, tiny extracellular space will produce an extracellular voltage drop and thus a radially varying transmembrane voltage. (3) The extracellular resistance is high enough to limit ion movement across the inner fiber membranes with which it is in series. Thus, changes in the size of extracellular clefts could change the flux of ions across inner fiber membranes even when no change occurs in the properties of these inner membranes.

Cell-to-Cell Communication

The cells of the lens exhibit extensive cell-to-cell coupling. This was first suggested by Cohen (31) from electron micrographs that showed ubiquitous membrane specializations (junctions) between cells of human lenses. Prior to this report, both Brindley (32) and Andrée (33) had found that localized damage to the lens caused its potential difference to fall in regions where damage had not occurred, a result that in retrospect supported the notion of electrical coupling. Duncan (18) placed two microelectrodes in the lens, one for passing current and one for measuring the potential difference between the inside of the lens and its bath. He found that, when current was passed, it caused a voltage change no matter where in the lens he placed the voltage-measuring electrode. These results showed that there was a direct pathway for current flow between the current electrode and the voltage electrode. The results of these current-passing experiments were explained by Duncan (18) as degeneration of the fiber membranes, but they are also consistent with a syncytial lens. Rae (12,13) showed that the idea of membrane degeneracy was incorrect by high-resolution potential profile measurements, which demonstrated that individual fibers maintain a voltage across their cell membrane, and by intracellular dye injection measurements (8,10), which showed that dye diffused in the cytoplasm of injected fibers diffused into adjacent fibers but did not cross fiber membranes when it was placed in the extracellular space. As a result of the work of this entire group of investigators, there is little doubt today that cells of the lens are extensively interconnected through specialized cell-to-cell communicating junctions.

Since the early experiments demonstrating fiber-fiber coupling, numerous investigators have published high-quality electron micrographs from freeze-fracture replicas of cell-to-cell fiber junctions (34–41), epithelial-epithelial junctions (35,42), and epithelial-fiber junctions (34,43). These micrographs show that the lens contains a large amount of gap junction material. In fact, the amount of "gap" junction that can be isolated from lenses is so large that numerous investigators have selected the lens as the tissue of choice for working out the details of gap junction chemistry (44–46).

Eisenberg and Rae (19) published the first report describing a method for quantitating the extent of cell coupling in lenses. The rationale for this method is as follows.

When current is passed from a microelectrode immersed in the saline bath and collected by a distantly located reference electrode, the current density is greatest right at the microelectrode tip, since all the current must pass through the small opening of the microelectrode. Consider the current electrode to be surrounded by an infinite number of concentric shells of saline (47). The surface area of the shells increases with the distance from the electrode tip. This means that the current density, the current flowing across a unit area of each shell, decreases with the distance away from the electrode tip. Since the resistivity of the saline is uniform, the voltage generated by current flow through the saline decreases progressively as one gets farther and farther away from the current electrode. This radial variation in the induced voltage can be measured with a second microelectrode by comparing the potential at its tip to the potential of the remotely located reference electrode. If measurements are made at different separations of the current and voltage microelectrode tips, the electrical field can be mapped. Since the measured voltage comes from current flow through the resistance of an electrolyte solution, the time course of the measured voltage response closely mimics the time course of the current, the primary distortion due to the finite frequency response of the voltage-measuring microelectrode and associated amplifiers. Therefore, a step of current produces a step of voltage; a high-frequency sinusoidal current produces a high-frequency sinusoidal voltage with little phase shift from the current sinusoid. This radially varying voltage decrement away from the current electrode is referred to as the *point source effect*. It is obvious that the magnitude of the voltage signal induced by the current flow will scale with the resistance of the solution; a larger voltage is generated when the solution resistance is higher.

If the current and voltage electrodes are now placed inside a single, spherical cell, the situation is somewhat more complex since now there is a surrounding membrane that must also be charged by the current flow. The cytoplasm of the cell behaves very nearly the same as the saline in the previous case, and one expects a point source effect in the cell cytoplasm. Therefore, a step of current will produce a complex voltage response. When the current is turned on, the membrane capacitance is discharged and provides essentially no impedance to current flow, but there will be a jump in the voltage due to the point source effect and then a slower change in voltage as the current charges the membrane capacitance. This jump is related almost entirely to the cytoplasmic resistance and not to membrane properties (48–50). If one applies the steps at a high enough frequency so that the capacitive reactance of the membranes becomes small in comparison with the point source resistance, there will be no discernible voltage induced across the membrane by current flow. The response to these high-frequency current steps will be a reasonably direct measure of the cyto-plasmic resistance of the cell. A similar result will occur, of course, with high-frequency sinusoidal currents. At high frequencies, the membrane capacitance shunts these currents so that no voltage is induced across the membrane. The in-phase sinusoidal voltages that are measured are due to voltage drops within the cell cytoplasm.

If the current and voltage microelectrodes are now placed within a spherical syncytium, the situation is more complicated, but analogous to the two previous cases. The resistance, which is analogous to the solution resistance or cytoplasmic resistance of the previous cases, is now composed of two resistances in series: the cytoplasmic resistance and the resistance due to the gap junctional channels. Current would flow from cell cytoplasm 1 to gap junction 1 to cell cytoplasm 2 to gap junction 2, and so on. Therefore, the point source effect should also occur in the spherical syncytium and should be scaled by the combined cytoplasmic and gap junctional resistances. Previous measurements have shown the cytoplasmic resistivity of lens fibers to be less than 5% of the total effective resistivity (51–53), and thus we expect the point source effect in a spherical syncytium to be dominated by the gap junctional properties. An increase in gap junction resistance, produced, for example, by closing some of the junctional channels, would result in a larger point source effect since the effective resistance to intracellular current flow would increase. Therefore the size of the voltage jump in step current experiments or the amplitude of the high-frequency sinusoidal voltage in sinusoidal current experiments is expected to increase as coupling between the cells decreases. However, this increase will not be a monotonic function of the uncoupling, since when two adjacent cells are totally uncoupled the induced voltage must clearly go to zero. However, cell to cell coupling in a spherical syncytium can be measured over a wide range of partially coupled states either by the magnitude of the voltage jump or by the magnitude of the high-frequency sinusoidal voltage response to current flow. This technique has been used extensively by Mathias et al. (15,30) to quantitate cell-to-cell communication in amphibian lenses.

The coupling that exists between cells in other tissues is highly labile. The communication can be disrupted by a wide variety of treatments (53,54). The major candidates for producing this uncoupling are H^+ and/or Ca^{2+}, with uncoupling occurring following an increase of one or both of these substances in cytoplasm (53,54). There is, of course, a possibility that the action of these substances is not direct and that other intermediates are required in the uncoupling process (55,56).

The ability of lens fibers to uncouple has been a matter of controversy. Goodenough (37) suggested that lens fibers are in constant communication and that this communication is unregulated. Bernardini and colleagues (57,58) argued that uncoupling was possible based on data that showed that the gap junctions could be crystallized in vitro or in rat lenses that had been freeze-thawed. This argument is based on original data from Peracchia that suggested that the communicating state of the gap junction is a noncrystalline state in freeze-fracture replicas, whereas the noncommunicating state is crystalline. Although crystalline junctions are generally accepted as uncoupled, there is some question of whether uncoupled junctions must be crystalline (59); nonetheless, considerable additional evidence for lens fiber uncoupling exists. Andrée (33) showed that amphibian lenses recover from dissection damage. Initially the resting voltage of the whole lens falls following microdissection damage and then the resting voltage returns to normal with time. This result is compatible with the notion that damaged fibers uncouple from their neighbors, but repair of the damaged fibers rather than uncoupling cannot be ruled out. Bernardini et al. (58) have shown recently that damaged lenses gradually recover their normal K^+ permeability. In that study, they also showed that if damaged lenses were incubated in a solution containing Procion yellow, damaged fibers were unable to exclude Procion yellow whereas immediately adjacent fibers could exclude the dye. These findings, which have also been seen by other investigators, strongly suggest that damaged fibers can uncouple from their neighbors.

Recently, Rae et al. (60), using the point source resistance to measure electrical coupling, have shown that 2,4-dinitrophenol causes a large degree of reversible uncoupling in surface fiber communication in frog lenses. Rae and Mathias (61) have also seen a smaller degree of uncoupling in frog lenses incubated in propionate Ringer's or in Ringer's containing 10^{-4} Ω ouabain. Recently, a large and reversible uncoupling has been seen with 1-heptanol (62). Thus, it seems reasonable to conclude that at least surface fiber junctions have the ability to uncouple and recouple. Gap junctions between more interior fibers are likely to also have the ability to couple and uncouple, but electrical studies have not been performed to localize the uncoupling in each of the treatments mentioned above.

At present, two studies have been published that show coupling between epithelial cells and fiber cells. Goodenough et al. (43) presented freeze-fracture replicas from chick lenses that showed gap junctions between the apical side of the epithelial cells and the outer membrane of surface fiber cells. Also in that study, they showed radioautographic patterns of diffusible metabolites compatible with epithelial-fiber coupling and clearly demonstrated fiber-fiber coupling.

More direct evidence for epithelial-fiber coupling in the frog lens has been published recently by Rae and Kuszak (63). In that study, a voltage-measuring microelectrode was placed in an epithelial cell while current was passed through a second microelectrode located in a fiber cell. The passage of current always produced a voltage change in the epithelial cell, a result that proves electrical coupling between fibers and epithelium. The intraepithelial location of the voltage electrode was verified by injecting the impaled cell with dye. This study further showed the coupling to be directly between an epithelial cell and an adjacent fiber cell rather than fiber-epithelial coupling only at the equator, with current reaching nonequatorial epithelial cells by passing through epithelial-epithelial coupling sites (34).

The cell-to-cell communication story in the lens has recently become more complicated by the report of Kuszak and colleagues (64,65) that many adjacent lens fibers in frog, rat, and chick lenses are fused and thus communication can occur through openings much larger than gap junction connexons. It is not yet clear whether these ubiquitous fusions form a continuous network from surface to nucleus, but future studies will have to concern themselves with assessing the relative extent to which communication occurs between fusion sites versus gap junction sites.

Ion Transport in the Lens

Na^+,K^+ Pump

Harris and coworkers showed early in the history of lens physiology that the lens had the ability to actively pump cations (66). When the lens was cooled, it gained Na^+ and lost K^+, and when it was rewarmed, it re-established its normal Na^+ and K^+ levels. Kinsey and Reddy (67), using a "chamber" that crudely isolated the anterior and posterior lens surfaces, found the pump to be located on the anterior side, presumably in the epithelial cells. This view was subsequently supported by studies of Paterson (68), Candia et al. (69,70), Yuge et al. (71), and others (72). Palva and Palkama (73), Unakar and Tsui (74), and Gorthy and Anderson (75) showed by histochemical means that the Na^+,K^+-ATPase was most concentrated on the lateral walls of the epithelial cells near their apical (fiber) end. Thus we suppose that the pump molecules are most concentrated there. However, it is well known from chemical studies that only about half of the Na^+,K^+-ATPase is associated with the epithelial cells; the rest is then obviously associated with the fiber mass (76). The spatial distribution of this fiber-associated Na^+,K^+-ATPase is not well described. To date, no quantitative ouabain-binding data or ouabain-binding radioautography has been published for the lens. So, it is conceivable that active ion pumping occurs throughout the fiber mass, but there are no compelling data to support this view. On the other hand, Neville et al. (77) have shown two different ATPases associated with the anterior surface of the rabbit lens and have presented data that suggest that at least some anterior surface fibers have the ability to actively pump ions. This view is shared by Gorthy and Anderson (75), who showed substantial Na^+,K^+-ATPase associated with the anterior end of surface fibers of the rat lens, near where the fibers entered the suture.

There is reasonable evidence to suggest that the pump is electrogenic; that is, the current produced by the pump is sufficient to hyperpolarize the lens resting voltage by about 10 mV. Numerous investigators have shown a rapid depolarization of the lens resting voltage following the application of ouabain or cold (78–81). Hightower and Kinsey (82) have shown that the resting voltage can overshoot its previous value when the lens is cooled and subsequently rewarmed, and this is a classic experiment to demonstrate electrogenicity. It therefore seems reasonable to think the pump is electrogenic.

It is also possible to estimate the average current produced by each epithelial cell. In the frog lens, the ouabain-induced resting potential change is about 10 mV for lenses whose input resistance is about 5000 Ω. The electrogenic current then must be (from Ohm's law)

$$J_p = \frac{10^{-2}\ V}{5 \times 10^3\ \Omega} = 2 \times 10^{-6}\ A$$

A frog lens of 0.2 cm radius must have about 10^5 epithelial cells, which contain half of the total lens pump sites (Na^+,K^+-ATPase). Therefore, half of the total electrogenic current, or 1×10^{-6} A, must come from these 10^5 cells. Thus each cell must produce about 10 pA of current. If the ATP turnover rate in these cells is similar to other cells where it has been measured (83,84), it is about 100 times per second. For a $3Na^+:2K^+$ pump coupling ratio, this produces about 100 charges per second. Since 10 pA = 6.3×10^7 charges per second, the number of pump sites per cell must be about 6.3×10^5. Although these numbers depend on several assumptions of unknown validity, it is reasonable to think that they are good estimates for pumping parameters in lens epithelial cells.

Localization of Conductances

Candia and coworkers (69) demonstrated that elevated K^+ ion depolarized the anterior surface but not the posterior surface of the lens. Delamere and Duncan (85) found that, for lenses preloaded with $^{42}K^+$, efflux was greater through the anterior surface than the posterior.

Using a two microelectrode method and applying steps of current, Eisenberg and Rae (19) were able to show that a simple 'giant single cell" model was not appropriate and they proposed that the majority of the membranes in the lens must have a very high specific resistance and allow little dc current flow. These studies demonstrated the necessity of considering the lens membranes to have different properties at different locations in the lens and the need for proper models for interpreting lens electrophysiologic data.

Eisenberg et al. (86) published a more realistic model for the linear equivalent circuit of the lens. This model assumes the lens to be composed of two membrane systems, an outer set of membranes at the lens surface and an inner set of membranes associated with fiber cells. All these cells are electrically coupled and surrounded by a narrow, tortuous extracellular space. This model is considerably more realistic than previous models but also has several possible failings. (1) It is unlikely that there are only two types of membranes. It is more likely that cells near the surface have one set of properties and that there is a gradual change in these properties with depth. (2) The model only addresses the situation in which anterior and posterior membrane properties cannot be easily distinguished, yet there is considerable evidence in the literature to suggest that anterior and posterior properties differ (69,85,87). (3) Some investigators now believe the lens resistive properties change with a change in transmembrane voltage (79,88,89). This voltage dependence, if real, is not included in the model, but for small signal impedance studies the nonlinearities are not an important factor.

Using this model, Mathias et al. (15) were able to show several important properties of the frog lens. First, the inner fiber cell membranes have a specific resistance about 1000 times greater than that of the surface cells (presumably the epithelial cells), thus confirming the speculations of Eisenberg and Rae (19). Second, the extracellular space has a high resistance associated with it, and even in the normal lens this resistance is high enough to affect current flow and thus ion movements across inner fiber membranes. Third, this high resistance of the inner fiber membranes, in series with the extracellular resistance, results in about half of all current flow (in small frog lenses) going through surface membrane. Therefore, even though more than 99% of all membrane surface is associated with inner fiber membranes, only half of the current flows across these membranes and out of the lens via the extracellular space, but this relation depends on the size of the lens. In these studies, Mathias et al. (15) used frequency domain methods. That is, rather than rectangular current pulses, they used a stochastic current (complex mixture of sinusoids) and determined the small signal impedance of the lens. These methods are well known in the electrical engineering literature to be methods of choice when dealing with complex circuits, and many physiologists have used impedance studies to characterize tissues. However, these methods are beyond the scope of this coverage. The reader should consult the papers of Eisenberg, Mathias, and Rae (15,16,21, 30,90), for more information about impedance analysis.

Using either input resistance or impedance measurements, it is possible to assess the effects of many compounds and interventions on lens permeation properties. The recent lens literature is filled with studies of such interventions (79,88,89,91-95). Drug studies can be done by simply adding the drug to the bath and determining resistance properties with and without the drug, for example. It is also possible in principle to localize the conductances for specific ions and, since conductance is related to permeability, also localize electrically detectable ionic permeabilities. These studies are performed by measuring lens electrical impedance for varying bath concentrations of the ion under study. As a permeable ion is removed and replaced with an impermeant ion of the same charge, the conductance of a membrane containing a population of channels for the permeant ion will fall and, concomitantly, the contributions of these channels to the lens impedance decreases. After this has been done for all important permeating ions, it should be possible to assess the locations of the conductances for specific ions and perhaps make some crude estimates of the actual value of the conductance. Using this approach, Mathias and colleagues (96-98) have determined that the surface cells of the frog lens have primarily K^+ conductance whereas the inner fiber membranes have primarily Na^+ and Cl^- conductances. These studies suggest that not only do the inner membranes have a much smaller total conductance than surface membranes, but also they have a different selectivity for Na^+, K^+, and Cl^-. Since these two populations of cells are electrically coupled, their cytoplasms must be at essentially the same voltage with respect to the bath. In accordance with arguments presented in the previous section, this means that an extracellular standing current flow must be produced even at rest and that the transmembrane voltage must be larger for cells nearer the lens surface. Studies that use simple input measurements without structural models cannot localize conductances since they include no local structures to which the conductance could be attributed.

It should be realized that the spatial parsing of conductances is very model dependent, as are all present studies that deal with the properties of individual channel types. Many present studies of lens input properties (94,95) interpret complex current-voltage relations as if the complexity arose from individual membrane channels. Although it is conceivable that channels could have such properties, it is almost certain that some of the properties arise from structural complexity, such as trapped extracellular space, ion accumulation and depletion, and spatial nonuniformity in conductances. This problem is made even more complex when one realizes that the voltage between the cytoplasm of the lens and its extracellular space cannot be controlled at the same value. If membranes contain processes that depend on transmembrane voltage, it is best to study them under "voltage clamp" conditions so that each cell has the same trans-membrane voltage, but because of the high resistance of the extracellular clefts, an extracellular voltage drop always occurs when current is passed. Therefore, even though it is possible to control the voltage of the lens cytoplasm in comparison to the bath, the transmembrane potential of fiber cells cannot be controlled. This point, which is well known by nerve, muscle, and cardiac electrophysiologists, has not been appreciated by all lens investigators (94,99).

Single Ion Channels

It is, of course, desirable to measure the properties of single ion channels directly rather than using circuitous impedance methods. Fortunately, it is now possible to do this using the technique of the "patch voltage clamp" (100–102). In this technique, a fire-polished microelectrode is gently pressed against the cell membrane. Following the successful application of a gentle suction, a small piece of membrane is sucked into the pipet tip and a chemical reaction occurs between the membrane and glass. This reaction results in the membrane being sealed to the glass, with a seal resistance in the range of a few to a few hundred gigohms. The pipet is connected to a sensitive circuit for measuring currents in the picoampere range. When single channels in the membrane open, ions move down their electrochemical gradient and produce a current that can be measured. To date, such currents have been measured by the authors from epithelial cells of grass frogs, bullfrogs, rats, rabbits, mice, and chick lenses and from anterior surface fibers of grass frog lens. Original records are shown in Fig. 7 for channels measured from an on-cell patch of frog lens apical epithelium. Several items of

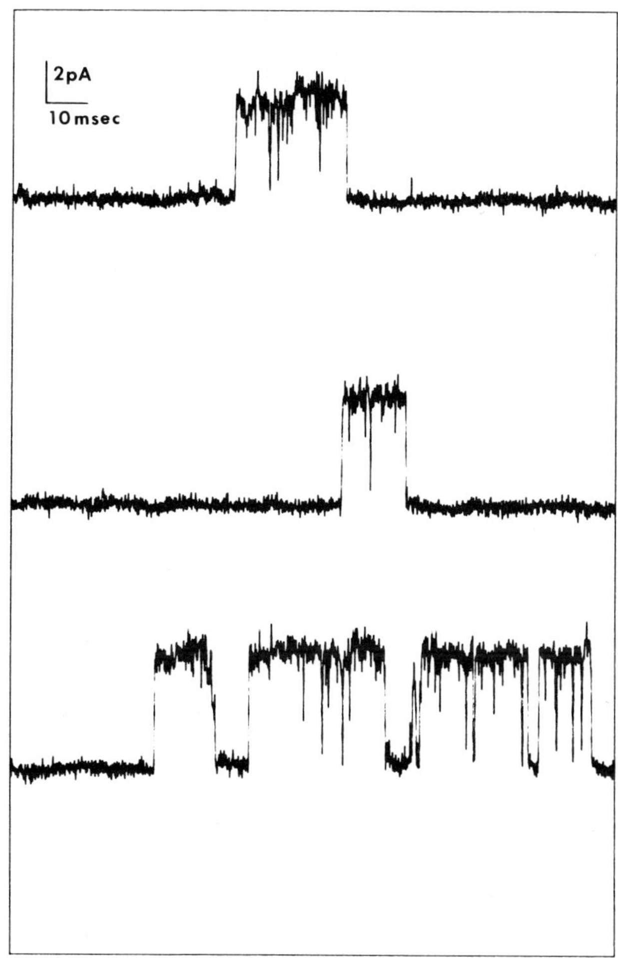

Figure 7 Single-channel currents recorded from the apical surface of the lens epithelium of the grass frog by use of the technique of patch voltage clamping. The records are from a nonselective cation channel recorded from an on-cell patch.

interest are already obvious from these preliminary single-channel studies. (1) The channels are not constantly open "tubes" through the membranes. Rather, they open and close in a stochastic manner. (2) This opening and closing rate seemingly depends on membrane voltage and the concentration of such substances as calcium in the cell. (3) There are numerous different kinds of channels of different size and selectivity. (4) The density of the channels is high.

In fact, it would seem that it is too high to be compatible with the life of the cell were all the channels to be open at once. These single-channel studies promise to be very useful for understanding mechanisms of lens ion permeation and volume regulation, but because of the bewildering array of processes that can be detected, it will be some time before the macroscopic and microscopic properties can be reconciled.

REFERENCES

1. T. Kuwabara. Exp. Eye Res., *20*: 427 (1975).
2. J. Kuszak, J. Alcala, and H. Maisel. Am. J. Anat., *159*: 395 (1980).
3. H. Maisel, C. V. Harding, J. R. Alcala, J. Kuszak, and R. Bradley. In *Molecular and Cellular Biology of the Eye* (H. Bloemendal, ed.), John Wiley and Sons, New York, 1981.
4. M. Sakuragawa, T. Kuwabara, J. H. Kinoshita, and H. N. Fukui. Exp. Eye Res., *21*: 381 (1975).
5. J. Kuszak and J. L. Rae. Exp. Eye Res., *35*: 499 (1982).
6. J. L. Rae, K. Truitt, and J. Kuszak. Invest. Ophthalmol. Vis. Sci., *24*: 1167 (1983).
7. Y. Yajima and T. Kuwabara. Invest. Ophthalmol. Vis. Sci., *20*: 134 (1981).
8. J. L. Rae. Invest. Ophthalmol., *13*: 147 (1974).
9. J. L. Rae and T. Stacey. Exp. Eye Res., *28*: 1 (1979).
10. J. L. Rae and J. E. Blankenship. Exp. Eye Res., *15*: 209 (1973).
11. J. L. Rae and T. Stacey. Dox. Ophthalmol. Proc. Ser., *8*: 233 (1976).
12. J. L. Rae. Exp. Eye Res., *19*: 227 (1974).
13. J. L. Rae. Exp. Eye Res., *19*: 235 (1974).
14. Y. Taura, T. Murata, and N. Akaike. Comp. Biochem. Physiol., *63*: 475 (1979).
15. R. T. Mathias, J. L. Rae, and R. S. Eisenberg. Biophys. J., *25*: 181 (1979).
16. J. L. Rae, R. S. Eisenberg, and R. T. Mathias. In *Red Blood Cell and Lens Metabolism* (Srivastava, ed.), Elsevier North Holland, Inc., 1980, p. 277.
17. J. L. Rae and R. T. Mathias. Unpublished observations, 1980.
18. G. Duncan. Exp. Eye Res., *8*: 406 (1969).
19. R. S. Eisenberg and J. L. Rae. J. Physiol. (Lond.), *262*: 285 (1976).
20. N. A. Delamere and G. Duncan. Exp. Eye Res., *31*: 637 (1980).
21. J. L. Rae, R. T. Mathias, and R. S. Eisenberg. Exp. Eye Res., *35*: 471 (1982).
22. M. J. Hogan, J. A. Alvarado, and J. E. Weddell. *Histology of the Human Eye*, Saunders, Philadelphia, 1971.
23. T. Wanko and M. A. Gavin. In *The Structure of the Eye* (G. K. Smelser, ed.), Academic Press, New York, 1961, p. 221.
24. T. Kuwabara. *Fine Structure of the Eye*, Security-Columbian, Boston, 1970.
25. N. S. Rafferty and E. A. Esson. J. Ultrastruct. Res., *46*, 239 (1974).
26. C. A. Paterson. Am. J. Physiol., *218*: 797 (1970a).
27. T. Yorio and P. J. Bentley. Exp. Eye Res., *23*: 601 (1976).

28. J. Kuszak, H. Maisel, and C. V. Harding. Exp. Eye Res., *27*: 495 (1978).

29. A. Huggert. Acta Ophthalmol. (Copenh.), *37*: 26 (1959b).

30. R. T. Mathias, J. L. Rae, and R. S. Eisenberg. Biophys. J., *34*: 61 (1981).

31. A. I. Cohen. Invest. Ophthalmol., *44*: 806 (1965).

32. G. S. Brindley. Br. J. Ophthalmol., *40*: 385 (1956).

33. G. Andrée. Pflugers Arch., *267*: 109 (1958a).

34. E. L. Benedetti, I. Dunia, and H. Bloemendal. Proc. Natl. Acad. Sci. USA, *71*: 5073 (1974).

35. E. L. Benedetti, I. Dunia, C. J. Bentzel, A. J. M. Vermorken, M. Kibbelaar, and H. Bloemendal. Biochim. Biophys. Acta, *457*: 353 (1956).

36. S. Okinami. Graefes Arch. Klin. Exp. Ophthalmol., *209*: 51 (1978).

37. D. A. Goodenough. Invest. Ophthalmol., *18*: 1104 (1979).

38. J. Kuszak, J. Alcala, and H. Maisel. Exp. Eye Res., *33*: 157 (1981).

39. C. Peracchia and L. Peracchia. J. Cell Biol., *87*: 708 (1980).

40. C. Peracchia and L. Peracchia. J. Cell Biol., *87*: 719 (1980).

41. J. Kuszak, J. L. Rae, B. Pauli, and R. Weinstein. J. Ultrastruct. Res., *81*: 249 (1982).

42. W. K. Lo and C. V. Harding. Invest. Ophthalmol. Vis. Sci., *20*: 129 (1981).

43. D. A. Goodenough, J. S. B. Dick, II, and J. E. Lyons. J. Cell Biol., *86*: 576 (1980).

44. B. J. Nicholson, M. W. Hunkapiller, L. E. Hood, J. P. Revel, and L. Takemoto. J. Coll Biol., *87*: 1539a (1980).

45. J. Alcala, N. Lieska, and H. Maisel. Exp. Eye Res., *21*: 584 (1975).

46. I. Dunia, C. S. Gosh, E. L. Benedetti, A. Zweers, and H. Bloemendal. FEBS Lett., *45*: 139 (1974).

47. R. S. Eisenberg and E. A. Johnson. In *Progress in Biophysics and Molecular Biology*, Vol. 20, Pergamon Press, New York, 1970, p. 1.

48. A. Peskoff and R. S. Eisenberg. J. Math. Biol., *2*: 277 (1975).

49. A. Peskoff and D. M. Ramirez. J. Math. Biol., *2*: 301 (1975).

50. X. Kevorkian and J. D. Cole. *Perturbation Methods in Applied Mathematics*, Springer-Verlag, New York, 1981.

51. H. Pauli and H. P. Schwan. IEEE Trans. Biomed. Eng., *11*: 103 (1964).

52. J. L. Rae and H. A. Germer. J. Appl. Physiol., *37*: 464 (1974).

53. W. R. Lowenstein. Physiol. Rev., *61*: 829 (1981).

54. D. C. Spray, J. H. Stern, A. L. Harris, and M. V. L. Bennett. Proc. Natl. Acad. Sci. USA, *79*: 441 (1982).

55. C. Peracchia, G. Bernardini, and L. L. Peracchia. J. Cell Biol., *91*: 124a (1981).

56. M. F. Johnson and F. Ramon. J. Physiol. (Lond.), *317*: 509 (1981).

57. G. Bernardini and C. Peracchia. Invest. Ophthalmol., *21*: 291 (1981).

58. G. Bernardini, C. Peracchia, and R. A. Venosa. J. Physiol. (Lond.), *320*: 187 (1981).

59. E. Raviola, D. A. Goodenough, and G. Raviola. J. Cell Biol., *87*: 273 (1980).

60. J. L. Rae, R. D. Thomson, and R. S. Eisenberg. Exp. Eye Res., *35*: 597 (1982).

61. J. L. Rae and R. T. Mathias. Unpublished observations, 1979.
62. J. L. Rae. Unpublished observations, 1982.
63. J. L. Rae and J. Kuszak. Exp. Eye Res., *36*: 317 (1983).
64. J. Kuszak and J. L. Rae. Invest. Ophthalmol., *22*: 285 (1982).
65. J. R. Kuszak, M. S. Macsai, J. L. Rae, and R. S. Weinstein. J. Cell Biol., *95*: 11a (1982).
66. J. E. Harris, L. B. Gehrsitz, and L. T. Nordquist. Am. J. Ophthalmol., *36*: 39 (1953).
67. V. E. Kinsey and V. N. Reddy. Invest. Ophthalmol., *4*: 104 (1965).
68. C. A. Paterson. Invest. Ophthalmol., *12*: 861 (1973).
69. O. A. Candia, P. J. Bentley, C. D. Mills, and H. Toyofuku. Nature (Lond.), *227*: 852 (1970).
70. O. A. Candia, P. J. Bentley, and C. D. Mills. Am. J. Physiol., *220*: 558 (1971).
71. T. Yuge, H. Takeda, A. Imamura, and M. Tani, Exp. Eye Res., *16*: 223 (1973a).
72. B. Becker and E. Cotlier. Invest. Ophthalmol., *1*: 642 (1962).
73. M. Palva and A. Palkama. Exp. Eye Res., *19*: 117 (1947b).
74. N. J. Unakar and J. Tsui. Invest. Ophthalmol., *19*: 378 (1980).
75. W. C. Gorthy and J. W. Anderson. Invest. Ophthalmol., *19*: 1038 (1980).
76. J. H. Kinoshita. Arch. Ophthalmol., *70*: 558 (1963).
77. M. C. Neville, C. A. Paterson, and P. M. Hamilton. Exp. Eye Res., *27*: 637 (1978).
78. J. L. Rae. Exp. Eye Res., *15*: 485 (1973a).
79. G. Duncan, N. A. Delamere, C. A. Paterson, and M. C. Neville. Exp. Eye Res., *30*: 105 (1980).
80. C. A. Paterson, M. C. Neville, R. M. Jenkins, and J. P. Cullen. Biochim. Biophys. Acta, *375*: 309 (1975).
81. K. R. Hightower and V. E. Kinsey. Exp. Eye Res., *24*: 587 (1977).
82. K. R. Hightower and V. E. Kinsey. Exp. Eye Res., *30*: 19 (1980).
83. J. F. Hoffman, J. H. Kaplan, and T. J. Callahan. Fed. Proc., *38*: 2440 (1979).
84. V. Harms and E. M. Wright. J. Membr. Biol., *53*: 119 (1980).
85. N. A. Delamere and G. Duncan. J. Physiol. (Lond.), *295*: 241 (1979).
86. R. S. Eisenberg, V. Barcilon, and R. T. Mathias. Biophys. J., *25*: 151 (1979).
87. O. A. Candia. Exp. Eye Res., *30*: 193 (1980).
88. L. Patmore and G. Duncan. Exp. Eye Res., *28*: 349 (1979).
89. L. Patmore and G. Duncan. Exp. Eye Res., *31*: 637 (1980).
90. R. S. Eisenberg and R. T. Mathias. CRC Crit. Rev. Bioeng., 1980, p. 203.
91. T. J. C. Jacob and G. Duncan. Exp. Eye Res., *31*: 505 (1980).
92. T. J. C. Jacob and G. Duncan. Exp. Eye Res., *34*: 445 (1982).
93. N. A. Delamere, G. Duncan, and C. A. Paterson. J. Physiol. (Lond.), *308*: 49 (1980).
94. N. A. Delamere, C. A. Paterson, and D. L. Holmes. Exp. Eye Res., *31*: 651 (1980).

95. N. A. Delamere and C. A. Paterson. Exp. Eye Res., *33*: 233 (1981).
96. L. Ebihara, R. T. Mathias, and J. L. Rae. Invest Ophthalmol., *22*: 101 (1982).
97. R. T. Mathias, J. L. Rae, and R. McCarthy. Invest. Ophthalmol., *22*: 101 (1982).
98. J. L. Rae and R. T. Mathias. Invest. Ophthalmol. Abstracts, *172*: (1980).
99. G. Duncan, L. Patmore, and P. B. Pynsent. J. Physiol. (Lond.), *312*: 17 (1981).
100. E. Neher, B. Sakmann, and J. H. Steinbach. Pflugers Arch., *375*: 219 (1978).
101. O. P. Hamill, A. Marty, E. Neher, B. Sakmann, and F. J. Sigworth. Pflugers Arch., *391*: 85 (1981).
102. J. L. Rae and R. A. Levis. Biophys. J., *41*: 226a (1983).

4
The Structure of Lens Proteins

PETER F. LINDLEY, MICHAEL E. NAREBOR, LESLEY J. SUMMERS
AND GRAEME J. WISTOW* / Birkbeck College, University of London, England

The cellular structure of all vertebrate eye lenses comprises long fiber cells that pack hexagonally in spherical laminae. However, the textures of these lenses range from very soft in avian species to very hard in the case of the shortsighted rat. The transparency and refractive power of the normal lens depend on a smooth refractive index gradient for visible light. These properties are achieved not only by the regular arrangement of the fiber cells but also the presence of stable intracellular lens-specific proteins, the crystallins (1–3). High concentrations of these proteins provide the medium of high refractive index necessary for the function of the lens, and the gradient of increasing refractive index from the cortex to the nuclear region is associated with increasing concentration of crystallins relative to water. Irregularities in the refractive index, which induce light scattering, are often found in certain types of cataract (4), and in senile nuclear cataract these are associated with increased degradation of lens proteins, cross-linking (including disulfide bonds), and aggregation of crystallins.

In vertebrates the crystallins account for some 90% of the soluble proteins of the lens and consist of three antigenically distinct families of polypeptides, called α, β, and γ. In avian and some reptilian lenses the γ-crystallins appear to be entirely replaced by a fourth antigenic protein called δ-crystallin. α-Crystallins are large macromolecular weight aggregates, about 800,000 d. There are two gene product subunits, molecular weight about 20,000 d, of related amino acid sequence (5). A proportion of these undergo posttranslational modification,

---

*Present affiliation: Laboratory of Molecular and Developmental Biology, National Eye Institute, Bethesda, Maryland.

which contributes to the considerable heterogeneity and polymerization of native α-crystallin. At least one model has been proposed describing the assembly of the subunits in the aggregates (6), but as yet no detailed structural information is available.

In contrast to the α-macromolecules, the γ-crystallins are a family of closely related monomers of low molecular weight, about 20,000 d. In between, the β-crystallins form aggregates of different sizes although the individual subunits are only slightly larger than those of the γ-crystallins and the aggregates never reach the size formed by the α-crystallins. In solution the mammalian crystallins are therefore comprised of a range of different-sized monomers and oligomers. There also exists a range of net charge at physiological pH, with the γ-crystallins the most positive and the α-crystallins the most negative; the β-crystallins exhibit a whole range of isoelectric points in between. This complex distribution of proteins of differing size and charge must interact with intracellular water, cell membranes, and the cytoskeleton, and also with each other, to achieve and maintain the refractive index gradient across the lens.

However, despite differences in charge and aggregation it now appears from amino acid sequence studies and molecular structural investigations that not only do the various γ-crystallins form a homologous group, but that, more surprisingly, there is also further homology with certain β-crystallins. Thus the γ- and β-crystallins may belong to a single protein superfamily accounting for a significant proportion of the total protein content of the mammalian lens.

It is the aim of this chapter to describe how three-dimensional single-crystal x-ray diffraction analyses in conjunction with amino acid sequence studies have reinforced the concept of a superfamily of γ- and β-crystallins. The molecular structure of bovine γ-II-crystallin will be described in detail, and possible roles for the γ-proteins in terms of their molecular structure, in the normal lens and in cataract, will be discussed. Molecular models for certain β-crystallins based on their homology with the γ-crystallins will be described, and an account will also be given of x-ray single-crystal structural studies currently being undertaken on eye lens proteins.

γ-CRYSTALLINS

The γ-crystallins are found mainly at the central core region of the mammalian eye lens (7,8), where the refractive index is a maximum and the water content a minimum. The core of the adult mammalian lens is formed from the original fetal structure, and since there is little or no protein turnover within it (9,10), the lifetimes of the γ-crystallin molecules are the same as that of the animal. It is therefore important for these proteins to have great stability in both their individual tertiary structures and their interactions with other cell components. Despite this requirement, γ-crystallins are rich in the sulfur-containing residues, cysteine and methionine, and aromatic residues (11,12), all of which are susceptible to oxidative processes. Thus, bovine γ-II contains 7 cysteine, 7 methionine, and 29 aromatic residues (10 phenylalanine, 15 tyrosine, and 4 tryptophan). The reactive sulfydryl groups in cysteines can interact with one another to form disulfide bridges, methionine can oxidize to polar sulfoxide, and photo-oxidation of aromatic residues can lead to a variety of products. Disulfide and other cross-links, higher oxidation states of sulfur, and unusual chromophores have all been reported in lenses afflicted with various forms of nuclear cataract (2) (see Ref. 13 and references cited therein). Furthermore, many senile and brunescent cataracts are initiated in the core region of the lens, and γ-crystallins have been identified in cross-linked aggregates with other lens components in cataractous human lenses (14). It is therefore apparent that studies of the spatial distribution and organization of these reactive residues within the overall structure of the γ-crystallin molecule may provide models for cataract formation. Such information can only be precisely provided by single-crystal x-ray diffraction studies. In this context it should be noted that the water content in crystals of the individual γ-crystallins is comparable to that in the actual eye lens, and although the crystalline state provides a slightly artificial environment in that the crystallin molecules interact only with one another, not other lens components, the structures of the γ-crystallin molecules are unlikely to be very much different from those adopted in vivo.

To date, crystallographic studies of γ-crystallins have been confined to those from bovine lenses where the fraction of these proteins relative to that of the combined α- and β-crystallins is approximately one-fifth. There is extensive amino acid sequence homology between the various monomeric bovine γ-crystallins (12,15), and single-crystal x-ray studies have shown that this also extends to structural homology.

Crystals of bovine γ-crystallins suitable for x-ray studies have been reported by Chirgadze et al. (16), Carlisle et al. (17), and Blundell et al. (18). X-ray structure analyses of bovine γ-II-crystallin at low resolution—5.5 Å (18) and 5.0 Å (19)—showed that the molecule was organized into two globular domains in close proximity to one another and each of radius approximately 25 Å. These studies have been extended to higher resolution—2.6 Å (20) and 1.9 Å (21)—to reveal the folding of the polypeptide chain and the individual amino acid residues. Further refinement of the structure of bovine γII at almost atomic resolution, 1.5 Å, is in progress (22). The structure of bovine γ-IIIb-crystallin has also been studied at low resolution, approximately 5.0 Å (23), and the analysis extended to medium resolution, 3.0 Å (24). At this resolution the polypeptide chain could be identified and the similarity of the molecule to that of γ-II observed. The structural analysis of bovine γ-IV is in progress (25). Crystallographic data for the bovine γ-crystallins are summarized in Table 1. Figure 1 shows single crystals of these proteins together with typical x-ray diffraction patterns.

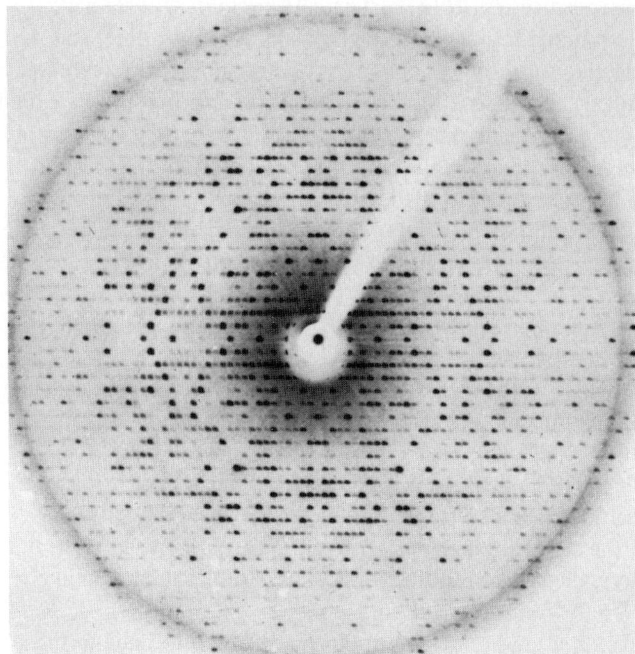

Figure 1　(A) Single crystals of bovine γ-crystallins and typical x-ray diffraction patterns recorded with a precession camera:　γ-II.

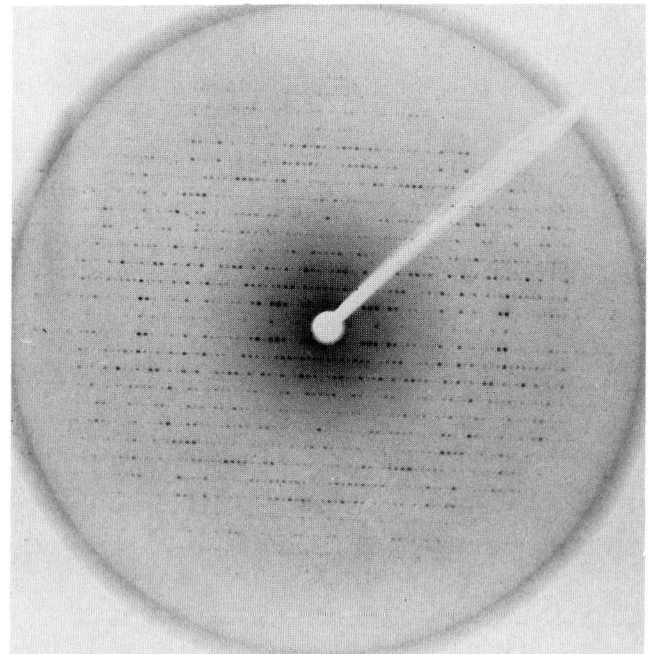

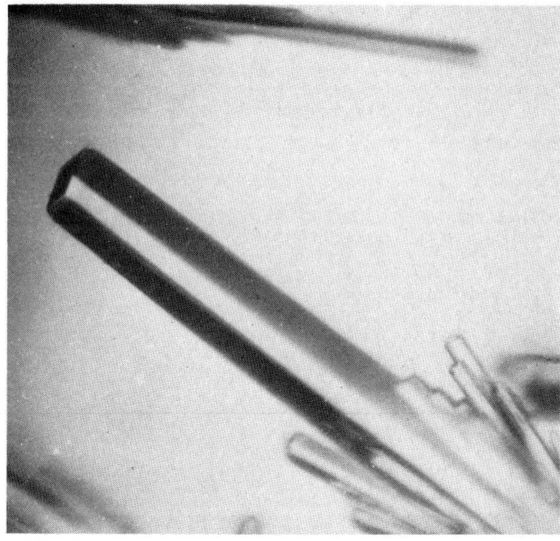

Figure 1 (B) Single crystals of bovine γ-crystallins and typical x-ray diffraction patterns recorded with a precession camera: γ-IIIb.

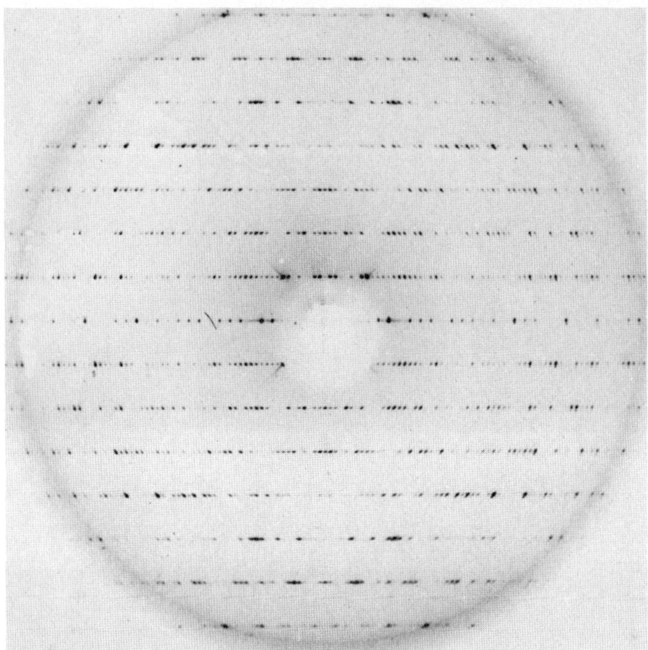

Figure 1 (C) Single crystals of bovine γ-crystallins and typical x-ray diffraction patterns recorded with a precession camera: γ-IV

Table 1   Crystallographic Data for Bovine γ-Crystallins

Crystal sytem	γ-II Tetragonal	γ-IIIb Orthorhombic	γ-IV Orthorhombic
Cell dimensions at ~20°C, Å			
a	57.8 (3)	58.4 (3)	35.1 (2)
b	57.8 (3)	69.8 (3)	46.2 (2)
c	98.7 (3)	117.7 (4)	186.2 (5)
Space group	$P4_1 2_1 2$	$P2_1 2_1 2_1$	$C222_1$ or $C222$
Number of molecules per asymmetric unit, Z	1	2	1

Structure of Bovine γ-II Crystallin

Overall Organization of the Molecule

The molecule of bovine γ-II-crystallin adopts a structure that has the highest
internal symmetry of any protein yet studied by x-ray diffraction techniques.
The molecule is a single-stranded polypeptide chain consisting of some 170
amino acids and organized, as shown in Fig. 2, into four remarkably similar
motifs, each consisting of approximately 40 residues. Within a motif the strand
of amino acid residues is folded into a pattern that has been likened to the space-
filling "Greek key" motif found on ancient pottery (26,27). This motif is
illustrated in Fig. 3A and can be seen to consist of four strands labeled a, b, c, and
d, running in opposite directions. As shown in Fig. 3B, the motifs in γ-II have
an extension between the a and b strands, which in the molecule structure folds
over, away from the loop between the b and c strands, to form an "ear"
(Fig. 3C).

In each domain two motifs 1 and 2 in the N-terminal domain and motifs 3
and 4 in the C-terminal domain, form two β-pleated sheets (see, for example,
Ref. 28), which then pack together in the shape of a wedge. A sheet does not
correspond precisely to a single motif but is derived from the a, b, and d strands
of one motif and the c strand of the second motif in the same domain. Figure
4A shows schematically the distribution of the strands of motif 1 between sheets
1 and 2 in the N-terminal domain. Figure 4B shows the complete α-carbon atom
skeleton of the amino acids in motifs 1 and 2 in this domain viewed almost
parallel to the sheets, that is, looking perpendicular to the end of the wedge in
Fig. 4A.

In Fig. 4B sheet 1 can be seen on the right-hand side of the diagram and
consists of strand c from motif 2, farthest away from the reader, and strands d,

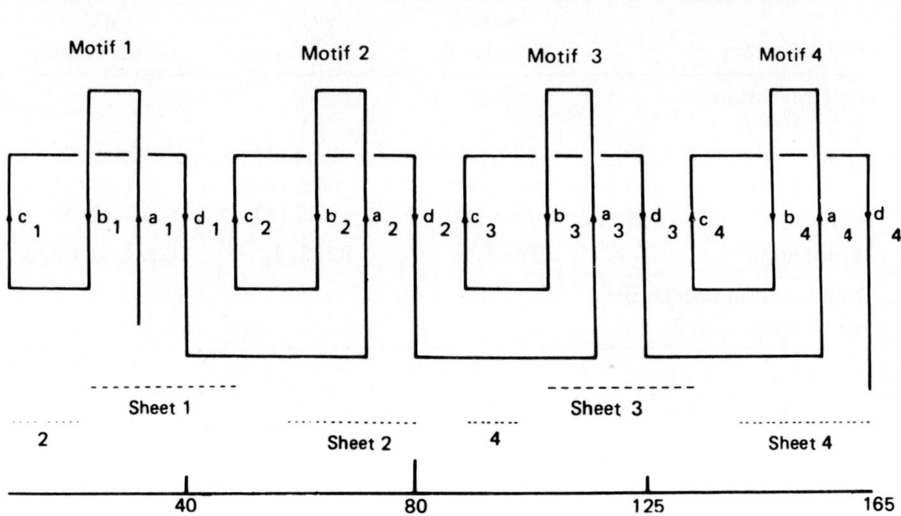

Figure 2 The overall organization of the molecule of bovine γ-II-crystallin showing the relation between the motifs, γ-pleated sheets, and domains (see Table 2).

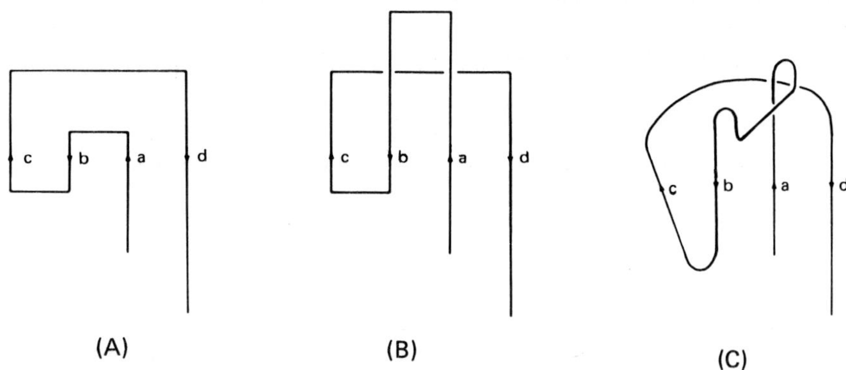

Figure 3 Schematic representations of the Greek key motif. (A) The four antiparallel strands a, b, c, and d in the idealized motif. (B) The extension between strands a and b in γ-II. (C) How this extension folds over in γ-II to form an ear.

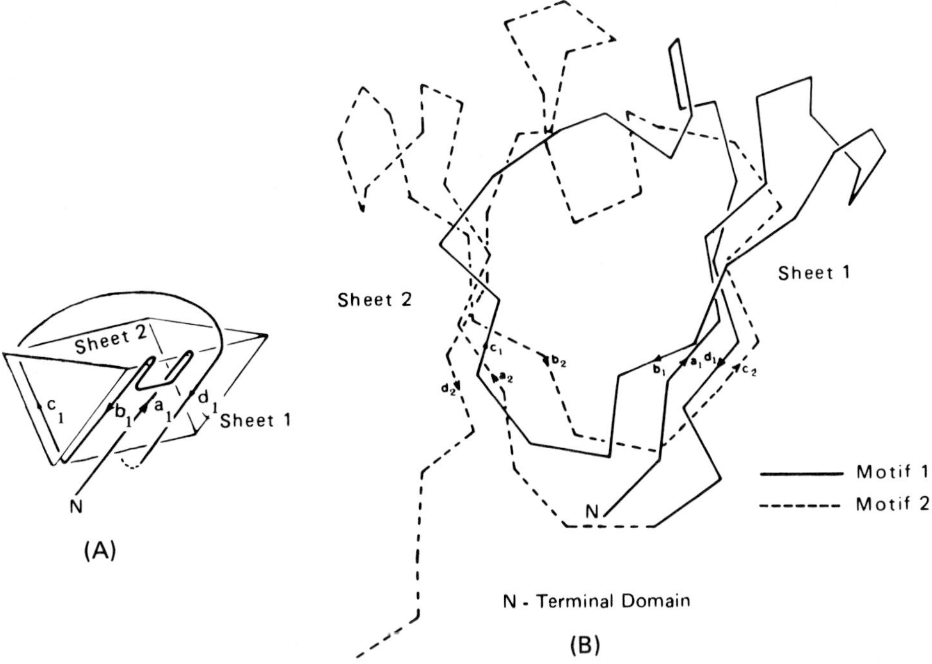

Figure 4   The relation between motifs, sheets, and domains. (A) A schematic drawing showing the distribution of a single motif between two sheets that comprise a wedge-shaped domain. (B) The α-carbon backbone of the amino acid residues of motifs 1 and 2 in the N-terminal domain viewed end-on to the wedge and hence almost parallel to β-sheets 1 and 2. Motif 1 is shown as a full line and motif 2 as a dashed line.

a, and b of motif 1, respectively, coming toward the reader; the ear between strands a and b is also clearly visible. Sheet 2 is on the left-hand side of Fig. 4B and comprises the c strand of motif 1, nearest the reader, and the d, a, and b (farthest away from the reader) strands of motif 2. The sides of the wedge thus consist of sheets 1 and 2. The b and c strands from a single motif form each end of the wedge, and the top is enclosed by the loops between the c and d strands of each motif, running in opposite directions to one another.

In a similar manner motifs 3 and 4 form sheets 3 and 4, which pack together in the C-terminal domain. Figure 5 shows that, in the molecule, the N- and C-terminal domains are then organized so that sheets 1 and 3 are on the outside surface of the molecule, whereas sheets 2 and 4 are on the inside and are partially in contact.

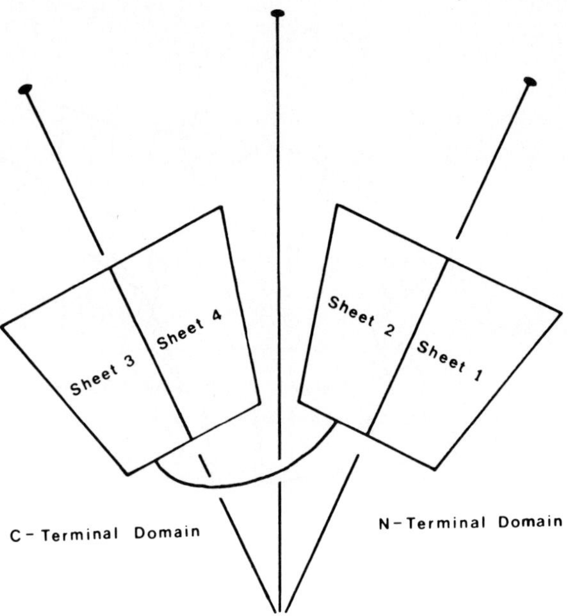

Figure 5   The organization of the β-sheets and domains in the bovine γ-II-crystallin molecule. Sheets 1 and 3 lie on the outside of the molecule, whereas sheets 2 and 4 are in partial contact.

Configuration of the Motifs

Although the four motifs differ considerably in amino acid composition (Table 2), the conformations of their amino acid α-carbon backbones are strikingly similar. Further, there are some residues common to all motifs that occur in structurally identical positions and appear to be crucial to the conformation of the motif. These two points will now be discussed in further detail.

Figure 6 shows the α-carbon backbone of motif 1 viewed approximately perpendicular to the plane formed by strands a, b, and d. Motifs 2, 3, and 4 are essentially the same except for some amino acid insertions in the loops between the c and d strands, which extend over the top of a domain wedge, and small conformational changes. However, as might be expected from Fig. 5, the similarity between motifs 1 and 3 on the outside of the bilobal γ-crystallin molecule or between motifs 2 and 4, which are partially in contact in the interdomain region, is greater than betwen two motifs in the same domain, that is, between 1 and 2 or 3 and 4. For motifs 2 and 4 the α-carbon atoms of 42 residues in topologically equivalent positions can be superimposed with a root mean square distance of only 1.05 Å, an exceedingly close fit; the agreement

Table 2 Amino Acid Sequence of Bovine γ-II-Crystallin[a]

Strands: a, b, c, d (each Greek key motif has four strands labeled a, b, c, and d).

Motif 1

1	2	3	4	5	6	7	8	9	10	11	12	13	14	15	16	17	18	19	20	21	23	24	25	26A	27	27B	28	29	30	31	32	33	34	35	36	37	38	39
Gly	Lys	Ile	Thr	Phe	Tyr	Glu	Asp	Arg	Gly	Phe	Gln	Gly	His	Cys	Tyr	Gln	Cys	Ser	Ser	Asn	Cys	Pro	Asn	Leu	Gln	Pro	Tyr	Phe	Ser	Arg	Cys	Asn	Ser	Ile	Arg	Val	Asp	Val

Motif 3

84	85	86	87	88	89	90	91	92	93	94	95	96	97	98	99	100	101	102	103	104A	105	106A	107	108	109	110	111	112	113	114	115	116	117	118	119A	119B	119C	120	121	122
Phe	Arg	Met	Arg	Ile	Tyr	Glu	Arg	Asp	Asp	Phe	Arg	Gly	Gln	Met	Ser	Glu	Ile	Thr	Asp	Asp	Cys	Pro	Ser	Leu	Gln	Asp	Arg	Phe	His	Leu	Thr	Glu	Val	Asn	Ser	Ile	Asn	Val	Leu	Glu

Motif 2

40A	41A	42A	43A	44A	45A	47	48A	49	50	51	51A	52	53A	54A	54B	54C	55A	56	56A	57	58	59	60	61	62	62A	63	64	65	66	67	68	69	70	71	72	73	74	75	76	77	78	79	80	81	82	83
Gly	Cys	Trp	Met	Ile	Tyr	Gln	Arg	Pro	Asp	Tyr	Gln	Gly	His	Gln	Tyr	Phe	Leu	Arg	Arg	Gly	Asn	Tyr	Pro	Asp	Tyr	Gly	Gln	Trp	Met	Gly	Phe	Asp	Asp	Ser	Ile	Arg	Ser	Cys	Arg	Leu	Ile	Pro	Gln	His	Thr	Gly	Thr

Motif 4

123	124	125	125A	125B	125C	126	127	128	129	130	131	132	133	134	135	136	137	138	139	140	141	142	143	144	145	146	147	148	149	150	151	152	153	154	155	156	157	158	159	160	161	162	163	164	165
Gly	Ser	Trp	Val	Leu	Tyr	Glu	Met	Pro	Ser	Tyr	Arg	Gly	Arg	Gln	Tyr	Leu	Leu	Arg	Pro	Gly	Glu	Tyr	Arg	Arg	Tyr	Leu	Asp	Trp	Gly	Ala	Met	Asn	Ala	Lys	Val	Gly	Ser	Leu	Arg	Arg	Val	Met	Asp	Phe	Tyr

[a] The numbering is that of Croft (12), and letters indicate insertions or replacements necessitated by the electron density map at 1.9 Å resolution. The sequence is divided into four approximately equal parts (see Fig. 2) corresponding to the four Greek key motifs, each with four strands labeled a, b, c, and d. Topologically equivalent residues are in columns, and the sequence has been ordered to highlight the similarity between motifs 1 and 3 and 2 and 4.

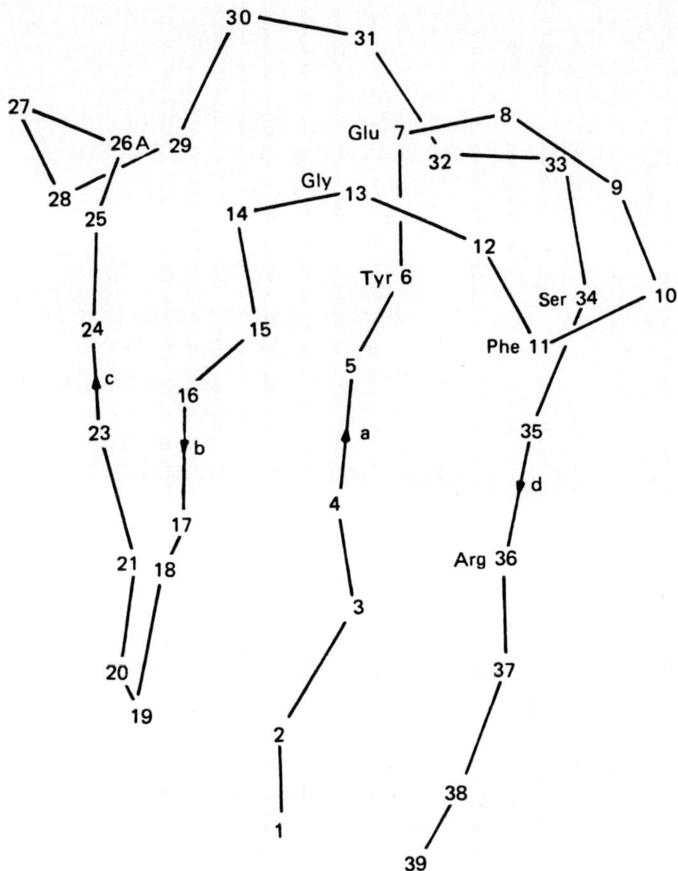

Figure 6 The α-carbon backbone of the amino acid residues comprising motif 1 viewed almost perpendicular to the plane containing strands a, b, and d. The folded ear formed by residues 8–12 inclusive is clearly seen at the top right-hand side of the figure.

between the α-carbon atoms of motifs 1 and 3 is slightly less but still very good, 1.74 Å for 36 residues.

In motif 1 the a strand of the Greek key consists of residues Lys 2, Ile 3, Thr 4, Phe 5, and Tyr 6 arranged in an extended chain conformation, whereas the b strand comprises residues His 14, Cys 15, Tyr 16, Gln 17, and Cys 18. These two strands run in opposite directions and are hydrogen bonded to one another. Glu 7 has a conformation (right-handed helical) that enables residues

Asp 8, Arg 9, Gly 10, Phe 11, and Gln 12 to fold over and form a turn between the a and b strands. In this part of the structure Gly 10 and Phe 11 assume a left-handed helical conformation so that the aromatic ring of Phe 11 is stacked against that of Tyr 6. After Gln 12 the chain has to complete the turn to fold down into the b strand. This fold is achieved through Gly 13, which has a conformation not normally permitted on sterical grounds for an amino acid residue with a side chain; Gly 13 is thus a key residue in this region of the motif structure. At the end of the b strand, residues Ser 19 and Ser 20 form a further turn into strand c. Only residues Asn 21 and Cys 23 in the c strand can be considered to run antiparallel to strand b, and these are not hydrogen bonded to it but are part of sheet 2 in the N-terminal domain. The polypeptide chain then assumes a rather irregular right-handed helical structure between Leu 24 and Tyr 28, but the following residues form a gentle loop around to strand d comprising residues Cys 32, Asn 33, Ser 34, Ile 35, Arg 36, Val 37, and Asp 38. Strand d is hydrogen bonded to strand a, runs in the opposite direction to it, and is part of the antiparallel right-handed β-pleated sheet formed from strands a, b, and d in motif 1 and the c strand of motif 2.

The importance of the structural role played by Gly 13 in motif 1 has already been indicated, and in motifs 2, 3, and 4 glycines are similarly found at topologically equivalent positions [Gly 52 (2), Gly 96 (3), and Gly 132 (4)]. Two further invariant residues are also found to have key structural roles in the motif conformation, and these are shown for motif 1 in Fig. 7. First, Tyr 6 (Tyr 45A, Tyr 89, and Tyr 125C in motifs 2, 3, and 4, respectively) has its aromatic ring stacked against that of the conservatively varied Phe 11 [Tyr 51 (2), Phe 94(3), and Tyr 130 (4)] on the turn between the a and b strands. Second, Ser 34 in strand d [Ser 73 (2), Ser 119A (3), and Ser 157 (4)] forms a hydrogen bond via its γ-oxygen atom to the peptide NH group of Phe 11. This hydrogen bond can be considered to pin the fold between strand a and b into position and therefore to hold the motif together as a structural unit. In each motif this important hydrogen bond appears to be protected from disruption by solvent by another amino acid residue attached to the d strand. Thus in motifs 1, 2, and 4, Arg 36, Arg 75, and Arg 159, respectively, are orientated so that their hydrophobic regions protect these hydrogen bonds and the hydrophilic guanidinium moieties are exposed to solvent. In motif 3, Asn 119C performs a similar function.

Thus very few invariant residues—Tyr 6, Gly 13, and Ser 34 (and Arg 36)— with the addition of the conservatively varied residues at Phe 11 and Glu 7 [Gln 47 (2), Glu 90 (3), and Glu 126 (4)] —appear to be essential for the motif structure. The hydrophobic or polar nature of other residues will be of importance when the sheets form a domain, but it is likely that all γ- (and β-) crystallins that contain these invariant residues will adopt structures containing similar Greek key motifs.

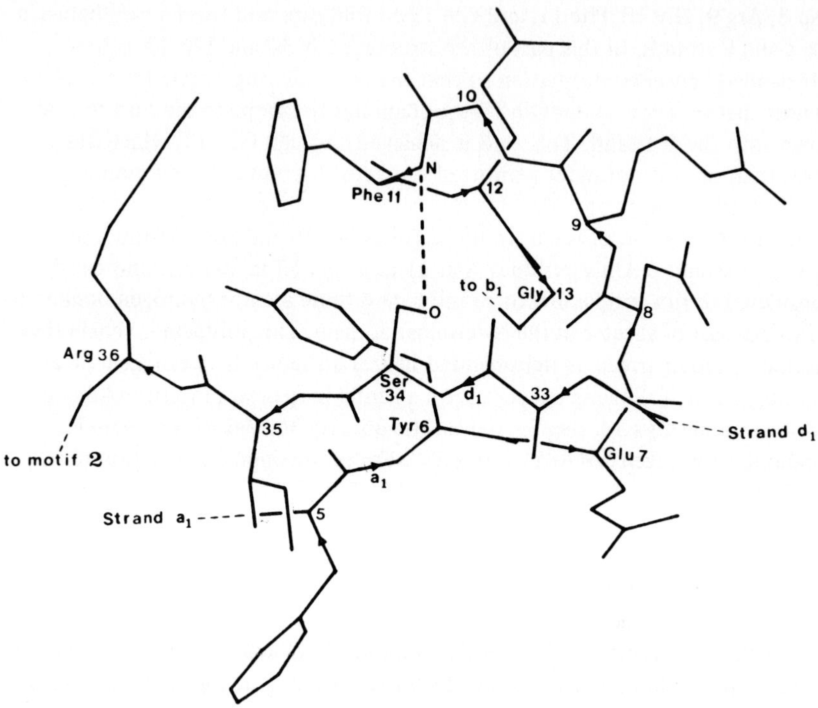

Figure 7 The region of motif 1 in the N-terminal domain containing amino acid residues invariant at topologically equivalent positions in the four motifs. The key residues are Tyr 6, Gly 13, Ser 34, and Arg 36. (Reprinted with permission from *Nature*, Vol. 289, No. 5800, pp. 771–777. Copyright 1981, Macmillan Journals Limited.)

Figure 8 shows the motif in the same orientation as Fig. 6 but complete with amino acid side-chain residues. A band of aromatic residues can be readily identified at a height of approximately two-thirds from the bottom of the motif. Within this band these residues are oriented so that the polar residues, for example, Tyr 6 and Tyr 16, lie on the outside surface of the motif, whereas the hydrophobic residues typified by Phe 5 and Phe 29 lie on the inside surface.

Finally, in the description of the motif structures and their participation in the corresponding β-sheets, Fig. 9 shows the intra-chain hydrogen bonding network within sheet 1 and involving strands a, b, and d from motif 1 and the c strand of motif 2. The N—O hydrogen bond distances range between 2.61 and 3.12 Å. Similar networks exist for the other three β-sheets, but in sheets

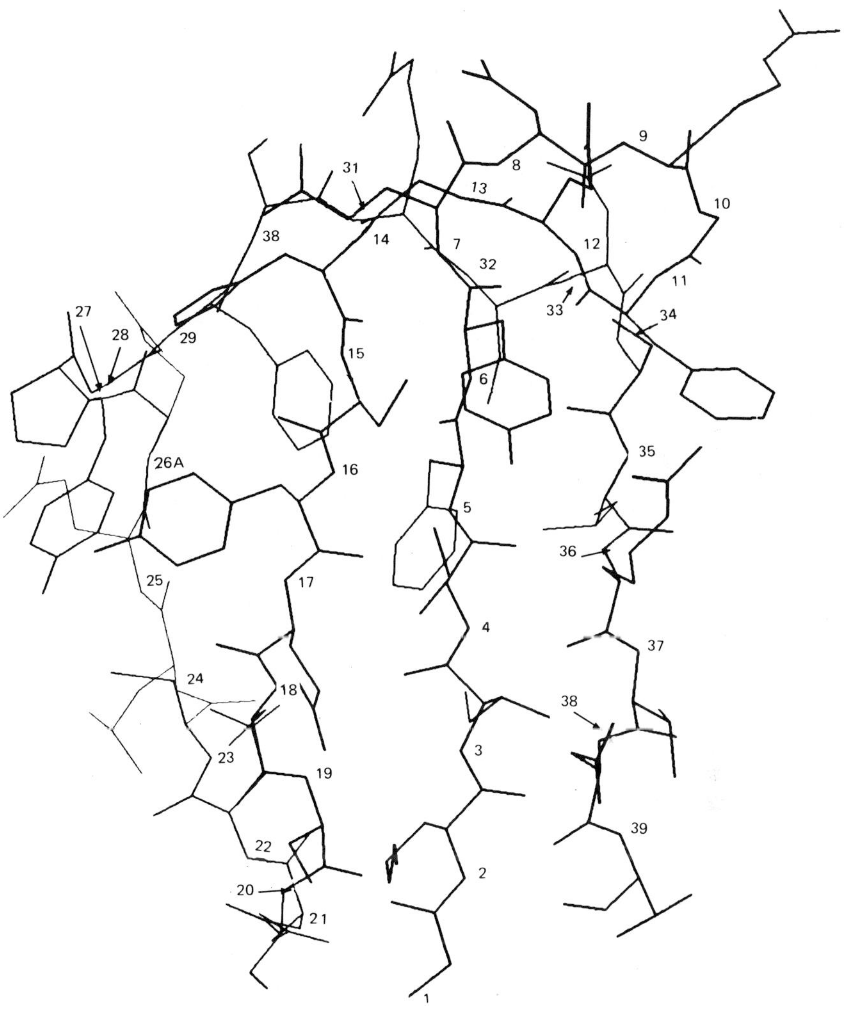

Figure 8 Motif 1 viewed in a similar direction to Fig. 6, but complete with amino acid residues. The sequence is that determined from electron density maps at 2.6 Å resolution and includes Asn 22, which is not apparent at higher resolution, 1.9 Å (see Fig. 6). (Reprinted with permission from *Nature*, Vol. 289, No. 5800, pp. 771–777. Copyright 1981, Macmillan Journals Limited.)

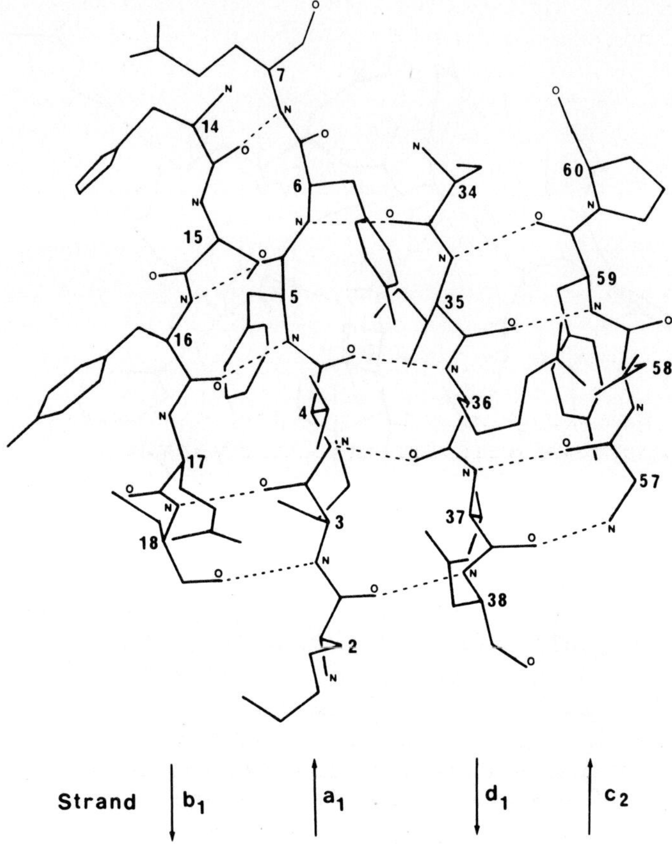

Figure 9   The interstrand hydrogen bonding network in sheet 1.

2 and 4, in the interdomain contact region of the molecule, only two hydrogen bonds (those equivalent to O35-N59 and O59-N35) join the c and d strands.

Organization of the N- and C-Terminal Domains

Each domain consists of two right-handed twisted β-pleated sheets (see, for example, Ref. 28a), where each sheet is formed from the c strand of one motif and the d, a, and b strands of the second motif in the same domain. When the two sheets pack together the bulky hydrophobic aromatic residues on the inside surfaces cause the domains to adopt a wedge-shaped structure (Fig. 4). The base of the wedge is formed by the turns between the b and c strands of the two motifs in the domain and the polypeptide chain connecting the d strand of one motif to the a strand of the second in that domain. The top of the wedge is enclosed by the loops, running antiparallel between the c and d strands of the two motifs. This arrangement introduces a pseudo-twofold axis between the sheets, as shown schematically in Fig. 5. Figure 10 shows stereo drawings of the N- and C-terminal domains complete with amino acid side-chain residues viewed almost perpendicular to the end faces of the wedges (Fig. 10A) and perpendicular to the sheets defining the sides of the respective wedges (Fig. 10B).

In general the hydrophobic residues are on the inside faces of the sheets and are orientated toward the centers of their respective domain cores. The hydrophilic and polar residues tend to be on the outside surfaces of the domains orientated toward the solvent. To illustrate this point, the regions of the domains directly between the sheets will be briefly described. In the hydrophobic domain cores rows of amino acids on the inside surfaces of the β-pleated sheets appear to form bilayers, as shown schematically in Fig. 11. The amino acid side chains in each row are listed in Table 3, together with the sheet, motif, and strand to which they are attached. A row consists of four residues on adjacent sheet strands (c$'$, d, a, and b) and runs almost perpendicular to the direction of the strands. In each row the outer residues are at the open ends of the domain wedge. These are exposed to solvent and tend to be hydrophilic or polar; the inner two residues are buried in the domain core, completely protected from solvent and are often Ile, Leu, or Val. The two rows from sheets 1 and 3 are almost linear, but in the rows from sheets 2 and 4 the tyrosine residues 54B and 135, respectively, are orientated toward the top of the wedge in the respective domains. However, in the N-terminal domain the α-carbon atom of Leu 55A (Leu 137 in the C-terminal domain) is almost in line with that of Trp 42A (Trp 125), whose bulky side chain completely fills the base of the wedge. The indole nitrogen atom of Trp 42A (Trp 125) forms a hydrogen bond with the peptide oxyten atom of Val 39 (Glu 122), which is at the end of strand d in motif 1. Thus residues 42A and 55A (125 and 137) could be considered as forming a partial second row from sheet 2 (4). The side chain of Leu 55A (Leu 137) extends into the space below Tyr 54B (Tyr 135) and is in contact with Val 37 (Val 120), Ile 44A (Leu 125B), and Ile 35 (Ile 119B).

Table 2 shows that very few of the residues in the hydrophobic core are invariant. The implication of this is that there are several combinations of amino acids that can pack together to fill a given spatial volume while preserving three-dimensional structural homology. Evolution, whether convergent or divergent, seems to have exploited these possibilities, and it is this variation combined with insertions in the loops between the c and d motif strands that makes recognition of structural homology very difficult purely from amino acid sequence information. The preservation of three-dimensional structure appears to override that of sequence.

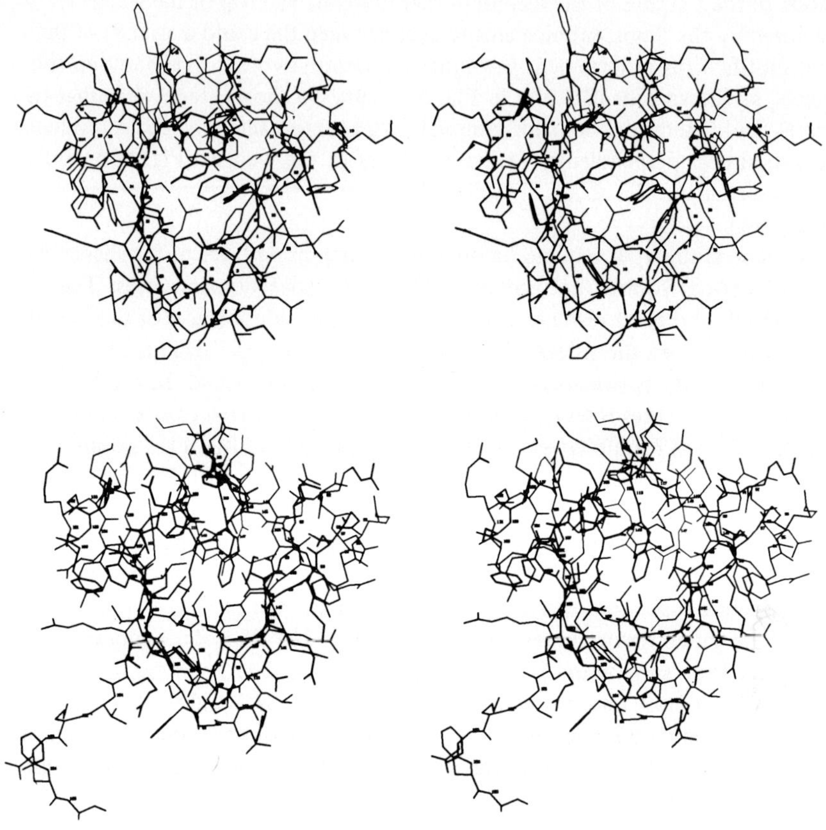

Figure 10A   Stereodiagrams of the domains of bovine γ-II-crystallin at 2.6 Å resolution. (A) Shows the N- and C-terminal domains viewed end on to the wedges.

Overall Structure of the γ-II Molecule

Figure 5 shows that the N- and C-terminal domains are connected by a short length of the polypeptide chain, which follows the surface of the molecule so that the β-pleated sheets comprising mainly motifs 1 and 3 lie on the outside of the molecule. The sheets comprising motifs 2 and 4 are adjacent to one another and are inclined at an angle of approximately 20° between strands. A further pseudo-twofold axis bisects the angle of approximately 40° between the intradomain diad axes. The twist between the domains when the strands of motifs 2 and 4 in sheets 2 and 4 are projected onto a mean plane is about 20°,

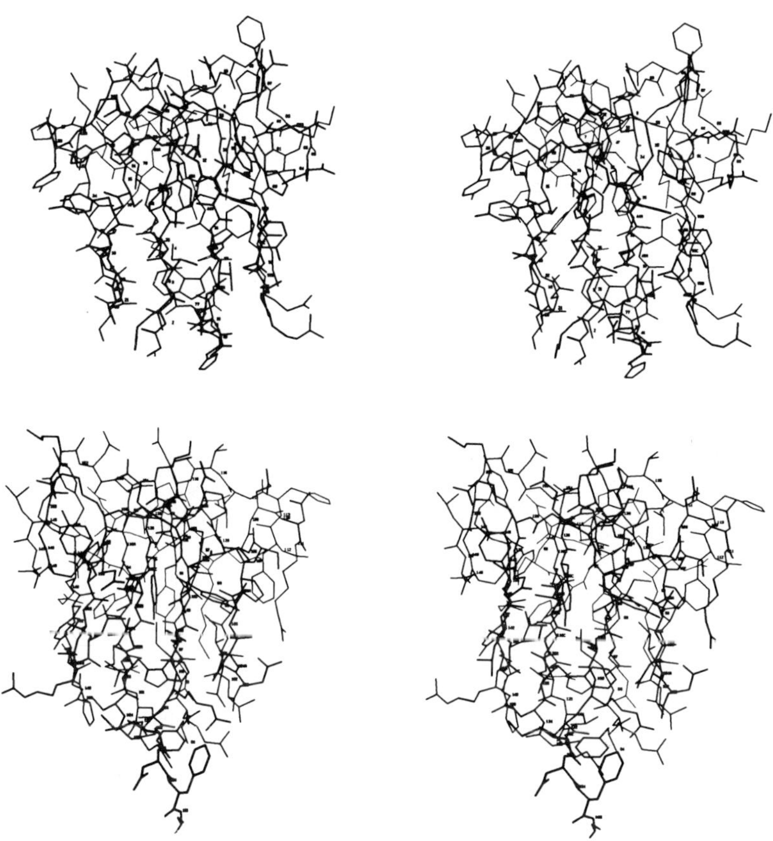

Figure 10B Shows the corresponding drawings viewed almost perpendicular to sheets 1 and 4, respectively. (Reprinted with permission from *Nature*, Vol. 289, No. 5800, pp. 771-777. Copyright 1981, Macmillan Journals Limited.)

similar to that between sheets in a single domain. Figure 12 is a stereodiagram of the complete molecule corresponding to the schematic representation of Fig. 5.

The connecting portion of the polypeptide chain comprises residues Ile 77 to Thr 83 inclusive and has an extended conformation, except for the turn at Gly 82 necessary to direct the chain into motif 3 at the beginning of the C-terminal domain.

The interdomain contact between the basal regions of sheets 2 and 4 is essentially hydrophobic, involving residues Met 43A and Phe 54C of motif 2 in the N-terminal domain and Leu 136 and Val 125A of motif 4 in the C-terminal domain (see Fig. 16). Other hydrophobic residues interact with the contact residues; Ile 77 and Val 161 both contact Phe 54C and Leu 136, and Ile 77 is in

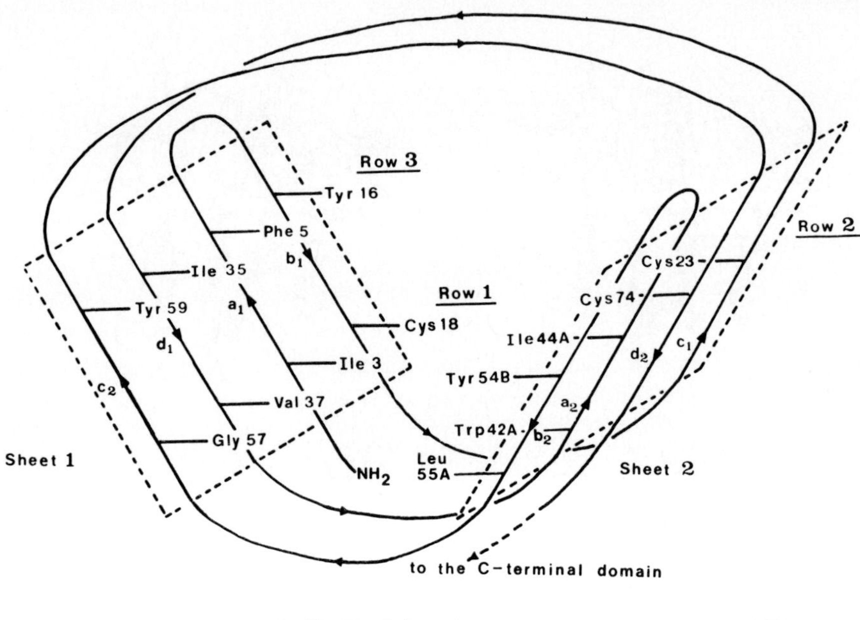

Figure 11 A schematic drawing of the hydrophobic core of the N-terminal domain showing the bilayer formed by rows of amino acid residues in sheets 1 and 2.

Table 3 The Hydrophobic Cores of the N- and C-Terminal Domains in γ-II

Row	β-Sheet	Residue (Motif and Strand)			
		N-Terminal domain			
1	1	Cys 18 (1b)	Ile 3 (1a)	Val 37 (1d)	Gly 57 (2c)
2	2	Cys 23 (1c)	Cys 72 (2d)	Leu 44A (2a)	Tyr 55A (2b)
3	1	Tyr 16 (1b)	Phe 5 (1a)	Ile 35 (1d)	Tyr 59 (2c)
		C-Terminal domain			
1	3	Ile 101 (3b)	Met 86 (3a)	Val 120 (3d)	Gly 140 (4c)
2	4	Cys 105 (3c)	Leu 158 (4d)	Leu 125B (4a)	Tyr 135 (4b)
3	3	Ser 99 (3b)	Ile 88 (3a)	Leu 119B (3d)	Tyr 142 (4c)

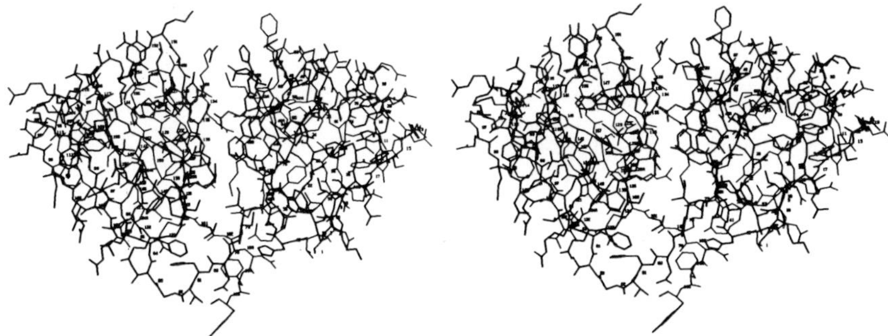

Figure 12   A stereodiagram of the complete molecule of bovine γ-II crystallin at 2.6 Å resolution. (Reprinted with permission from *Nature*, Vol. 289, No. 5800, pp. 771–777. Copyright 1981, Macmillan Journals Limited.)

contact with Met 43A. Otherwise, the contact region is surrounded by hydrophilic side chains interacting via their aliphatic or aromatic components while polar groups are exposed to the solvent.

The C-terminal portion of the polypeptide chain can be considered to comprise residues Val 161 to Tyr 165 inclusive, and extends away from the C-terminal domain so that Tyr 165 is in the vicinity of the N-terminus. In the electron density synthesis computed at 1.9 Å resolution, the C-terminal tyrosine is ill defined and may well be undergoing large thermal librations or occupying several positions in a disordered structure.

In the molecule as described above, 122 residues (70%) have conformations that can be assigned to the β-sheet category, whereas 34 residues (19%) can be classified as α-helical. Spectroscopic studies (29,30) are in agreement with a large β-sheet content but indicate no detectable α-helix. In the molecule there are 7 glycines and 11 other residues with left-handed α-helical conformations that may give circular dichroism of the opposite sign, effectively compensating for much of the ellipticity due to right-handed α-helical residues.

The high symmetry of the molecule and structural similarity of the motifs is emphasized in Fig. 13, a view of the α-carbon backbone almost parallel to the interdomain pseudo-twofold axis. The similarity in the three-dimensional structure of the motifs could be taken to suggest that the protein evolved divergently in two stages. First, a gene duplication of a protein of about 40 amino acid residues to a protein, approximating to a single domain, and a second gene duplication to yield the two domain structure. Such a model is consistent with the observation that the two domains resemble each other more closely than do the two motifs within a single domain.

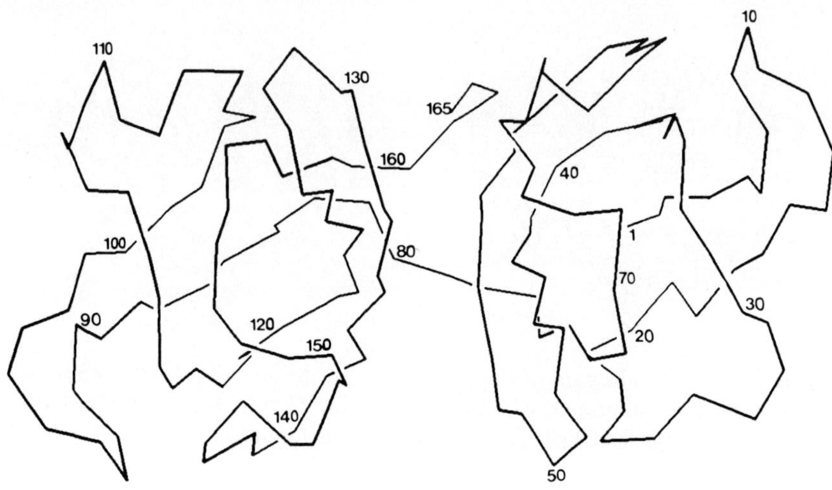

Figure 13   The α-carbon backbone of the amino acid residues of bovine γ-II-crystallin viewed parallel to the pseudo-twofold axis between the N- and C-terminal domains. The high structural symmetry of the molecule is clearly shown. (Reprinted with permission from *Nature*, Vol. 289, No. 5800, pp. 771–777. Copyright 1981, Macmillan Journals Limited.)

Structures of Bovine γ-IIIb- and γ-IV-Crystallins

The structure of bovine γ-IIIb-crystallin has been determined by the method of multiple isomorphous replacement to a resolution of 3.0 Å (24). The asymmetric unit in the crystal contains two independent molecules, and these are arranged approximately head to tail. Examination of the polypeptide chain folding in any one molecule shows that the N- and C-terminal domains are remarkably similar and comprise two right-handed twisted β-pleated sheets made up from simple Greek key motifs. The domains are related by a pseudo-twofold axis orientated at approximately 45° with respect to one another, and within each domain the sheets are also related by a pseudo-diad axis. Figure 14 shows the α-carbon backbone of one γ-IIIb molecule viewed in a direction corresponding approximately to that of Figs. 5 and 12.

The structure appears to be almost identical at this resolution to that of bovine γ-II, as would be expected from the homology between the respective amino acid sequences. This is estimated at approximately 85% from the partial sequence of γ-IIIb (31). A full structural comparison of the two proteins must await the high-resolution structural analysis of γ-IIIb.

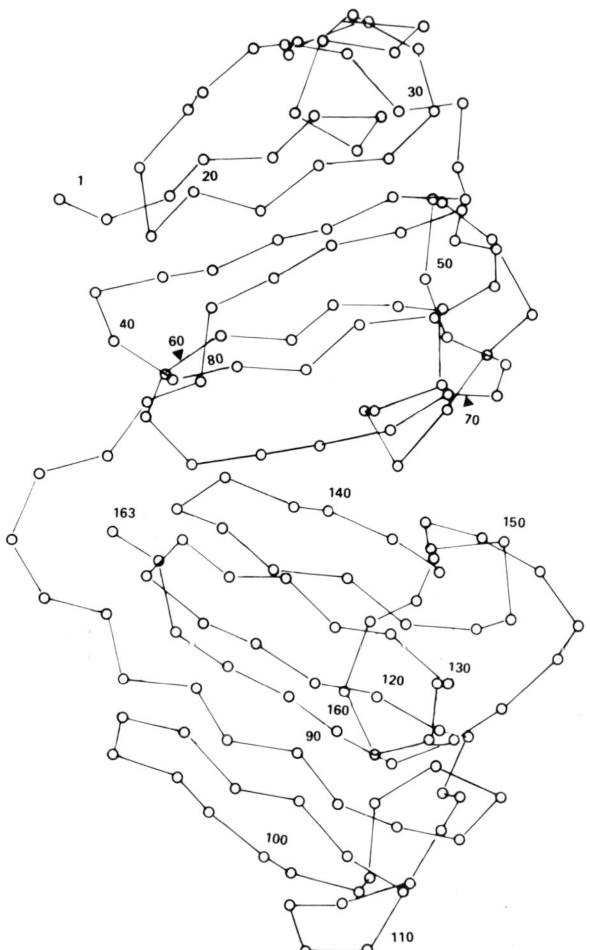

Figure 14 The α-carbon backbone of the amino acid residues of bovine γ-IIIb-crystallin viewed in a similar orientation to Fig. 12. (Reprinted in part with kind permission from *FEBS Letters*, Vol. 131, No. 1, pp. 81–84. Copyright 1981, Elsevier-North Holland Biomedical Press.)

Crystals of bovine γ-IV-crystallin have been reported (18), and a structural analysis is in progress. This analysis should show that the molecule also has the motif and domain structure of γ-II- and γ-IIIb-crystallins. The main aim of these studies is to compare the nature of the molecular surfaces of the different γ-crystallin molecules.

Molecular Surfaces and Interactions in γ-Crystallins

As yet only detailed studies of the nature of the molecular surface and molecular interactions for bovine γ-II have been reported (21,32), but differences in the balance between polar and nonpolar surface residues may well be responsible for the different physical properties of the homologous family of γ-crystallins. For example, the γ-crystallins as a class show a marked temperature-dependent solubility, with the different members exhibiting a range of solubilities.

The surface of the γ-II molecule is largely polar and hydrophilic, and many surface residue side chains appear well ordered in the high-resolution structure of the molecule. The preponderance of acid (Asp and Glu) and basic (Arg, Lys, and His) side chains is greater for the C-terminal domain than the N-terminal domain. Thus, in addition to the chain termini, the C-terminal domain comprises 14 basic and 13 acidic residues, compared with 8 basic and 7 acidic for the N-terminal domain; all these residues may be charged at pH 7. Within each domain the charges are nearly balanced, and since the majority of the charged residues are arginines and aspartates there will be little change in the number of charged groups over medium pH ranges. Over 50% of the ionic side chains form ion pairs, and since arginines and aspartates have less conformational flexibility than glutamates and lysines, this may contribute to the extremely well ordered surface skeleton of alternating charged groups. It has been noted (33) that enzymes in thermophilic bacteria have abundant surface ion pairs that contribute to thermal stability, and in a similar way an "exoskeleton" of organized charged groups in γ-crystallins could be important in holding together the polypeptide strands that form each domain. An example of an extended network of charged residues centered on Asp 92 is shown in Fig. 15. Six alternately negatively and positively charged residues, Glu 141, Arg 143, Asp 93, Arg 95, Asp 92, and Arg 144, form a network on one molecule. However, in the crystal lattice the chain may extend farther than this, since Arg 144 is close to Asp 103 of a symmetry-related molecule. Several other intermolecular ion pairs are also present in the crystal lattice.

In addition to these intermolecular ionic interactions, several other different types of intermolecular contact occur in crystals of bovine γ-II, but the majority are polar, with many of the residues usually regarded as hydrophobic (Phe and Met), probably involved because of their polarizability. There are few direct van der Waals contacts between nonpolarizable atoms. A detailed study suggests that interactions between sulfur-containing residues and/or residues containing delocalized π-orbital systems may also be of importance as regards crystallin-crystallin interactions in the lens (21).

Distribution of Sulfur in γ-Crystallins

Two general observations should be made with respect to the distribution of sulfur in protein molecules. First, an intracellular protein of molecular weight

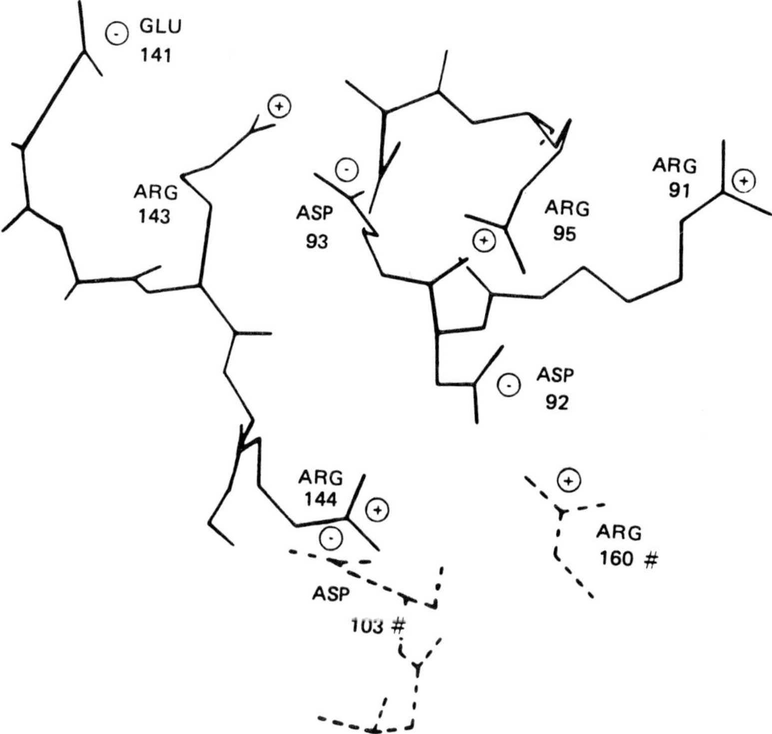

Figure 15 An extensive network of charged groups on the surface of the γ-II molecule, centered on Asp 92. These and similar interactions may be partially responsible for the stability of the γ-crystallin molecule. (Reprinted with permission from *Nature*, Vol. 289, No. 5800, pp. 771–777. Copyright 1981, Macmillan Journals Limited.)

around 20,000 d would be expected to contain about two cysteine residues (34). Second, several protein structures have been found in which there are chains of alternating sulfur-containing and aromatic residues (35) or arginine residues (36). In such structures it has been proposed that interactions between the polarizable sulfur atoms and the delocalized π-electron density may contribute to protein stability (37).

The distribution of cysteine residues in bovine γ-crystallins is shown in Fig. 16. Bovine γ-II has seven sulfydryl groups, only one of which, Cys 105 (topologically equivalent to Cys 23), is in the C-terminal domain. The γ-IIIb and γ-IV have fewer cysteine residues than γ-II but more than would be expected for intracellular proteins of their molecular weight.

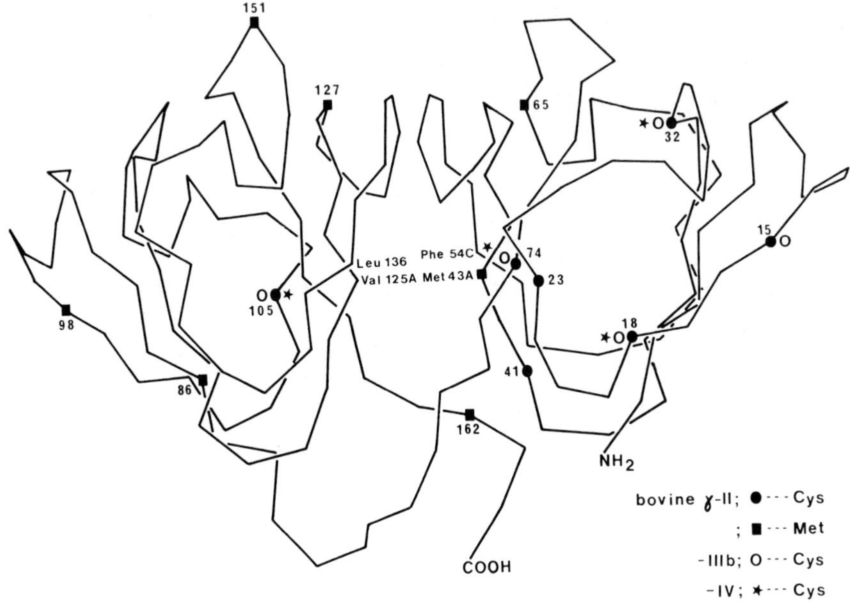

Figure 16   The distribution of the cysteine residues in bovine γ-crystallins. The diagram shows the α-carbon backbone of the molecule of γ-II-crystallin viewed in the same orientation as Fig. 12. In addition, the positions of the methionine residues in γ-II are shown, as well as the four residues comprising the hydrophobic patch between sheets 2 and 4 of the N- and C-terminal domains, respectively.

Although in γ-II the distribution of cysteine residues is asymmetric between the two domains, the overall distribution of sulfur atoms is more even; there are five methionine residues in the C-terminal domain and only two in the N-terminal domain (Fig. 16). All sulfur atoms in the molecule are found to be near other sulfur atoms, aromatic or arginine residues. For example, in the N-terminal domain both tryptophan residues are associated with Cys 32 and Cys 74, all of which are buried in the hydrophobic core of this domain. Cys 32 has Trp 64 and Phe 29 on either side, with Phe 5 underneath toward the base of the wedge. Cys 74 is in contact with Trp 42A, which fills the base of the wedge and also Phe 5. Extended molecular orbitals involving sulfur atoms and delocalized π-electron systems throughout the molecule can be envisaged.

Possible Roles for γ-Crystallins in the Transparent Lens and in Cataract

The γ-crystallins are predominant in the nuclear regions of those vertebrate lenses that have a steep refractive index gradient and a relatively low water content. The surface of the γ-II molecule appears to have several sites for noncovalent interactions with other crystallins but probably has a low tendency to bind water, since many of the ionic and polarizable surface residues are involved in intramolecular interactions, as described previously. However, because many of the surface polar side chains are fixed by such intramolecular interactions, they could have a major effect in the local ordering of water molecules.

Small hydrophobic surface patches may also play a role in ordering surface water, and both types of surface region will contribute, but in a different way, to the solubility and potential for surface interactions with other cell components. These may be critical for any role that these monomeric proteins have as spacers between other more closely interacting cell components in the nuclear region of the lens. High-resolution studies and comparisons with other members of the γ-crystallin family and, indeed, other crystallins are required to substantiate these proposals.

With respect to cataract it is established that the increased light scattering of senile nuclear cataracts is associated with protein aggregation (38–41), together with chemical cross-linking, including disulfide bridges and methionine oxidation (42–45) and unidentified covalent cross-links (46,47). Alteration of the surface properties of the γ-crystallins might well be expected to lead to supcraggregation of other crystallins, giving rise to light scattering. Such alterations could be caused by chemical modification (48), and the resulting noncovalent aggregation could then be followed by covalent cross-linking.

The high-resolution structure of bovine γ-II enables an assessment to be made of the accessibility of the cysteine residues. Thus, Cys 15 is readily accessible to solvent and could cross-link with other molecules with little disruption to the γ-II molecule if the oxidation state of the lens became sufficiently high. Cys 41A and Cys 105 (on the C-terminal domain) are also accessible, the former particularly so if the domains become separated or cleaved, whereas Cys 32 and Cys 74 are buried, and any participation in intermolecular disulfide cross-linking would involve gross disruption of the molecule and presumably denaturation. The proximity of Cys 18 and Cys 23 indicates that these residues are more likely to take part in intramolecular cross-linking. Indeed, evidence for such a link is found in the structure of bovine γ-II (22), although this linkage may have arisen due to the age and oxidation state of the crystals used for the x-ray data collection. The formation of such an intramolecular disulfide bond would tend to stabilize the molecule, and it is interesting to speculate whether this type of mechanism helps to prevent premature destructive oxidation of lens proteins in senescence or cataract.

Cross-linking between the exposed C-terminal tyrosine residues caused by photo-oxidation may be relevant to brunescent cataract (49).

The quantity and distribution of the sulfur-containing and aromatic residues may be of importance with regard to ultraviolet (UV) radiation. It has been suggested that the extended molecular orbitals formed by these residues may facilitate electron flows in cytochromes and other electron transfer proteins. Such mechanisms could also protect internal aromatic residues in crystallins from the effects of UV radiation (50,51).

β-CRYSTALLINS

β-Crystallins have properties intermediate between α-and γ-crystallins. They are heterogeneous, with at least five polypeptide subunits involved in aggregate formation in the range 50,000–200,000 (52,53). In all vertebrate classes, however, one particular subunit predominates (54), and this has been designated βBp (55). Driessen et al. (56) determined the amino acid sequence for the monomer subunit of βBp-crystallin, molecular weight around 23,000 d, from bovine lenses. Unexpectedly this β-crystallin was found to be partially homologous (approximately 25%) with bovine γ-II-crystallin and in a similar manner appeared to show an internally duplicated sequence of amino acids. On this basis Driessen et al. (56) proposed that the presentday β- and γ-crystallins arose from an intragenic duplication of an ancestral βγ-crystallin gene.

The proposal of structural similarity between β- and γ-crystallins has been considered further by Driessen et al. (57), Siezen (58), and Wistow et al. (59) in terms of β-crystallin internal homology and, in addition, secondary and tertiary structure, charge stabilization, location of sulfydryl groups, and intermolecular interactions. Supporting evidence for structural similarity can be found in optical rotatory dispersion and circular dichroism spectroscopic studies of β- and γ-crystallins, which show that these proteins contain a high proportion of β-sheet structure with little or no α-helix (29,60). However, the most striking evidence is that provided by Wistow et al. (59), who have used interactive computer graphics to build a molecular model of the bovine β-Bp-crystallin protomer based on the positions of the atoms of bovine γ-II-crystallin. Further, by considering the nature of the molecular surface and the chain extensions at the N- and C-termini, these workers have been able to propose possible modes of aggregation of βBp subunits.

The amino acid sequence of a β-crystallin from a mouse lens has also been reported (61), and predictive computer graphics techniques have again enabled a model for this crystallin to be devised (62). The bovine βBp- and murine pMβCr1-crystallin model-building studies and their implications will now be described.

Structure Prediction and Molecular Models of β-Crystallins

The computer graphics system used for these studies was an Evans and Sutherland Picture System II linked to a PDP11/60 computer. The program FRODO (63), modified by Jones and Tickle, was used to construct molecular models, and the program MIDAS (written by Tickle) was used to display the resulting structures. The general strategy adopted involving the following stages.

1. Sequence alignment. As shown in Table 4, the amino acid sequences of the β-crystallins were aligned against that of bovine γ-II, motif by motif, so that the amino acid residues that appear to be crucial with respect to the motif conformation (for example Tyr 6, Gly 13, Ser 34, and Arg 36 in motif 1 of γ-II) were in their appropriate positions and in such a way that insertions or deletions would minimize the disruption of the backbone γ-II structure.
2. Construction of the main polypeptide chain backbones. The main-chain conformations of the β-crystallins were then modeled with interactive computer graphics using the atomic coordinates of γ-II as a guide. Networks of hydrogen bonds within each β-sheet (see Fig. 9) were maintained in a similar manner to those observed in bovine γ-II-crystallin.
3. The hydrophobic core. Residues of the hydrophobic domain cores were next added to the main-chain models and adjusted so that, as far as possible, they were close-packed in the cores while maintaining acceptable geometries and so that there were no nonbonded atomic contacts significantly less than the sum of the respective van der Waals radii.
4. Surface residues. The surface residues were added in positions close to those occupied by the biologically equivalent residues of the γ-II molecule. Some adjustments were made to optimize interactions according to van der Waals radii criteria and to form ion pairs.
5. Termini. The N- and C-termini residues were added finally, in conformations suggested by secondary structure prediction techniques (64,65).

The implications of the model-building studies are that both β-crystallins show the same fourfold internally repetitive structure of four Greek keys as found in bovine γ-II-crystallin. Table 4 shows that of the invariant amino acid residues that appear to play a crucial role in the motif conformation, Gly 13 in motif 1 (52, 96, and 132 in motifs 2, 3, and 4, respectively) and Ser 34 (73, 119A, and 157) are completely conserved, whereas Tyr 6 (45A, 89, and 125C) is conservatively varied. In β-Bp-crystallin, Arg 36 in motif 1 is replaced by an entirely hydrophobic but still bulky leucine residue. In both β-crystallins the residues equivalent to Glu 7, Phe 11, and Val 37 of motif 1 in γ-II and their topologic equivalents in the other three motifs are also conservatively varied, allowing an overall conformation close to that found in γ-II-crystallin.

Table 4 Comparison of Amino Acid Sequences of Bovine βBp and Murine pMβCr1 Crystallins with Bovine γ-II-Crystallin[a]

N-terminus

pMβCr1	Ac-	Met Leu Tyr Leu(-10) Val Leu Phe Leu Val Pro Phe Asn Ser(-1)
βBp	Ac-	Ala Ser Asn His Glu Thr Gln Ala Gly Lys Pro Gln Pro Leu Asn

MOTIF 1

	← strand a →	← strand b →	← strand c →
pMβCr1	Ile Gln Ile Ihr Ile Tyr Asp Gln Glu Asn Phe Gln Gly	Lys Arg Met Glu Phe Thr Ser	Ser Cys Pro Asn
βBp	Pro Lys Ile Ile Ile Phe Glu Gln Glu Asn(10) Phe His Gly	His Ser Gln Glu Leu Asx Pro(20)	Gly Asx Pro Cys
γ-II	Gly Lys Ile Thr Phe Tyr Glu Asp Arg Gly Phe Gln Gly	His Cys Tyr Gln Cys Ser Ser	Asn Cys Pro Asn

MOTIF 3

pMβCr1	Ser Lys Ile Thr Asn Phe Glu Lys Glu Asn Phe Ile Gly	Arg Gln Trp Glu Ile Cys Asp Asp	Tyr Pro Ser
βBp	His(84) Lys Ile Thr Leu Tyr Glu(90) Asn Pro Asn Phe Thr Gly Lys	Lys Met Glu Val(100) Ile Asp Asp Asp	Val Pro Ser
γ-II	Phe Arg Met Arg Ile Tyr Glu Arg Asp Asp Phe Arg Gly	Gln Met Ser Glu Ile Thr Asp Asp	Cys Pro Ser

MOTIF 2

pMβCr1	Gly Ala Trp Ile Gly Tyr Glu His Thr Ser Phe Cys Gly	Gln Gln Phe Ile Leu Glu	Arg Gly Glu Tyr Pro Arg
βBp	Gly Pro Trp Val Gly Tyr Glu Gln Ala Asn(50) Cys Lys Gly [40A]	Glu Gln Phe Val Phe Glu	Lys Gly Glu Tyr Pro(60) Arg
γ-II	Gly Cys Trp Met Ile Tyr Gln Arg Pro Asp Tyr Gln Gly	His Gln Tyr Phe Leu Arg	Arg Gly Asn Tyr Pro Asp

MOTIF 4

pMβCr1	Gly Ala Trp Val Cys Tyr Gln Tyr Pro Gly Tyr Arg Gly	Tyr Gln Tyr Ile Leu Glu Cys Asp His His		Gly Gly Asp Tyr Lys His
βBp	Gly Thr Trp Val Gly Tyr Gln Tyr Pro Gly Tyr(130) Arg Gly [123]	Leu Gln Tyr Leu Leu Gln	Lys	Gly(140) Asp Tyr Lys Asp
γ-II	Gly Ser Trp Val Leu Tyr Glu Met Pro Ser Tyr Arg Gly	Arg Gln Tyr Leu Leu Arg	Pro	Gly Glu Tyr Arg Arg

C-terminus

pMβCr1	Ile Gln Gln
βBp	Ile Arg Asp Met Gln(165) Trp His Gln Arg Gly Ala Phe His Pro Ser Ser
γ-II	Val Met Asp Phe Tyr

[a]The bovine sequence is arranged to show the equivalent residues of the four motif folding units in vertical columns (see Table 2). Residues in the β-crystallins are then aligned for maximum homology with the γ-II sequence.

There are several interesting compensatory amino acid changes in the β-crystallins with respect to bovine γ-II. One example is that, in both β-crystallins, Tyr 62 in strand c of motif 2 in γ-II is replaced by a tryptophan residue. Without other changes this bulky residue would either protrude into solvent or considerably disturb the nearby conserved Trp 64. However, insertions in the loop between the c and d strands of this motif in the β-crystallins allow Trp 62 to be turned around into a more internal environment and to be protected from solvent by the amino acid residues on the lengthened loop; these changes thus appear to be concerted.

Chain Termini in β-Crystallins

Compared to the γ-crystallins, bovine βBp crystallin has polypeptide chain extensions at both the N- and C-termini, and it has been proposed that these may be involved in oligomer formation as described in the next section. However,

```
                                                   ←———— strand d ————→
Val Ser Glu     Arg Asn Phe Asp                    Asn Val Arg Ser Leu Lys     Val Glu Cys
Leu Lys Glu     Thr Gly Val Glu                    Lys Ala Gly Ser Val Leu     Val Gln Ala
                            30
Leu Gln Pro         Tyr Phe Ser                    Arg Cys Asn Ser Ile Arg     Val Asn Val


Leu Gln Ala Met Gly Trp Phe Asn Asn                Glu Val Gly Ser Met Lys     Ile Gln Cys
Phe His Ala His Gly Tyr Gln Glu                    Lys Val Ser Ser Val Arg     Val Gln Ser
        110                                                                            120
Leu Gln Asp Arg Phe His Leu Thr                    Glu Val Asn Ser Ile Asn     Val Leu Glu

                                                                         ←——— inter-domain ———→
                                                                                region
Trp Asp Ala             Trp Ser Gly Ser Asn Ala Tyr His Ile Glu Arg Leu Met Ser Phe Arg Pro Ile Cys Ser Ala Asn His Lys Glu
Trp Asp Ser             Trp Thr Ser Ser         Arg Arg Thr Asp Ser Leu Ser Ser Leu Arg Pro Ile Lys Val Asp Ser Gln     Glu
                                                                70                                            80
Tyr Gly Gln             Trp Met Gly                     Phe Asp Asp Ser Ile Arg Ser Cys Arg Leu Ile Pro Gln His Thr Gly     Thr


Trp Pro Glu             Trp Gly Ser His         Ala Gln Thr Ser Gln Ile Gln Ser Ile Arg Arg
Ser Gly Asp             Phe Gly                 Ala Pro Gln Pro Gln Val Gln Ser Val Arg Arg
                                                150                                      160
Tyr Leu Asp             Trp Gly                 Ala Met Asn Ala Lys Val Gly Ser Leu Arg Arg
```

in the murine pMβCr1 molecule, the C-terminus terminates at the surface of the
C-terminal domain and there is only one extension, albeit of some 13 residues,
at the N-terminus. This extension is largely hydrophobic in nature and contains
sequences of leucine, valine, and phenylalanine amino acid residues characteristic
of membrane anchors and signal peptides, which dissolve in the membrane
hydrophobic lipid fraction. However, Inana et al. (62) have argued that the
N-terminal extension is unlikely to correspond to a signal peptide; not only is
it considerably shorter than most of those already reported, but also there is no
reason the protein should be exported or indeed should pass through any
organelle membrane. Instead, these authors have proposed that the N-terminal
extension acts as a membrane anchor, with the chain adopting an α-helical
conformation between residues –13 and –5, which is terminated by the proline
residue at –4. The presence of four leucine residues, which are strongly helix
inducing, at –12, –10, –8, and –6, support this idea, and the hydrophilic residues
Asn –2 and Ser –1 could act as a flexible link, allowing the hydrophobic region
access to the lipid bilayer.

Considering the overall morphology of eye lenses, some crystallins, although intracellular, must interact with the very extensive membrane surface. The N-terminal extension of murine pMβCr1-crystallin can be envisaged as linking the bilobal structure to the membrane by interacting with it, although not penetrating through to the polar head groups on the extracellular surface.

Aggregation of β-Crystallins

In solution β- and γ-crystallins differ in their ability to self-associate so that γ-crystallins are monomeric but β-crystallins, with the exception of βs-crystallin (66), always aggregate. The principal subunit of bovine β-crystallin has been isolated under dissociating conditions and crystallized as a reassociated βBp dimer (67). Wistow et al. (59) have been able to propose two possible models for this dimer by considering, first, differences between hydrophobic patches on the surfaces of the predicted structure for the βBp subunit and the γ-II molecule and, second, the nature of the chain termini extensions in the β-crystallin.

With regard to hydrophobic patches, Table 4 shows that Tyr 6 in motif 1 of γ-II is replaced in βBp by the more hydrophobic Phe 6; indeed, in the eight equivalent motifs in γ-II and βBp, this is the only place where Tyr 6 is not conserved. In addition, Arg 36 in motif 1 of γ-II, which appears to shield the hydrogen bond between Ser 34 and Phe 11, is replaced by a hydrophobic leucine residue in βBp, and again this only occurs in this motif. Finally, Ile 4 in βBp replaces the more hydrophilic Thr 4 in γ-II. As shown in Fig. 17, four residues— Ile 4, Phe 6, Phe 11, and Leu 36—form a hydrophobic patch on motif 1 in βBp that provides a potential site for intersubunit contact. On the C-terminal domain in βBp a second potential site arises, with the replacement of Arg 87 and Glu 100 in motif 3 on the outside surface of the γ-II molecule by the uncharged Thr 87 (at pH 7) and Val 100, respectively.

As stated previously, the major differences between the monomeric γ-II molecule and the βBp protomer are the existence of the termini chain extensions in the latter. Secondary structure prediction methods (64,65) indicate that neither the 15 residues in the N-terminal extension nor the 14 C-terminal residues have a strong tendency to form regular β-sheet or α-helical structure. However, six residues in the N-terminal chain—Ser −14 to Glu −9—could form a short helix followed by a reverse turn through Ala −8 and Gly −7. Wistow et al. (59) have noticed that this type of structure is similar to the N-terminal arms of spherical virus coat proteins (68,69) and lactate dehydrogenase (70,71), which are also involved in oligomer formation. Indeed, these authors have proposed that a helical region extending from the N-terminus and ending in a turn or coil conformation at or around a common glycine residue, Gly −7 in βBp, may be a structural feature enhancing the stability of certain oligomeric proteins.

By considering the surface hydrophobic patches on the βBp protomer that do not appear on the monomeric bovine γ-II-crystallin and the possible role of the

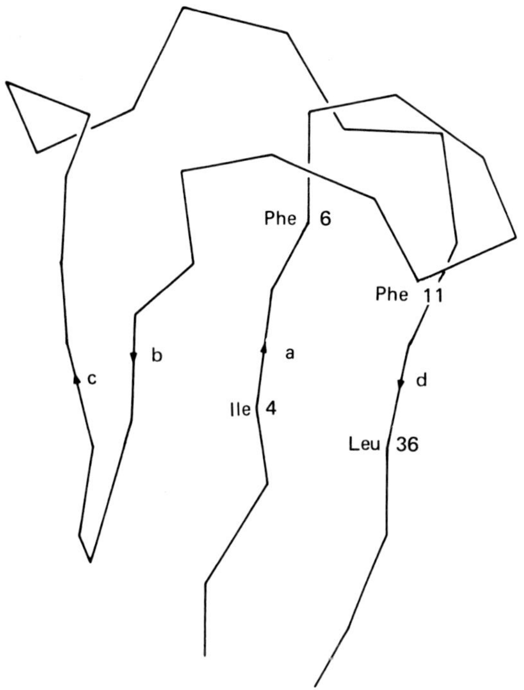

Figure 17 The hydrophobic patch of aminoacid residues on motif 1 in the N-terminal domain of bovine βBp-crystallin.

chain termini in oligomer formation, Wistow et al. (59) have used interactive computer graphics to propose two models for the βBp dimer. These are shown schematically in Fig. 18 and consist of a compact model where the protomer domains are approximately related by three mutually perpendicular twofold axes and an extended model involving isologous interaction of the N-terminal domains and possessing twofold symmetry.

In the extended model the interaction between the two protomers is analogous to the interaction between the domains within a single protomer, and the model allows for the formation of higher aggregates via the free C-terminal arm and other types of β-crystallin subunit. In the compact model the number of interactions between protomers is maximized. Thus, for example, Phe 172 in the C-terminal arm of one protomer is in contact with the hydrophobic patch on motif 1 of the other protomer. Other contacts involve the N-terminal arm of one molecule interacting with the hydrophobic patch at Val 100 of the other, and the connecting lengths of polypeptide chain between

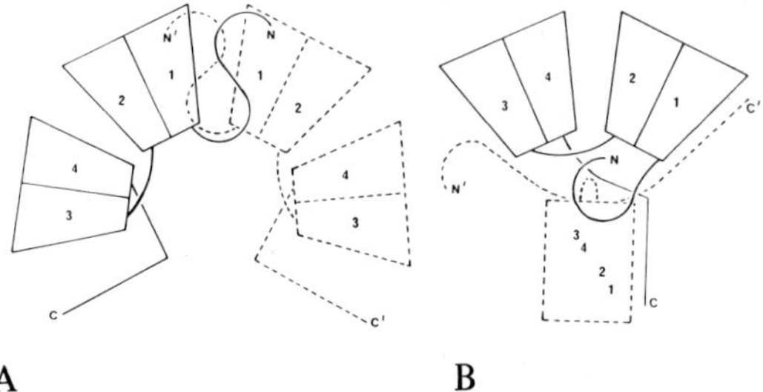

Figure 18   Schematic diagrams showing possible packing of βBp protomers to form dimers. Diagram (A) shows the extended model involving an isologous interaction of the N-terminal domains. Diagram (B) represents a more compact model in which the number of interactions between protomers is maximized and where the protomer domains are approximately related by three mutually perpendicular twofold axes. In both diagrams one protomer is shown with a full line and the other as a dashed line.

the domains of each protomer lie close to one another so that there is a possibility of forming a two-stranded section of antiparallel β-sheet involving residues 78–81 of both molecules.

Although further biochemical and spectroscopic studies may enable one or the other of these models to be preferred, only a single-crystal three-dimensional x-ray structure analysis will give detailed information on the quaternary interactions in βBp-crystallin.

X-ray Studies of β-Crystallins

Single crystals of the bovine βBp dimer suitable for x-ray diffraction studies have been grown by the hanging-drop vapor diffusion technique (67). These are shown in Fig. 19, together with a typical x-ray diffraction pattern. The crystals become rapidly disordered if diffraction experiments are carried out at room temperature or at +5°C; x-ray work is currently undertaken at –5°C. The protein crystallizes in the orthorhombic system with a = 154.2 (3), b = 165.6 (3), and c = 78.4 (2) Å. The conditions limiting possible reflection indicate space group $C222_1$ or C222. Assuming that the volume of the unit cell occupied by solvent is

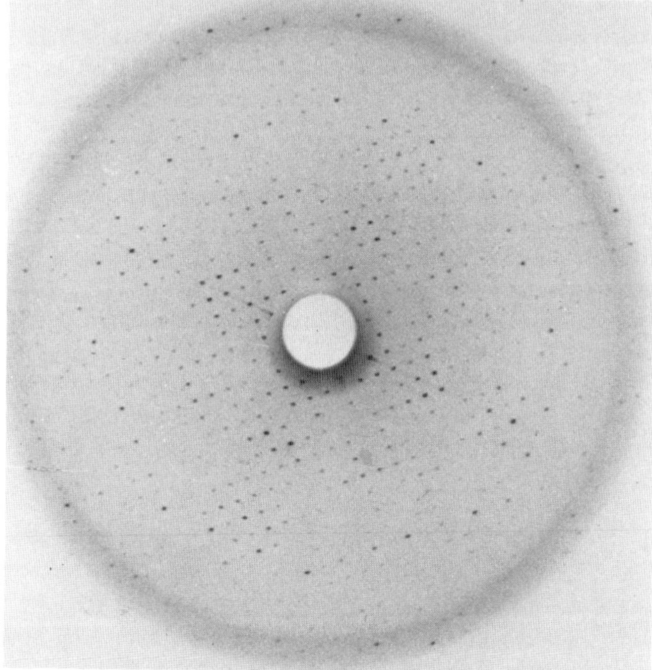

Figure 19 Crystals of bovine βBp-crystallin with a precession photograph of the hk0 zone showing the pseudo-fourfold symmetry.

approximately 50%, the asymmetric repeating unit in the crystal must contain two independent dimer entities, each of molecular weight around 46,000 d, although these may be associated in the crystal. The diffraction photographs also reveal a pseudo-tetragonal cell with the same c axis as the orthorhombic cell but with an a axis of length 113.1 Å and parallel to the ab face diagonal of the orthorhombic cell. This implies that the two dimer βBp molecules in the asymmetric unit may be related by approximate twofold symmetry.

Further diffraction studies are under way (72), and a set of x-ray intensity data has been collected for the native protein crystals to a resolution of 3.5 Å. A search for a heavy atom derivative has proved successful, and it is intended to undertake the x-ray analysis using both the method of isomorphous replacement and that of molecular replacement, using the known structure of bovine γ-II-crystallin. These studies should not only confirm the predicted structure for the βBp monomer subunit but in addition, by determining the relative orientations of the two subunits in the dimer, should provide for the first time experimental details of quaternary interactions between crystallin molecules.

δ-CRYSTALLINS

δ-Crystallin is a tetrameric protein of molecular weight 200,000 d that appears to replace entirely the γ-crystallin components of certain avian and reptilian lenses (73–76). Lenses in which the protein is a major component have a relatively high water content and are thus easily deformed, a requirement for their extensive accommodatory function. Circular dichroism and laser Raman spectroscopic studies (77,78) show that δ-crystallin has a large proportion of α-helix—up to 70%. It must therefore be entirely different in molecular structure from the α-, β-, and γ-crystallins in mammalian lenses, where the predominant structure is β-sheet. Whether the accommodatory properties of these lenses are partly fulfilled by protein molecules moving with respect to one another or individual components within molecules moving with respect to one another, or both, is not known. In the latter case, structures with high proportions of α-helix, where the hydrogen bonding is mainly intrachain, might be more flexible than predominantly β-sheet proteins, where the hydrogen bonding is interchain.

Electrophoretic studies on polyacrylamide gels in the presence of SDS have shown the molecule to be a tetramer with four subunits, each of molecular weight approximately 50,000 d (79). These subunits in chick δ-crystallin have been shown to be similar on the basis of tryptic peptides (80), but the extent of the identity must await a full amino acid sequence study. In newly synthesized δ-crystallin from embryonic chick, quail, and turkey lenses, two subunits of slightly different molecular weight have been identified—a major component of 48,000 d and a minor one of 50,000 d (73,76). The relative proportion of the

Table 5 Amino Acid Compositions of δ-Crystallins

Amino Acid	Moles of amino acid per subunit			
	Chick	Quail	Turkey	Bovine γ-II[a]
Thr	34	41	29	5 (13)
Ser	47	37	42	12 (30)
Pro	13	14	13	8 (20)
Gly	26	23	21	15 (38)
Val	34	37	29	7 (18)
Ile	39	41	29	8 (20)
Leu	77	74	59	11 (28)
Ala	39	32	29	2 (5)
Cys	—	—	—	7 (18)
Met	9	9	8	7 (18)
Tyr	4	5	4	15 (38)
Phe	9	9	8	10 (25)
Trp	—	—	—	4 (10)
Asp	34	32	29	20 (50)
Glu	60	60	55	18 (45)
Lys	34	32	29	2 (5)
His	4	5	4	3 (8)
Arg	17	18	17	20 (50)

[a]The figures in parentheses are the moles of amino acid per mole of bovine γ-II-crystallin scaled up by a factor of 2.5 for comparison with the δ-crystallin.

two subunits appears to depend on the conditions of culture (81–83). There is also evidence for two genes for chick and turkey δ-crystallin, although subunit heterogeneity could possibly be caused by posttranslational processing (84). Amino acid composition studies have shown that chick, quail, and turkey δ-crystallins have similar compositions (Table 5). In comparison with bovine γ-II, the δ-crystallin molecule contains a far larger proportion of leucine, alanine, and valine residues, amino acids that tend to induce α-helix formation. There appears to be no tryptophan, and the amounts of sulfur-containing and aromatic residues are small.

Crystals of δ-crystallin isolated from adult turkey (and quail) lenses suitable for x-ray diffraction experiments have been grown by vapor diffusion techniques

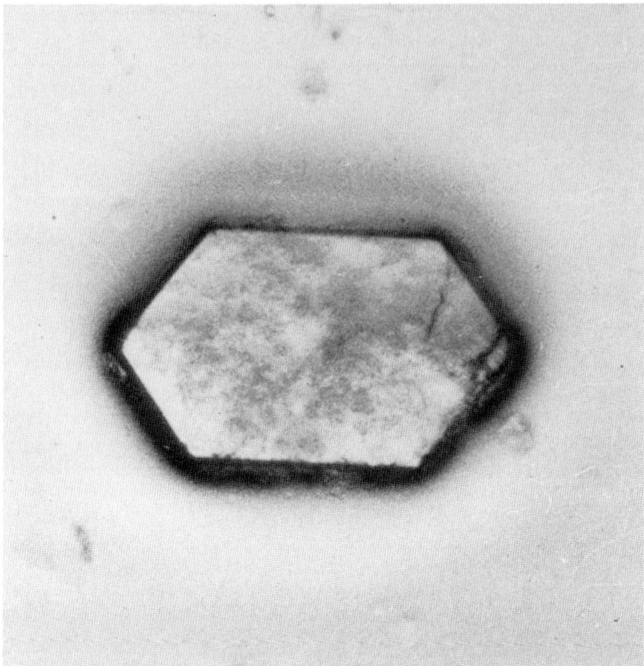

Figure 20 A crystal of turkey δ-crystallin showing the hexagonal platelet habit.

(85). The protein crystallizes in the form of hexagonally shaped plates that belong to the orthorhombic system (Fig. 20). At room temperature the x-ray diffraction patterns show considerable crystal disorder, but two distinct unit cells have been identified. One crystal has shown a unit cell with dimensions a = 99.9 (2), b = 135.4 (3), and c = 140.0 (3) Å, with space group symmetry $P2_12_12_1$. The volume of this cell could accommodate four tetramers of molecular weight 200,000 d, assuming a water content of around 40% by volume. No restrictions would be placed on the composition of the tetramers due to crystallographic symmetry; that is, they could consist of identical subunits, a combination of two different subunits, or indeed, four nonidentical subunits. However, most crystals at room temperature grow with orthorhombic cells having a slightly smaller b dimension, 133.4 (3) Å, and a halved c dimension, 69.1 (2) Å. The conditions limiting possible reflection in the x-ray diffraction patterns indicate space group $P2_12_12$. In this cell only two tetrameric molecules can be accommodated, assuming a water content of 40% by volume, and these must lie on the crystallographic twofold axes. That is, the tetramers themselves must possess twofold symmetry so that the subunit composition must be either A_4 with four identical subunits, or A_2B_2 with equal numbers of two different subunits disposed about a twofold axis. Although this observation does not rule out the possibility of different combinations of nonidentical subunits in the larger cell, the fact that most crystals have a cell where the tetramers must possess at least twofold symmetry strongly suggests that, in the bulk of δ-crystallin from adult turkey lenses, the maximum heterogeneity possible would be two sets of extremely similar subunits. The large $P2_12_12_1$ cell would then be explained in terms of a different packing of the molecules in the crystal lattice, with the tetramers moving slightly away from the twofold axis. This view is reinforced by recent x-ray work carried out at -5°C.

At temperatures of -5°C, crystals of turkey δ-crystallin obtained by the addition of PEG crystallize with space group $P2_12_12$ and yield x-ray diffraction patterns extending to a resolution of at least 2.5 Å; the crystalline disorder is considerably less than at room temperature (Fig. 21). However, the situation is further complicated in that it has been found (86) that the unit cell c dimension is particularly sensitive to the crystal wetness. Very wet crystals give unit cells in which c = 80.0 (2) Å, but a second cell with c = 70.0 (2) Å can often be obtained as the crystals dry out, and values as low as 62.5 (2) Å have been observed with very dry crystals. This represents a decrease in the unit cell volume of some 22%. In several cases crystals have undergone topotactical changes during the recording of x-ray diffraction patterns. At the commencement of these experiments with wet crystals the diffraction patterns show a c parameter of 80 Å, but as the crystals dry out the pattern corresponding to c = 70 Å is superimposed and subsequent exposures show only this pattern; the crystals do not change in external shape or appearance. This phenomenon is most simply explained by the δ-crystallin molecules coming closer together within the crystal lattice to give a higher packing density as the loss of solvent water proceeds. These rearrangements may be closely similar to those that occur in vivo as the avian lens carries out its accommodatory functions.

Further x-ray structural work is in progress; data collection for the native crystals and three heavy atom derivatives for use in the method of isomorphous replacement is under way. Initially these studies should establish a model for the subunit quaternary structure and, at higher resolution, the topology of the polypeptide backbone chain, which will indicate whether high symmetry of folding has occurred for this predominantly α-helical protein. These studies will be supplemented by amino acid sequence analysis, which should not only indicate regions of secondary structure but also regions of homology (87). A comparison of the structural role of this low thiol containing protein from the central region of avian and reptilian lenses with that of the γ-crystallins found in mammalian lenses may give considerable insight into the relation between these proteins and the different optical and mechanical properties of the respective lenses.

Lindley et al.

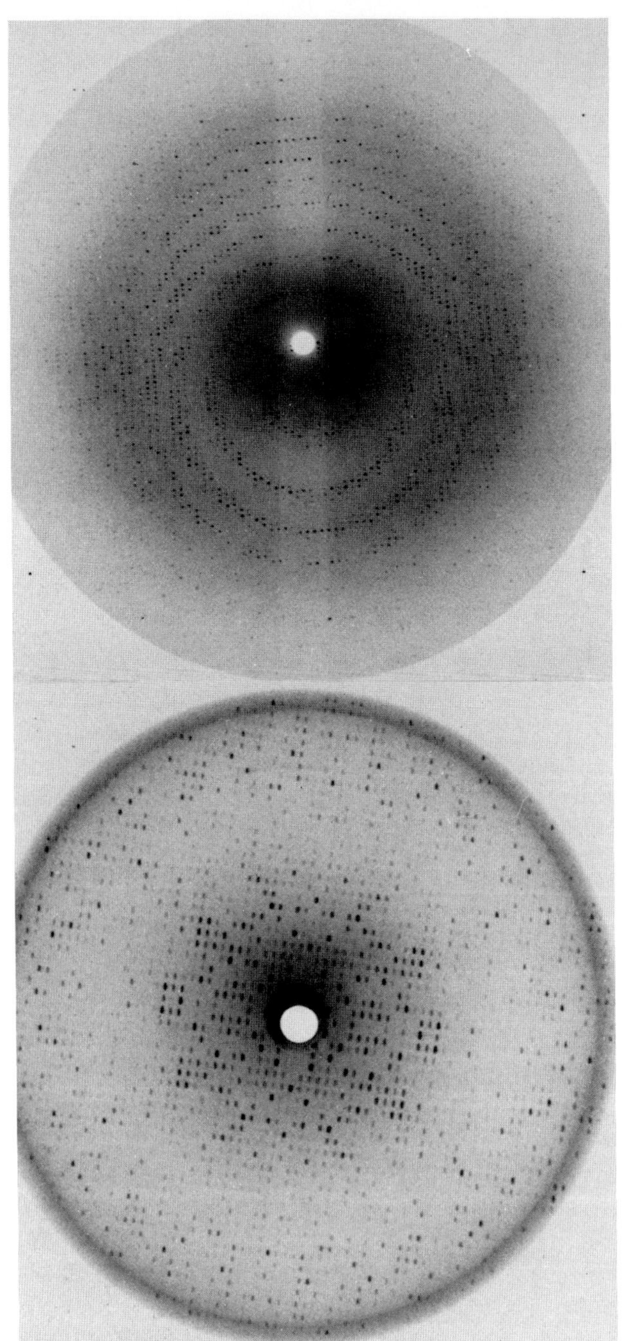

Figure 21 X-ray diffraction patterns of turkey δ-crystallin taken at −5°C: (A) A precession photograph of the hk0 zone. (B) A 2° oscillation photograph taken on an Arndt-Wonacott camera at the synchrotron radiation source at Daresbury (UK) by permission of the Science and Engineering Research Council (UK). The limit of the resolution is better than 2.5 Å.

SUMMARY

Structural analyses of crystallin proteins using x-ray techniques have so far demonstrated the remarkably symmetrical structures of bovine γ-II and γ-IIIb, which may confer stability on these components in the healthy lens. Studies of the nature of the molecular surface of bovine γ-II-crystallin and of the distribution of sulfur-containing and aromatic residues have also suggested possible roles for the monomeric γ-proteins both in the normal transparent lens and in cataract.

Molecular model-building techniques together with amino acid sequence studies and supported by ultraviolent and circular dichroism spectroscopic studies have indicated a superfamily of lens proteins involving the γ-crystallins and the oligomeric β-crystallins. It is likely that this concept will be reinforced by the results of x-ray structure analyses now in progress and further amino acid sequence studies on γ- and β-crystallins. During the preparation of this chapter, a partial amino acid sequence of a frog γ-crystallin (88) and two rat γ-crystallins (89) were reported, all of which show repetitive homology and conserved amino acid residues typified by bovine γ-II-crystallin.

As yet no single-crystal structural work has been undertaken on α-crystallins. Problems of heterogeneity and aggregate formation may severely complicate the formation of crystals suitable for x-ray analysis, but this type of study is at an early stage. However, the amino acid sequences of both the A_2 (90) and B_2 (91) subunit chains are known, and although there is no apparent sequence homology between these two subunits and those of γ- and β-crystallins, it has been proposed (58) that similar principles of folding units, domains, hydrophobic cores, and charge stabilization found in these proteins may also be applicable to the α-crystallins.

The molecular structures and models so far obtained are the first steps in understanding at the molecular level the interactions of crystallins that stabilize the intracellular transparent matrix of the lens fiber cells. As further crystallin molecular structures and aggregates are defined, our knowledge of the complex interactions between these proteins, water, and other cell components, will be enhanced and may eventually lead to an understanding of cataract formation and of the different physical properties possessed by the various vertebrate lenses. Much work remains to be undertaken, but we can look forward to progress in this area of eye lens research with optimism and excitement.

ACKNOWLEDGMENTS

We would like to thank several colleagues at Birkbeck College for discussions, help, and encouragement with some of the research described in this chapter, in particular Dr. Christine Slingsby and Professor Tom Blundell. We would also like to thank the Medical Research Council, UK (Grant No. 76/862) for supporting the structural studies on bovine γ-II-crystallin, βBp-crystallin, and turkey δ-crystallin.

REFERENCES

1. J. J. Harding and K. J. Dilley. Exp. Eye Res., *22*: 1–73 (1976).
2. R. C. Augusteyn. In *Mechanisms of Cataract Formation in the Eye Lens* (G. Duncan, ed.), Academic Press, London, 1981, pp. 71–115.
3. H. Bloemendal. In *Molecular and Cellular Biology of the Eye Lens* (H. Bloemendal, ed.), John Wiley and Sons, New York, 1981, pp. 1–47.
4. B. T. Philipson: Invest. Ophthalmol., *8*: 258–270 (1969).
5. W. W. De Jong. In *Molecular and Cellular Biology of the Lens* (H. Bloemendal, ed.), John Wiley and Sons, New York, 1981, pp. 221–278.
6. J. G. Bindels, R. J. Siezen, and H. J. Hoenders. Ophthalmol. Res., *11*: 441–452 (1979).
7. J. Papaconstantinou. Biochim. Biophys. Acta, *107*: 81–90 (1965).
8. C. Slingsby and L. R. Croft. Exp. Eye Res., *17*: 369–376 (1973).
9. C. F. Wannemacher and A. Spector. Exp. Eye Res., *7*: 623–625 (1968).
10. K. J. Dilley and R. Van Heyningen. Doc. Ophthalmol. Pres., *8*: 171 (1976).
11. I. Bjork. Exp. Eye Res., *3*: 254–261 (1964).
12. L. R. Croft. J. Biochem., *128*: 961–970 (1972).
13. J. J. Harding. In *Molecular and Cellular Biology of the Lens* (H. Bloemendal, ed.), John Wiley and Sons, New York, 1981, pp. 327–365.
14. W. H. Garner, M. H. Garner, and A. Spector. Biochem. Biophys. Res. Commun., *98*: 439–447 (1981).
15. C. Slingsby and L. R. Croft. Exp. Eye Res., *26*: 291–304 (1978).
16. Y. N. Chirgadze, S. V. Nikonov, M. B. G. Garber, and L. S. Reshetnikova. J. Mol. Biol., *110*: 619–624 (1977).
17. C. H. Carlisle, P. F. Lindley, D. S. Moss, and C. Slingsby. J. Mol. Biol., *110*: 417–419 (1977).
18. T. L. Blundell, P. F. Lindley, D. S. Moss, C. Slingsby, I. J. Tickle, and W. G. Turnell. Acta Crystallogr., *B34*: 3653–3657 (1978).
19. Y. N. Chirgadze, Y. V. Sergeev, N. P. Fomenkova, S. V. Nikonov, and Y. V. Lunin. Proc. Acad. Sci. USSR, *250*: 762–765 (1980).
20. T. L. Blundell, P. F. Lindley, L. Miller, D. S. Moss, C. Slingsby, I. J. Tickle, W. G. Turnell, and G. J. Wistow. Nature, *289*: 771–777 (1981).
21. G. J. Wistow, P. F. Lindley, L. Miller, D. S. Moss, C. Slingsby, W. G. Turnell, and T. L. Blundell. J. Mol. Biol., *170*: 175–202 (1983).
22. L. J. Summers, G. J. Wistow, M. E. Narebor, D. S. Moss, P. F. Lindley, C. Slingsby, T. L. Blundell, H. Bartunik, and K. Bartels. In "Peptide and Protein Reviews", *3*: 147–168 (1984).
23. Y. N. Chirgadze, V. D. Oreshin, Y. V. Sergeev, S. V. Nikonov, and Y. V. Lunin. FEBS Letts., *118*: 296–298 (1980).
24. Y. N. Chirgadze, Y. V. Sergeev, N. P. Fomenkova, and V. D. Oreshin. FEBS Lett., *131*: 81–84 (1981).

25. P. F. Lindley, C. Slingsby, H. White, and T. L. Blundell. Unpublished results, 1983.
26. J. S. Richardson. Nature, *268*: 495–500 (1977).
27. J. S. Richardson. In *Advances in Protein Chemistry*, Vol. 34 (C. B. Anfinsen, J. T. Edsall, and F. M. Richards, eds.), Academic Press, New York, 1981, pp. 167–339.
28. L. Stryer. *Biochemistry*, Freeman, San Francisco, 1975, pp. 32–35.
28a. C. Chothia, M. Levitt, and D. Richardson. Proc. Natl. Acad. Sci. USA, *74*: 4130–4134 (1977).
29. J. Horwitz. Exp. Eye Res., *23*: 471–481 (1976).
30. J. S. Zigler Jr., J. Horwitz, and J. H. Kinoshita. Exp. Eye Res., *31*: 41–55 (1980).
31. L. R. Croft. CIBA Found. Symp., *19*: 207–226 (1973).
32. G. Wistow. Ph.D. Thesis, University of London, England (1982).
33. J. E. Walker, A. J. Wonacott, and J. I. Harris, Eur. J. Biochem., *108*: 581–586 (1980).
34. R. C. Fahey, J. S. Hunt, and G. C. Windham. J. Mol. Evol., *10*: 155–160 (1977).
35. R. S. Morgan, C. E. Tatsch, R. H. Gushard, J. M. McAdam, and P. K. Warme. Int. J. Pept. Protein Res., *11*: 209–217 (1978).
36. R. S. Morgan and J. M. McAdam. Int. J. Pept. Protein Res., *15*: 177–180 (1980).
37. B. L. Bodner, L. M. Jackman, and R. S. Morgan. Biophys. J., *17*: 579 1977).
38. B. T. Philipson and P. P. Fagerholm. CIBA Foundation Symp., *19*: 45–63 (1973).
39. R. J. W. Truscott and R. C. Augusteyn. Exp. Eye Res., *24*: 159–170 (1977).
40. J. A. Jedziniak, D. F. Nicoli, H. Baram, and G. B. Benedek. Invest. Ophthalmol. Vis. Sci., *17*: 51–57 (1978).
41. A. Spector, M. H. Garner, W. H. Garner, D. Roy, P. Farnsworth, and S. Shyne. Science, *204*: 1323–1326 (1979).
42. L. J. Takemoto and P. Azari. Exp. Eye Res., *23*: 1–7 (1976).
43. R. J. W. Truscott and R. C. Augusteyn. Biochim. Biophys. Acta, *492*: 43–52 (1977).
44. A. Spector and D. Roy. Proc. Natl. Acad. Sci. USA, *75*: 3244–3246 (1978).
45. M. H. Garner and A. Spector. Proc. Natl. Acad. Sci. USA, *77*: 1274–1277 (1980).
46. R. H. Buckingham. Exp. Eye Res., *14*: 123–129 (1972).
47. L. J. Takemoto and P. Azari. Exp. Eye Res., *24*: 63–70 (1977).
48. J. J. Harding. In *Ageing of the Lens* (F. Regnault, O. Hockwim, and Y. Courtois, eds.), Elsevier, Amsterdam, 1980, pp. 71–80.
49. S. Garcia-Castineiras, J. Dillon, and A. Spector. Exp. Eye Res., *26*: 461–476 (1978).
50. D. V. Bent and E. J. Haydon. J. Am. Chem. Soc., *97*: 2612–2619 (1975).
51. S. Arian, M. Benjamini, J. Feitelson, and G. Stein. Photochem. Photobiol., *12*: 481–487 (1970).

52. P. Herbrink, H. Van Westreenen, and H. Bloemendal. Exp. Eye Res., *20*: 541–548 (1975).

53. J. G. Bindels, A. Koppers, and H. J. Hoenders. Exp. Eye Res., *33*: 333–343 (1981).

54. J. S. Zigler and J. B. Sidbury. Comp. Biochem. Physiol., *55*: 19–24 (1976).

55. P. Herbrink and H. Bloemendal. Biochim. Biophys. Acta, *336*: 370–382 (1974).

56. H. P. C. Driessen, P. Herbrink, H. Bloemendal, and W. W. De Jong. Exp. Eye Res., *31*: 243–246 (1980).

57. H. P. C. Driessen, P. Herbrink, H. Bloemendal, and W. W. De Jong. Eur. J. Biochem., *121*: 83–91 (1981).

58. R. J. Siezen. FEBS Lett., *133*: 1–8 (1981).

59. G. Wistow, C. Slingsby, T. L. Blundell, H. Driessen, W. De Jong, and H. Bloemendal. FEBS Lett., *133*: 9–16 (1981).

60. G. Armand, E. A. Balazs, and M. Testa. Exp. Eye Res., *10*: 143–150 (1970).

61. G. Inana, T. Shinohara, J. V. Maizel, Jr., and J. Piatigorsky. J. Biol. Chem., *257*: 9064–9071 (1982).

62. G. Inana, J. Piatigorsky, B. Norman, C. Slingsby, and T. L. Blundell. Nature, *302*: 310–315 (1983).

63. T. A. Jones. J. Appl. Crystallogr., *11*: 268–272 (1978).

64. P. Y. Chou and G. D. Fasman. Biochemistry, *13*: 222–245 (1974).

65. J. Garnier, D. J. Osguthorpe, and B. Robson. J. Mol. Biol., *120*: 97–120 (1978).

66. A. F. van Dam. Exp. Eye Res., *5*: 255–266 (1966).

67. C. Slingsby, L. R. Miller, and G. A. M. Berbers. J. Mol. Biol., *157*: 191–194 (1982).

68. S. C. Harrison, A. J. Olsen, C. E. Schutt, F. K. Winkler, and G. Bricogne. Nature, *276*: 368–373 (1978).

69. C. Abad-Zapatero, S. S. Abdel-Meguid, J. E. Johnson, A. G. W. Leslie, I. Rayment, M. G. Rossmann, D. Suck, and T. Tsukihara. Nature, *286*: 33–39 (1980).

70. M. J. Adams, G. C. Ford, R. Koekoek, P. J. Lentz, A. McPherson, M. G. Rossmann, I. E. Smiley, R. W. Schevitz, and A. J. Wonacott. Nature, *227*: 1098–1103 (1970).

71. J. J. Holbrook, A. Liljas, S. J. Steindal, and M. G. Rossman. In *The Enzymes*, Vol. 2, 3rd ed. (P. D. Boyer, ed.), Academic Press, New York, 1975, pp. 191–192.

72. C. Slingsby, L. R. Miller, P. F. Lindley, and T. L. Blundell. Unpublished results, 1983.

73. M. Rabaey. Exp. Eye Res., *1*: 310–316 (1962).

74. R. M. Clayton. In *The Eye*, Vol. 5 (H. Davson, ed.), Academic Press, New York, 1979, pp. 339–494.

75. D. S. McDevitt and L. R. Croft. Exp. Eye Res., *25*: 473–481 (1977).

76. L. A. Williams and J. Piatigorsky. Eur. J. Biochem., *100*: 349–357 (1979).

77. J. F. R. Kuck, Jr., E. J. East, and N. T. Yu. Exp. Eye Res., *23*: 9–14 (1976).

78. J. Piatigorsky, J. Horwitz, and R. T. Simpson. Biochim. Biophys. Acta, *490*: 279–289 (1977).
79. J. Piatigorsky, P. Zelenka, and R. T. Simpson. Exp. Eye Res., *18*: 435–446 (1974).
80. J. Piatigorsky. J. Biol. Chem., *251*: 4416–4420 (1976).
81. J. Piatigorsky and T. Shinohara. Science, *196*: 1345–1347 (1977).
82. T. Shinohara and J. Piatigorsky. Nature, *270*: 406–411 (1977).
83. J. Piatigorsky. Exp. Eye Res., *27*: 227–237 (1978).
83a. R. Reszelbach, T. Shinohara, and J. Piatigorsky. Exp. Eye Res., *25*: 583–593 (1977).
84. S. P. Bhat, R. E. Jones, M. A. Sullivan, and J. Piatigorsky. Nature, *284*: 234–238 (1980).
85. E. Narebor, C. Slingsby, P. F. Lindley, and T. L. Blundell. J. Mol. Biol., *143*: 223–225 (1980).
86. E. Narebor, C. Slingsby, P. F. Lindley, and T. L. Blundell. Unpublished results, 1983.
87. J. M. Nickerson and J. Piatigorsky. FEBS Lett., *144*: 289–292 (1982).
88. S. I. Tomarev, A. S. Krayer, K. G. Skryabin, A. A. Bayer, and G. G. Gause, Jr. FEBS Lett., *146*: 315–318 (1982).
89. J. M. Moormann, J. T. der Dunnen, H. Bloemendal, and J. G. G. Schoenmakers, Proc. Natl. Acad. Sci. USA, *79*: 6876–6880 (1982).
90. F. J. van der Ouderaa, W. W. De Jong, and H. Bloemendal. Eur. J. Biochem., *39*: 207–222 (1973).
91. F. J. van der Ouderaa, W. W. De Jong, A. Hilderink, and H. Bloemendal. Eur. J. Biochem., *49*: 157–168 (1974).

5

Biochemistry of Lens Plasma Membranes and Cytoskeleton

JOSE ALCALA AND HARRY MAISEL / Wayne State University School of
Medicine, Detroit, Michigan

The vertebrate eye lens is a uniquely specialized organ that functions to maintain
in optic focus points of visual attention of varying distances of origin. To accomplish its function, the lens must possess transparency and elasticity. Transparency allows the lens to refract light in a manner proper to visual acuity. In
vertebrate species capable of accommodation, elasticity confers upon the lens
the remolding attribute proper to the accommodative reflex (1). Among species
capable of accommodation, such as the human, the two properties have an
inverse relation of importance to lens function; when one is present but the
other is impaired, lens function is defeated. Although phylogenetic processes
have established a variety of accommodative mechanisms in existing vertebrate
species (1), lens transparency is common to all vertebrate organisms. Transparency, therefore, is the invariant imperative to lens function.

Transparency is primarily a physical phenomenon, and as such, all ontogenetic morphologic, biochemical, and physiologic processes of the lens must be
viewed as essential to its manifestation. Trokel (2) provided the definition of
the physical basis for lens transparency that has guided most recent investigations into its biologic manifestation. In Trokel's view, cytoplasmic proteins are
held responsible for small-particle light scatter and the plasma membranes are
held responsible for large-particle light scatter. Transparency is thus achieved in
the lens, via the minimization of light scattering, by the existence of structural
regularity at two levels: structural regularity of cytoplasmic (soluble) proteins
minimizes the small-particle light scatter, and uniform alignment of the fiber
cells' plasma membranes minimizes the large-particle light scatter. Transparency,
therefore, is created in the lens by ontogenetic processes that organize regularity
at two structural levels: the fiber cytoplasmic matrix (cytoskeleton and soluble

protein) and the fiber cell plasma membrane. Lasser and Balazs (3) provided the first conclusive biochemical definition of these two major lens fractions.

HISTORICAL PERSPECTIVE

The first systematic biochemical fractionations of the lens (4,5) revealed the existence of three soluble lens proteins: the α-, β-, and γ-crystallins. The insoluble lens residue was generically designated by Mörner (4), as "albuminoid." Early studies on albuminoid reported it to be largely made up of an insoluble form of α-crystallin (for a review, see Ref. 6). Later investigations, utilizing various electrophoretic and immunologic methods, reported species differences in the crystallin makeup of albuminoid, although in most it was reported to consist of a mixture of all three crystallins (for a review, see Ref. 7).

The fractionation of albuminoid utilizing concentrated urea solutions (8) led to the realization that the composition of the lens water-insoluble fraction was more complex than had been reported by the earlier studies. Dische et al. (9) first showed that the urea-insoluble portion of the lens water-insoluble fraction largely consisted of the fiber cell plasma membranes with adherent protein material Dische (10) believed represented the crystallins detected in albuminoid by earlier investigators. Waley (8) had redefined albuminoid to represent the urea-soluble portion of the lens water-insoluble residue and concluded that the urea-insoluble portion was an artifact formed upon dilution of the lens homogenate, but Harding (11) warned that the urea-insoluble portion was generated artifactually by oxidation during extraction in atmospheric oxygen. Kramps et al. (12) later modified the urea extraction of lens albuminoid by the introduction of reducing agents to obviate the generation of artifactual urea-insoluble protein. Elsewhere, Wanko and Gavin (13) had observed filamentous and chainlike elements on sectioned calf lenses, and Maisel and Perry (14) recovered these structures from high-speed supernatants of chick lens water-soluble fractions. The smooth filaments (12–14 nm in diameter) and filamentous chains, consisting of globular protein particles (12–15 nm in diameter) attached to a filament backbone (7–9 nm in diameter), were believed to be constituents of the lens water-insoluble fraction and to represent the cytoskeleton of the lens fiber cells.

Finally, Lasser and Balazs (3), through a combined ultrastructural and biochemical study, provided the first clear biochemical definition of the lens water-insoluble fraction, ending the need for its vague and confusing generic designation as albuminoid as originated by Mörner (4). Through the centrifugation of the calf lens homogenate in its own fluid under a nitrogen atmosphere, these investigators demonstrated that the lens water-insoluble residue largely consisted of the fiber cells' plasma membranes and cytoskeleton. The preponderance of crystallins in this fraction, as observed by previous investigators, was held to be the result of their incomplete removal by insufficient washing of the residue. The urea-soluble portion of the lens water-insoluble fraction was redefined as consisting of the cytoskeletal elements of the fiber cells complexed with insolubilized lens crystallins, and the urea-insoluble portion as consisting of the fiber cells' plasma membranes. Evidence obtained in that study of a physical association between these three major lens fiber components and of the preponderance of cell junctions on the fiber plasma membranes opened the way for active investigations into the cytochemical and biochemical basis of the ultrastructural regularity of the lens cytoplasmic matrix and of the uniform alignment of lens fiber plasma membranes uncovered by Trokel's (2) observations on the physical basis of lens transparency.

LENS BIOCHEMICAL FRACTIONS

The vertebrate lens has the lowest water content and highest protein concentration of any soft tissue known in nature. Some fish (trout) have lenses with the highest protein content (50%); bird (pigeons) lenses have the lowest (20%), and human lenses are intermediate (30%) in protein content (15). About 80–90% of lens protein is soluble in water or aqueous buffers, referred to as the lens *water-soluble fraction*, and is comprised primarily of three groups of proteins generically designated as "lens crystallins" (16): the α-, β-, and γ-crystallins of all vertebrate lenses and the δ-crystallin of reptilian and avian lenses (1).

Although the three groups of soluble crystallins make up about 80–90% of the lens dry weight, the remainder (10–20% of lens dry weight) is made up of various structural proteins collectively referred to as the lens *water-insoluble fraction* and previously generically designated as albuminoid. Treatment of the lens water-insoluble fraction with 7–8 M urea under reducing conditions yields a *urea-insoluble fraction*, consisting primarily of fiber cell plasma membranes (Fig. 1), and a *urea-soluble fraction*, consisting partly of proteins from the fiber cell cytoskeleton (16). The remainder of the urea-soluble fraction consists of insolubilized lens crystallins, the proportion of which increases with aging (for a review, see Ref. 7).

About 70–80% of the lens water-insoluble fraction is urea soluble (12,17,18) under reducing conditions, and 20–30% is recoverable as the plasma membrane-enriched urea-insoluble fraction. Lens fiber plasma membranes have been estimated to constitute about 0.4–1.0% of the wet weight of the lens (dry weight as percentage of the wet weight) (17,19,20).

LENS PLASMA MEMBRANES

Biologic membranes consist primarily of lipids and proteins and a small amount of carbohydrate in the form of glycolipid and glycoproteins (21). The main

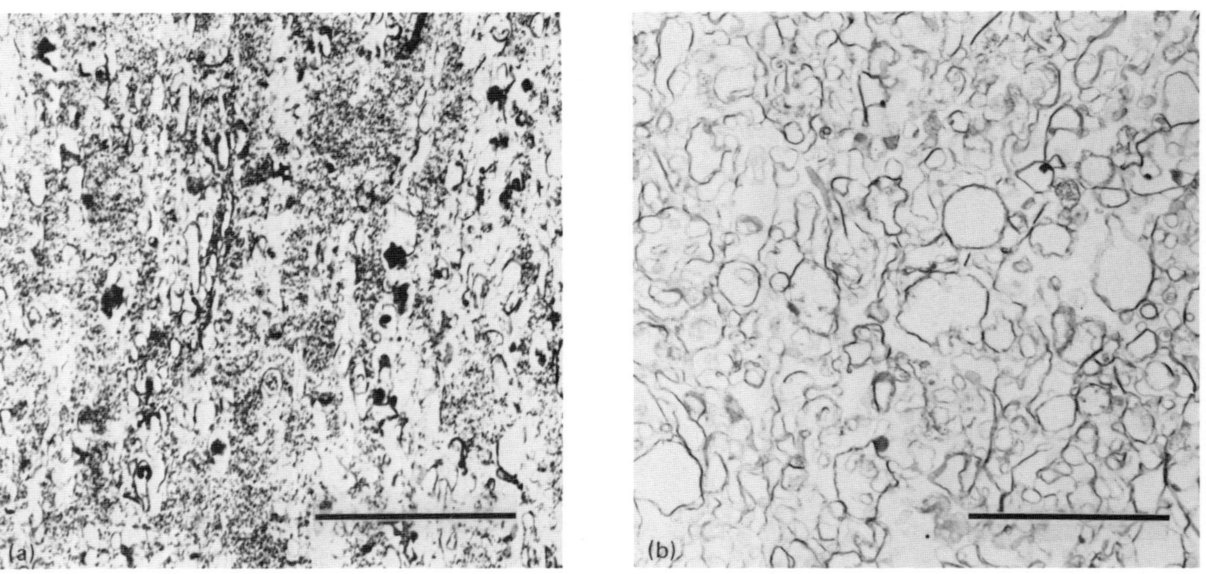

Figure 1 Transmission electron micrograph of the chick lens (a) water-insoluble fraction (WIF), and (b) fiber cell plasma membranes prepared by the urea method ($\times 27,000$); bar = 1 μm.

structural components of membranes are the phospholipids, which form an extended bilayer through alignment of their hydrophobic fatty acid chains to form the core of the bilayer and their hydrophilic head groups facing its inner and outer surfaces. The proteins embedded in the lipid bilayer result in a complex set of protein-lipid interactions. Thus, the core of the bilayer consists of the hydrocarbon chains of lipid molecules and the hydrophobic segments of the polypeptide chains of some membrane proteins. *Intrinsic* membrane proteins (21) are operationally defined as those proteins that can only be removed from the membrane by detergents or other chaotropic agents and are thus partially or completely buried in the lipid bilayer of the membrane. *Extrinsic* membrane proteins (21), on the other hand, are operationally defined as those proteins that can be removed from the membrane with high ionic strength solutions or metal chelating agents (e.g., EDTA), and are thus attached to the surface of the lipid bilayer or to other membrane proteins via calcium or ionic bridges.

As a solid organ, the lens possesses the unique characteristic of being made up of a single differentiated cell type, the lens fiber cell (22). The anterior monolayer of embryonal lens epithelial cells proliferate and move toward the lens equator, where the continuous differentiation of new fiber cells occurs (23). Differentiation of epithelial cells into fiber cells at the lens equator entails extensive cellular elongation and its consequent 1000-fold increase in the elaboration of new plasma membrane by the elongating fiber cells (24). Differentiation of lens fiber cells also entails the loss of the cell nucleus and all intracellular membranous organelles (22,23). The bulk (94%) of the lens is thus made up of a single cell type, possessing a single cellular membrane, the plasma membrane (25).

Methods of Isolation

There are currently several different methods of isolating lens fiber cell plasma membranes based on either density characteristics or solubility properties of the membranes, or both. The most commonly used methods and the basic steps in the procedures are summarized in Table 1. All the methods utilize the decapsulated lens (i.e., the lens fiber cell mass) as starting material; decapsulation of the lens results in the removal of the lens epithelium as well as most of the equatorial lens fiber region along with the lens capsule (31). Alternatively, specific lens fiber regions (e.g., cortex or nucleus) are utilized as starting material when study of aging or developmental changes in the biochemical composition of the membranes is desired (18).

The "primary treatment" in Table 1 refers to the homogenization of the decapsulated lens and steps designed for the removal of the lens-soluble proteins, generating the thoroughly washed lens water- (or buffer-) insoluble fraction. The lens water-insoluble fraction is known to consist of a plasma

Table 1 Methods of Isolating Lens Fiber Plasma Membranes from
Decapsulated Lenses

Primary treatment	Secondary treatment	References
Homogenization in its own fluid as a medium and centrifugation, 200,000 × g	Homogenization in 7 M urea or 1% deoxycholate and centrifugation, 70,000 × g	3
Repeated homogenization in buffer and centrifugation, 20,000–100,000 × g	Repeated homogenization in 7 or 8 M urea and centrifugation, 70,000–100,000 × g	17–19, 26
Homogenization in 0.25 M sucrose; underlayering by 2.4 M sucrose and centrifugation, 75,000 × g	Repeated suspension in 8 M urea buffer; underlayering by 1.6 M sucrose and centrifugation, 75,000 × g	27
Repeated homogenization in buffer, and 7 M urea and centrifugation, 10,000 × g	Homogenization in 0.1 M NaOH and centrifugation, 10,000 × g	28
Homogenization in 7 M guanidine HCl, citraconic anhydride, and dialysis against 0.2% NH_4OH	Repeated washing in 0.005 M phosphate buffer and centrifugation, 135,000 × g	29
Repeated homogenization in buffer and centrifugation, 7000–12,000 × g	Sucrose gradient centrifugation at maximum speed; membranes collected at 1.16–1.14 g/cm^3 density interface	30

membrane-cytoskeleton complex and insolubilized lens crystallins (Fig. 1) (3).
The "secondary treatment" (Table 1) refers to steps designed for the removal
of the cytoskeleton and insolubilized crystallins through either solubilization of
these components (chemical) or their separation by gradient centrifugation
(physical), generating the lens fiber cell plasma membrane fraction (Fig. 1). The
various methods compiled in Table 1 can be grouped on the basis of the treatments as the urea methods (3,17–19,26–28), the acylation method (29), and the
gradient centrifugation method (30).

Only a few studies on lens epithelial cell plasma membranes have been conducted, and their isolation has been performed either by the traditional differential centrifugation method for isolating cellular membranes (32) or by the gradient centrifugation method (33).

Protein Composition

Studies on membrane proteins comprise an analysis by sodium dodecyl sulfate-polyacrylamide gel electrophoresis (SDS-PAGE) under reducing conditions, which dissociates proteins into their constituent polypeptide subunits (34). Lens fiber cell plasma membranes isolated from diverse vertebrate species have been resolved electrophoretically into a large number of polypeptides ranging in molecular weight from 10,000 to 250,000 d (Fig. 2) (16), and a number of these components have been singled out by various investigators for particular study (35).

The major polypeptide components of vertebrate lens fiber cell plasma membranes have been compiled in Table 2. A component has been assumed to be in all vertebrate lenses if it has been found in the lens membranes of representatives of more than three classes of animals. If comparative information was lacking, only the species in which it was first demonstrated has been noted. Assignation to a specific class of animals denotes either the absence of the component in other classes or the demonstration of its class specificity. The references noted include reports of the presence of components of a similar or identical molecular weight in the species under study, whether or not a definitive identification was attempted. The molecular weight range provided for a particular component reflects the differences in its estimation by the various investigators.

Initial analyses of lens epithelial cell plasma membranes by SDS-PAGE described a large number of polypeptides ranging in molecular weight from 17,000 to 250,000 d (16,33), but no follow-up studies have been conducted on any of these components. Consequently, only the much studied polypeptide components of lens fiber cell plasma membranes will be reviewed.

Intrinsic Proteins

Original SDS-PAGE analyses of the isolated lens fiber plasma membranes demonstrated a predominant polypeptide component of about 25,000 to 27,000 d molecular weight (17,19), originally designated main intrinsic polypeptide (MIP; Table 2) (36). Its molecular weight, variously estimated as 27,500 (17), 27,000 (18), 26,500 (36), 26,000 (26), and 25,500 (19) d, was generally agreed to be closer to 26,000 d (26,37,38); hence, its later designation as MP26 (Fig. 2) (39). MP26 has been demonstrated to be the main or only protein component of isolated bovine (40) and chick (41) lens fiber cell junctions (see Lens Fiber Cell Junctions).

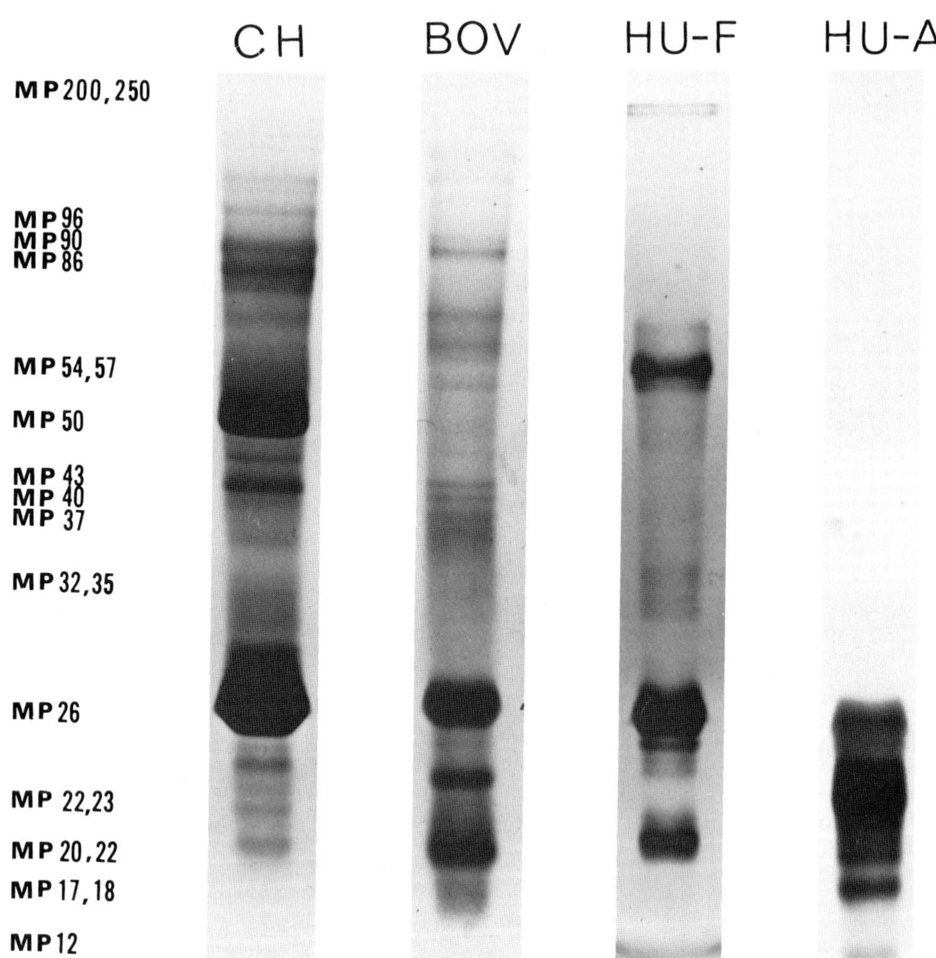

Figure 2   Silver-stained SDS-PAGE gels of isolated lens fiber cell plasma membranes from chick (CH), bovine (BOV), fetal human (HU-F), and adult human (HU-A) lenses; MP = membrane polypeptide. The membrane polypeptides were designated as in Table 2.

MP26:   Lens MP26 was originally shown to resist the treatment of lens membranes with 8 M urea (19) and to be fully soluble in chloroform-methanol (17,36), suggesting that a significant portion of the protein is disposed within the lipid bilayer of the membrane. Immunocytochemically, MP26 was shown to be a transmembrane protein (42), spanning the lipid bilayer of the membrane.

Table 2 Polypeptide Composition of Vertebrate Lens Fiber Plasma Membranes[a]

Membrane component	Identification	Source	References
MP200, 250	Spectrin	All	17, 18, 95, 96
MP220–225	Fibronectin	All	18, 95, 84, 85
MP96–130	α-Actinin	All	35, 84, 186
MP90–107	Na^+,K^+-ATPase[b]	All	97, 98, 115
MP86–95	Beaded-chain filament protein	All	35, 14, 99, 192
MP54–57	Vimentin	All	17, 100, 101
MP50	δ-Crystallin	Chick	60, 102, 103
MP43	Extrinsic protein	Human	18, 86, 87, 95
MP40	Actin	All	84, 104, 198
MP37	G3PD[c]	Bovine	91
MP32, 35	EEP[d]	All	49, 52, 54, 55
MP26	MIP[e]	All	17, 43, 52, 82
MP24	MIP	Frog	43, 82
MP22–23	MIP	Mammals	29, 43, 52, 81
MP22	Extrinsic protein	Human	92
MP20–22	α-Crystallin	Mammals	17, 18, 104
MP17–18	Unknown	Mammals	17, 19, 82
MP12	Unknown	Human	18, 189

[a]See text for explanation. Nomenclature of Bloemendal et al. (39).

[b]Sodium potassium activated adenosine triphosphatase.

[c]Glyceraldehyde 3-phosphate dehydrogenase (107).

[d]EDTA-extractable protein (49).

[e]Main intrinsic polypeptide (36).

Consistent with this conclusion, proteolytic digestion of intact lens membranes from diverse vertebrate species results in the cleavage of MP26 to a 20,000–22,000 d portion (core) not accessible to further proteolytic digestion (43). The 22,000 d core is assumed to represent the hydrophobic portion of the protein buried in the lipid bilayer of the membrane and has been shown to be nearly identical in chick, bovine, and human lenses by tryptic-peptide mapping (43). *Biochemical properties:*   Amino acid analyses of MP26 from diverse sources and isolated by different methods demonstrate a very similar amino acid composition and a low polarity (Table 3). A polarity below 40% is characteristic of

Table 3 Comparison of Amino Acid Compositions of Lens Fiber Plasma Membrane MP26[a]

Amino acid	Chick[b]	Calf[c]	Human[d]
Aspartic acid	5.8	4.9	4.7
Threonine	4.0	4.5	5.1
Serine	6.4	7.8	8.5
Glutamic acid	8.4	6.3	5.3
Proline	8.2	6.7	4.4
Glycine	12.7	14.6	11.6
Alanine	14.0	13.1	12.8
Valine	4.8	7.4	7.2
Cysteine	1.6	0.0	0.0
Methionine	0.5	1.6	3.6
Isoleucine	2.6	2.3	3.0
Leucine	16.0	12.0	13.8
Tyrosine	0.7	2.7	3.2
Phenylalanine	4.6	4.7	7.6
Lysine	1.5	2.8	1.6
Histidine	3.0	3.1	2.5
Arginine	5.4	5.3	5.7
Polarity[e]	35%	35%	33%

[a]Values expressed as mol%.

[b]Values from Maisel and Alcala (20) rounded off to one decimal.

[c]Data from Kibbelaar and Bloemendal (26).

[d]Data from Broekhuyse (44).

[e]Sum of mol% of polar amino acids, according to Capaldi and Vanderkooi (45).

intrinsic membrane proteins (45). Circular dichroism spectra of isolated MP26 and of sonicated membrane suspensions suggest that it possesses 58% α-helical structure (46); proteolytic cleavage of MP26 to its 22,000 d derivative does not appear to alter this conformation.

The presence of multimeric aggregates of MP26 in preparations of isolated chick lens fiber membranes fractionated by SDS-PAGE has been reported (47), and these have been interpreted as indicative of the presence of multimeric forms of MP26 in the membranes. Although multimeric aggregation of MP26 from mammalian lenses has been demonstrated to be artifactually generated by various conditions and treatments (48), such artifactual aggregation has never been observed to occur with chick lens MP26 (27) (authors' observations). The molecular (multimeric) structure of the MP26 protein in the intact membrane is not known at present (see Lens Fiber Cell Junctions).

The electrophoretic behavior of lens MP26 under varying conditions in SDS-PAGE has been extensively studied. Heating MP26 samples in SDS solution at 100°C for 3 min to solubilize the protein prior to electrophoresis caused the loss of the protein as an insoluble residue (49). Samples of MP26 from bovine (48) and human (37) lenses were also demonstrated to form high-molecular-weight aggregates upon heating at 100°C for 1 min in SDS solutions; the aggregates are too large to enter the acrylamide gel. Recent observations, however, indicate that high concentrations (1–5%) of the reducing agent β-mercaptoethanol in the electrophoresis solubilizing solution may be a prerequisite to the severe aggregation of human MP26 upon heating (29); neither heating nor the β-mercaptoethanol alone is capable of causing the aggregation. Broekhuyse and Kuhlmann (50) also demonstrated a change in the MP26 band to a double band dependent on the absence of a reducing agent during electrophoresis. The observed change of MP26 from a single to a double band in the absence of the reducing agent in that study has been confirmed to occur for the chick lens MP26 (51). The precise reasons for the observed differences in the electrophoretic behavior of MP26 from diverse sources is not known at present. The reasons are all the more puzzling given the observed similarities in molecular weights (52), amino acid compositions (Table 3), conformational properties (53), and immunologic affinities (54,55) of MP26 isolated from diverse vertebrate species. Speculation has centered on the putative glycoprotein nature of MP26 (48) and the known anomalous behavior of glycoproteins when subjected to SDS-PAGE (34).

Initial tests of calf and chick lens MP26 by periodic acid-Schiff staining (PAS) on SDS-PAGE and by gas chromatography failed to detect any carbohydrate associated with the protein (17,36). Positive PAS staining of the MP26 polypeptide gel band from the bovine (48) and human lens (29), however, was later reported. The positive staining of MP26 with glycoprotein stains, however, appears to be misleading. It has been demonstrably obtained with lens MP26 even in the absence of periodate treatment (56). Positive PAS staining of the MP26 gel band may be due to the now demonstrated comigration and presence of lipids (glycolipids?) with the polypeptide in SDS-PAGE (17), generated by peroxidation of polyunsaturated fatty acids (yielding malonaldehyde) (57) or by the hydrolytic release of fatty aldehydes from plasmalogens (58). The observed PAS staining on the tracking dye front in SDS-PAGE gels of bovine (48) and human (29) lens membranes supports this conclusion; the tracking dye front in SDS-PAGE gels does not normally bind protein stains and is the region of electrophoretic migration of excess SDS and membrane lipids (59).

Proteolipid nature: Initial analyses of lens fiber membranes by SDS-PAGE demonstrated the association of lipid with the MP26 band in the gels (17,60). This evidence, and the original observations that MP26 is fully soluble in solutions of chloroform-methanol (17,36), suggested that at least some of this lipid was tightly bound to the protein. The recovery of the MP26 protein from the original spot (point of application) in silica gel H plates following the two-dimensional thin-layer chromatography of lipid extracts of bovine, chick, and human lens fiber plasma membranes (61), further suggests that lipid is tightly bound to MP26. As the origin spot has been determined to represent about 1–2% of lens lipid (61,62), these observations support the conclusion that a portion of this lipid may be tightly bound to the protein. The observed behavior and characteristics of lens MP26 qualifies this protein to be classified as a *proteolipid* (63). Proteolipids are a class of proteins characterized by their solubility in chloroform-methanol (63).

The presence of an undetermined amount of lipid in the MP26 gel band in SDS-PAGE (17) introduces a significant uncertainty regarding the behavior of the protein moeity in the band and one that has been largely ignored. The occasional observation (64) of lower molecular weight components suggests that the lipid associated with the protein-SDS complexes in the gel band may contribute to the observed mobility of the complexes. Likewise, the reported occasional observations of multimeric forms (47) and the anomalous behavior of MP26 in SDS-PAGE (48,50), following various procedures and treatments, may be due to partial delipidation of this proteolipid, leading to its hydrophobic aggregation. *Immunologic properties*:   Initial immunologic studies on MP26 reported it to be organ, tissue, and class specific (65). Its developmental pattern of appearance in chick and calf lenses traced by specific immunofluorescence demonstrated it to be a protein specific to the lens fiber cell (38,66). Lens MP26 is the most specific marker for lens fiber plasma membrane differentiation (67). SDS-PAGE analyses have confirmed the absence of MP26 in the lens epithelial cell membranes (33,38). The apparent class specificity of MP26 was reciprocally confirmed utilizing an antiserum to the mammalian (bovine) protein in reaction with the chicken lens fiber membranes (68), and its organ and tissue specificities verified in reactions with an antiserum to the calf lens protein (38).

The apparent antigenic specificities of MP26, however, appear to have been misleading. Friedlander (69) first reported the cross-reactivity of an antiserum to calf lens MP26 in immunofluorescent staining of chick lenses. Later, Bouman and Broekhuyse (54) demonstrated the cross-reactivity of an antiserum to chick lens MP26 in reactions with the mammalian (calf, pig, and sheep) lens protein, and the reciprocal reactivity of antisera to the mammalian proteins. The cross-reactivity was made evident only with the presence of 4% polyethylene glycol in the agarose utilized for immunodiffusion (54). The reason for discrepancies in the reported antigenic behavior of MP26 is not known at present. At least part of the cross-reactivity of antisera to MP26 may be explainable by the lipid tightly associated with the protein as isolated by SDS-PAGE (17) acting as a hapten.

Biosynthesis: The biosynthesis of MP26 has been studied in the organ-cultured chick lens (70). Synthesis of MP26 was sustained only by fiber cells of the outer and deep lens cortex, and no synthesis was detected in the lens epithelial or nuclear fiber cells. The absence of synthesis of MP26 by epithelial cells helps to explain the previously reported absence of this protein in lens epithelium (33,38). Synthesis of MP26 by deep cortical fiber cells suggests that there is continued membrane elaboration by the terminally differentiating fiber cells. As terminally differentiating fiber cells lack a nucleus, the observation also suggests the existence of long-lived messenger-RNA responsible for the synthesis of this protein in the cortical fiber cells. The isolation of calf lens MP26 messenger-RNA and polyribosomes has been reported (71) and the cell-free synthesis of this polypeptide fully effected. Polyribosomes directing the synthesis of MP26 were reported to be strongly associated with the lens plasma membrane-cytoskeleton complex (71).

The in vitro synthesis of bovine lens MP26 (72) indicates that this membrane protein is synthesized without a cleavable leader sequence and inserts into membranes cotranslationally. No proteolytic processing or glycosylation appears to accompany its membrane insertion (72). Chymotryptic cleavage of membrane-inserted MP26 demonstrated that its N-terminal faces the cytoplasmic side of the lipid bilayer of the membrane (72). Elsewhere, analyses of the cyanogen bromide peptides of the trypic cleavage products of intact bovine lens MP26 suggest that both the N-terminal and C-terminal ends of this protein face the cytoplasmic side of the membrane (73).

MP22–23:   The human lens fiber membrane main intrinsic protein consists of two polypeptides of 26,000 and 22,000 d (MP26, MP22–23; Table 2) molecular weight (18,29,37), which undergo a reciprocal gradual change in relative abundance from a predominance of MP26 at birth to equality (74) or a predominance of MP22 after midlife (18). The two polypeptides have been demonstrated to share lability to heat-induced aggregation in SDS (37) and to have very similar amino acid compositions (75). MP22 is believed to be derived from MP26 by posttranslational modifications, as evidence by demonstrated immunochemical affinities (68,76) and by its apparent in vitro proteolytic derivation from MP26 (43,77,78). Interestingly, a similar interconversion has been reported for the Philly and Nakano strains in the mouse but at the time of cataract formation (79,80).

Questions have been raised concerning the equivalency of the in vitro proteolysis of MP26 to its putative in vivo proteolysis (78). The in vitro proteolysis of MP26 yields a product that differs by about 1–2000 d in molecular weight from its putative in vivo product (43,78). The reason for the difference in molecular weight is not known at present.

The main intrinsic protein of monkey lens fiber plasma membranes also appears to consist of two polypeptides, MP26 and MP22 (55), and the appearance of MP22 in aged bovine lens fiber membranes has been noted (81). The appearance of MP22 in the aged Siberian tiger lens (authors' observations) is suggestive that this may be a general feature of aging mammalian lens fiber plasma membranes. It has been suggested that MP22 is probably derived from MP26 by posttranslational cleavage as part of either an age-dependent (29) or developmentally relevant (37) process. The suggestion that MP22 arises as a degradation product of MP26 (81), however, is consistent with its primary appearance in aging and pathologic (cataractous) lenses.

MP24:   The frog lens fiber membrane main intrinsic protein has been noted to consist of only a 22,000–24,000 d (MP24; Table 2) molecular weight component (43,82). Whether the MP24 polypeptide of the frog lens is a primary gene product or arises by a process similar to that envisioned for the human MP22 remains to be determined.

Aside from MP26, MP22, MP24, and MP90–107 (Na^+,K^+-ATPase; see Enzymes), no other lens fiber plasma membrane polypeptides have been clearly identified as intrinsic membrane proteins (Table 2). Some controversy has existed concerning a 34,000 d ("MP34") membrane polypeptide (30), alternatively classified as an intrinsic (35) and extrinsic (44) protein of the lens fiber plasma membranes. As this protein is readily extractable by EDTA solutions from lens fiber membranes isolated by the urea method (44), it will, be considered an extrinsic protein for this chapter.

Extrinsic Proteins

Only a few polypeptides of the lens fiber plasma membranes have been unequivocably demonstrated to be extrinsic membrane proteins and have been singled out for particular study by various investigators. A number of components coextracted in the lens urea-soluble fraction along with proteins of the lens cytoskeleton are assumed to be extrinsic membrane proteins based on information about them in other cells and tissues, where they have been well characterized.

MP220-225:   The cell attachment glycoprotein *fibronectin* is a cell surface component that mediates adhesion of cells to their extracellular matrix (83). Fibronectin resolves in SDS-PAGE as two closely spaced bands corresponding to 220,000–225,000 d molecular weight, and the proteins from different sources are all immunologically indistinguishable and are very similar in amino acid composition (83). It appears to have some kind of association with the intracellular cytoskeleton as demonstrated by the coalignment of their respective bundles observed by immunnofluorescence. Although it can directly bind to actin, it is believed that an intermediary transmembrane protein mediates its interaction with the cell's cytoskeleton (83).

The presence and synthesis of fibronectin in explanted bovine lens epithelial cells has been reported (84,85). Its appearance in the cultures coincides with conversion of the cells from an epitheloid to a fibroblastlike configuration (85). Polypeptides in the molecular weight range of this protein abound in lens fiber plasma membrane preparations (Table 2) (17,18,60) but constitute less than 0.5% of the membrane protein recovered by SDS-PAGE. It is more likely that the urea treatment of the membranes results in its coextraction with the urea-soluble protein, a conclusion supported by its elution from affinity (gelatin-Sepharose) chromatography columns by 8 M urea (83). Several polypeptides in this molecular weight range are consistently recovered in the urea-soluble fraction.

MP43:   A 43,000 d (MP43; Table 2) polypeptide, originally isolated from the high-molecular-weight protein aggregates of the water-insoluble fraction of cataractous human lenses (86), has been localized in the fiber plasma membranes of fetal, young, and adult normal human lenses by specific immunofluorescence (87). MP43 is probably linked to the membranes via calcium, as it was found to be released into the lens water-soluble fraction upon homogenization of the lenses in the presence of the calcium chelator EGTA (86). The polypeptide was demonstrated to be specific to the fiber plasma membranes, as it was shown to be absent in epithelial cells (87). Its developmental pattern of distribution was shown by immunofluorescence to be initially within the cytoplasm of differentiating (elongating) fetal and equatorial adult fiber cells but to become increasingly confined to the fiber plasma membranes with fiber cell maturation (87). At least part of the protein isolated as MP43 appears to contain nonmuscle actin (88), and antibody to this protein has been shown to cross-react with actin (89).

MP32, 35:   Originally designated as "MP34" (33), this 34,000–38,000 d group of polypeptides was initially thought to constitute the main intrinsic protein of the lens fiber plasma membranes (30). However, its predominance in the membranes appears to have been misleading, following the demonstration of the heat lability of MP26 (35), and there is general agreement now of the predominance of MP26 in the lens fiber plasma membranes (39). The "MP34" protein is the main constituent of the lens epithelial cell membranes (33) but is now recognized to be a minor component of the lens fiber plasma membranes, comprising about 5–6% of the fiber membrane protein (17,49).

Broekhuyse and Kuhlmann (49) first demonstrated that the "MP34" protein was extractable with the calcium chelator EDTA, following water treatment of lens fiber membranes isolated by the urea method, hence its later designation as EEP (EDTA-extractable protein). Its prior molecular weight estimations of 37,500 (19), 35,000 (11), and 34,000 d (30) appears to have been misleading, as the isolated protein resolved in SDS-PAGE into two components of 32,000 and 35,000 d (MP32 and 35; Table 2) (49). Part of the protein is calcium precipitable, and thin-layer filtration demonstrated that no oligomers are formed in the

absence of detergent. The two polypeptides are antigenically different but may
share a determinant with soluble γ-crystallin (90). Its amino acid composition is
similar to that of the lens crystallins in polarity (50%) but different from them
in composition (49). Some differences between the proteins isolated from
diverse vertebrate sources have been detected by immunochemistry and isoelectric focusing (52,54,55). Although no detectable EEP is extractable from human
lenses, low levels are obtainable from monkey lenses (55).

More recently, Lenstra et al. (91), utilizing fiber membranes isolated by the
gradient centrifugation method, have reported the resolution of bovine lens
"MP34" by SDS-PAGE into several components designated as MP35, MP36.5,
and MP37, one (MP37) of which has been identified as the enzyme
glyceraldehyde 3-phosphate dehydrogenase (see Enzymes). It is not clear,
however, which of these components correspond to the MP32 and MP35 of EEP.

MP22:   A 22,000 d (MP22; Table 2) polypeptide has been extracted with the
calcium chelator EGTA from human lens fiber plasma membranes prepared by
the acylation method (92). Although it comigrates in SDS-PAGE with the MP22
polypeptide, which is believed to be a degradation product of MP26, the two
differ significantly in amino acid composition. Moreover, although it shares
antigenic reactivity with soluble γ-crystallin, as does bovine EEP, it also differs
significantly from EEP in amino acid composition (92).

A third MP22 polypeptide has been detected by the same group of investigators (76) by immunochemical analyses of the MP22 SDS-PAGE band of
normal human lenses. Although referred to as "intrinsic" (93), this MP22
protein has been demonstrated to be immunologically distinct from the intrinsic
MP22 believed to be a degradation product of MP26 (76). Support for the
conclusion of unique MP22 species has been obtained by the demonstration of
the independent synthesis of MP22 and MP26 in the organ-cultured human lens
(94). The MP22 protein displays a developmental pattern of distribution (93)
identical to that of the extrinsic MP43 polypeptide also discovered by the same
group of investigators (86). The probable relation of these polypeptides are
unknown at present. Thus there appear to be three distinct MP22-associated
polypeptides: one, an intrinsic protein thought to be a degradation product of
MP26 (29); a second, presumably intrinsic (93), but a product of independent
synthesis (94); and a third, believed to be extrinsic by virtue of its EGTA
extractability (92).

MP10-12, 17-18:   A few polypeptides in the molecular weight range 10,000–
20,000 d have been occasionally described for the lens fiber plasma membranes
but have not been singled out for particular study. A 17,000–18,000 d (MP17-
18; Table 2) polypeptide of unknown identity has been described in fiber plasma
membranes isolated from fish (82), frog (82), bovine (19,49), and human (18)
lenses. The bovine lens MP17 has been noted to decrease slightly upon heating of
fiber membrane preparations for SDS-PAGE (50), especially in the absence of

reducing agents on electrophoresis. A 10,000–12,000 d (MP10–12; Table 2) polypeptide has been described in isolated human lens fiber plasma membranes (18,95).

Others: A number of other components have been demonstrated to be or are assumed to be extrinsic components of the lens fiber plasma membranes (Table 2). Among those under active study are MP200, 250, lens spectrin (96); MP96–130, α-actinin (35); MP86–95, the beaded-chain filament protein (99); MP54–57, vimentin (100,101); MP50, chick δ-crystallin (102,103); MP40, lens actin (84); MP37, glyceraldehyde 3-phosphate dehydrogenase (91); and MP20–22, mammalian α-crystallin (17,104).

Some of these membrane components (Table 2) are also components of the lens fiber cell cytoskeleton (spectrin, α-actinin, the beaded-chain filament protein, vimentin, and actin) and are reviewed in that section in greater detail (see Lens Cytoskeleton). Components known to be enzymes are discussed in the section on enzymes (see Enzymes), and crystallin components are included in a discussion on the interactions of lens membranes with crystallins and the cytoskeleton (see Interactions Among Lens Membranes, Cytoskeleton, and Crystallins).

Enzymes

It is generally believed that with very few exceptions all biologic enzymes so far detected would also be found in the lens if the appropriate methods were employed (105). Among enzymes known to be specifically (but not exclusively) associated with plasma membranes in other tissues are 5-nucleotidase, various specific ATPases, alkaline phosphatase, leucine aminopeptidase, and adenylcyclase (106). The assignation of particular enzymes to lens plasma membranes reviewed here is not in terms of the physiologic demonstration of the presence of the enzymes in the lens but with their specific localization by immunochemical, cytochemical, and biochemical (SDS-PAGE) means in the lens epithelial and fiber plasma membranes. At the present time only two enzymes have been localized to the lens plasma membranes, although many others are assumed to be present by virtue of their known association with the lens water-insoluble fraction.

Membrane-bound enzymes may be classified under the same two categories as all other membrane proteins, that is, as intrinsic and extrinsic components of the membranes (21), with some qualifications. Intrinsic enzymes generally cannot be removed from the membrane without their inactivation and are often known to require specific lipids for their optimal activity. Extrinsic enzymes can be removed from the membrane by salt solutions and do not require lipid for their activity; however, extrinsic enzymes may encompass soluble enzymes, which become secondarily associated with the membrane.

MP37: Jedziniak and coworkers (107) first purified a membrane-bound form of the enzyme glyceraldehyde 3-phosphate dehydrogenase (G3PD; EC 1.2.1.12) from human normal and cataractous lenses. The enzyme has been identified independently by Bloemendal and coworkers (91) as the 37,000 d polypeptide (MP37; Table 2) that forms part of the group of polypeptides formerly designated as MP34. The absence of the enzyme from membranes isolated by the urea methods is believed to be due to its loose association with the membranes (91). Its loose association with the membrane qualifies MP37 to be classified as an extrinsic enzyme of the lens fiber plasma membranes. Whether it is secondarily associated with the membrane or constitutes a permanent component is not known at present. In this regard, Russell and Kinoshita (108) report that G3PD, isolated with the β-crystallin peak by G200 column chromatography from mouse or bovine lenses, rapidly associates with isolated plasma membranes (108). The specific function of this glycolytic enzyme in the lens fiber plasma membranes is not known at present (91).

MP90-107:   The most studied enzyme in lens biochemistry, Na^+,K^+-ATPase (EC 3.6.1.3), was initially localized in the rat lens epithelial plasma membranes by ultrastructural cytochemical methods (109,110). Enzymic reaction was initially localized only on the lateral membranes; the apical and basal membrane surfaces were negative (110). Subsequently, ultrastructural cytochemical localization was demonstrated on the apical rat lens epithelial membranes (111) and on the anterior and posterior pole membranes of cortical fiber cells (111, 112) as well. This last observation verified earlier conclusions from physiologic studies in rabbit lenses of the presence of Na^+,K^+-ATPase enzymatic activity in these same lens fiber regions (113). The catalytic subunit of the enzyme has been identified as a 90,000 d (MP90) polypeptide by radiolabeled phosphorylation of isolated chick lens fiber plasma membranes and analysis by SDS-PAGE (115) and its identity verified by peptide mapping and specific immunochemistry (97). Finally, the specific isolation of Na^+,K^+-ATPase has been effected from chick lens epithelium (115) and pig lens fiber cell plasma membranes (98). The smaller glycoprotein subunit (MP51) of the enzyme was also identified in the last study (98). The requisite use of detergents for the extraction of Na^+,K^+-ATPase from lens membranes (98,115) qualifies MP90-107 to be classified as an intrinsic enzyme of the lens fiber plasma membranes.

Others: Leucine aminopeptidase (116), acid phosphatase (117), and Ca^{2+}, Mg^{2+}-ATPase (118) are among other enzymes believed to be either membrane associated or membrane bound in the lens. No specific membrane polypeptides, however, have yet been associated with these enzymes by SDS-PAGE of isolated lens plasma membranes. The protease leucine aminopeptidase has been detected by specific immunoelectrophoresis in the water-insoluble fraction of human lenses (116), suggesting the presence in that fraction of a membrane-bound form of the enzyme. The presence of acid phosphatase in lenses of rats in which cataractogenesis was initiated through galactose feeding (119), as well as in lenses of normal young and aged rats (117,119), has been demonstrated by ultrastructural cytochemistry. The enzyme was localized in extracellular sites between epithelial cells and between cortical fibers in the entire intercellular space between the fibers (119). The existence of a lens Ca^{2+},Mg^{2+}-ATPase has been demonstrated in the epithelial, cortical, and nuclear lens areas of mammalian lenses by physiologic methods (118).

Glycoproteins

Membrane glycoproteins are largely intrinsic proteins, embedded in the lipid bilayer of the membrane and possessing covalently bound carbohydrate chains that face the outside of the membrane (21). Dische (9,10) first demonstrated that lens carbohydrate was associated with the lens plasma membranes.

The lens fiber plasma membrane fraction is known to consist of 3–4% carbohydrate (44,120), which has been found to consist of fucose, mannose, galactose, glucose, *N*-acetylgalactosamine, *N*-acetylglucosamine, and traces of sialic acid. The most abundant (50%) carbohydrate is galactose (44,120). Gas-liquid chromatographic analysis of bovine lens fiber membranes isolated by the acylation method (121) revealed that mannose and fucose are associated with the nonlipid fraction of the membranes, probably glycoproteins, and all hexosamines and 50–60% of bound hexoses occur in membrane glycolipids. The most specific identification of lens fiber membranes glycoproteins to date, utilizing the galactose oxidase-tritiated sodium borohydride labeling of galactose-containing proteins, has demonstrated polypeptides MP103, MP69, MP50, MP32, MP26, and MP22 in rabbit lenses (122) and polypeptides MP128, MP103, MP82, MP71, MP35, and MP22 in mouse lenses (123) as glycoproteins.

Lipid Composition

Almost all the lens lipid is found in the fiber cell plasma membranes (24). Lipids make up from 47% (bovine and chick) (17,19,60) to 55% (human) (18) of the lens fiber cell plasma membranes. Lens lipids consist of phospholipids, cholesterol, and glycosphingolipids (Table 4) (44). The major phospholipids are phosphatidylcholine, phosphatidylethanolamine, phosphatidylserine, phosphati-dylinositol, and sphingomyelin, and a minor fraction consisting of phosphatidyl-glycerol, phosphatidic acid, diphosphatidylglycerol, lysophosphatidylcholine, lysophosphatidylethanolamine, and plasmalogens (44,124). The only sterol detected by gas-liquid chromatography in the lens is cholesterol (125). Glyco-sphingolipids are known to be found in the lens (127,128) but have not been widely studied.

Phospholipids

Phosphatidylethanolamine is the major membrane phospholipid in the rabbit (124), bovine (24,124), and chick (61) lenses, but sphingomyelin predominates in human (24,62,125) lenses. The fatty acid profiles of lens phospholipids show a

Table 4 Lipid Composition of Vertebrate Lenses

Lipids	References
Phospholipids	24, 44, 61, 62, 124, 125
Phosphatidylcholine	
Phosphatidylethanolamine	
Phosphatidylserine	
Phosphatidylinositol	
Phosphatidylglycerol	
Phosphatidic acid	
Diphosphatidylglycerol	
Lysophosphatidylcholine	
Lysophosphatidylethanolamine	
Plasmalogens	
Sphingomyelin	
Mono- and dimethylphosphatidylethanolamine	
Di- and triphosphoinositide	
Cholesterol	24, 44, 61, 125, 126
Glycosphingolipids	127, 128

predominance of C_{16}, C_{18}, and C_{24}, mostly saturated and monounsaturated fatty acids for most vertebrate species studied (44,124,125). Polyunsaturated fatty acids have been reported to be very low in concentration in the lens (129). The presence of arachidonic acid (20:4) in bovine lenses, however, has been reported (44).

Cholesterol

Cholesterol is the only sterol found in the normal lens (125). The overall cholesterol/phospholipid molar ratios are relatively high (1.0–3.5) in most vertebrate lenses studied (24,61,125). The human lens plasma membranes apparently attain the highest cholesterol/phospholipid ratios known in nature (24). Cholesterol esters, not normally found in plasma membranes, have been found in low concentration in normal human lenses (125). The relatively high cholesterol content of lens fiber plasma membranes is widely believed to contribute, along with membrane proteins, to creating a more rigid (less fluid) membrane (35,44,61,125).

Glycosphingolipids

Glycosphingolipids are present in the lenses of various species studied (bovine, pig, rat, and human) (44,127,128). Gangliosides detected in pig and bovine lenses consist mostly of hematoside (G_{M3}), G_{M1} in rat lenses, and both G_{M1} and G_{M3} predominate in human lenses (128). The long-chain bases of the sphingolipid ceramide fraction of human lenses consist primarily of sphinganine and 4-sphinganine (127) and are more complex than that of the bovine lens sphingolipids (44,125), which consist of only 4-sphinganine. Human lens sphingolipids are the only sphingolipids known to consist of primarily saturated sphingosine bases (127). As glycolipids are known to be asymmetrically localized in the outer half of the membrane lipid bilayer and to have special structural and functional roles (e.g., as receptors) (130), their presence in lens fiber plasma membranes is intriguing.

Biosynthesis

The lens largely synthesizes all its own membrane lipids, with the exception of essential fatty acids, the synthesis occurring primarily in the epithelium, equator, and outer cortex, the lens nucleus displaying essentially no activity (44). Human lens fiber differentiation is associated with the increase synthesis of phosphatidylinositol (24). Bovine lens fiber differentiation, however, is associated with a decrease in phosphatidylcholine and phosphatidylinositol but increases in phosphatidylserine and sphingomyelin (131). The sphingomyelin concentration also increases in the transition from fetal to adult lens (132), presumably through increased synthesis. Chick lens fiber differentiation is associated with the increase synthesis of phosphatidic acid and the decreased turnover and increased synthesis of phosphatidylinositol, phosphatidylcholine, and phosphatidylethanolamine (133,134), resulting in phospholipid accumulation in the elongating fiber cells. The addition of the methylation inhibitor 3-deazaadenosine to the explanted chick lens epithelial cells inhibits both the methylation of phosphatidylethanolamine to phosphatidylcholine and the in vitro elongation (differentiation) of the cells (135).

Studies on the incorporation of [^{14}C]stearic acid into the organ-cultured rabbit lens (136) demonstrated the incorporation of label into fatty acids of 20 carbons and longer, revealing the presence of a fatty acid elongation system in the lens. The incorporation of label into cholesterol in the same study (136) also reveals that this fatty acid can serve as a carbon source for cholesterol synthesis in the lens. Studies of sterol synthesis in the rat lens in vitro (126) have demonstrated that the lens possesses the capacity to synthesize all its cholesterol at early ages but may acquire an extralenticular source for at least one-half of its normal requirements at later ages.

LENS FIBER CELL JUNCTIONS

Gap junctions are rigid plaques of plasma membranes, made up of membrane proteins and lipids, which interlock adjacent cells (137), narrowing the intercellular space between the outer leaflets of the two participating cells from its normal 20+ nm to 2–3 nm. The proteins, embedded in the lipid bilayers of the adjacent membranes, stretch across the intercellular space and interlock the cells (137) and form channels, allowing the direct exchange of ions and small molecules (less than 1200 d) between the adjacent cells (137). Gap junctions are believed to be the structural entities effecting the electrotonic and metabolic coupling between adjacent cells displaying these forms of functional communication (137).

Intercellular junctions occur in both epithelial and fiber cells in the lens (25). The epithelial junctions are believed to be gap junctions (138) and to differ morphologically and biochemically from the fiber cell intercellular junctions (138), but as they have not been isolated, very little is known concerning their composition. Remarkably, although gap junctions constitute about 3–4% of the membrane surface of the cells in most tissues where they occur, junctions constitute from 50 to 60% of the surface plasma membrane of the lens fiber cells (40,139). This constitutes the greatest amount of surface plasma membrane devoted to intercellular junctions known in nature. Lens fiber cell junctions consist of aggregates of 9 nm intramembrane particles arranged in a noncrystalline, random, pleomorphic pattern (40,41). Lens fiber cell junctions differ from gap junctions and from the lens epithelial junctions in that they retain a disordered aggregation (noncrystalline) of their intramembrane particles upon fixation for ultrastructural study (40,35,138). Crystallization of the intramembrane particles of gap junctions has been correlated with a high-resistance state (closed channels), as contrasted to the noncrystallized (disordered) aggregation of particles correlated with a low-resistance state (open channels) (137). Lens fiber cell junctions are believed to aid in the structural establishment of lens transparency (140) and in allowing the exchange of ions and small metabolites (40,138) between the largely metabolically inactive lens fiber cells.

Recently, on the basis of morphologic and biochemical criteria (141–143), several investigators have concluded that lens fiber cell junctions are not "typical" gap junctions and should not be classified as such. The evidence centers on observed differences between liver and lens fiber cell junctions, and between their proteins (142). The occasional observation in lens freeze-fracture preparations of structures displaying exclusively square lattice arrays of their intramembranous particles (142) contrasts with the hexagonal pattern observed in isolated liver gap junctions. Orthogonal arrays, however, if they do occur in the lens, have been demonstrated to be unrelated to the lens fiber cell junctions (144) and to be a frequent result of intrinsic proteolysis upon storage of lens tissue at 37°C without fixation; square arrays are never observed if lens tissue is fixed immediately (41,140,144). The other line of evidence concerns the previous observed lack of immunochemical cross-reaction of antisera to the liver and lens fiber cell junction proteins (141–143). Very recent immunochemical evidence, obtained with affinity-purified antibodies to the liver protein (145) in reaction with the lens junctional protein, however, demonstrate structural homology between the proteins.

Methods of Isolation

Lens fiber cell junctions were first isolated as the detergent-insoluble fraction following the treatment of lens fiber plasma membranes with 1.0% sodium deoxycholate (64). Currently, the most commonly employed method involves a modification to the method developed by Goodenough (40). The decapsulated lenses are homogenized (41) in a $NaHCO_3$-EDTA buffer and centrifuged, the resulting pellet resuspended in a Tris-EDTA-$CaCl_2$ buffer and recentrifuged to recover the lens water-insoluble fraction. The latter fraction is then treated first with 4 M urea and then with 7 M urea to remove the lens cytoskeleton and insolubilized crystallins, subjected to a sucrose step gradient (25, 41, and 45%), and the purified membranes collected at the 41–25% interface. The isolated membranes are treated with 0.2% sodium deoxycholate and the purified junctions recovered through a second step-gradient centrifugation (40,41). An alternate method (146) utilizes the treatment of lens fiber plasma membranes, isolated by the gradient centrifugation method, with 0.1% Sarkosyl and centrifugation to recover a pellet rich (95%) in junctions.

The isolated lens fiber cell junctions have been analyzed ultrastructurally by transmission electron microscopy following sectioning of the pellet, freeze-fracture replication (Fig. 3), and negative staining (40,41,146) and biochemically by SDS-PAGE (Fig. 3) (40,41,61,146) for protein composition and by thin-layer chromatography for lipid composition (61).

Protein Composition

Finbow et al. (147) have recently reviewed the independent lines of evidence that suggest that a 26,000 d polypeptide constitutes the primary or only protein of gap junctions. The presence of higher molecular weight components in junctional preparations is believed to be the result of contamination and lower molecular weight components are held to represent proteolysis of the 26,000 d protein (147). Correlated electron microscopy and x-ray diffraction had yielded a model of the molecular structure of gap junction connexons consistent with this evidence (148).

For the lens, MP26 (Table 2) has been shown to be the main or only protein component of isolated bovine (40) and chick (41) lens fiber cell junctions and has been localized immunocytochemically on the junctions (42,149). Moreover, lens fiber cell junctions isolated to homogeneity display a single polypeptide component in SDS-PAGE (61). A degree of structural homology between the 26,000 d proteins of liver cells and lens fiber cells has been uncovered by

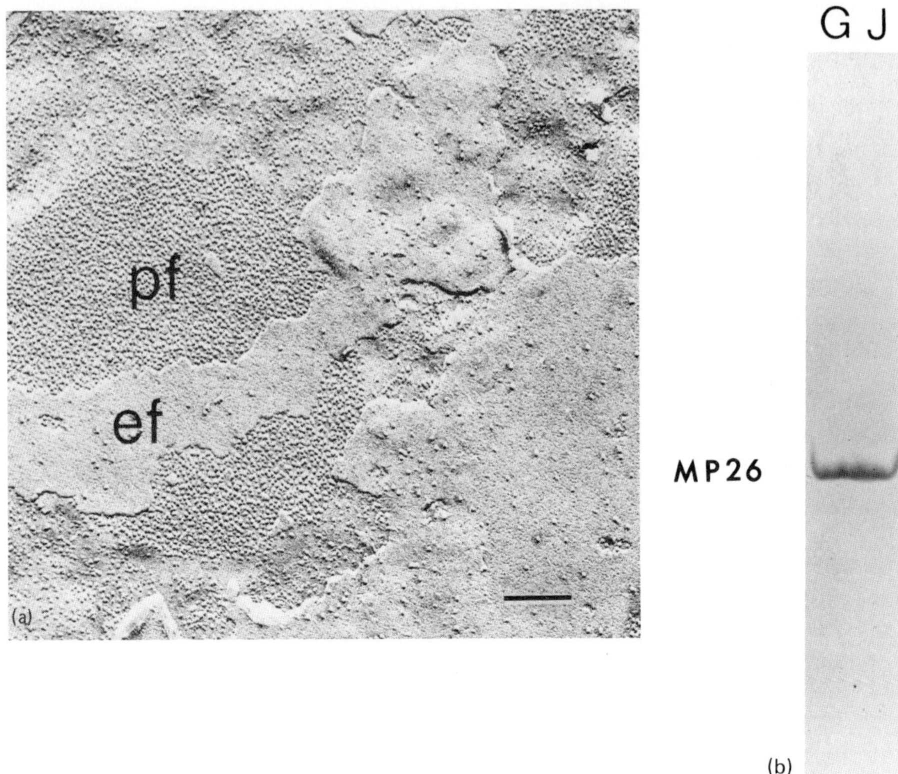

Figure 3 (a) Freeze-fracture replica of an intact chick lens fiber cell surface membrane. The particle-rich areas represent fiber cell junctions; pf = protoplasmic face; ef = exoplasmic face (X 55,500); bar = 0.2 μm. (b) Coomassie bluestained SDS-PAGE gel of isolated chick lens fiber cell junctions (GJ); MP = membrane polypeptide.

sensitive immunochemical procedures (145). Partial sequencing of the proteins at the N-terminus, however, demonstrated no particular homology between these proteins other than a strongly hydrophobic character (150).

The exact molecular organization of gap junction connexons (intramembrane particles) is not known at present. The model proposed by x-ray diffraction suggests a hexameric structure (148). The uniform size of lens fiber cell junction particles (41), their pattern of crystallization (137), and rotary-shadowing visualization (151) suggest that lens junctional particles are made up of either tetramers (152) or hexamers (151) of MP26. Tryptic peptide analysis of multimeric aggregates of MP26 in SDS-PAGE gels of isolated chick lens fiber membranes, and chemical cross-linking of lens membrane components, demonstrate extensive nearest-neighbor interactions of MP26 with itself in the intact junctional complex (47,146). The conventional interpretation of the freeze-fracture replicas of gap junctions may be misleading. The idea has been advanced that gap junctional intramembrane particles might not be exclusively made up of protein (153) and that lipid may contribute significantly to the observed morphology of junctional particles. The proteolipid nature of lens MP26 constitutes support for this conclusion (see Lens Plasma Membranes).

Lipid Composition

Very few studies on the lipid composition of gap junctions have been performed, presumably due to the low yields obtained in junctional preparations. As the inherent detergent resistance of gap junctions allows for their isolation from nonjunctional membrane, isolated gap junctions partly reflect their nonextracted lipid composition (154). The liver gap junctions display a higher cholesterol and lower phospholipid content than that of the plasma membranes from which they are derived (154,155). The unavoidable detergent extraction in the isolation procedures is believed (154,155) to be accountable for the lower phospholipid content of the isolated gap junctions. Remarkably, gap junctions isolated by different methods display a lack of sphingomyelin (137,154,155).

Analysis of the lipid composition of isolated chick lens fiber cell junctions demonstrated that the junctions contain about 50% of the lens fiber membrane protein and lipid (61). Isolated lens fiber cell junctions are similar to gap junctions in overall lipid composition, but unlike gap junctions, the lens junctions contain sphingomyelin. The cholesterol/phospholipid molar ratio is 2.1 for total fiber membranes but 3.1 for the chick lens fiber cell junctions (61). The major phospholipid in chick lens fiber cell junctions is sphingomyelin (40%) but phosphatidylethanolamine (39%) in the total fiber membranes. The junctions were found to contain about 57% of the total fiber cholesterol and about 53% of the total fiber sphingomyelin (61). The high cholesterol content in isolated junctions, in spite of the expected detergent effect, is suggestive of its tight binding to the junctional protein; the sphingomyelin content of the isolated lens junctions, likewise, is suggestive of a strong interaction between this phospholipid and, possibly, junctional cholesterol. It appears, however, that phosphatidylinositol, phosphatidylethanolamine, and to a lesser extent phosphatidylcholine, may have been preferentially extracted from the junctions by the detergents employed in the isolation procedure, accounting for the lowered phospholipid content of the isolated junctions (61).

The high cholesterol and sphingomyelin content suggests that lens fiber cell junctions constitute highly rigid membrane regions conferring significant constraints to the movement of their intramembrane particles (connexons) in the plane of the membrane, a conclusion supported by the results of recent physical studies indicating a higher degree of rigidity (order) in lens plasma membranes (156,157) than in any other membranes known in nature. These observations help to explain the resistance to the crystallization of their intramembranous particles (40,137), observed so far only in lens fiber junctions.

The importance of cholesterol to the integrity of lens fiber junction structure is exemplified by the hereditary mouse cataracts (79,80,158,159). There is a marked decrease in cell junctions during cataract formation in the mouse (159) and a concomitant degradation of MP26 to a 23,000 d polypeptide (79,80). Studies on the lipid synthesis in the Philly mouse lens (158) with radioactively labeled acetate demonstrate a high specific activity but lower cholesterol content at all ages, suggestive of a failure to incorporate cholesterol in the membranes.

Interactions with Calcium and Calmodulin

Interactions with Calcium

The astronomical size and number of junctions in lens fiber cells (139), the evidence that the fiber cells are electrically coupled (160) and that Procion dyes and small metabolites diffuse readily from one fiber cell to another (138), and the observation that lens fiber cell junctional particles do not crystallize upon fixation with glutaraldehyde (40) have led to the hypothesis that the lens fiber junctions remain permanently in a noncrystalline, low-resistance (open channels) state (40). The hypothesis implies that lens fiber cell junctions are physiologically unregulated (40). The implications of the hypothesis have led to active controversy, as several conditions have been found in which lens fiber junctional particles can be made to crystallize (152,161–163); no direct evidence of electrical uncoupling as a result of these treatments, however, has been obtained, although a calcium-dependent impediment to the movement of Procion yellow dye between fiber cells subjected to injury has been demonstrated (163). Recently, Rae et al. (164) have demonstrated the reversible electrical uncoupling of fiber cells in the frog lens.

The most commonly employed methods for the crystallization of lens fiber cell junctional particles has been the presence of Ca^{2+} ions along with isolated junctions in the incubating medium (161). A hypothesis (161) has been proposed for the regulation of the functional state of gap junctions (open or closed channels) based on the direct binding of divalent cations (Ca^{2+}, Mg^{2+}, Sr^{2+}, and Ba^{2+}) or H^+ to the protein of the gap junctional particles, leading to a narrowing of the junctional channels and to the crystallization of the particles. More direct evidence for this process in the lens has been lacking due to the generally believed conclusion of the existence of a high intracellular calcium-buffering capacity in the lens as a result of the extremely high soluble protein concentration of this organ. Support for this conclusion is found in recent observations that 75% of the total calcium in the lens is associated with soluble protein (165) although a significant proportion (20%) is membrane bound.

Interactions with Calmodulin

Calmodulin is a ubiquitous Ca^{2+}-binding protein that modulates many calcium-dependent regulatory processes in cells (166). The specific binding of calmodulin to the isolated calf lens fiber cell junctions (167) and its specific binding to chick lens MP26 (168) have been recently demonstrated. MP26 has been demonstrated, by measurements of intrinsic fluorescence and circular dichroism (169), to undergo changes in its conformation as a result of calmodulin binding. Moreover, evidence that trifluoperazine, a calmodulin inhibitor, affects lens fiber cell junction structure (170) further supports the conclusion that lens fiber cell junctions may be regulated by a calcium-dependent process.

LENS CYTOSKELETON

The cytoskeleton of biologic cells comprises cytoplasmic filamentous structures, some of which are contractile, involved in maintaining the structural integrity of cells by providing internal structural support (171,172), localized contractility (173), and structural support for the plasma membrane (174) and for membrane receptors (175). Cytoskeletons are composed of filamentous structures classified by their diameter as microfilaments (5 nm), intermediate-sized filaments (7–11 nm), thick filaments (12–15 nm), and microtubules (25 nm).

Microfilaments consist of two protofilamentous strands of globular actin molecules twisted around each other in helical fashion. Actin, the protein of microfilaments, can occur in either a globular soluble form (G-actin) of 45,000 d molecular weight or polymerized into long double-stranded helical (F-actin) microfilaments (176). Three different forms of the actin protein have been identified by isoelectric focusing: α-actin, found exclusively in cardiac and skeletal muscle; and β- and γ-actins, found in a variety of nonmuscle cells (177–179). Actin microfilaments in nonmuscle cells are involved in many contractile cellular processes, including cell division, cell movement, cytoplasmic streaming, and changes in cell shape (for a review, see Ref. 173). The strong gelation property of actin is believed to confer upon it a cytoskeletal role in addition to its known contractile role in cells (180).

Intermediate-sized filaments (intermediate filaments) consist of protofilamentous strands of globular protein organized in helical array that in sections display a hollow core of 2–3 nm diameter (181). Five different classes can be distinguished by biochemical and immunologic criteria (172,181): the epithelial cytokeratins (40,000–68,000 d), the muscle desmins (55,000 d), the neurofilaments (68,000–210,000 d), the astrocyte glial fibrillary acidic protein (50,000–55,000 d), and the vimentin filaments (57,000 d), typical of cells derived from embryonal mesenchyme. However, recent studies with monoclonal antibodies suggest that all types of intermediate filaments share antigenic determinants in common (182). Furthermore, more than one class of intermediate filament can exist in the same cell (181). Intermediate filaments are believed to function primarily in a cytoskeletal role in cells (172).

The major recognized thick filaments comprise the nonmuscle myosin (183) and spectrin (184). Nonmuscle myosin (200,000 d) has been detected, isolated, and biochemically identified in a variety of cells (183) and is found in intracellular actin stress-fiber bundles (185). It is believed to function in intracellular movement and cell motility. Spectrin, the major extrinsic protein of the plasma membrane of the red blood cell (184), is composed of two subunits of 220,000 and 240,000 d, referred to as band 1 (α) and band 2 (β), respectively (184). It has been recently detected in a wide variety of cells (96). In association with actin and other proteins, it is believed to mediate cytoplasmic filament-membrane interactions and to confer form or structural integrity to the plasma membranes of cells (96).

Microtubules consist of concentric arrays of 13 protofilaments made up of alternating globular protein subunits of α- and β-tubulin, each of about 50,000 d molecular weight (for a review, see Ref. 171), and microtubule-associated proteins (MAP), which act as promoters of tubulin assembly into microtubules. Microtubules are ubiquitous organelles that function in a variety of cell processes, including cell division, intracellular transport, ciliary and flagellar motion, secretion, cell elongation, and shape changes (171).

Cytoskeletal Components of Lens Cells

It has become increasingly evident that lens cells contain many of the cytoskeletal elements present in other nonmuscle cells (Table 5) (35,186), as well as a lens-specific cytoskeletal component, the beaded-chain filament (3,14). Microfilaments (187), intermediate filaments (187), and microtubules (188) have been identified ultrastructurally in the intact lens. Myosin in cultured calf lens epithelial cells (84) and spectrin in chick lens epithelium (96) have been detected by immunofluorescence. The lens beaded-chain filaments, apparently unique to the lens fiber cells (14), have been observed ultrastructurally in the intact lens (189).

Methods of Isolation

Lens cytoskeletal components can be either destabilized (solubilized) or stabilized (retained) in the lens water-insoluble fraction depending on the ionic

Table 5 Cytoskeletal Components and Proteins of Lens Cells

Cytoskeletal component	Cytoskeletal protein	References
Microfilaments	Actin	187, 193, 195–197
Intermediate-sized filaments	Vimentin	84, 187, 201–203
Microtubules	Tubulin	84, 188, 205
Beaded-chain filaments	Backbone protein	3, 35, 14, 189, 205
Stress fibers[a]	Myosin, α-actinin, tropomyosin	84, 186, 209
Membrane skeleton	Spectrin	96

[a]Also contain actin.

conditions used for the initial homogenization of the lens (71,190,191). Cyto-skeleton-stabilizing buffer conditions are a prerequisite for the further fractionation of the lens cytoskeletal proteins from the lens water-insoluble fraction (membrane-cytoskeleton complex; see Introduction) (71,190,191). The most frequently used procedures (71,191) for the isolation of lens cytoskeletal proteins involves homogenization of the lens in a high ionic strength buffer containing Mg^{2+} and centrifugation to recover the water-insoluble fraction (pellet). Following repeated washing and centrifugation, to remove as much of the soluble crystallins as possible, the water-insoluble fraction is extracted with 6–8 M urea and centrifuged (71,190–192). The 6–8 M urea extract obtained may be used for the chromatographic separation of the lens cytoskeletal (urea-soluble) proteins, including actin, vimentin, and spectrin (190,192).

The polypeptide separation obtained by DEAE-52 column chromatography of the chick lens urea-soluble cytoskeletal proteins (192) is shown in Fig. 4. Actin elutes before application of the 15–65 mM NaCl gradient and can be recovered in relatively pure form. Contaminating proteins can be separated from the actin fraction by chromatography on a hydroxylapatite column (192). The ability of intermediate filaments to repolymerize upon removal of the urea by dialysis is utilized to isolate an enriched fraction of intermediate filaments from peaks 6 and 7 (Fig. 4). Higher molecular weight cytoskeletal proteins (e.g., spectrin) elute at the end of the salt gradient and in the column "push."

The specific isolation of lens actin has been accomplished by a variety of methods. The G form of the protein can be isolated from the lens water-soluble fraction by affinity chromatography utilizing DNase-I linked to agarose, and the F form has been isolated by urea column chromatography from the lens urea-soluble fraction (193). Fractions enriched in actin can also be prepared from acetone extracts of the lens water-insoluble fraction (192,194).

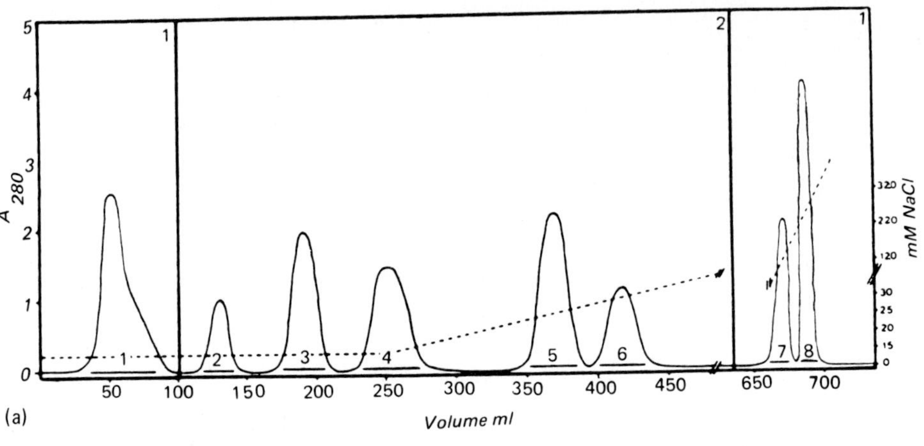

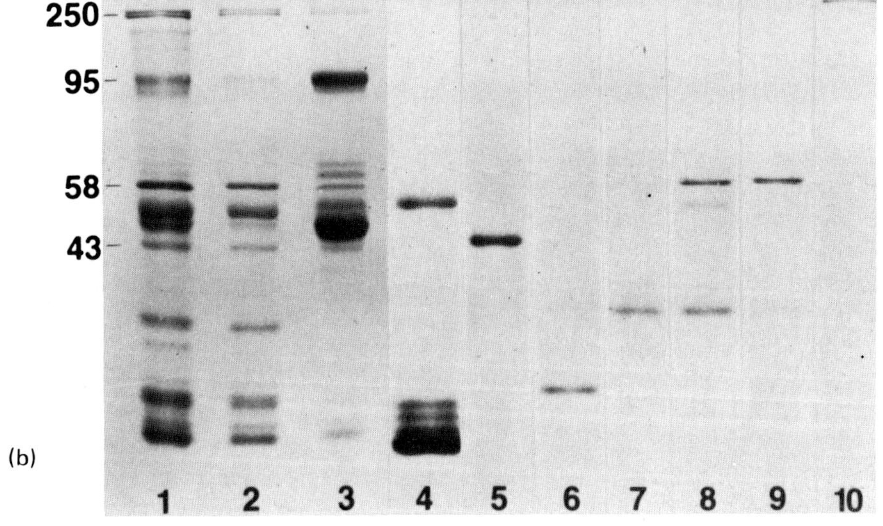

Figure 4 (a) Elution profile of the DEAE-52 column separation of the chick total lens 8 M urea-soluble fraction (USF). The solid line denotes the A_{280} absorbance and the dashed line denotes the salt (NaCl) gradient. (b) Coomassie blue-stained SDS-PAGE gels of (1) chick total lens 8 M urea-soluble fraction; (2) 10,000 × *g* supernatant of (1), applied to the DEAE-52 column; (3) 10,000 × *g* pellet of (1); (4) peak 1; (5) peak 2; (6) peak 3; (7) peak 4; (8) peak 5; (9) peak 6; (10) peaks 7 and 8. Molecular weights in kilodaltons. (Reproduced with the permission of Academic Press, Ltd.)

Distribution and Biochemical and Immunological Properties

Microfilaments: The distribution of lens actin microfilaments, studied ultrastructurally (187) in vertebrate lenses, is suggestive of different functional patterns. In nonaccommodating (mice and rats) lenses the filaments are grouped in bundles around the cell nucleus, in cellular processes, and throughout the cytoplasm of the epithelial cells. In anteriorly flattened accommodating lenses (frog, squirrel, and human), the filaments are also scattered throughout the cytoplasm of the epithelial cells but are particularly accumulated on either side of the epithelial-fiber cell junction, where they form a dense lattice closely associated with the lens fiber plasma membrane. It has been hypothesized that the role of microfilaments in the lens, depending on their pattern of distribution, is either to structurally support a spherical shape (nonaccommodating lenses) or to provide a contractile force to return the flattened anterior lens surface to the accommodated state (accommodating lenses) (187).

The presence of actin (Table 5) in the lens has been confirmed cytochemically by specific immunofluoresence (195) and labeling with heavy meromyosin (196), and biochemically by isoelectric point (197) and peptide mapping (193). Lens actin is the typical nonmuscle type consisting mainly of β- and γ-actin (192,197). Its isoelectric point is near 5.5 (197) and its amino acid composition similar to that of muscle actin (193). Although actin constitutes about 10.7% of the total protein in newborn rat lens epithelium and the β-/γ-actin ratio is 2.2, it constitutes only about 1.1% of the total protein of newborn and adult rat cortical lens fiber cells with a β-/γ-actin ratio of 1.2 (197). Actin was reported to be absent from rat lens nuclear fiber cells (197), although it has previously been detected by immunofluorescence in the rat lens nucleus (198).

Antibodies prepared against chick lens actin cross-react with muscle actin and with lens actin of different vertebrate species, including human lens actin (192). Immunologic studies show that actin is particularly concentrated at the short sides of the hexagonal lens fiber cells (192,198). A similar localization has also been demonstrated by the use of phallotoxin (199). Although actin is concentrated along the length of the lens fiber plasma membrane, it is also found more centrally in the fiber cells (192). In chick and calf lenses the plasma membrane association of actin is found in both cortical and nuclear fiber cells (192,198, 199). In rat lens nuclear fiber cells, however, actin is observed only in a cytoplasmic position, the fiber membrane regions appearing devoid of actin (198). Following injury to the lens epithelium, actin-containing stress fibers are detected in those cells that migrate into the wound area (200), and their disappearance is noted following the wound repair.

Intermediate Filaments: Lens intermediate filaments (Table 5) are typical 7–11 nm structures found primarily in epithelial and cortical fiber cells and largely absent from the lens nucleus (23). Although lens cells are epithelial derivatives, lens intermediate filaments are of the vimentin class (typical of mesenchymally derived cells) as determined by immunologic properties (84), amino acid composition, and isoelectric point (201). Its molecular weight in the lens has been variously estimated as 54,000 (202) and 57,000 (203) d and was first purified from the water-insoluble fraction of bovine cortical fiber cells by ion-exchange chromatography on DEAE-52 and carboxymethylcellulose (202). Lens vimentin comprises at least three isoelectric variants between pH 5.3 and 5.5, and its amino acid composition is enriched in glutamic acid and leucine (202). Similar results have been obtained following the isolation of vimentin from pig lenses (201). The intermediate filament protein appears to be phosphorylated (192).

Antigenic determinants of the glial fibrillary acidic (GFA) protein (Table 5) have been detected in chick lens epithelium with an antiserum specific to that protein (204), although this may be indicative of the observation that all different types of intermediate filaments share common antigenic determinants (182). Biosynthesis of vimentin has been observed to occur in the organ-cultured chick lens only by epithelial and superficial cortical fiber cells (70), an observation that correlates closely with the morphologic distribution of intermediate filaments in the lens (23). Although intermediate filaments have been found to be distributed similarly to actin microfilaments (186) within lens epithelial cells, their distribution within lens fiber cells differs in the more central localization of intermediate filaments and the peripheral (membrane association) localization of microfilaments (199). The function of intermediate filaments in the lens is not known at present.

Beaded-Chain Filaments: The beaded-chain filaments (Table 5), cytoskeletal elements apparently unique to the lens (3,14,35), consist of globular protein particles 12–15 nm in diameter, attached to a filament backbone 7–9 nm in diameter (14). Beaded-chain filaments are restricted in distribution to the lens cortical and nuclear fiber cells and are absent from the lens epithelium (189).

Immunochemical and SDS-PAGE analyses of isolated preparations have demonstrated two major urea-soluble proteins to be specifically associated with the lens beaded-chain filaments (205). Although most extensively studied in the chick lens, they have been observed to be present in the lenses of animals from all vertebrate classes examined (206). The first represents a polypeptide of 86,000–95,000 d, varying with the species (86,000–130,000 d), and displaying an isoelectric point of about 5.0. The second represents polypeptides of 43,000–47,000 d and displaying a series of isoelectric variants from pH 5.0 to 5.8. In the chick lens, the 86,000 and 43,000 d polypeptides are only found in elongating and mature fiber cells and are absent from the epithelial cells (205). The studies on the biosynthesis of these polypeptides in the organ-cultured chick lens corroborate this pattern of distribution, as their synthesis was not observed in the epithelial cells but was initially observed in elongating cortical fiber cells (70).

Antibodies produced against the chick 86,000 and 43,000 d proteins cross-react with homologous proteins of bovine and human lenses (Fig. 5). No cross-reaction was observed with other tissues, such as muscle, skin, or liver. It has been reported that actin forms the backbone of the bovine lens beaded-chain filaments (35), although this does not appear to be the case for the chick lens (194). The chick lens 43,000 d beaded-chain filament protein has been found to differ from actin by immunologic properties (205), by its greater molecular weight, and by differences in amino acid composition and isoelectric point (194). The beaded-chain filament proteins have been demonstrated to be phosphorylated (192).

The lens beaded-chain filaments (14) are believed to be associated with the plasma membrane and to function either as structural organizers of α-crystallin or as structural support for lens polyribosomes (35).

Microtubules: Microtubules (Table 5) have been demonstrated to be restricted in distribution to epithelial and cortical fiber cells in a wide variety of lenses, including the human lens (188). Their presence in the nucleus of the human lens has been reported (207), although not confirmed. Although microtubules in the lens are believed to be primarily associated with developmentally related events, they are more abundant in the mature fiber cells of the more movable cortical lens areas, where they are believed to function in a cytoskeletal role to maintain the elongated and curved shape of the fiber cells (188).

The microtubule protein, tubulin, has been demonstrated in cultured epithelial cells by immunofluorescence (84), as well as in epithelial and super-ficial cortical fiber cells of the intact chick lens (205) and by SDS-PAGE in the developing chick lens (205). Tubulin, however, has not yet been isolated from the lens.

Others: Spectrin (Table 5), the protein of the red blood cell membrane skeleton, has been detected by immunofluorescence in chick lens epithelium (96). Immunoautoradiographic analysis of SDS-PAGE gels of chick lens epithelial proteins demonstrated a correspondence between the lens urea-soluble 250,000 d membrane polypeptide and the α (band 1) subunit of erythrocyte spectrin (96).

Myosin, α-actinin, and tropomyosin (Table 5) have been detected in cultured (elongating) calf lens epithelial cells (84) by immunofluorescence in association with actin-containing stress fiber bundles.

The Role of the Cytoskeleton in Lens Fiber Cell Differentiation

Differentiation of lens epithelial cells into lens fiber cells involves proliferation, elongation of the cells, the elimination of cell nuclei and of all membranous organelles, and separation of the cells from the surrounding epithelium and capsule (22). Elongation of the cell bodies appears to be the most important factor in the differentiation of lens fiber cells (22). The morphologic and bio-chemical aspects of lens fiber cell elongation have been extensively studied both

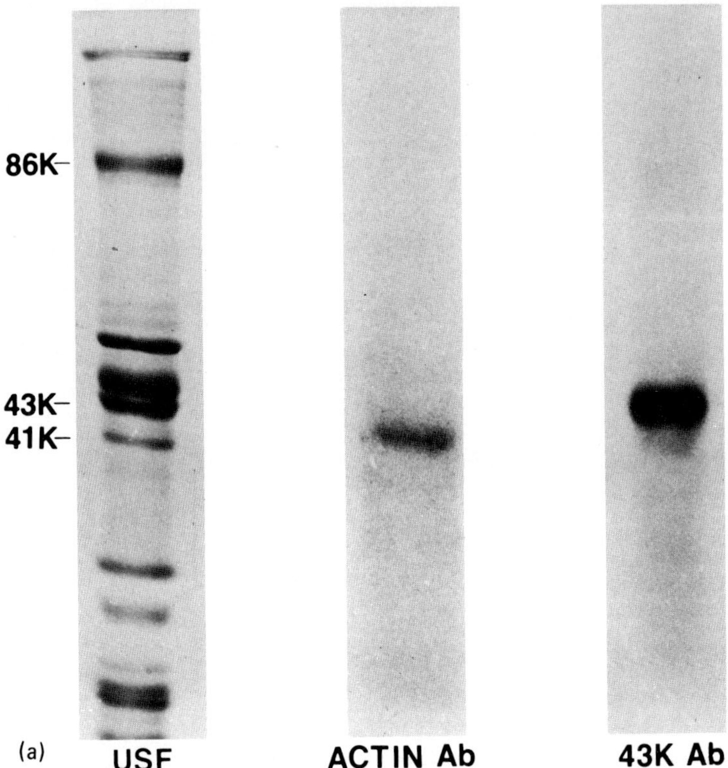

Figure 5   (a) Immunoblot analyses of antisera prepared against chick lens actin
(41,000 d polypeptide) and the beaded-chain filament (43,000 d) polypeptide.
USF: Coomassie blue-stained SDS-PAGE gel of the chick total lens 8 M urea-
soluble fraction. Actin Ab: Immunoblot of USF reacted with the antiserum
prepared against lens actin. 43K Ab: Immunoblot of USF reacted with the
antiserum prepared against the chick lens beaded-chain filament, 43,000 d poly-
polypeptide. The immunoblots were visualized by the peroxidase-antiperoxidase
(PAP) method.

in vivo (197,208) and in vitro (84,85,186,195,209–214). As cellular motility and
shape changes are known to involve a redistribution of cytoskeletal filamentous
cell components (185), the study of lens fiber cell elongation has centered on
the role of microfilaments, intermediate filaments, and microtubules in this
process (186,195).

Cultured lens epithelial cells elongate spontaneously into long, fiberlike cells
by a process believed to reflect the in vivo elongation of lens fiber cells (215).
The in vitro elongation of lens epithelial cells has been analyzed by specific

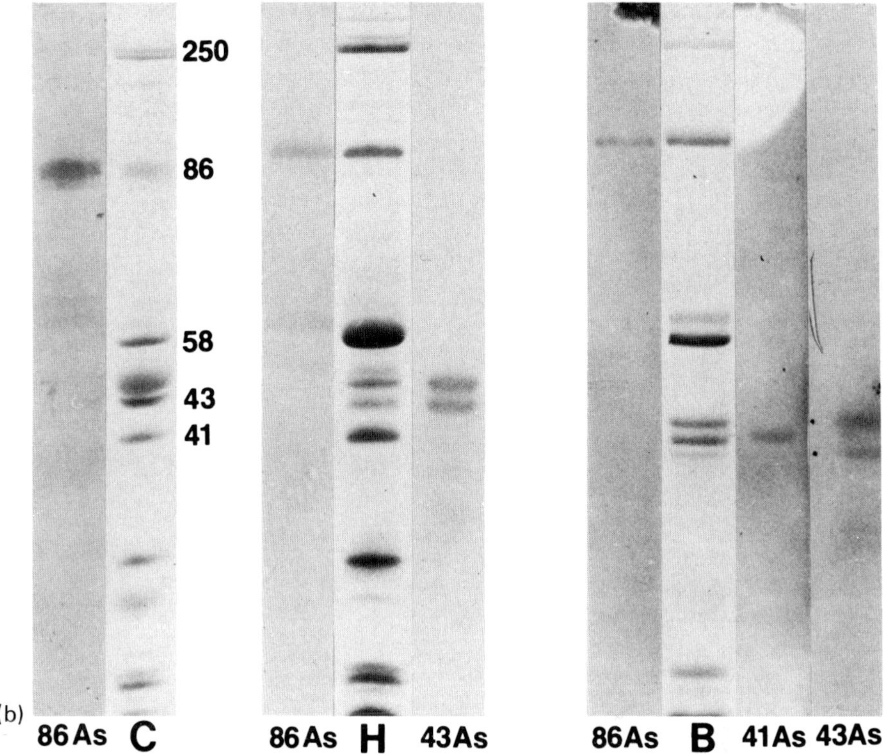

Figure 5 (b) Immunoblot analyses of antisera prepared against chick lens actin (41 As) and the beaded-chain filament 43,000 (43 As) and 86,000 d (86 As) polypeptides in reaction with total lens 8 M urea-soluble fractions from chick (C), human (H), and bovine (B) lenses. (C) H and B denote the Coomassie blue-stained SDS-PAGE gels of the corresponding 8 M USF fractions. 86 As, 43 As, and 41 As denote the immunoblots of the USF fractions reacted with the corresponding antisera. The immunoblots were visualized by the peroxidase-antiperoxidase method. Note that the antiserum prepared against the chick lens 86,000 d polypeptide cross-reacts with a corresponding polypeptide in the human and bovine lenses but the antiserum prepared against the chick lens 43,000 d polypeptide cross-reacts with two components of the human and bovine lenses. (Reproduced with the permission of Academic Press, Ltd.)

immunofluorescence and transmission electron microscopy and the molecular events studied by SDS-PAGE and isoelectric focusing (186). The involvement of microfilaments, intermediate filaments, and microtubules in the elongation of lens epithelial cells in culture has been demonstrated by specific immunofluorescence

(84,85,186,195,209). In particular, the primary event in the flattening and elongation of lens epithelial cells in culture involves the redistribution of the assembly of microfilaments into, first, a "geodesic dome" pattern in a perinuclear position and, later, into linear stress fiber bundles in close association with the plasma membrane (209) as the final flattening and elongation of the cells takes place. The elongation of the lens epithelial cells in culture is associated with a shift from G-actin to F-actin, an increase in total actin, primarily through increased synthesis of β-actin, a decrease in vimentin relative to actin, a decrease in tubulin synthesis, and the initial synthesis of myosin (209). The stress fiber bundles in elongated cultured lens epithelial cells have been demonstrated to contain actin, myosin, tropomyosin, and α-actinin (84, 186,209). The believed correspondence of in vitro lens epithelial cell elongation to the in vivo elongation of lens fiber cells has been questioned (209) on the grounds that stress fibers do not occur in lens fibers in vivo and on evidence that the in vitro elongation of epithelial cells involves the adaptation of the cells to the culture conditions and their transformation into fibroblastlike cells (210).

At least some of the features of the elongation of lens epithelial cells in vitro are similar to the elongation of lens fiber cells in vivo (197,208). Thus, the elongation of lens fiber cells in bovine lenses has been demonstrated to involve an increase in F-actin (208) but a decrease in the amount of vimentin. Changes in the relative amounts of β- and γ-actins appear to accompany lens fiber cell elongation in both bovine (208) and rat lenses (197). These observations are in agreement with observations on the intact chick lens (192). In the chick lens, there is an observed increase in the ratio of actin to vimentin when the poly-peptide pattern of epithelial and superficial cortical fiber cells are compared (192). Furthermore, in the chick lens there is the initial appearance of the 43,000–47,000 d polypeptide associated with the beaded-chain filaments.

Microtubules in the lens are believed to be associated with two develop-mentally distinct events: microtubules are known to be associated with the interkinetic migration of cell nuclei and the mitosis of cells in the lens placode (211). In addition, microtubules have been observed to be oriented parallel to the direction of cell elongation in chick lenses (212), and treatment with anti-microtubule agents was observed to disrupt elongation of cultured lens epithelial cells, suggesting their role in such elongation (213). However, cell elongation has recently been demonstrated in vitro, in the absence of micro-tubules, following treatment with the anti-microtubule drug nocodazole (214). Moreover, microtubules do not appear to increase in number in initially (actively) elongating fiber cells but instead appear to increase in terminally elongating cells deep in the bow region of the lens equator (22). This evidence and previous evidence of a decrease tubulin synthesis in the elongation process of cultured lens epithelial cells (209) support the conclusion that microtubules do not play an active role in lens fiber cell elongation.

INTERACTIONS AMONG LENS MEMBRANES, CYTOSKELETON, AND CRYSTALLINS

Important molecular interactions between the crystallin and noncrystallin components of the lens became the subject of intensive investigations, following their first demonstration by Maraini and Fasella (216). The in vitro demonstrations (216,217) that lens crystallins could bind to lens fiber plasma membranes were followed by the Lasser and Balazs (3) observations of strong interactions between the fiber cells' cytoskeleton, crystallins, and plasma membranes. The idea, first advanced by Maraini and Fasella (216), that such interactions in the intact lens may play a role in determining the structural organization of the various cytoplasmic components of the lens fiber cell, has just begun to be actively pursued (35).

Interactions Among Lens Membranes and Crystallins

Initial analyses of isolated lens fiber plasma membranes by SDS-PAGE demonstrated the presence of (soluble) crystallins in the membranes (17,19,30,60). Prior reports that insolubilized crystallins formed part of the lens water-insoluble fraction (for a review, see Ref. 7), and that a particular form of aged α-crystallin freely bound to lens fiber ghosts (217), raised the possibility that these crystallins were contaminants of purified lens fiber plasma membranes (17,39, 49). A stark crystallin and class specificity, however, was noted in that only α-crystallin polypeptides in mammalian lenses (17–19,30) and δ-crystallin polypeptides in avian lenses (60), and no other crystallin polypeptides, were consistently detected electrophoretically and immunologically in the purified membranes. Significantly, any soluble lens crystallin contamination of the isolated membranes would have been expected to be representative of the total soluble crystallin composition of the lens (i.e., β- and γ-crystallins as well). The consistent observations of the presence of specific lens crystallins in otherwise purified preparations of fiber plasma membranes, even after their extensive washing with buffers, treatment with 8 M urea, and mild proteolytic digestion (30), suggest that crystallins constitute a significant component of the lens fiber plasma membranes. An identical association of hemoglobin with the membrane in intact erythrocytes has been recently demonstrated (218).

MP50

The isolation of δ-crystallin with fiber membranes from a region of the chick lens cortex (outer cortex) (102) where no cytoplasmic (soluble) δ-crystallin is detected ruled out the possibility that this component constituted a methodologic contaminant and confirmed the original observations. The immunologic identity of the membrane component with that of soluble δ-crystallin was demonstrated in that study (102). As δ-crystallin was detected immunologically only following total detergent solubilization of the membranes, it was thought to be an intrinsic membrane protein. This last conclusion, however, appears to have been misleading.

Zelenka et al. (27) attempted to demonstrate that soluble δ-crystallin significantly contaminates plasma membranes isolated from total chick lens cortex otherwise devoid of δ-crystallin as isolated by them. The use of sodium citrate, a known chelator of calcium, at high ionic strength in the isolation buffer used in that study, however, raised the possibility that the absence of δ-crystallin in the membranes was explainable if its association with the membranes was dependent on calcium. Chick lens fiber membranes isolated under calcium-chelating conditions do show a marked depletion of δ-crystallin, but membranes isolated in excess calcium retain significant amounts of δ-crystallin (103). δ-Crystallin has been thus shown to be an extrinsic protein of the chick lens fiber plasma membranes and its linkage to the membranes to be calcium dependent. Although both subunits (48,000 and 50,000 d) of δ-crystallin are observed to be associated with the membranes of fetal chick lenses, only the higher molecular weight subunit (MP50; Table 2) is associated with the adult chick lens fiber membranes (27,102).

MP20–22

α-Crystallin (MP20–22; Table 2) polypeptides are routinely recovered from isolated mammalian lens fiber plasma membranes (17–19,30) and their presence confirmed immunologically only following total detergent solubilization of the membranes (18). The crystallin specificity of the interaction has been demonstrated in experiments that show that only the newly synthesized α-A_2 chains of α-crystallin and no other chains of this crystallin interact with the membranes (104). The report also demonstrated that no aging or maturation processes as originally believed (217) are required for the interaction. The presence of an intact MP26 protein in the membranes appears not to be a prerequisite to the association, as its cleavage to MP22 does not reduce the interaction (104). A similar incorporation of newly synthesized α-crystallin polypeptide into reticulocyte plasma membranes has been reported (219).

Interactions Among Lens Membranes and Cytoskeleton

It is now a well-established observation in many cells and tissues that the cell's cytoskeletal elements form specific associations with the cytoplasmic surface of plasma membranes (220). Furthermore, a submembranous "membrane skeleton" has been analytically defined in the circulating red blood cells (for a review, see Ref. 174).

A number of components of the lens fiber cell cytoskeleton have been identified as interacting with the plasma membranes. Early ultrastructural analyses of isolated lens fiber plasma membranes demonstrated the attachment of microfilaments to the membranes (30), and subsequently newly synthesized lens actin (MP40; Table 2) (104) was shown to become associated with isolated lens fiber plasma membranes. Its unique association with the fiber plasma membranes has been demonstrated by specific immunofluorescence (198,199). Intermediate filaments are found throughout the cytoplasm of epithelial and cortical fiber cells and do not appear to have a membrane association as judged by immunofluorescence (199). However, electron microscopic and biochemical evidence has suggested an intimate association between intermediate filaments and the plasma membrane (101). The newly synthesized intermediate filament protein, vimentin (MP47–52; Table 2), has been demonstrated to become associated with lens fiber plasma membranes (101). The specific association of these components to the fiber plasma membranes resists their subsequent treatment with urea and does not appear to depend on an intact MP26 (104). The lens beaded-chain filaments have been demonstrated to be associated with the plasma membrane ultrastructurally (35) and its protein (MP83–95; Table 2) localized to a submembranous position by specific immunofluorescence (192). The presence of spectrin (MP200, 250; Table 2), and α-actinin (MP97; Table 2) in close association with the lens membranes has been demonstrated by specific immunocytochemistry (35,96).

Interactions Among Lens Cytoskeleton and Crystallins

Unique macromolecular interactions have been demonstrated to exist between the lens cytoskeleton and the soluble lens proteins (crystallins) (3,35). This interaction is believed to be instrumental in organizing the macromolecular architecture of the crystallins within the fiber cells (3,35,206,221). Initial ultra-structural observations of Lasser and Balazs (3) demonstrated the structural association of spherical particles (12–14 nm) of α-crystallin with a filamentous backbone not unlike the beaded-chain filaments first described by Wanko and Gavin (13) and later isolated by Maisel and Perry (14). However, the proportion (2–5%) of α-crystallin consistently recovered with the lens water-insoluble fraction, and the consistent coisolation of actin together with α-crystallin in the urea-soluble fraction (35), is also indicative of an association between the crystallins and actin microfilaments. Moreover, both actin and α-crystallin are retained by affinity columns of DNase-I and are coeluted only by high concentrations of guanidine hydrochloride (35). Finally, a stochiometric relation appears to exist in the proportions of newly synthesized α-crystallin and actin incorporating into the membrane-cytoskeleton complex following translation of lens polyribosomes (221).

Lens polyribosomes specific for MP26 and attached to an actin filamentous backbone, isolated from the bovine lens membrane-cytoskeleton complex (water-insoluble fraction) (35,71), have been described as "beaded-chain" filaments and have been demonstrated to have a close structural association with the plasma membrane (35). However, the cytoskeletal beaded-chain filaments of the chick lens have been demonstrated not to bind heavy meromyosin (189) and antibodies to actin shown not to cross-react with the beaded-chain filament backbone protein (205). Moreover, the spherical particles on the cytoskeletal beaded-chain filaments have been demonstrated to be smaller than polyribosomes (3,14). This evidence supports the conclusion that the "beaded-chain" filaments associated with lens polyribosomes may represent unique structures, unrelated to the cytoskeletal beaded-chain filaments (205). We propose that, to avoid confusion and to differentiate them from the actin-polyribosome complex, the term "beaded-chain filament" be reserved for the cytoskeletal structures.

Possible Intermediaries in the Interactions

The effectiveness of agents (Table 1; urea and guanidine hydrochloride) known to disrupt hydrogen bonding in the dissociation of the cytoskeleton from lens fiber plasma membranes is strongly suggestive that specific and sensitive protein conformations are involved in their association. As these agents solubilize the major lens cytoskeletal elements (222), it may be concluded that specific conformations of lens cytoskeletal proteins are essential to this interaction. The evidence obtained in other cells and tissues, however, indicates the existence of specific intermediary molecules mediating the association between the plasma membrane and cytoskeleton (220). Observations of the binding of hemoglobin to erythrocyte membranes (218) also indicate that the erythrocyte band 3 polypeptide may serve as an intermediary linkage of this cytosolic component to the membranes.

A number of identified extrinsic proteins of lens fiber plasma membranes have been suggested as possible intermediary molecular linkages of the lens cytoskeleton and/or crystallins to the membranes. Aside from spectrin (MP200, 250; Table 2), α-actinin (MP96–130; Table 2), and actin (MP40; Table 2), which are known to perform this function in other cells and tissues (96,218), the MP43 polypeptide is believed to be involved in this function (86). Its immunofluorescent localization on the membranes and its biochemical coextraction with high-molecular-weight aggregates of lens crystallins (86), suggest that it may be involved as an intermediary molecular linkage of soluble crystallins (or cytoskeleton) to the lens fiber plasma membranes.

The nature of the specific association of polypeptides of α-crystallin to the lens fiber plasma membranes is at present not very clear. α-Crystallin polypeptides are routinely observed to coextract with the EEP protein (MP32, 35; Table 2) (49,90), indicating some sort of association. Thus, EEP is believed to be involved as an intermediary in the association of this crystallin with the lens fiber plasma membranes (90). Newly synthesized α-crystallin polypeptides, however, are believed to become inserted into the lipid bilayer of the membrane, as their insertion resists subsequent urea and proteolytic treatments of the membranes (104,221). Observations on the insertion of newly synthesized crystallins into the plasma membranes are relevant to elucidation of the possible mechanism by which the cytoskeleton may influence the structural organization of soluble crystallins in the lens. Initially, newly synthesized α-A_2 subunits incorporate rapidly into the membranes, and with prolonged incubation, the α-B_2 and β-B_{1a} crystallin incorporate (101,221). The observations suggest that the initially incorporating α-A_2 chains of α-crystallin may serve as nucleation sites for further assembly of lens crystallins (101).

AGING CHANGES IN LENS MEMBRANES AND CYTOSKELETON

The proteins of the vertebrate lens are probably the longest lived proteins known in nature (223). The lens never sheds any of its cells, and new fiber cells, following their terminal differentiation, gradually undergo the loss of their nuclei and full complement of protein-synthesizing organelles (23). Protein synthesis in the lens, therefore, is restricted to the epithelium and superficial layers of fiber cells and is virtually absent in the lens nuclear fiber cells (224). In long-lived vertebrate species, such as the human, proteins of the lens nuclear fiber cells attain the same age as the individual and accumulate a number of age-related modifications. Some of the best studied to date involve increased aggregation and insolubilization, deamidation and C-terminal degradation, disulfide and nondisulfide cross-linkage, racemization of aspartic acid residues, photo-oxidation of tryptophan, development of nontryptophan fluorescence, and pigmentation (for a review, see Ref. 225).

Changes in Lens Membranes

Proteins

Polypeptide changes in the protein composition of lens fiber plasma membranes have been studied in bovine (81) and human (18,74,95) lenses. The major age-dependent change in human fiber membranes consists of the gradual conversion of MP26 to MP22 throughout the life span (18,74). The content of vimentin (MP54–57; Table 2) by human membranes steadily decreases with aging (18,95). The α-crystallin (MP20–22; Table 2) content of human membranes also decreases steadily, beginning with lenses 30 years and older (18,95). The MP43 (Table 2) protein increases in human fiber membranes in an age-dependent fashion (18,95). The human MP10–12 (Table 2) protein is absent in the membranes from fetal lenses, appears first at birth, and rises linearly in its relative weight fraction throughout the life span (18,95). Aging bovine lens fiber plasma membranes display a decrease in the EEP protein (MP32, 35; Table 2) and display a similar conversion of MP26 to MP22 to that in the human lens fiber membranes (81). The gradual conversion of MP26 to MP22 in human lenses gains special relevance with respect to the function and structural integrity of its fiber cell junctions, and to their lability during the development of senile cataracts; such a correlation has been demonstrated for the Nakano mouse cataract (159).

Aging human lenses display increased oxidation of methionine and cysteine residues of intrinsic membrane proteins (225). Whereas in the young (12–20 years old) human lens most of the methionine and cysteine are in the unoxidized state, in the old lens (60–65 years old) only the membrane intrinsic polypeptides and the urea-soluble high-molecular-weight insolubilized crystallins complexed with extrinsic components of the membranes are oxidized (225).

Lipids

Higher protein/lipid ratios are found upon aging in bovine (226) and human lens fiber membranes (227). Aging of fiber membranes is also associated with higher cholesterol/phospholipid molar ratios in both bovine (226) and human (125) lenses. The cholesterol/phospholipid molar ratio rises progressively from lens equator, cortex, and nucleus (226), and the concentration of sphingomyelin increases in parallel fashion (24,125). Unusually high concentrations of sphingomyelin are attained in the human lens upon aging (72–75% of phospholipid) (24). Concomitantly, there is an almost complete loss of phosphatidyl-ethanolamine and phosphatidylcholine with aging in human lenses (24). Unusually high concentrations of lysophospholipids are already present in the 70-year-old human lens (24). The human lens fiber plasma membranes have been reported to contain very low levels of polyunsaturated fatty acids (129) and to undergo no significant changes upon aging. However, significant increases in polyunsaturated (22:4 and 22:6) fatty acids in the human lens with aging have been reported recently (228).

Thus, the aging of human lens fiber plasma membranes is associated with higher protein/lipid and cholesterol/phospholipid ratios and increasing concentrations of sphingomyelin, presumably not through synthesis but by the loss of membrane phospholipids. The presence of phospholipase and lysophospholipase activities in all lens parts has been demonstrated (229). The most remarkable and potentially significant aging changes in the lipid composition of the membranes appear to be the loss of phospholipids and the conversion of the membranes into less fluid, more rigid structures through the retention of cholesterol, sphingomyelin, and intrinsic proteins. Although lipid peroxidation does not appear to be a major factor in the aging of lens fiber plasma membranes, its potential to disrupt important protein-lipid interactions and the molecular interactions of the membrane with the cytoskeleton and crystallins remains to be determined.

Structural Changes

The age-dependent changes in the composition and distribution of the protein and lipid components of the lens fiber plasma membranes should be reflected in concomitantly gradual but profound changes in the molecular architecture of the membrane and in the membranous process system interconnecting neighboring fiber cells. As the lens never sheds any of its cells, the continued differentiation of new fiber cells at the lens equator results in the internalization of fiber cells of increasing maturation and age. Thus, fiber cells of all stages of maturation can be observed in the adult lens (22). Ultrastructural studies (140,230) of the fiber cell surfaces in the intact lens demonstrate the progressive loss of the hexagonal cross-sectional fiber shapes, of their interlocking devices, and of their ordered alignment with increasing maturation and in the older nuclear fiber cells.

Changes in Lens Cytoskeleton

Many striking changes in the occurrence or distribution of lens cytoskeletal components have been demonstrated to be the result of aging (189,231). Microtubules are present in lens epithelium and are particularly abundant in the equatorial and cortical fiber cells but are held to be totally absent from the lens nuclear fiber cells, denoting a loss of microtubules with aging (188). Indeed, microtubules are markedly reduced in all areas where they are expected to be plentiful in aged human lenses (188). Microfilaments and intermediate filaments are observed in epithelial cells of human lenses at all ages (189). These and the initial appearance of beaded-chain filaments are also observed in superficial cortical fiber cells of the 24-year-old lens (189). The number of intermediate filaments decreases, and the beaded-chain filaments become more coiled and aggregated in the deep cortical fiber cells of the 24-year-old lens (189). The 80-year-old human lens fiber cells contain only aggregated beaded-chain filaments (189). Both microfilaments and intermediate filaments are reported to increase in the epithelial cells of aged rat and mouse lenses (231).

Many of the morphologic changes in the lens cytoskeleton observed with aging are reflected in the biochemical changes observed in cytoskeletal proteins (95,191,197,198,232). Actin, the protein of microfilaments, has been reported to be absent from rat lens nuclear fiber cells (197), although it has been demonstrated to be present by immunofluorescence (198). A striking change in its pattern of distribution in nuclear fiber cells, however, was observed in that study, in that it displayed an exclusively submembranous localization in the older nuclear fiber cells (198). Striking changes in the urea-soluble protein pattern of the avian and mammalian lenses are observed when comparing the cortical and nuclear fiber cells (191,232). In particular, significant reductions are observed in the content of actin and intermediate-filament polypeptides, and more strikingly in all high-molecular-weight cytoskeletal polypeptides, including those corresponding to spectrin (191,232). In the human lens, all cytoskeletal proteins are readily discernible up to the 14-year-old lens (95). Thereafter, there is a rapid loss of these components even from the cortex (95).

Thus, one of the most striking changes in the lens fiber cell cytoskeleton upon aging is the loss of intermediate filaments (189), so that in the 80-year-old lens, intermediate filaments are found only in the epithelial cells. The loss of intermediate filaments with age in the human lens is reflected in the disappearance of its corresponding polypeptide from the gel electrophoretic pattern of its urea-soluble protein (18,95), suggesting that the protein is degraded with age. Evidence for the presence in the lens of a Ca^{2+}-activated protease specific for the protein of intermediate filaments has been demonstrated (233). A Ca^{2+}-activated thiol proteinase specific for intermediate filaments has been found in many other cells and tissues (234).

CONCLUSIONS

The multiplicity of methods developed for the isolation of lens fiber cell plasma membranes (Table 1) have arisen from the investigators' desires to obtain "clean" lens fiber plasma membranes (i.e., totally free of lens soluble proteins and cytoskeletal elements). This approach, however, has led to the development of some methods that result in the total destruction of the important molecular interactions known to exist between the soluble proteins, cytoskeleton, and the plasma membranes in the normal lens. Important membrane molecular components involved in these interactions, and serving as anchoring intermediates between the cytoskeleton and plasma membrane, may be removed by the more chemically drastic treatments. On the other hand, residual soluble proteins and cytoskeletal components, contaminating the membranes and resulting from their incomplete removal, can result in misleading conclusions on the number and nature of the components involved in these interactions. As pathologic processes in the lens (i.e., cataractogenesis) may be caused by (or reflect) alterations in these interactions, proper interpretation of putative causes may be compromised.

Evidence is accumulating of topographic and consequently functional heterogeneity on the lens fiber cell surface membrane. The localization of Na^+,K^+-ATPase in the apical (polar) regions of lens fiber plasma membranes has been demonstrated by ultrastructural cytochemistry. Topographic membrane specialization of the apical fiber membrane is suggested by the presence of special features on the polar fiber regions of the rat lens (112); the presence of elaborate foldings of plasma membrane, microvillar processes, and cytochemically demonstrable ouabain-sensitive Na^+,K^+-ATPase is suggestive of the regional specialization of plasma membrane of apical lens fiber regions (112). Morphologic studies of lens fiber cell differentiation have revealed the progressive elaboration of regional membrane specializations (140) in the form of a variety of interlocking devices, the establishment of different membrane domains, the structural modulation of developing cell junctions, and the elaboration of extensive fiber membrane surface area as junctions.

Biochemically, the lipid compositional differences observed between lens fiber cell junctions and total fiber plasma membranes and the well-characterized unique protein composition of lens fiber cell junctions suggest that the lens fiber junctions constitute unique plasma membrane regions. Moreover, the findings of a detergent-resistant lipid fraction that interacts strongly with the MP26 protein in lens fiber junctions and of differences in the lipid composition of membrane detected in different lens biochemical fractions (235) are suggestive of the local accumulation of specific lipids in different fiber membrane areas and of a differential effect of aging changes in lens proteins and lipids upon the structure and function of the lens fiber plasma membrane. The evidence so far accumulated supports the conclusion that the lens fiber plasma membrane is not a unitary entity but a mosaic of different protein-lipid domains. Thus, considerable initial progress has been made in knowledge of the topographic distribution of proteins and lipids in the lens fiber plasma membrane, and on the organizational protein-lipid interactions in the intact membrane.

Differentiation of lens fiber cells entails the rapid synthesis of plasma membrane proteins and lipids, and evidence is accumulating to the effect that, following their insertion into the membrane, membrane constituents (proteins and lipids) may undergo further reassortments and modifications (35,37,71,74, 93,135). These postsynthetic modifications presumably could be related to the establishment of topographic heterogeneity in the lens fiber plasma membrane and to its now demonstrated interactions with the fiber cell's cytoskeleton. Proper interpretation of this evidence, however, awaits a more complete biochemical definition of the functional and structural mosaicism of the lens fiber cell membrane surface as exemplified by topographic domains specialized for membrane-cytoskeleton interactions, enzymes, interlocking devices, and intercellular junctions.

The evidence available at present suggests that the membrane-associated crystallins and cytoskeletal proteins have a role in the macromolecular organization of the intracellular constituents of the lens fiber cell. The lens-specific beaded-chain filaments are the cytoskeletal components most likely to fulfill this role. The remaining cytoskeletal components (microfilaments, intermediate filaments, microtubules, stress fibers, and the membrane skeleton) probably perform functions in the lens not unlike those functions that have been well characterized for them in other cells and tissues (see Lens Cytoskeleton), and are most probably adapted to the different accommodative mechanisms of vertebrate lenses. As current models of the structural basis of lens transparency do not exclude an ontogenetic role for the preferential association between lens crystallins and the plasma membrane-cytoskeleton complex (236), the following possible models may be suggested.

1. Specific membrane-associated cytoskeletal proteins may serve as nucleation sites for the further organizational polymerization of their corresponding cytoskeletal components, and further interaction between the crystallins and these cytoskeletal components in turn affect the structural organization of the lens soluble proteins.
2. Specific membrane-associated crystallin and cytoskeletal proteins may serve as nucleation sites for the further, independent, structural organization of these two lens components.

A third and potentially relevant model is suggested by the recent identification of spectrin as an important cytologic component of lens cells. Does there exist a subsurface, membrane skeleton in lens fiber cells akin to that of the red blood cell, and a separate cytoplasmic cytoskeleton? And, what are the organizational interactions between these two, and between them and the soluble proteins?

The evidence available at present can be used to support all these possibilities. Now that several putative anchorage (intermediary) molecules have been tentatively identified, the possibility exists for a future investigational sorting among these possible models.

ACKNOWLEDGMENTS

We wish to thank Mr. Jon Lawniczak for his assistance in the preparation of the silver-stained slab gels of the lens membranes and Dr. Jerome R. Kuszak for the use of the freeze-fracture replica of the lens fiber cell junctions. The previously unpublished portions of the authors' work were supported by grants EY 01855 to J.A. and EY 01417 to H.M. from the National Eye Institute, National Institutes of Health.

REFERENCES

1.   W. W. De Jong. In *Molecular and Cellular Biology of the Eye Lens* (H. Bloemendal, ed.), John Wiley and Sons, New York, 1981, Ch. 6, p. 221.
2.   S. Trokel. Invest. Ophthalmol. Vis. Sci., *1*: 493 (1962).
3.   A. Lasser and E. A. Balazs. Exp. Eye Res., *13*: 292 (1972).
4.   C. T. Mörner, Hoppe Seylers Z. Physiol. Chem., *18*: 61 (1894).
5.   E. L. Burky and A. C. Woods. Arch. Ophthalmol., *57*: 464 (1928).
6.   S. G. Waley. In *The Eye*; Vol. 1, (H. Davson, ed.), Academic Press, London, 1969, p. 299.
7.   J. J. Harding and K. J. Dilley. Exp. Eye Res., *22*: 1 (1976).
8.   S. G. Waley. Exp. Eye Res., *4*: 293 (1965).

9. Z. Dische, M. A. Hairstone, and G. Selmenis. Protides Biol. Fluids, *15*: 123 (1967).

10. Z. Dische. Exp. Eye Res., *11*: 338 (1971).

11. J. J. Harding. Exp. Eye Res., *8*: 147 (1969).

12. H. A. Kramps, H. J. Hoenders, and J. Wollensak. Biochim. Biophys. Acta, *434*: 32 (1976).

13. T. Wanko and M. A. Gavin. J. Biophys. Biochem. Cytol., *6*: 97 (1959).

14. H. Maisel and M. M. Perry. Exp. Eye Res., *14*: 7 (1972).

15. P. Racz, K. Tompa, and I. Pocsik. Exp. Eye Res., *29*: 601 (1979).

16. H. Bloemendal. In *Molecular and Cellular Biology of the Eye Lens* (H. Bloemendal, ed.), John Wiley and Sons, New York, 1981, Ch. 1., p. 1.

17. J. Alcala, N. Lieska, and H. Maisel. Exp. Eye Res., *21*: 581 (1975).

18. J. Alcala, J. Valentine, and H. Maisel. Exp. Eye Res., *30*: 659 (1980).

19. R. M. Broekhuyse and E. D. Kuhlmann. Exp. Eye Res., *19*: 297 (1974).

20. H. Maisel and J. Alcala. In *The Red Blood Cell and the Lens Metabolism* (S. Srivastava, ed.), Elsevier, New York, 1980, p. 213.

21. S. J. Singer. Annu. Rev. Biochem., *43*: 805 (1974).

22. T. Kuwabara. Exp. Eye Res., *20*: 427 (1975).

23. H. Maisel, C. V. Harding, J. Alcala, J. Kuszak, and R. Bradley. In *Molecular and Cellular Biology of the Eye Lens* (H. Bloemendal, ed.), John Wiley and Sons, New York, 1981, Ch. 2, p. 49.

24. R. M. Broekhuyse. In *The Human Lens in Relation to Cataract* (K. Elliot and D. Fitzsimons, eds.), Elsevier, Amsterdam, 1973, p. 135.

25. E. L. Benedetti, I. Dunia, C. J. Bentzel, A. J. M. Vermorken, M. Kibbelaar, and H. Bloemendal. Biochim. Biophys. Acta, *457*: 353 (1976).

26. M. A. Kibbelaar and H. Bloemendal. Exp. Eye Res., *29*: 679 (1979).

27. P. Zelenka, R. Reszelbach, and J. Piatigorsky. Biochim. Biophys. Acta, *556*: 447 (1979).

28. P. Russell, W. G. Robison, Jr., and J. H. Kinoshita. Exp. Eye Res., *32*: 511 (1981).

29. D. Roy, A. Spector, and P. N. Farnsworth. Exp. Eye Res., *28*: 353 (1979).

30. H. Bloemendal, A. Zweers, F. Vermorken, I. Dunia, and E. L. Benedetti. Cell Differ., *1*: 91 (1972).

31. J. W. McAvoy. Differentiation, *17*: 85 (1980).

32. H. Roelfzema, R. M. Broekhuyse, and J. H. Veerkamp. Exp. Eye Res., *18*: 579 (1974).

33. A. J. M. Vermorken, J. M. H. C. Hilderink, I. Dunia, E. L. Benedetti, and H. Bloemendal. FEBS Lett., *83*: 301 (1977).

34. C. Tanford and J. A. Reynolds. Biochim. Biophys. Acta, *457*: 133 (1976).

35. L. Benedetti, I. Dunia, F. C. S. Ramaekers, and M. A. Kibbelaar. In *Molecular and Cellular Biology of the Eye Lens* (H. Bloemendal, ed.), John Wiley and Sons, New York, 1981, Ch. 4, p. 137.

36. R. M. Broekhuyse, E. D. Kuhlmann, and A. L. H. Stols. Exp. Eye Res., *23*: 365 (1976).

37. J. Horwitz, N. P. Robertson, M. M. Wong, J. S. Zigler, and J. H. Kinoshita. Exp. Eye Res., *28*: 359 (1979).

38.   R. M. Broekhuyse, E. D. Kuhlmann, and H. J. Winkens. Exp. Eye Res., *29*: 303 (1979).

39.   H. Bloemendal, A. J. M. Vermorken, M. Kibbelaar, I. Dunia, and E. L. Benedetti. Exp. Eye Res., *24*: 413 (1977).

40.   D. A. Goodenough. Invest. Ophthalmol. Vis. Sci., *18*: 1104 (1979).

41.   J. R. Kuszak, J. Alcala, and H. Maisel. Exp. Eye Res., *33*: 157 (1981).

42.   J. Alcala, P. Waggoner, R. Bradley, and H. Maisel. In *Immunology and Immunopathology of the Eye* (A. M. Silverstein and G. R. O'Connor, eds.), Masson, New York, 1979, Ch. 57, p. 319.

43.   L. J. Takemoto, J. S. Hansen, and J. Horwitz. Comp. Biochem. Physiol., *68B*: 101 (1981).

44.   R. M. Broekhuyse. In *Mechanisms of Cataract Formation in the Human Lens* (G. Duncan, ed.), Academic Press, New York, 1981, Ch. 6, p. 151.

45.   R. A. Capaldi and G. Vanderkooi. Proc. Natl. Acad. Sci. USA, *69*: 930 (1972).

46.   J. Horwitz and D. Bok. Invest. Ophthalmol. Vis. Sci. (Suppl.), *24*: 235 (1983).

47.   L. J. Takemoto and J. S. Hansen. Exp. Eye Res., *32*: 781 (1981).

48.   M. M. Wong, N. P. Robertson, and J. Horwitz. Biochem. Biophys. Res. Commun., *84*: 158 (1978).

49.   R. M. Broekhuyse and E. D. Kuhlmann. Exp. Eye Res., *26*: 305 (1978).

50.   R. M. Broekhuyse and E. D. Kuhlmann. Exp. Eye Res., *28*: 615 (1979).

51.   H. Maisel and J. Alcala. (Unpublished observations).

52.   A. A. Bouman, A. L. M. De Leeuw, and R. M. Broekhuyse. Exp. Eye Res., *32*: 491 (1981).

53.   J. Horwitz and M. M. Wong. In *The Red Blood Cell and the Lens Metabolism* (S. Srivastava, ed.), Elsevier, New York, 1980, p. 229.

54.   A. A. Bouman and R. M. Broekhuyse. Exp. Eye Res., *32*: 299 (1981).

55.   A. A. Bouman, A. L. M. de Leeuw, E. D. Kuhlmann, and R. M. Broekhuyse. Exp. Eye Res., *33*: 309 (1981).

56.   D. Sas, P. Keeling, K. Johnson, and R. Johnson. J. Cell Biol., *87*: 95a (1980).

57.   W. G. Niehaus, Jr., and F. Wold. Biochim. Biophys. Acta, *196*: 170 (1970).

58.   J. T. Dodge and G. B. Phillips. J. Lipid Res., *8*: 667 (1967).

59.   R. Mira y Lopez and P. Siekevitz. Anal. Biochem., *53*: 594 (1973).

60.   H. Maisel, J. Alcala, and N. Lieska. Doc. Ophthalmol., *8*: 121 (1976).

61.   J. Alcala, M. Katar, and H. Maisel. Curr. Eye Res., *9*: 569 (1983).

62.   J. S. Andrews and T. Leonard-Martin. Invest. Ophthalmol. Vis. Sci., *21*: 39 (1981).

63.   J. Folch-Pi and M. B. Lees. J. Biol. Chem., *191*: 807 (1951).

64.   I. Dunia, C. Shen Ghosh, E. Benedetti, A. Zweers, and H. Bloemendal. FEBS Lett., *45*: 139 (1974).

65.   J. Alcala and H. Maisel. Exp. Eye Res., *26*: 219 (1978).

66.   P. R. Waggoner and H. Maisel. Exp. Eye Res., *27*: 151 (1978).

67.   H. Rink. Biophys. Struct. Mech., *9*: 95 (1982).

68. J. S. Zigler, Jr., and J. Horwitz. Invest. Ophthalmol. Vis. Sci., *21*: 46 (1981).
69. M. Friedlander. In *Immunological Approaches to Embryonic Development and Differentiation*, Part II, Vol. 14, (A. A. Moscona and A. Monroy, eds.), Academic Press, New York, 1980, p. 321.
70. M. Bagchi, M. Ellis, and H. Maisel. Curr. Eye Res., *2*: 75 (1982).
71. F. C. S. Ramaekers, A. M. E. Selten-Versteegen, E. L. Benedetti, I. Dunia, and H. Bloemendal. Proc. Natl. Acad. Sci. USA, *77*: 725 (1980).
72. D. L. Paul and D. A. Goodenough. J. Cell Biol., *96*: 636 (1983).
73. L. J. Takemoto, J. S. Hansen, B. J. Nicholson, M. Hunkapiller, J.-P. Revel, and J. Horwitz. Biochim. Biophys. Acta, *731*: 267 (1983).
74. D. Roy. Biochem. Biophys. Res. Commun., *88*: 30 (1979).
75. D. Roy, L. Rosenfeld, and A. Spector. Exp. Eye Res., *35*: 113 (1982).
76. M. H. Garner, D. Roy, and A. Spector. Exp. Eye Res., *34*: 781 (1982).
77. J. Horwitz and M. M. Wong. Biochim. Biophys. Acta, *622*: 134 (1980).
78. R. M. Broekhuyse and E. D. Kuhlmann. Exp. Eye Res., *30*: 305 (1980).
79. P. Russell, and J. H. Kinoshita. Invest. Ophthalmol. Vis. Sci. (Suppl.), *19*: 150 (1980).
80. T.-S. Hu, P. Russell, and J. H. Kinoshita. Exp. Eye Res., *35*: 521 (1982).
81. A. A. Bouman, A. L. M. de Leeuw, and R. M. Broekhuyse. Exp. Eye Res., *31*: 495 (1980).
82. M. Katar, J. Alcala, and H. Maisel. Ophthalmic Res., *13*: 206 (1981).
83. E. Ruoslahti, E. Engvall, and E. G. Hayman. Coll. Res., *1*: 95 (1981).
84. F. C. S. Ramaekers, M. Osborn, E. Schmid, K. Weber, H. Bloemendal, and W. W. Franke. Exp. Cell Res., *127*: 309 (1980).
85. Y. Courtois, C. Arruti, D. Barritault, J. Tassin, M. Olivic, and R. C. Hughes. Differentiation, *18*: 11 (1981).
86. A. Spector, M. H. Garner, W. H. Garner, D. Roy, P. N. Farnsworth, and S. Shyne. Science, *204*: 1323 (1979).
87. P. N. Farnsworth, A. Spector, J. R. Lozier, S. E. Shyne, M. H. Garner, and W. H. Garner. Exp. Eye Res., *32*: 257 (1981).
88. D. Roy, E. Wada, and A. Spector. Invest. Ophthalmol. Vis. Sci. (Suppl.), *24*: 235 (1983).
89. P. Farnsworth, B. Lubit, and D. McCafferty. Invest. Ophthalmol. Vis. Sci. (Suppl.), *24*: 235 (1983).
90. A. A. Bouman, A. L. M. De Leeuw, E. F. J. Tolhuizen, and R. M. Broekhuyse. Exp. Eye Res., *29*: 83 (1979).
91. J. A. Lenstra, A. J. M. van Raaij, and H. Bloemendal. FEBS Lett., *148*: 263 (1982).
92. D. Roy. Exp. Eye Res., *32*: 533 (1981).
93. P. N. Farnsworth, S. E. Shyne, M. Garner, D. Roy, and A. Spector. Curr. Eye Res., *2*: 81 (1982).
94. W. H. Garner, S. R. Leff, and A. Spector. In *The Red Blood Cell and the Lens Metabolism* (S. Srivastava, ed.), Elsevier, New York, 1980, p. 367.
95. P. J. Ringens, H. J. Hoenders, and H. Bloemendal. Exp. Eye Res., *34*: 201 (1982).

96. E. A. Repasky, B. L. Granger, and E. Lazarides. Cell, *29*: 821 (1982).
97. L. J. Takemoto, J. S. Hansen, and L. E. Hokin. Exp. Eye Res., *35*: 337 (1982).
98. P. C. Sen and D. R. Pfeiffer. Biochim. Biophys. Acta, *693*: 34 (1982).
99. H. Maisel and M. Ellis. Curr. Eye Res., *3*: 369 (1984).
100. F. C. S. Ramaekers, I. Dunia, E. L. Benedetti, and H. Bloemendal. Proc. Int. Soc. Eye Res., *1*: 42 (1980).
101. F. C. S. Ramaekers, I. Dunia, H. J. Dodemont, E. L. Benedetti, and H. Bloemendal. Proc. Natl. Acad. Sci. USA, *79*: 3208 (1982).
102. J. Alcala, H. Maisel, and N. Lieska. Exp. Cell Res., *109*: 63 (1977).
103. J. Alcala, H. Maisel, M. Katar, and M. Ellis. Exp. Eye Res., *35*: 379 (1982).
104. H. Bloemendal, T. Hermsen, I. Dunia, and E. L. Benedetti. Exp. Eye Res., *35*: 61 (1982).
105. O. Hockwin and C. Ohrloff. In *Molecular and Cellular Biology of the Eye Lens* (H. Bloemendal, ed.), John Wiley and Sons, New York, 1981, Ch. 9, p. 367.
106. J. W. DePierre and M. L. Karnovsky. J. Cell Biol., *56*: 275 (1973).
107. J. Jedziniak, J. Rokita, M. Meys, and L. Arredondo. Invest. Ophthalmol. Vis. Sci. (Suppl.), *22*: 157 (1982).
108. P. Russell and J. H. Kinoshita. Invest. Ophthalmol. Vis. Sci. (Suppl.), *22*: 157 (1982).
109. M. Palva and A. Palkama. Exp. Eye Res., *19*: 117 (1974).
110. M. Palva and A. Palkama. Exp. Eye Res., *22*: 229 (1976).
111. N. J. Unakar and J. Y. Tsui. Invest. Ophthalmol. Vis. Sci., *19*: 630 (1980).
112. W. C. Gorthy and J. W. Anderson. Invest. Ophthalmol. Vis. Sci., *19*: 1038 (1980).
113. M. C. Neville, C. A. Paterson, and P. M. Hamilton. Exp. Eye Res., *27*: 637 (1978).
114. L. J. Takemoto, J. S. Hansen, and L. E. Hokin. Biochem. Biophys. Res. Commun., *100*: 58 (1981).
115. H. Maisel and J. Alcala. Invest. Ophthalmol. Vis. Sci. (Suppl.), *24*: 269 (1983).
116. A. Taylor, M. Daims, J. Lee, and T. Surgenor. Curr. Eye Res., *2*: 47 (1982).
117. W. C. Gorthy. Exp. Eye Res., *27*: 301 (1978).
118. K. R. Hightower, V. Leverenz, and V. N. Reddy. Invest. Ophthalmol. Vis. Sci., *19*: 1059 (1980).
119. N. J. Unakar, G. Price, and J. Tsui. Ophthalmic Res., *14*: 83 (1982).
120. I. Kabasawa and H. N. Fukui. Jpn. J. Ophthalmol., *21*: 348 (1977).
121. A. V. Johnson, W. A. Szarek, and D. J. Walton. Exp. Eye Res., *35*: 391 (1982).
122. R. Garadi, F. J. Giblin, and V. N. Reddy. Invest. Ophthalmol. Vis. Sci., *22*: 553 (1982).
123. R. Garadi, V. N. Reddy, P. Kador, and J. Kinoshita. Invest. Ophthalmol. Vis. Sci. (Suppl.), *24*: 235 (1983).

124. G. L. Feldman, T. W. Culp, L. S. Feldman, C. K. Grantham, and H. T. Jonsson, Jr. Invest. Ophthalmol., *3*: 194 (1964).

125. E. Cotlier, Y. Obara, and B. Toftness. Biochim. Biophys. Acta, *530*: 267 (1978).

126. R. J. Cenedella. J. Lipid Res., *23*: 619 (1982).

127. R. V. P. Tao and E. Cotlier. Biochim. Biophys. Acta, *409*: 329 (1975).

128. C. P. Sarkar and R. J. Cenedella. Biochim. Biophys. Acta, *711*: 503 (1982).

129. L. Rosenfeld and A. Spector. Exp. Eye Res., *35*: 69 (1982).

130. T. Yamakawa and Y. Nagai. Trends Biochem. Sci., *3*: 128 (1978).

131. R. M. Broekhuyse and J. H. Veerkamp. Biochim. Biophys. Acta, *152*: 316 (1968).

132. H. Roelfzema, R. M. Broekhuyse, and J. H. Veerkamp. Exp. Eye Res., *23*: 409 (1976).

133. P. S. Zelenka. Invest. Ophthalmol. Vis. Sci. (Suppl.), *18*: 66 (1979).

134. P. S. Zelenka. Invest. Ophthalmol. Vis. Sci. (Suppl.), *19*: 53 (1980).

135. P. S. Zelenka, D. C. Beebe, and D. E. Feagans. Science, *217*: 1265 (1982).

136. B. Alberts-Jackson and F. T. Bunch. Curr. Eye Res., *2*: 233 (1982).

137. C. Peracchia. Int. Rev. Cytol., *66*: 81 (1980).

138. D. A. Goodenough, J. S. B. Dick, II, and J. E. Lyons. J. Cell Biol., *86*: 576 (1980).

139. J. Kuszak, H. Maisel, and C. V. Harding. Exp. Eye Res., *27*: 495 (1978).

140. J. Kuszak, J. Alcala, and H. Maisel. Am. J. Anat., *159*: 395 (1980).

141. E. L. Hertzberg, D. J. Anderson, M. Friedlander, and N. B. Gilula. J. Cell Biol., *92*: 53 (1982).

142. G. Zampighi, S. A. Simon, J. D. Robertson, T. J. McIntosh, and M. J. Costello. J. Cell Biol., *93*: 175 (1982).

143. D. L. Paul and D. A. Goodenough. J. Cell Biol., *96*: 625 (1983).

144. J. Kistler and S. Bullivant. FEBS Lett., *111*: 73 (1980).

145. O. Traub and K. Willecke. Biochem. Biophys. Res. Commun., *109*: 895 (1982).

146. L. J. Takemoto and J. S. Hansen. Biochem. Biophys. Res. Commun., *99*: 324 (1981).

147. M. Finbow, S. B. Yancey, R. Johnson, and J.-P. Revel. Proc. Natl. Acad. Sci. USA, *77*: 970 (1980).

148. D. L. D. Caspar, D. A. Goodenough, L. Makowski, and W. C. Phillips. J. Cell Biol., *74*: 605 (1977).

149. D. Bok, J. Dockstader, and J. Horwitz. J. Cell Biol., *92*: 213 (1982).

150. B. J. Nicholson, M. W. Hunkapiller, L. E. Hood, J.-P. Revel, and L. Takemoto. J. Cell Biol., *87*: 200a (1980).

151. J. R. Kuszak, J. L. Rae, B. U. Pauli, and R. S. Weinstein. J. Ultrastruct. Res., *81*: 249 (1982).

152. C. Peracchia and L. L. Peracchia. J. Cell Biol., *87*: 719 (1980).

153. A. J. Verkleij and P. H. J. T. Ververgaert. Biochim. Biophys. Acta, *515*: 303 (1978).

154. D. Henderson, H. Eibl, and K. Weber. J. Mol. Biol., *132*: 193 (1979).

155. W. H. Evans and J. W. Gurd. Biochem. J., *128*: 691 (1972).

156. L.-K. Li, A. Spector, U. Cogan, and D. Schachter. Exp. Eye Res., *34*: 145 (1982).

157. J. S. Puskin and M. B. Wiese. Exp. Eye Res., *35*: 251 (1982).

158. J. S. Andrews, T. Leonard-Martin, and P. Kador. Invest. Ophthalmol. Vis. Sci. (Suppl.), *24*: 31 (1983).

159. M. Tanaka, P. Russell, S. Smith, S. Uga, T. Kuwabara, and J. H. Kinoshita. Invest. Ophthalmol. Vis. Sci., *19*: 619 (1980).

160. G. Duncan. In *The Human Lens in Relation to Cataract* (K. Elliot and D. Fitzsimons, eds.), Elsevier, Amsterdam, 1973, p. 99.

161. C. Peracchia and L. L. Peracchia. J. Cell Biol., *87*: 708 (1980).

162. G. Bernardini and C. Peracchia. Invest. Ophthalmol. Vis. Sci., *21*: 291 (1981).

163. G. Bernardini, C. Peracchia, and R. A. Venosa. J. Physiol. (Lond.), *320*: 187 (1981).

164. J. L. Rae, R. D. Thomson, and R. S. Eisenberg. Exp. Eye Res., *35*: 597 (1982).

165. K. R. Hightower, F. J. Giblin, and V. N. Reddy. Proc. Int. Soc. Eye Res., *1*: 28 (1980).

166. W. Y. Cheung. Science, *207*: 19 (1980).

167. E. L. Hertzberg and N. B. Gilula. Cold Spring Harbor Symp. Quant. Biol., *46*: 639 (1982).

168. M. J. Welsh, J. C. Aster, M. Ireland, J. Alcala, and H. Maisel. Science, *216*: 642 (1982).

169. S. J. Girsch and C. Peracchia. Invest. Ophthalmol. Vis. Sci. (Suppl.), *24*: 27 (1983).

170. C. Peracchia, G. Bernardini, and L. L. Peracchia. J. Cell Biol., *91*: 124a (1981).

171. L. A. Amos. In *Microtubules* (K. Roberts and J. S. Hyams, eds.), Academic Press, New York, 1979.

172. E. Lazarides. Nature, *283*: 249 (1980).

173. M. Clark and J. Spudich. Annu. Rev. Biochem., *46*: 797 (1977).

174. S. E. Lux. Nature, *281*: 426 (1979).

175. J. Prives, A. B. Fulton, S. Penman, M. P. Daniels, and C. N. Christian. J. Cell Biol., *92*: 231 (1982).

176. I. Blikstad, F. Markey, L. Carlsson, T. Persson, and U. Lindberg. Cell, *15*: 935 (1978).

177. J. I. Garrels and W. Gibson. Cell, *9*: 793 (1976).

178. R. C. Whalen, G. S. Butler-Browne, and F. Gros. Proc. Natl. Acad. Sci. USA, *73*: 2018 (1976).

179. P. A. Rubenstein and J. A. Spudich. Proc. Natl. Acad. Sci. USA, *74*: 120 (1977).

180. T. D. Pollard. J. Supramol. Struct., *5*: 317 (1976).

181. W. W. Franke, E. Schmid, C. Grund, and B. Geiger. Cell, *30*: 103 (1982).

182. R. M. Pruss, R. Mirsky, M. C. Raff, R. Thorpe, A. J. Dowding, and B. H. Andeston. Cell, *27*: 419 (1981).

183. K. Fujiwara and T. D. Pollard. J. Cell Biol., *71*: 848 (1976).

184. F. H. Kirkpatrick. Life Sci., *19*: 1 (1976).

185. E. Lazarides. J. Cell Biol., *68*: 202 (1976).

186. F. C. S. Ramaekers and H. Bloemendal. In *Molecular and Cellular Biology of the Eye Lens* (H. Bloemendal, ed.), John Wiley and Sons, New York, 1981, Ch. 3, p. 85.

187. N. S. Rafferty and W. Goossens. Exp. Eye Res., *26*: 177 (1978).

188. T. Kuwabara. Arch. Ophthalmol., *79*: 189 (1968).

189. R. h. Bradley, M. E. Ireland, and H. Maisel. Acta Ophthalmol. (Copenh.), *57*: 461 (1979).

190. H. Bloemendal. CRC Crit. Rev. Biochem., *12*: 1 (1982).

191. H. Maisel, J. Alcala, N. Lieska, and N. Rafferty. Ophthalmic Res., *9*: 147 (1977).

192. M. Ireland and H. Maisel. Curr. Eye Res., *7*: 961 (1984).

193. M. A. Kibbelaar, A. M. E. Selten-Versteegen, I. Dunia, E. L. Benedetti, and H. Bloemendal. Eur. J. Biochem., *95*: 543 (1979).

194. M. Ireland and H. Maisel. Ophthalmic Res., *14*: 428 (1982).

195. M. O. Lonchampt, M. Laurent, Y. Courtois, P. Trenchev, and R. C. Hughes. Exp. Eye Res., *23*: 505 (1976).

196. M. Ireland, H. Maisel, and R. H. Bradley. Ophthalmic Res., *10*: 231 (1978).

197. G. Y. Mousa and J. R. Trevithick. Exp. Eye Res., *29*: 71 (1979).

198. M. A. Kibbelaar, F. C. S. Ramaekers, P. J. Ringens, A. M. E. Selten-Versteegen, L. G. Poels, P. H. K. Jap, A. L. Van Rossum, T. E. W. Feltkamp, and H. Bloemendal. Nature, *285*: 506 (1980).

199. F. C. S. Ramaekers, L. G. Poels, P. H. K. Jap, and H. Bloemendal. Exp. Eye Res., *35*: 363 (1982).

200. S. R. Gordon, E. Essner, and H. Rothstein. Cell Motility, *4*: 343 (1982).

201. N. Geisler and K. Weber. FEBS Lett., *125*: 253 (1981).

202. N. Lieska, J. Chen, H. Maisel, and A. E. Romero-Herrera. Biochim. Biophys. Acta, *626*: 136 (1980).

203. D. Barritault, Y. Courtois, and D. Paulin. Biol. Cellulaire, *39*: 335 (1980).

204. J. Hatfield, R. Skoff, H. Maisel, and L. Eng. J. Cell Biol., *98*: 1895 (1984).

205. M. Ireland and H. Maisel. Invest. Ophthalmol. Vis. Sci. (Suppl.), *22*: 147 (1982).

206. R. Bradley and H. Maisel. Experientia, *34*: 470 (1978).

207. P. N. Farnsworth, S. S. Shyne, S.-J. Caputo, A. V. Fasano, and A. Spector. Exp. Eye Res., *30*: 611 (1980).

208. F. C. S. Ramaekers, T. R. Boomkens, and H. Bloemendal. Exp. Cell Res., *135*: 454 (1981).

209. F. Ramaekers, P. Jap, G. Mungyer, and H. Bloemendal. Curr. Eye Res., *2*: 169 (1982).

210. J. A. Lenstra, M. W. A. C. Hukkelhoven, A. A. Groenveld, R. A. M. M. Smits, P. J. J. M. Weterings, and H. Bloemendal. Exp. Eye Res., *35*: 549 (1982).

211. T. L. Pearce and J. Zwaan. J. Embryol. Exp. Morphol., *23*: 491 (1970).
212. B. Byers and K. R. Porter. Proc. Natl. Acad. Sci. USA, *52*: 1091 (1964).
213. J. Piatigorsky, H. DeF. Webster, and M. Wollberg. J. Cell Biol., *55*: 82 (1972).
214. D. C. Beebe, D. E. Feagans, E. J. Blanchette-Mackie, and M. E. Nau. Science, *206*: 836 (1979).
215. J. Piatigorsky. Ann. N.Y. Acad. Sci., *253*: 333 (1975).
216. G. Maraini and P. Fasella. Exp. Eye Res., *10*: 133 (1970).
217. P. G. Bracchi, F. Carta, P. Fasella, and G. Maraini. Exp. Eye Res., *12*: 151 (1971).
218. M. Sayare and M. Fikiet. J. Biol. Chem., *256*: 13152 (1981).
219. A. J. M. Vermorken, M. A. Kibbelaar, J. M. H. C. Hilderink, and H. Bloemendal. Biochem. Biophys. Res. Commun., *88*: 597 (1979).
220. B. Geiger, A. H. Dutton, K. T. Tokuyasu, and S. J. Singer. J. Cell Biol., *91*: 614 (1981).
221. F. C. S. Ramaekers, A. M. E. Selten-Versteegen, and H. Bloemendal. Biochim. Biophys. Acta, *596*: 57 (1980).
222. H. Maisel. Exp. Eye Res., *25*: 595 (1977).
223. J. S. Zigler Jr. and J. Goosey. Trends Biochem. Sci., *6*: 133 (1981).
224. C. F. Wannemacher and A. Spector. Exp. Eye Res., *7*: 623 (1968).
225. H. J. Hoenders and H. Bloemendal. In *Molecular and Cellular Biology of the Eye Lens* (H. Bloemendal, ed.), John Wiley and Sons, New York, 1981, Ch. 7, p. 279.
226. R. M. Broekhuyse, E. D. Kuhlmann, J. Bijvelt, A. J. Verkleij, and P. H. J. T. Ververgaert. Exp. Eye Res., *26*: 147 (1978).
227. R. M. Broekhuyse, E. D. Kuhlmann, and P. H. K. Jap. Ophthalmic Res., *11*: 423 (1979).
228. D. Chand and S. D. Varma. Invest. Ophthalmol. Vis. Sci. (Suppl.), *24*: 31 (1983).
229. R. M. Broekhuyse and F. J. Daemen. In *The Eye*, Vol. 2, (F. Snyder, ed.), Plenum Press, New York, 1977, p. 145.
230. H.-E. Hoyer. Cell Tissue Res., *224*: 225 (1982).
231. N. Rafferty, W. Goosens, and A. Roth. Ophthalmic Res., *11*: 276 (1979).
232. S. Nasser, R. Bradley, J. Alcala, and H. Maisel. Exp. Eye Res., *30*: 109 (1980).
233. M. Ireland and H. Maisel. Invest. Ophthalmol. Vis. Sci. (Suppl.), *24*: 29 (1983).
234. W. J. Nelson and P. Traub. J. Cell Sci., *57*: 25 (1982).
235. J. Alcala, M. Katar, and H. Maisel. Proc. Int. Soc. Eye Res., *2*: 44 (1982).
236. M. Delaye and A. Tardieu. Nature, *302*: 415 (1983).

6

Lens Metabolism

HONG-MING CHENG AND LEO T. CHYLACK, JR.* / Harvard Medical School, Boston, Massachusetts

The term "metabolism" is defined as the "sum of the processes in the building up and destruction of protoplasm; specif: the chemical changes in living cells by which energy is provided for initial processes and activities and new material is assimilated to repair the waste" (1). Although this is an excellent general definition, a considerably more restricted discussion of anabolic and catabolic processes in the lens is covered in this chapter. Much of what is written in the preceding chapters of this book legitimately might be considered "metabolism," so this chapter focuses on control mechanisms of glycolysis, the major source of ATP in the lens, and the interaction of this pathway wtih the hexose monophosphate shunt, the sorbitol pathway, and oxidative phosphorylation. We will try to present evidence supporting actual and potential associations between a metabolic disorder and lens opacification. Motivating most lens scientists is a wish to understand how normal metabolism contributes to lens clarity and how abnormal metabolism leads to cataract formation.

As in many other tissues, the metabolic processes of the lens are concerned with fueling cell division and growth, active transport of cations, amino acids, and myoinositol, and protein and lipid synthesis.

The primary purpose all metabolic processes subserve in the lens is to maintain lens clarity (i.e., the integrity of lens structure so that a macroscopically uniform refractive index within a given zone remains unperturbed). Failure of lens metabolism ultimately results in the formation of opacities or cataracts. This has been amply illustrated with experimental cataract models, such as the "hypoglycemic" cataract (2,3), cataracts induced with oxidants (4-6), and hereditary cataracts (e.g., Nakano and Philly mouse) (7,8). An overstimulated metabolism also causes cataracts; such is the case for the diabetics. Far less clear,

*Brigham and Women's Hospital, and Massachusetts Eye and Ear Infirmary, Boston, Massachusetts

however, are the cataractogenic factors in the human aging lens. Most likely they involve the decompensation of already diminished metabolic processes in the aging lens. For example, substantial loss of hexokinase (HK) in the cortex-nucleus has been reported in the clear, aging human lens (9). This loss of HK could conceivably retard the glycolytic rate due to the unique position of HK in regulating glucose flux. Such a decrease may affect the aging lens's ability to control ionic balance, to provide sufficient cofactors, such as NADPH, and to provide the carbon skeleton for various biosyntheses.

Rather than being static, the character of lens metabolism evolves with age; there is a decline in glycolytic rate with age in the rat lens (10). In 1961, Kinoshita et al. (11) demonstrated that young calf lenses could sustain normal ATP, Na^+, and K^+ contents and amino acid incorporation into protein in the absence of oxygen as long as sufficient glucose was present. Many other studies (12–15) have supported the belief that the Embden-Meyerhof pathway, by which glucose is metabolized, is the principal energy-producing pathway in the young lens. Trayhurn and van Heyningen (16–19) showed quite clearly that, as the calf lens ages, aerobic metabolism assumes relatively greater significance than anaerobic glycolysis in the production of ATP. The control of sugar metabolism in the crystalline lens has been the subject of two review articles (20,21).

Glucose is transported into the lens by facilitated transport (22), heretofore believed to be an insulin-independent transport process. Recently, Coulter et al. (23) have demonstrated, with a sensitive radioimmunoassay system, a level of insulin in primary aqueous humor equal to 3% of that in serum. This level increases 30% after feeding (compared with a 175% increase in plasma insulin) and is sustained for a longer period in aqueous than in serum. Insulin receptors exist in the lens, and insulin is a known mitogen (24), but it is not known whether insulin influences sugar metabolism in the lens.

The possibility of reversing the age-related decline in glycolysis, by bypassing the hexokinase bottleneck, was studied recently (9); it was found that HK and phosphofructokinase (PFK) activities in epithelium remained constant with age and did not decrease with cataract formation; both enzymes decreased, however, in the cortical-nuclear fraction. ATP levels were significantly lower in cataractous than in clear lenses and decreased rapidly in cataracts, but not clear lenses, incubated in isotonic media. Cataracts incubated in media containing glucose-6-phosphate (G6P) or fructose-1,6-diphosphate (FDP) produced significantly more ATP than with glucose in the medium (Table 1). This suggested that HK and PFK together constitute a bottleneck to full ATP synthesis from glucose in cataractous lenses. That this occurs in the face of a normal HK and PFK activity in the epithelium suggests that oxidative phosphorylation rather than glycolysis might be the source of ATP in the G6P- and FDP-supplemented lens.

Another intriguing study of the effects of age on lens metabolism was done by Orhloff et al.; they investigated the conversion of enzymatically active lens

Table 1 Production of ATP in Incubated Human Cataractous Lenses[a]

Incubation time (hr)	Addition	ATP (μmol/g lens)	P<
6	Glucose	0.67 ± 0.39 (6)	0.4
	G6P	0.37 ± 0.24 (6)	0.5
	FDP	0.68 ± 0.33 (10)	0.2
	Glucose/FDP	0.66 ± 0.13 (8)	0.4
	None	0.47 ± 0.23 (6)[b]	—
6	Glucose/ADP	0.67 ± 0.31 (6)	0.6
	G6P/ADP	0.90 ± 0.09 (6)	0.05
	FDP/ADP	1.07 ± 0.33 (15)	0.01
	ADP	0.57 ± 0.32 (7)[b]	—
18–24	Glucose/ADP	1.15 ± 0.23 (6)	0.3
	G6P/ADP	1.60 ± 0.62 (8)	0.05
	FDP/ADP	2.23 ± 0.94 (7)	0.01
	ADP	0.98 ± 0.28 (7)[b]	—

[a]Values are mean ± SD. Number of lenses in parentheses.
[b]Background values for statistical comparisons.

enzymes to enzymatically inactive, immunologically reactive proteins (25). Using quantitative immunologic titration and enzyme-specific antibodies, inactive aldolase and PFK molecules could be detected in 6-year-old rabbit lenses. They also found changes in the heat sensitivity and kinetic properties of enzymes in the aging lens (26,27). The posttranslational changes that lead to enzyme inactivation have not been elucidated; for nonenzymatic lens proteins nonenzymatic glycosylation is one such example. The effects of such deactivation of HK and PFK might be relevant to the age-related decrease in overall lens glycolysis.

LENS EPITHELIUM AEROBIC METABOLISM

An interesting study of human lens epithelium in normal and cataractous lenses (28), using a Nikon 20x dipping-cone objective, revealed little variation in cell size among transparent lenses aged 40–70 years (mean size = 257 μm^2, with diameters ranging from 12.5 to 22.5 μm and cell density = 3900 ± 220 cells per mm^2). In cataractous lenses from a slightly older population (52–92 years), the cell diameter and density of the epithelium varied greatly by cataract type. The central epithelium was separated from the capsule in many lenses by dense subcapsular vacuolization, but the peripheral epithelium appeared entirely normal in these cataracts. The authors speculated that the subcapsular lens edema was due to insufficient epithelial transport function.

In the older human lens there is a decrease in epithelial intracellular organelles and a condensation of chromatin (29) which suggests decreased metabolic activity. The dense lamina on the internal aspect of the nuclear envelope may indicate also diminished nuclear activity. With aging there is a steady accumulation of cytoplasmic granules (30), which may be crystallin protein complexes. In addition, there is an increase in cytoplasmic filaments with aging (31). In senility, epithelial cells assume a shriveled appearance (32). However, Perry et al. (29) showed that aged human epithelial cells in culture reenter the proliferative phase and show chromatin, nucleoli, and polysome contents more typical of much younger cells. To supply the energy for this increased metabolic activity, many more mitochondria are formed and the number of cristae per mitochondrion increases. However, with continued in vitro maintenance these apparently rejuvenated epithelial cells undergo variable differentiation, so that very few retain the morphologic characteristics of young cells. That even a transient rejuvenation is possible is not surprising to someone who has studied the biochemical characteristics of the young and aged human lens epithelium—many enzymes retain normal activities, even in the advanced immature cataractous lens.

Other metabolic pathways, such as oxidative phosphorylation, the citric acid cycle, the hexose monophosphate shunt, glycogenolysis, and the sorbitol pathway, are also present in the lens. These pathways are under delicate controls and interact with one another. Thus, the overall metabolic activity of the lens probably never reaches a steady state.

The complex nature of lens metabolism is further illustrated by the presence of large quantities of odd, small molecules, such as inositol, taurine, and organophosphates such as α-glycerophosphate, phosphocreatine, glycerol-3-phosphoryl-ethanolamine, and glycerol-3-phosphorylcholine (33). Most of these are without clearly defined purpose or function. Nevertheless, progress in the past decade has elucidated the major metabolic pathways in the lens: glycolysis, the hexose monophosphate shunt, the sorbitol pathway, and their auxiliary pathways. Only in the mitochondria of epithelium are critic acid cycle and oxidative phosphorylation activity of major importance. Studies on the interaction of these pathways are expedited by the application of nuclear magnetic resonance (NMR) spectroscopy to the lens. This new technology is noninvasive and allows an overall view of phosphorus and carbon metabolism on a real-time basis.

The following is a review of our current understanding of major metabolic pathways in the crystalline lens.

GLYCOLYSIS

The lens derives most of its energy (more than 70%) from anaerobic glycolysis, in which 1 mol of glucose is utilized to generate 2 mol of ATP (Fig. 1). This pathway is controlled by three kinases: HK, PFK, and pyruvate kinase (PK). The

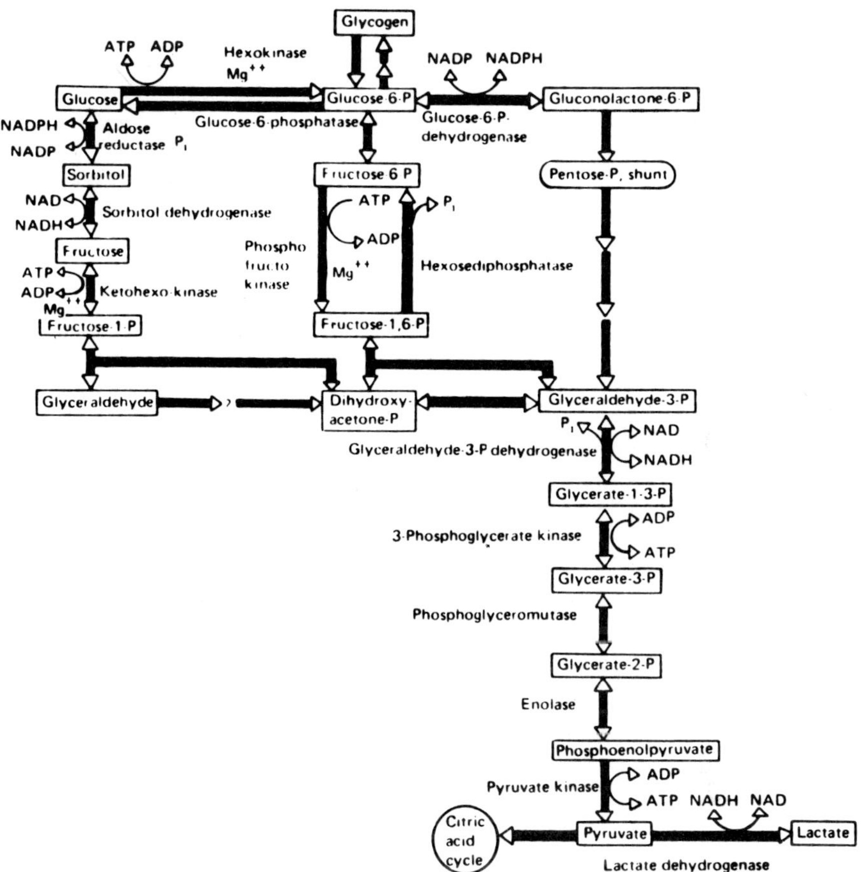

Figure 1 Major metabolic pathways in the crystalline lens.

regulation of glycolysis by these three enzymes is tightly coordinated and is based on the lens's need for ATP. In this regard, the lens is similar to other tissues. However, several characteristics of the enzymes, particularly those of hexokinase, are unique in that they may be the basis of the diminished energy production in the aging lens. The properties of these enzymes are described below.

Hexokinase (HK) (34–36)

There are at least four HK isozymes in mammalian tissues (13). Only types I and II are found in the lens. Type I exists in an active and latent form. The latter is bound to membranes and is released with such detergents as Triton X-100 and Tween 80.

In the human lens both types I and II are present in cells cultured from very young infant lenses (34); However, in the aging lens, the presence of type II isozyme, although demonstrable with a thermodeactivation study, cannot be detected with starch gel electrophoresis. Kinetically, the two forms differ in their affinity toward glucose. Type I has a K_m for glucose of 0.07 mM and type II, 1 mM; thus, the former is preferentially activated at physiologic levels of glucose. Nevertheless, type II comprises 70% of total soluble HK in the lens. In lenses incubated in high-glucose medium, glucose-6-phosphate accumulates, suggesting activation of type II HK isozyme.

The regulatory role of HK is based largely on the small amount of enzyme present in the lens relative to other glycolytic enzymes (37) and its sensitivity to the feedback inhibition by G6P. Green et al. (38) were the first to note that the rate of glycolysis in the lens was limited by the amount of hexokinase present. Both Harris et al. (39) and Patterson and Bunting (40) observed the same phenomenon; however, it was Pirie (41) who introduced the term "pacemaker" for this regulatory function of HK. Marked increases in free glucose, but not lactate, occurred when ambient glucose concentration was raised. In contrast to the abundant literature emphasizing the low activity of HK in the lens, there is a single publication by Gonzalez et al. (42) documenting a twofold increase in HK in the lens of rats with 4 weeks of alloxan diabetes. Similar increases occur in intestinal mucosa (43) and kidney cortex (44), and the exact purpose served by this increase is not clear. In the case of the lens, however, such an increase might decrease the availability of glucose to aldose reductase and result in less sorbitol formation—a definite benefit to the lens. In spite of the limit on the maximum flux of glucose through HK, there is variation of HK activity due to product inhibition by G6P. The decrease in G6P inhibition of HK is apparent when the hexose monophosphate shunt is activated by oxidants. The activated shunt can consume more than three times the normal output of G6P from HK. This is discussed further in the Hexose Monophosphate Shunt section. The lactate production rate does not decrease during shunt activation.

The two HK isozymes differ in their sensitivity to heat inactivation: type II is deactivated completely after 30 min at 37°C in the absence of glucose; type I is stable. This deactivation can be prevented by including as little as 2.0 mM glucose in the incubating medium (Fig. 2). Heat inactivation of HK may be the basis of the "hypoglycemic" cataracts observed in human infants (45–48). The opacities are typically lamellar, possibly reflecting the transient nature of hypoglycemic episodes and the slow recovery of HK activity. Rat lens incubated in media containing no glucose rapidly consumed endogenous glucose. ATP levels dropped to 40% of control levels during a 4 h incubation without glucose, and with both substrates depleted the total HK activity began to decrease

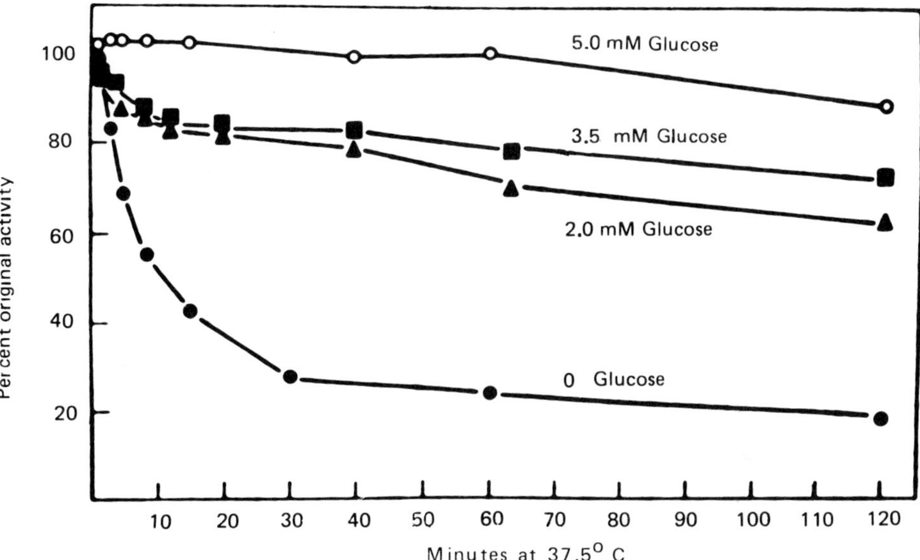

Figure 2 Heat sensitivity of rat lens hexokinase.

abruptly. By 8 h the hexokinase activity level was approximately 20% of control levels (Fig. 3). This resulted in further loss in K^+ ion and gain of Na^+ ion and water. Glucose deficiency affected HK more than the rest of glycolysis as shown by incubating homogenates of heat-treated lens with glucose or G6P and then measuring lactate production. The decrease in lactate production from glucose occurred concurrently with the loss of HK activity; there was only a gradual loss of lactate production if G6P was supplied. This suggested that HK was more susceptible to heat stress than other glycolytic enzymes. That hexokinase was primarily involved was shown by producing the identical cataract and metabolic changes in the glucose-deprived lens in the lens incubated with 2-deoxyglucose, a competitive inhibitor of hexokinase. By increasing the level of glucose, the cataract formed in the presence of 2-deoxyglucose could be prevented.

Neither endogenous free glucose nor glycogen could sustain the lens in the face of glucose deprivation. There appeared to be no alternative exogenous energy-yielding substrate in the lens. In a study by Barber et al. (49) of rat lenses incubated in a glucose-free balanced salt solution, rapid net proteolysis began when ATP levels were depleted. Glucose deficiency and deprivation are major experimental cataractogenic stresses; it remains to be seen if the same occurs in human lenses.

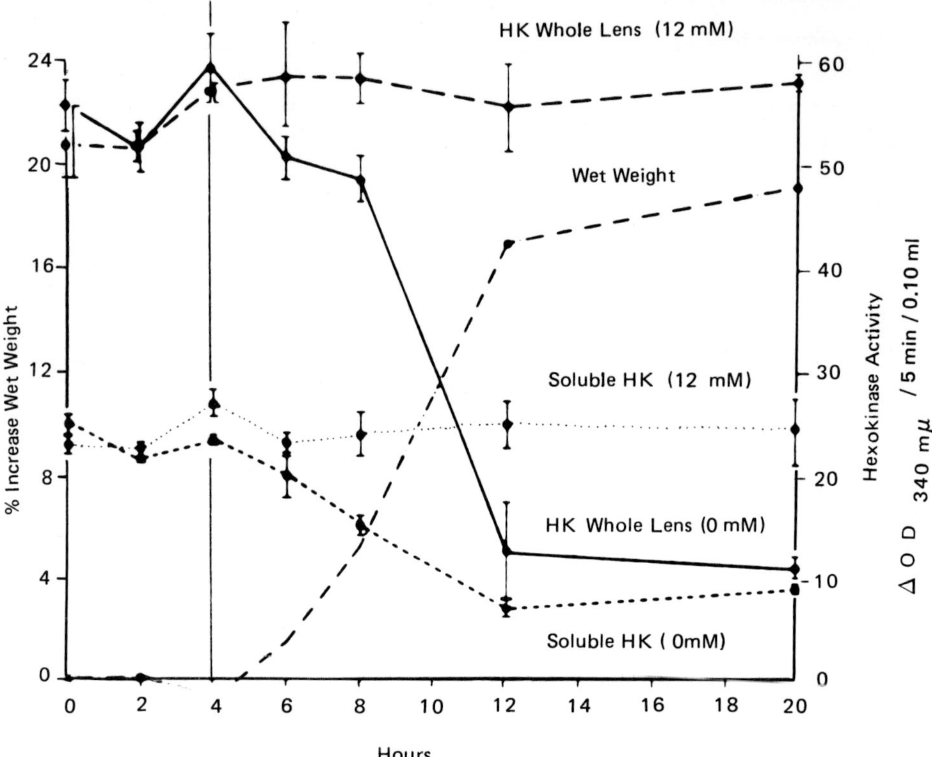

Figure 3   Decrease of hexokinase activity in the "hypoglycemic" lens.

Hockwin et al. (50) studied hexokinase in 5-year-old bovine lenses. They also found two forms with separate K_m values for glucose (K_{m_1} 6.2-6.7 $\times$ 10^{-5} M and K_{m_2} = 1.0-1.4 $\times$ 10^{-3} M) but postulated that the type 2 form may be an allosteric version of type 1 because (1) the behavior of type 1 can be described by the Hill equation with n = 2, and (2) the second K_m value disappeared when the ionic strength of buffer was increased. They proposed that type 2 might be a "stationary intermediate towards enzyme inactivation" developing from type 1 in the process of aging. This interpretation would conflict with that of Chylack et al., since they were able to demonstrate type II (type 2) only in very young lens cells. Hypoglycemic lenses showed a hazy lens surface, due to the destruction of the epithelial layer, and gradually developed cortical and nuclear opacities. In the rat lens model system the structure of the cataract resembled that of the human hypoglycemic cataract.

This heat inactivation can also be partially arrested with the inclusion of EDTA, suggesting that the heat sensitivity is a metal-mediated oxidation. HK is known to be oxidized and deactivated by diamide, a thiol oxidant (51). This raises an intriguing question: whether the loss of HK observed in the human aging lens cortex and nucleus is a result of oxidative stress. Decreased energy output could certainly result if extensive loss of HK occurred.

Phosphofructokinase (52–54)

PFK is present in the lens in two interconvertible forms: one dominates at pH 7.05 or lower (PFK-1), and the other above 7.05 (PFK-II). PFK-II is believed to be the functional form because of its extreme sensitivity to metabolic effectors.

PFK is apparently a fine-tuning mechanism for the control of glycolysis: it is inhibited by H^+, ATP, and Ca^{2+}, but activated by cAMP, ADP, AMP, NH_4^+, K^+, Pi, SO_4^{2-}, G6P, fructose-6-phosphate (F6P), and FDP. The final PFK activity therefore depends upon the combined effect of these effectors. However, the major effectors appear to be ATP, H^+, G6P, F6P, and possibly cAMP. Under normal physiologic conditions, the lens contains 2–3 mM ATP (55) and has an intracellular pH of 6.9 (33). Based on kinetic studies, this means that PFK would not be active unless it is activated by G6P and/or F6P, the products of HK catalysis. Wolfe et al. (56) have shown that PFK reaction rates measured as a function of F6P concentration can be described by the Hill equation:

$$V = \frac{V_{max}\,[F6P]^n}{[F6P]^n + K_s}$$

Values of K_m for F6P increase with increasing ATP concentration, and values of the Hill coefficient remain constant at 2 for all ATP concentrations studied. ATP inhibition is competitive with F6P and can be reversed by cyclic AMP. The cAMP level in rat lens is approximately 0.2 μM (57).

A mathematical model describing the velocity and modulation of reaction rates has been proposed by Wolfe et al. (56):

$$V = \frac{V_{max}\,[F6P]^n}{[F6P]^n + \left\{ K^\circ_{F6P\text{-}0.5} \left[\dfrac{1 + ([ATP]/K_{iATP})^m}{1 + ([cAMP]/K_{acAMP})} \right] \right\}^n}$$

The constants in this equation were obtained by curve fitting and show that ATP effectively controls PFK activity. One can use this equation to predict the PFK rate as a function of F6P, ATP, and cAMP concentrations and compare theoretical expectations with experimental results (56) with very good agreement. However, it should be noted that control at PFK step is more stringent than that at HK, since G6P accumulates when the lens is incubated in high-glucose medium (58). Coordination between HK and PFK is therefore mediated by the concentration of G6P and F6P.

Purified PFK is a tetramer with a molecular weight of 400,000 and elutes in the α-crystallin fraction. It is one of the largest enzymes in the lens: most other enzymes are eluted within the heavy β fraction. With gel filtration, Wolfe has shown that PFK undergoes enzyme-dependent aggregation (57), and work in other tissues—rat liver (59) and rabbit muscle (60)—shows that the affinity of F6P for PFK increases with increasing enzyme aggregation.

The development and further testing of the validity of mathematical models such as that just given will help to define which of the many factors regulating PFK (and perhaps HK and PK) activity are crucial under homeostatic conditions and which are altered in degenerative or disease states.

PFK monomers are biologically inactive but can be reaggregated in the presence of F6P at alkaline pH. On the other hand, PFK tetramers tend to dissociate into monomers at acidic pH and in cold. This dissociation can be prevented by ATP or F6P. In fact, PFK is one of a handful of cold-sensitive enzymes known to exist in nature. Cheng and Chylack have shown (51) that rat lens PFK is cold labile at acidic pH even in the presence of sulfate and inorganic phosphate, two known positive effectors. The inactivation is irreversible but can be prevented by ATP. PFK is also very unstable at 37°C at pH 7.15 in a desalted Tris-Cl buffer. Sufficient ATP (2–3 mM) and free sulfate (4–5 mM) are present in the lens to protect PFK against cold lability or heat lability at pH 7.40, but recent NMR studies showing the intracellular pH of the lens to be 6.95 (33) raise doubts about the security of PFK. Should the concentration of ATP or sulfate drop below the above-mentioned concentrations, this may lead to irreversible deactivation of PFK. In fact, this may have happened in the cortex of the adult human lens, where there is little HK or PFK activity (9).

Hockwin et al. (26) have shown that heat deactivation of PFK from the bovine nucleus can be mimicked by 5 krad of x-irradiation; this suggests that x-irradiation may be a cooperative risk factor in the aging process of the lens.

Pyruvate Kinase (PK) (61)

There are two major species of PK isozymes, muscle (M) and liver (L) types, in mammalian tissues. The L isozymes are indifferent to FDP activation, whereas the M isozymes are strongly affected by FDP and amino acids.

Lens PK is a variant of M_2 isozyme; it is totally inhibited by 2.5 mM ATP. Its affinity toward phosphoenolpyruvate (PEP) can be altered by preincubating the enzyme with FDP; for example, the K_m for PEP is changed from 0.267 to 0.037 mM with preincubation for PK isolated from human lenses. Activation of PK therefore depends upon PFK activity; both enzymes are controlled by the lens ATP contents.

Lens PK is also inhibited by L-amino acids, such as alanine, isoleucine, methionine, proline, threonine, phenylalanine, valine, and cysteine; this inhibition is counteracted by serine. The biological significance of amino acid inhibition and counterinhibition is unknown at present.

KETOHEXOKINASE (62)

An alternative pathway by which glucose might bypass the HK bottleneck and enter glycolysis is via the sorbitol pathway to fructose and then via ketohexokinase to fructose-1-phosphate (F1P) and aldolase to dihydroxyacetone phosphate (DHAP) and glyceraldehyde. The DHAP is metabolized as in glycolysis; the glyceraldehyde may be phosphorylated by triose kinase to glyceraldehyde-3-phosphate (GA3P) or oxidized to glycerate via aldehyde dehydrogenase and NAD. Glycerate could be phosphorylated to glycerate 2-P by glycerate kinase and then metabolized by glycolytic enzymes. Orhloff (62) has looked for all these enzymes in the older bovine lens but was unable to detect any glycerokinase or glycerate kinase. Ketohexokinase was known to be present (61), and aldehyde dehydroxygenase was found in whole lens and even the nucleus of very old lenses. So if fructose is to be an energy source for the lens, it must pass via ketohexokinase to F1P and then via aldolase to DHAP, which then is converted to glyceraldehyde-3-P by triosephosphate isomerase. Of course, the activity of ketohexokinase is usually low in the young rat lens and probably inadequate to support the lens in a glucose-free environment, as shown by Chylack (2,3).

THE HEXOSE MONOPHOSPHATE SHUNT

The hexose monophosphate shunt (HMPS) is depicted diagrammatically in Fig. 1. Not all the enzymes in the shunt have been demonstrated rigorously to be present in the lens; however, it is generally assumed to be a complete pathway in the lens that can consume about 14% of the glucose metabolized by the lens (63). The first two enzymes of the pathway, glucose-6-phosphate dehydrogenase (G6PD) and 6-phosphogluconic acid dehydrogenase (6PGD), have been studied extensively in the lens (64,65). The shunt competes with glycolytic enzymes for G6P and serves a reductive rather than a biosynthetic or oxidative (energy-yielding) role, contrary to other tissues. In the first two reactions, NADPH is produced, and it is this compound that is in demand by many other pathways in the lens (e.g., sorbitol pathway and glutathione reductase).

A major advance in our ability to study the HMPS was the development by Nisselbaum and Green (66) and the application to the lens by Giblin and Reddy (67) of a cycling assay system in which thiazolyl blue is employed as a terminal electron acceptor. NADPH reduces phenazine ethosulfate to NADP and phenazine ethosulfate (reduced). The electron is transferred from reduced phenazine ethosulfate to thioazolyl blue with the formation of formazan blue. The assay is simple and precise and accurately measures amounts of coenzyme as low as 0.01 nmol. Using this assay Giblin and Reddy found unexpectedly high levels of NADPH and NADP in lens epithelium (25 times higher than in cortex or whole lens). They speculated that these high levels were related to the high level of GSH and the high activity of GSH reductase in lens epithelium. The purpose served would be to maintain several key enzymes (e.g., HK) in a reduced state in an environment that is more oxidative than the rest of the lens.

This HMPS is activated whenever there is a demand for NADPH, for example, from aldose reductase or from glutathione reductase. This pathway is, therefore, closely associated with sugar cataractogenesis and the detoxification of H_2O_2 and related compounds in the lens.

Little is known regarding the component enzymes of the shunt. The first enzyme, G6PD, is stabilized by NADP(H) and is regulated by the NADP/NADPH ratio (64,65). This enzyme is present in excess in the lens. An experimental model employing steroids as G6PD inhibitors has shown that the correlation between G6PD activity and GSSG reduction is not linear; an 80% loss of G6PD activity still allows the generation of enough NADPH for more than 50% of the level of GSSG reduction in the controls (68). Thus, unless the enzyme loss is total or nearly total, the residual activity is probably not a limiting factor in the mechanism that affects disulfide reduction.

Activation of HMPS in conjunction with the activation of the sorbitol pathway has been known since the early 1960s. In general, the shunt is activated linearly with increasing glucose up to 50–60 mM, that is, at least 10 times the basal activity. Activation of the shunt with thiol oxidants, such as diamide, t-butylhydroperoxide (tBHP), and H_2O_2, however, is less. Giblin et al. (69) showed 7.7 times control with tBHP (Fig. 4), as if tBHP is removed from the medium, the activity of the HMPS reverts to control levels. At maximum stimulation the shunt turns over 100% of the lens NADP to NADPH every 48 sec. This results in the reduction of oxidized glutathione (GSSG) at the rate of 26% of the total GSH in the lens per hour. In this same study the distribution of HMPS activity in the lens was measured: 65% of total was present in the equatorial cortex (which made up 45% of total lens weight), 15% in the middle cortex, and 20% in the nucleus. The shunt, glutathione peroxidase, glutathione reductase, and NADPH combine to protect the lens against oxidative stress. However, there is a limit to the protective capacity of this system; this is shown by the inability of the lens to mobilize more than 50% of the HMPS capacity to offset the oxidative stress posed by 0.25 mM tBHP. The accumulation of GSSG was interpreted to mean that much of the unused shunt capacity was located in the inner lens and geographically unavailable to offset the oxidative stress of an exogenous oxidant.

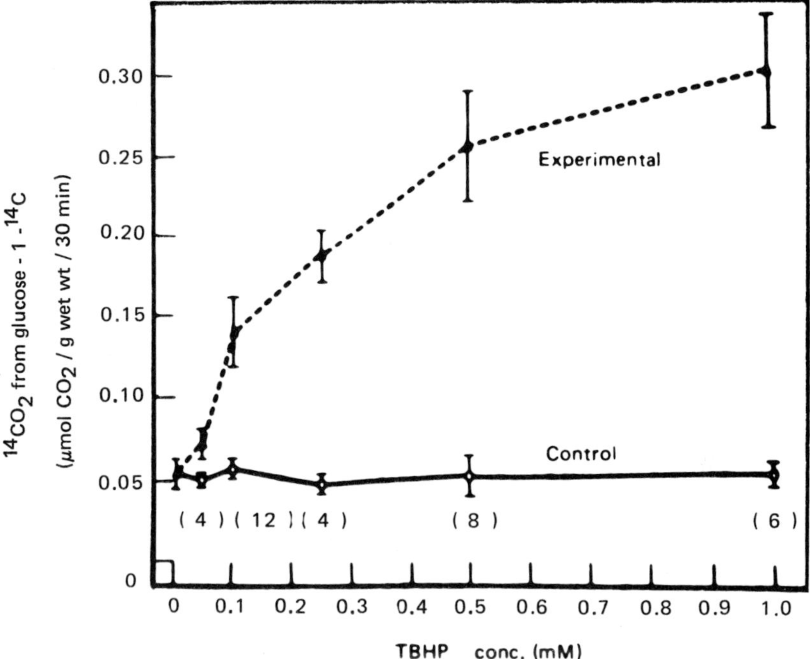

Figure 4 Activation of the lens hexose monophosphate shunt with *t*-butyl-hydroperoxide (tBHP).

In another elegant study, Giblin et al. (70) showed that the HMPS and GSH combine to offset an oxidative stress of 0.5 μmol of H_2O_2 per lens per hour. In this experiment GSSG did not accumulate, suggesting that the capacity of the HMPS near the lens surface was indeed sufficient to protect the lens.

The following sequence of reactions was proposed to be involved in protecting the lens against oxidative stress:

$$H_2O_2 + 2GSH \rightarrow GSSG + 2H_2O$$

$$GSSG + NADPH + H^+ \rightarrow 2GSH + NADP$$

$$HMPS \rightarrow NADPH$$

Increasing the concentration of oxidants tends to inhibit the shunt activity, suggesting damage to the shunt. Thus, in the lens, the HMPS seems limited in its ability to detoxify thiol oxidants, particularly H_2O_2. Detailed studies have shown that the shunt generates 0.4 μmol NADPH per rabbit lens every 3 hr in the presence of 0.05–0.07 mM H_2O_2 with a concomitant H_2O_2 removal of 1.46 μmol per lens every 3 hr (i.e., 27% of H_2O_2 is removed through the gluta-thione peroxidase-reductase-HMPS mechanism).

THE SORBITOL PATHWAY SUGAR CATARACTS

There are two enzymes in this pathway: aldose reductaste (AR), which requires NADPH, and polyol dehydrogenase (PD), which requires NAD^+ (Fig. 1). An excellent detailed review of the role of AR in the etiology of diabetic cataracts has recently been published (71). It covers in great depth the scientific developments leading to the strong association between AR activity and diabetic cataract formation. There is abundant evidence that AR in many animal lenses has a K_m of approximately 30 mM. When glucose levels are elevated, glucose is reduced by AR and NADPH to sorbitol, which accumulates within the cell whose cell membrane is impermeable to it. Galactose, another aldose, is similarly converted to galactitol (dulcitol). Sorbitol, but not galactitol, can be oxidized by the second enzyme of the pathway PD to fructose, which slowly diffuses out of the cell or, where ketohexokinase exists, may be converted to F1P. It is the rapidity with which sorbitol and galactitol are produced from their hexose precursors and the impermeability of the cell membrane to the sugar alcohol that create a hypertonic milieu within the cell; this leads to cellular swelling, rupture, and disintegration. Crabbe et al. (72) question the importance of sorbitol accumulation as the primary factor in sugar cataractogenesis, preferring to regard the depletion of NADPH as more important. They cite older data by Patterson and others (73–75) in which streptozotocin-diabetic rats fed a diet high in sugar and unsaturated fat show less extensive lens opacification even though sorbitol levels remain high. Unfortunately, they have no data on NADPH/NADP ratios in diabetic and nondiabetic lenses to substantiate this speculation.

That AR is important in sugar cataract formation is suggested by strong direct and indirect evidence. Varma and Kinoshita showed that congenitally hyperglycemic mice do not get "sugar" cataracts; the aldose reductase activity in this mouse lens is only a tenth that in the rat lens (76). Insufficient polyol is formed in these lenses to create any internal hyperosmolarity and osmotic stress. All other enzyme levels tested were similar to those in rat lens.

Inhibitors of AR (ARI) were developed and found to be effective in delaying or preventing sugar cataract formation. Varma and others (71,77) present a detailed account of the role of flavonoids as ARI. One of the first inhibitors to be tested was tetramethylene glutaric acid (78). Later in vivo testing of AY22284, another ARI, was found to be successful in delaying sugar cataract formation (79). Fukushi et al. showed that sorbinil, a new, extremely potent ARI could prevent, not just delay, cataracts in diabetic rats (80) (Fig. 5). A more recent study by Hu et al. (81) showed that sorbinil was as effective as removing galactose from the diet (82–84) in reversing the sugar cataract already formed in the lens. In another study, Gillis and Chylack (85) demonstrated that fasting normal, nondiabetic rats lead to a significant drop in lens fructose and sorbitol,

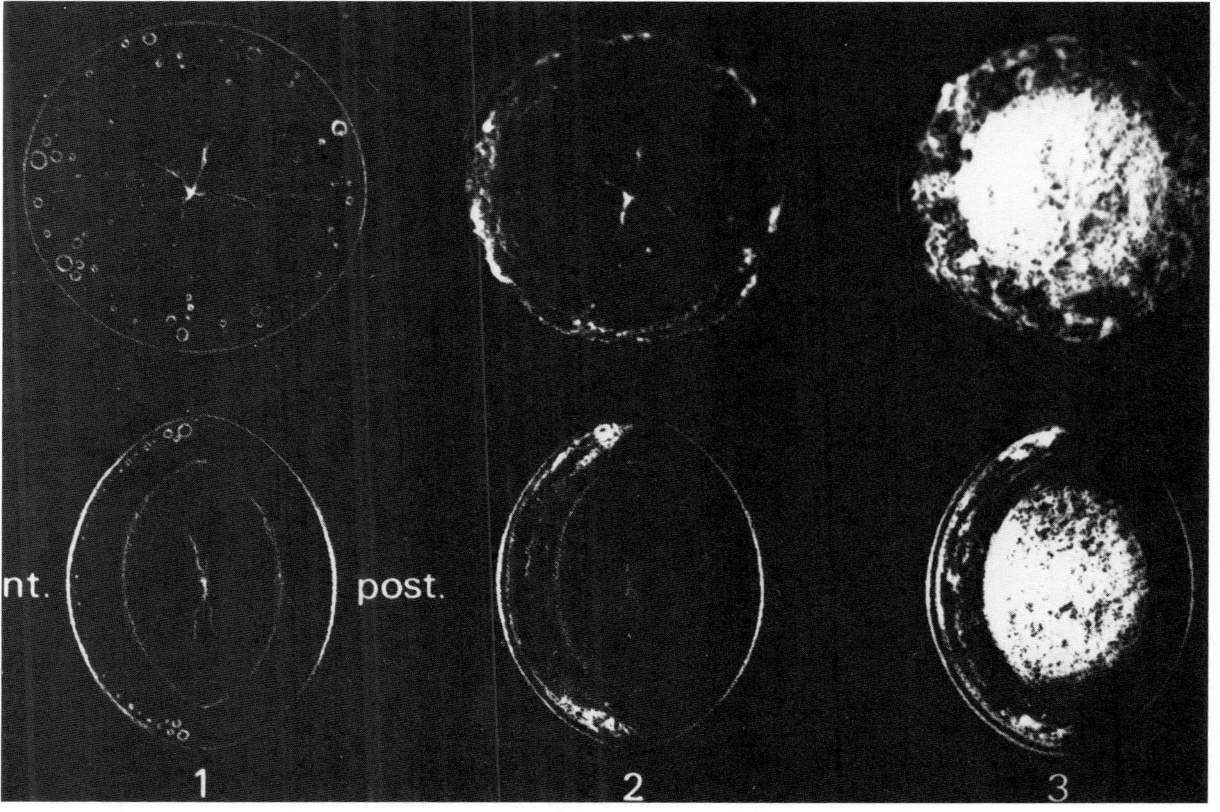

Figure 5 Progression of lens opacification in the diabetic rats: (1) 3 weeks, (2) 5 weeks, and (3) 6 weeks. Sorbinil treatment prevents cataract formation in these rats.

Table 2 Specific Radioactivity of NaB^3H_4 Incorporation in Galactosemic Rats Treated With and Without Sorbinil

Experiments	Days with Galactose feeding	Specific radioactivity (cpm/mg protein)		
		50% Galactose	50% Galactose + sorbinil[a]	Control
I	7	16,200 ± 600[b]	18,600 ± 400	—
	14	24,800 ± 400	23,300 ± 700	12,400 ± 400
	21	25,100 ± 600	26,000 ± 800	—
	28	30,000 ± 900	31,200 ± 800	11,700 ± 300
II	7	17,800 ± 1000[c]	17,200 ± 800	—
	28	36,200 ± 2300	41,400 ± 2100	15,200 ± 500

[a]None receiving sorbinil treatment had cataract or vacuoles developed in the lenses.

[b]The radioactivity incorporation is expressed as counts per minute (cpm) per milligram protein content (Lowry's protein determination). Each value shown is the mean of duplicate analyses from paired lenses.

[c]Each value shown in experiment II is the mean of analyses on three pairs of lenses using $[^3H]NaBH_4$ with a specific radioactivity different than that in experiment I.

suggesting that the metabolism of fructose by the lens may be another method of reducing the sorbitol accumulation.

In a report by Stevens et al. (86), it was suggested that nonenzymatic glycosylation of lens proteins, rather than sorbitol production and accumulation, was the principal cause of experimental diabetic cataracts. However, this suggestion was refuted by Chiou et al. (87) in a study of lenses from rats on a normal and high-galactose diet treated with and without sorbinil (Table 2). All galactosemic rats not on sorbinil had cataracts; none receiving sorbinil had cataracts. There was no difference in the extent of nonenzymatic glycosylation in lenses of galactosemic rats treated with and without sorbinil. It was concluded that nonenzymatic glycosylation was not responsible for the sugar-induced cataracts.

The extension of ARI from successful animal application to human use has been complicated by the findings of Kador et al. (88) and Jedziniak et al. (89). The former found significant differences in the susceptibility of human lens AR, human placental AR, and rat lens AR to inhibition by structurally distinct aldose reductase inhibitors. They correctly pointed out that clinical efficacy of AR inhibition must be based on testing an AR inhibitor on the enzyme of the particular target tissue. Furthermore, Jedziniak et al. conducted a detailed characterization of the sorbitol pathway in the human normal and diabetic lens (Table 3). They found that the amount of AR in human lenses was only 16–30% of the level in rat and rabbit lenses, two representative animal lenses with very

Table 3 Sorbitol Pathway in the Human Lens

	AR			Lens conc		PD (Sorbitol → fructose)			Lens conc
	K_m*	K_i*	V_{max}†	(mM)		K_m*	K_i*	V_{max}†	(mM)
Glucose:	200				Sorbitol:	1.45		0.19	
Nondiabetic lens				0.7-2.2	Nondiabetic lens				0.3-2.4
Diabetic lens				3.0-4.5	Diabetic lens				1.7-9.5
NADPH	0.06			0.02	NAD$^+$	0.06			0.57
NADP$^+$		0.07		0.02	NADH		0.02		0.21
+SO$_4$			4.8						
−SO$_4$			1.2						

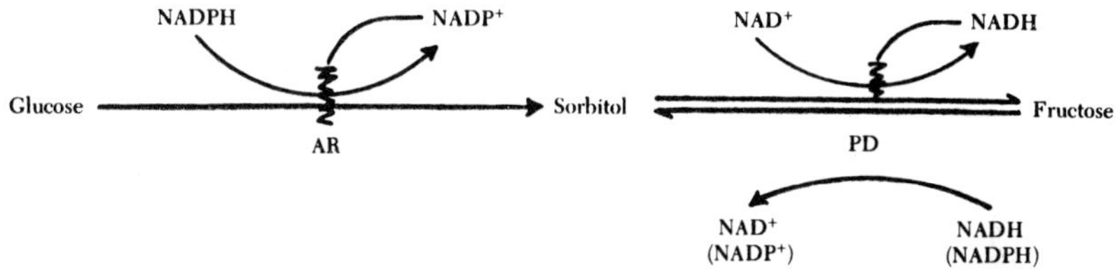

	PD (Fructose → sorbitol)			Lens conc
	K_m*	K_i*	V_{max}†	(mM)
Fructose:	40		1.9	
Nondiabetic lens				0.4-1.4
Diabetic lens				1.2-12.0
NADH	0.02			0.21
NAD$^+$				0.57
(NADPH)	0.27			0.02
(NADP$^+$)				0.02

*K_m + K_i expressed as mM.

†V_{max} μmol/min/lens.

low and very high AR activity. In addition they found a K_m for glucose of 200 mM for human lens AR. This means that glucose is converted to sorbitol in human lens only at very high levels of glucose, as might occur in poorly controlled diabetes. They also found AR to be confined primarily to the epithelium of the adult lens and in this region to be two to three times lower than the level in this region of the juvenile lens. The K_m for NADPH for human lens AR is 0.06 mM, and the concentration in the lens was 0.02 mM, enough to exert some inhibitory effect on AR. The authors concluded that the level of AR in juvenile lenses was sufficient to cause osmotic stress, but questioned whether such stress could develop in adult lens.

An even more surprising difference between the sorbitol pathway in the human and animal lens is the extremely high level of PD activity in the former. In the fructose to sorbitol direction the activity of PD in human lens is nearly 80 times more active than in the animal lens. Its K_m for fructose is 40 mM, and for NADH is 0.02 mM. These data suggest that sorbitol may be formed from fructose in the human lens.

It was concluded that human lens AR was capable, under optimal conditions, of producing at most 6–7 μmol sorbitol per lens every 24 hr. In young lens epithelium, at this rate of sorbitol production, an osmotic stress of 280–570 mOsM would be created every 24 hr per lens epithelium. Significantly less stress would be developed in the adult human lens epithelium.

Two studies (90,91) shed light on the ability of the human lens to produce and accumulate sorbitol in the face of 35.5 mM glucose; human diabetic and nondiabetic cataracts were incubated in high-glucose medium immediately after extraction. One initial finding was the markedly reduced level of AR in cataractous lenses. Even with this reduced amount of AR, clear and cataractous lenses accumulated significant levels of sorbitol and fructose. Considerable swelling occurred in control lenses (in 5.5 mM glucose) but even more occurred in high-glucose lenses. In fact, the swelling was sufficient to spontaneously rupture many of the high-glucose lenses. Alrestatin (AY22884 = 1,3-dioxo-1H-benz[de]-isoquinoline-2-[3H] acetic acid), an ARI, at a final concentration of 4×10^{-4} M, completely blocked net sorbitol accumulation, reduced fructose accumulation, and reduced the number of lenses rupturing in high-glucose medium. These data suggested that there was sufficient sorbitol production, even in the older lens, to produce a severe osmotic stress.

In a study of lens growth, cell hydration, and protein composition of neonatal rats fed a high-galactose diet (92), it was found that sorbinil preserved the normal features of all these parameters. Normal lens growth and fiber ultra-structure were preserved.

Laurent et al. (93) have studied the thickness and metabolism of lens capsules from 5-week-old prediabetic and 12-week-old diabetic *kk* mice. These mice have genetically determined diabetes that develops after 5 weeks of age. During the prediabetic state, the anterior lens capsules were 21% thicker ($p < 0.01$) than capsules from Swiss control mice. These changes occurred before the onset of elevated blood glucose and were believed to be genetic in origin. However, this difference increased to 103% in 12-month-old diabetic mice, suggesting that the duration of diabetes also played a role in determining capsule thickness. These data showed that basement membrane thickening occurs in diabetes both in vascular and avascular tissues. Incorporation of [^{3}H]proline into hydroxy[^{3}H]proline was greater in *kk* diabetic lens capsule and suggested an increase in collagen biosynthesis as the basis for increased capsular thickness, but there was also decreased noncollagenous glycoprotein synthesis. The authors infer from this that increased protein synthesis in general is not responsible for the increased basement membrane thickness.

In a study focused more on the effects of galactosemia than on lens protein synthesis, Kador et al. (94) demonstrated a depression of the synthesis of lens crystallins but not noncrystallin proteins. This effect was correlated with the influx of Na^+ and the loss of K^+ from the lens. There was a marked decrease in dry weight of galactosemic lenses, presumably due to the depressed crystallin synthesis. Removal of galactose from the diet led to a gradual (17 days) recovery of crystallin synthesis and a cessation of crystallin leak (5 days). These results were similar to those of Shinohara and Piatigorsky (95) and Piatigorsky et al. (96), who described altered crystallin synthesis in the chick lens in response to changes in the intracellular concentrations of Na^+ and K^+ and crystallin leak from cultured cataractous Nakano lenses.

Another elegant study on the effects of galactosemia on the rat lens (83) revealed marked drops in GSH, taurine, and other free amino acids; the rate of synthesis of GSH was not decreased. The biochemical changes occurring in the lens during the recovery phase (when galactose has been removed from the diet) are outlined and indicate that recovery is delayed long after the dulcitol level has returned to the level in control lenses. This is apparently due to the persistently high Na^+ concentration and delayed recovery of K^+. It was striking to note the reversal of a totally opaque, mature, galactose cataract to one in which only a pinhead nuclear opacity remained.

Another interesting development in the field of sugar cataract formation is the role of the sugar in altering tertiary protein structure. Liang and Chakrabarti (97) studied with circular dichroism (CD) of isolated, purified α-crystallin from bovine lenses treated with glucose and G6P (Fig. 6). They found three positive and two negative CD bands in the near-ultraviolet (UV) region. The positive ellipticities of these near-UV CD bands increased greatly upon incubation with glucose and G6P. The secondary structure did not change. The increase was not due to glycosylation. At higher glucose or G6P concentrations, the proteins aggregated. These data suggest that sugars can modify the tertiary structure of lens proteins and may contribute to the cataractous process in diabetics and galactosemics.

A recent study by Cheng et al. (98) suggested an antioxidant role for ARI. In addition to their well-known ability to block sorbitol production, they also reduce NADPH consumption. By preserving the NADPH pool, ARI increase the reserve of reducing equivalents in the lens and enhance its ability to overcome oxidative stress. If such work is further substantiated by work with human lenses, it may mean that ARI will be of use to nondiabetics and diabetics alike; the former will benefit from its antioxidant properties and the later from both the antioxidant and antisorbitol accumulation properties.

The past two decades have witnessed a continuing dedication by the lens community to the basic research on sugar cataract formation and the sorbitol

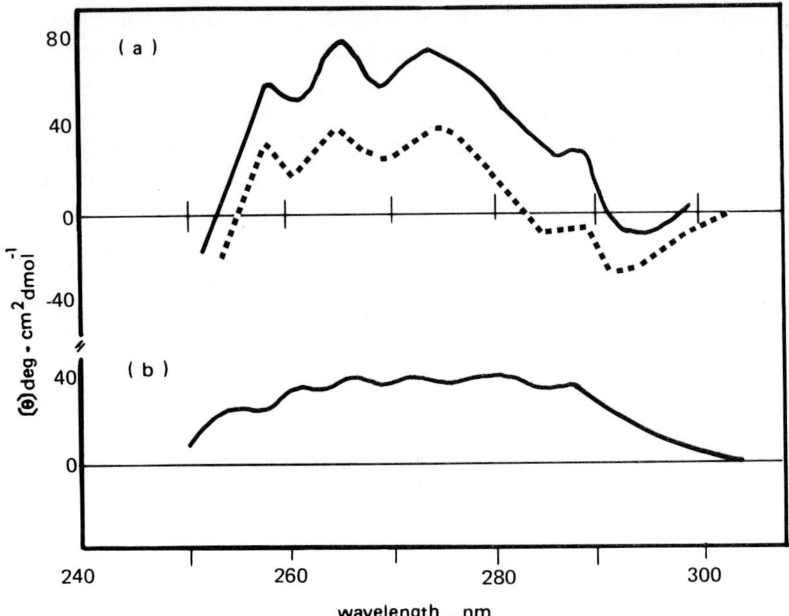

Figure 6   Circular dichroism (CD) spectra of bovine lens α-crystallin. (a) Near-UV CS increase of α-crystallin caused by incubation with (——) and without (· · ·) G6P. (b) Difference CD spectrum of α-crystallin incubated with and without G6P.

pathway. Complementing this has been research on other cataracts that suggests that cataract formation is a multifactorial process. Sorbitol production in diabetic lenses and galactitol production in galactosemic lenses are definitely one factor in the etiology of cataracts in these populations. Whether they are the primary causal factors remains to be determined. There has been sufficient confidence in the role of these mechanisms in diabetic cataracts to stimulate the pharmaceutical industry to develop more potent ARI. Also, there has been considerable effort expended in testing those drugs for toxicity to humans, ease of corneal penetration, or lenticular absorption after oral administration. All of this has been aimed at the successful development of an anticataract drug, either oral or topical, for use in diabetics. There has also been great interest in the role of aldose reductase in other complications of diabetes mellitus (i.e., diabetic polyneuropathy, nephropathy, and retinopathy). The successful treatment of diabetic complications with aldose reductase inhibitors will be a major contribution to society and one of the most significant applications of basic research for the public good in the past 50 years.

The biologic function of this pathway is unclear, although it has been proposed (1) to serve as a secondary ATP-generating system and (2) to serve as an osmotic buffer that reduces the hypertonic effect imposed on the lens from high-glucose levels in the aqueous humor during hyperglycemia (99). Proposal 1 is based on the discovery of enzymes capable of further metabolizing fructose such as ketohexokinase. Proposal 2 is based on an observation that rapid changes in extracellular osmolarity (50 mOsM) occurring as glucose increases from 5.5 to 55.5 mM can significantly dehydrate the lens. This type and magnitude of osmolar stress occurs in hyperglycemic, hyperosmolar coma, a serious well-known complication of diabetes mellitus in humans. A similar study by Jacob and Duncan (100) measured the effect of a 40 mOsM osmotic shock on amphibian lens conductance. That an increase of -80 mV in electrical conductance of the amphibian lens can occur in response to a 50 mOsM increase in osmolarity (from added glucose) was demonstrated. When the lens was returned to the isosmotic control medium, further swelling occurred. When isosmotic medium containing 40 mM glucose (and 20 mM NaCl removed) was used, the same increase in baseline electrical conductance occurred. Similar increases were observed with nonmetabolizable sugars, such as 3-O-methyl-glucose. These data suggest that the small fluctuations in external glucose levels, as occur in diabetes, may cause significant changes in lens membrane permeability due to the "double osmotic shock" the lens experiences. The dehydration from glucose-derived osmotic stress can be partially offset by intracellular sorbitol production (99,100,101). Blocking sorbitol production with an aldose reductase inhibitor eliminates its protective effect. These data suggest that the sorbitol pathway may be a means of counteracting the effects of fluctuating glucose-derived osmotic stress in the lens and that cyclic activity is its normal mode of action. Sustained activity of the sorbitol pathway is abnormal and results in the well-known sequence leading to either diabetic or galactosemic cataract formation.

GLUTATHIONE METABOLISM

Like other tissues (Fig. 7), the lens contains high concentrations of glutathione (GSH). The highest GSH level appears to be in the epithelium; in the rabbit lens, for example, there is 64 μmol GSH per gram tissue, which is six times the whole lens concentration (102).

There is apparently a slow turnover of GSH; Reddy et al. (103) reported a rate of 1.4%/hr. GSH turnover, essentially the synthetic activity, requires approximately 11% of ATP generated from glycolysis. Enzymes responsible for GSH synthesis (glutathione synthetase and cysteinylglycine synthetase), and for GSH degradation (γ-glutamyltranspeptidase and cysteinylglycine dipeptidase) have been identified and characterized (104).

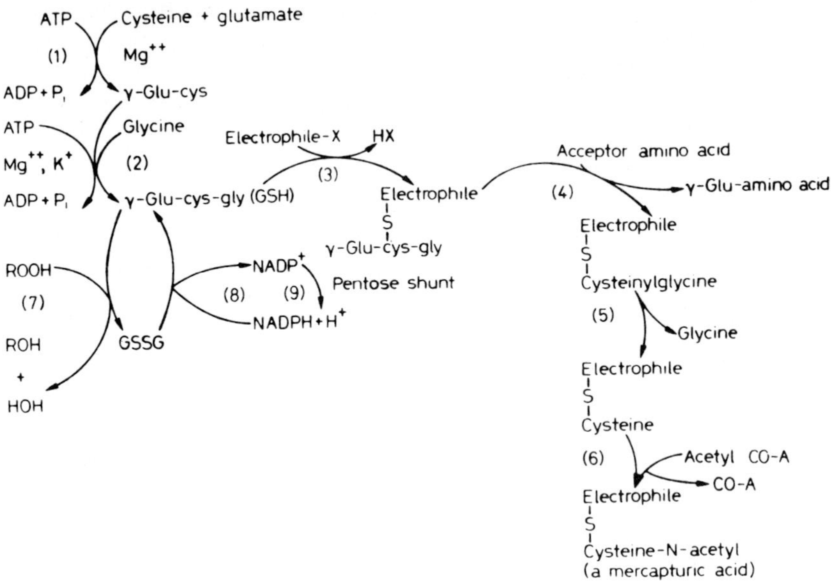

Figure 7 Pathways of glutathione metabolism.

Normal lenses maintain a steady-state concentration of GSH; however, this begins to drop in lenses undergoing cataract formation. This has been found to be true in almost all experimental cataracts and also in human senile cataracts. The disappearance of GSH may be due to its diffusion though damaged cell membranes or its formation of GSH-protein mixed disulfides. In any case, the loss of GSH will ultimately affect many changes in lens structure and function. This can be illustrated by an examination of the multiple role of GSH in the lens.

1. Sulfhydryl protection. It has been shown that crystallins isolated from human senile cataracts can be reduced in vitro by incubating the proteins with a GSH-maintaining system (105). Reduction of disulfide bonds and maintenance of existing sulfhydryls by GSH are probably the most important functions of GSH.

2. γ-Glutamyl cycle. GSH may participate in amino acid transport as a γ-glutamyl donor in the γ-glutamyl cycle. The presence of components of this cycle in the lens has been reported (106,107). The mechanism involves the transfer of the γ-glutamyl group, through the catalysis of γ-glutamyl transpeptidase, to a receptor amino acid. The conjugate is then converted to the free amino acid and 6-oxoproline by lactamase.

The enzyme γ-glutamyltranspeptidase has been histochemically localized in rabbit lens, ciliary process, and cornea; it has been purified from the bovine lens epithelium and appears to be membrane bound. It readily splits GSH and GSSG, but requires an acceptor amino acid for the γ-glutamyl moiety.

3. Ion transport. Maintenance of Na^+ and K^+ concentrations in the lens is mediated through Na^+,K^+-ATPase, which is extremely sensitive to thiol oxidation. It has been demonstrated that GSH oxidants, such as diamide, azoester, t-butylhydroperoxide, and H_2O_2, all cause a decrease in ^{86}Rb uptake (4,102,108,109). The close association between lens GSH content and ionic pump activity is further illustrated by the use of 1-chloro-2,4-dinitrobenezene, which conjugates GSH through glutathione-S-transferase; a linear correlation between the GSH level and the ^{86}Rb uptake was observed (110).

4. Detoxification of H_2O_2. Removal of H_2O_2 by glutathione peroxidase is coupled to glutathione reductase and the hexose monophosphate shunt. In the rabbit lens, the rate of H_2O_2 removal is 1.46 μmol per lens every 3 hr at 37°C, with a concomitant increase of HMPS activity of 0.4 μmol per lens every 3 hr (70). Combining the GSH peroxidase and reductase mechanism and catalase activity, the lens is apparently capable of detoxifying substantial amounts of H_2O_2 present in the aqueous humor. GSH peroxidase is a selenoenzyme containing 4 mol selenium per mole enzyme. It has a molecular weight of 140,000 d, and apparent K_m values for glutathione, t-butylhydroperoxide, cumene hydroperoxide, and H_2O_2 are 2.9, 0.54, 0.65, and 0.045 mM, respectively (111). Since the aqueous humor contains 0.06 mM H_2O_2 (112), it is expected that glutathione peroxidase remains constantly activated. Selenium-deficient rats lost 85% of the GSH peroxidase activity, and the defect led to cataract formation in these rats (113). Whether selenium deficiency is a causative factor in human senile cataract is not clear (114).

5. Removal of xenobiotics. In addition to the glutathione peroxidase-reductase system, there is a separate detoxification system, (glutathione-S-transferase, in conjunction with γ-glutamyltranspeptidase), which seems to be responsible for the removal of xenobiotics (104). Glutathione-S-transferase has been purified from bovine lens (115,116); it has a molecular weight of 49,000 d and is a dimer. Unlike the rat liver enzyme, the lens GSH S-transferase does not have any GSH peroxidase activity. Nevertheless, it remains unclear whether there is a linear correlation between the loss of GSH and the loss of cellular functions, or whether there is a certain threshold level of GSH below which cellular functions decline.

NUCLEAR MAGNETIC RESONANCE STUDIES OF THE LENS

The application of NMR spectroscopy to lens research is a recent development. The advantages of this technique over conventional biochemical methods are: (1) NMR allows real-time measurements of lens metabolism; (2) NMR is noninvasive; therefore, the lenses, following NMR studies, can be processed further for other complementary studies; and (3) NMR permits an overall view of lens metabolism because of its nondiscriminating detection of resonance signals. Furthermore, information regarding the behavior of water in the lens can be obtained by measuring the relaxation times (spin-lattice, T_1, and spin-spin, T_2). This can be done only with NMR spectroscopy, because the relaxation times are a measurement of the approach to equilibrium with the thermal environment (T_1) and with the spin system (T_2) following a radiofrequency pulse.

Some of the studies employing NMR technology are described below.

The State of Water in the Lens

The relaxation times of water in living tissues have been found to be shorter than those in pure water. This is also true in the crystalline lens. Further, Neville et al. (117) found that these relaxation times were dependent on free-water content. Racz et al. (118) later measured this parameter and showed that, at temperatures below $-9°C$, the relaxation times of nonfrozen (or bound) water were identical for both normal and totally opaque lenses; however, above $-9°C$, the relaxation times increased in the opaque lens, suggesting an increase in the ratio of free to bound water and/or decreased bonding sites for water in the opaque lens. These findings were confirmed by Pope et al. (119). Recently, Lerman et al. (120) also showed two types of water in the lens that do not exchange rapidly, confirming the work by Racz. Further, they found that lowering the temperature in the presence of acrylamide induced a significant change of T_2 in one fraction of lens water; acrylamide is known to cross-link with lens proteins and may affect protein-water interactions as a result. These results indicate that lens water can be divided into free and protein-bound fractions and that the two fractions do not exchange rapidly. Further, there appears to be more free water in a totally opaque cataractous lens, and the ratio of free to bound water can be altered chemically. It should be noted that these data (i.e., the T_1 and T_2 values) also provide the basis for proton imaging of normal and cataractous lenses, a technique to characterize the cataractogenic process by changes in the state of water in the lens.

Lens Metabolism as Revealed by NMR

$[^{31}P]$NMR is a relatively insensitive technique incapable of detecting naturally abundant ^{31}P if the concentration of the phosphorus compound is lower than 0.2 mM. Nevertheless, it allows the detection of intra- and extracellular P_i, and

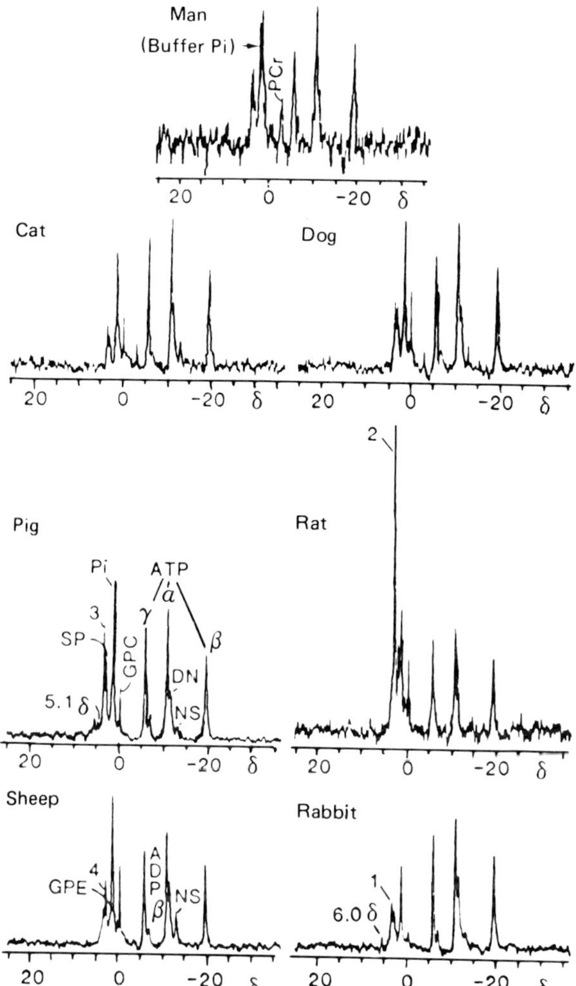

Figure 8 [^{31}P]NMR spectra of mammalian lenses. (P_i: inorganic phosphate; PCr: creatine phosphate; SP: sugar phosphates; GPC: glycerol-3-phosphoryl-choline; DN: dinucleotides; NS: nucleoside diphosphosugars; and GPE: glycerol-3-phosphorylethanolamine.)

intracellular organophosphates, such as α-glycerophosphate, sugar phosphates, α,β, and γ-ATP, α- and β-ADP, AMP, phosphocreatine, and NADP(H) + NAD(H), in the mammalian lens. Kopp et al. (121) have shown that the distribution of these organophosphates varies from species to species, and that the cat lens is closest to the human lens in organophosphate content (Fig. 8).

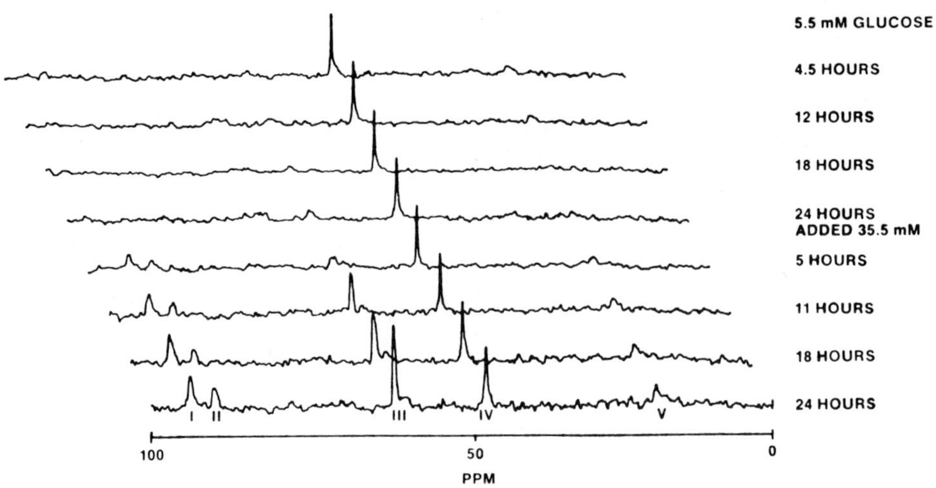

Figure 9 [^{31}C]NMR spectra showing accumulation of sorbitol in the lens incubated in high-glucose medium. (I and II: β- and α-glucose, respectively; III: sorbitol; IV: standard; and V: lactate.)

[^{31}C]NMR, on the other hand, utilizes ^{13}C-enriched substrates, such as [^{13}C]glucose and [^{13}C]tryptophan, to compensate for the lack of naturally abundant ^{13}C compounds (Fig. 9). NMR spectroscopy is ideally suited for the study of dynamic metabolism in tissue because it documents the continuous change of metabolic events. Based on [^{31}p]NMR spectroscopy, it was found that α-glycerophosphate (αGP) increased in lenses incubated in 30 mM galactose (122). This was subsequently found in lenses incubated in 35.5 mM glucose (123). In both cases, there was an increased carbon flow through glycolysis. The increased glycolytic rate, coupled with the NADH generated from the sorbitol pathway, was responsible for the increased αGP. The use of an ARI, sorbinil, was found to obliterate the increase (123). Thus, in theory, a topical NMR spectrometer that successfully detects αGP signals in the lens in situ could be used to diagnose the efficacy of aldose inhibitors. Another new finding is the intralenticular pH, which turns out to be approximately 6.9 in the normal lens (33,124). It rises to 7.3 when the lens is deprived of glucose, and drops to 6.4 in medium containing Ca^{2+}. The change in lens pH will modify enzyme activities in the lens; PFK is one such pH-dependent enzyme. The mechanism with regard to how the lens regulates is internal pH is still a mystery.

Sorbitol

The slow turnover rate of sorbitol in the "hyperglycemic" lens was nicely demonstrated by a $[^{13}C]$NMR study (125), in which the lenses were incubated in 35.5 mM $[^{13}C]$glucose for 12 hr to allow accumulation of sorbitol; this was followed by a return to 5.5 mM for the next 73 hr, and during this time, sorbitol returned to background levels. It is apparent that synthesis occurs approximately six times as rapidly as depletion.

It should be noted that, at present, the application of NMR to lens research has not yet reached its fullest potential. There is a dual role for NMR technology in biomedical research. (1) It can be used as a tool for basic research either by itself or in conjunction with conventional biochemical assays; and (2) it can be developed into a clinically applicable diagnostic procedure, such as, proton imaging, chemical shift imaging, or surface coil (topical) imaging or spectroscopy. Most of the studies to date utilizing NMR spectroscopy of the lens have used only a small number of the many resources of NMR technology. We should expect to see a fast expansion of NMR research of the lens in the next few years.

IONIC TRANSPORT

Several studies have shown that ^{86}Rb uptake, which occurs via the K^+ pump, decreases if the lens is treated with oxidants such as tBHP, H_2O_2, diamide, and azoester, or with free radicals generated from oxygen with photodynamic systems (4,103,109,110,126). The decreased ^{86}Rb uptake does not seem to be reversible with sulfhydryl reagents, such as dithiothreitol. Thus, in theory, this uptake should serve as a good index of the lens's prior exposure to oxidation. Nevertheless, several investigations of the ionic pump activity in the human lens show conflicting results. (1) Kobayashi et al. (127) found that Na^+,K^+-ATPase activity declined in human senile cataracts, and that the extent of loss seemed to correspond with the severity of cortical opacification. (2) Paterson et al. (128), on the other hand, found no correlation between the enzyme activity and the type of lens, cataractous or not. There is now evidence indicating that only the ion transport but not the ATP-hydrolyzing activity of Na^+,K^+-ATPase is sensitive to oxidation; therefore, measurement of the inorganic phosphate released does not fully indicate the ionic pump activity in the human aging lens. Nevertheless, Pasino and Maraini (129,130) investigated ^{86}Rb uptake in the human aging lens and cataract and found no loss of ^{86}Rb uptake with cataract formation. An increase in ^{86}Rb efflux, however, showed that membrane integrity had been compromised in the cataracts. Disturbance to the ionic pump as an important cataractogenic factor was illustrated by the Nakano mouse cataract model and the x-ray cataract model (131).

Other than Na^+,K^+-ATPase, a Mg^{2+}- and Ca^{2+}-dependent ATPase has been found on the outer surfaces of bovine lens epithelial cells (132). Hamilton et al. (133) found that Na^+,K^+-ATPase (ouabain-sensitive) was sensitive to Ca^{2+} inhibition; whereas Ca^{2+} seemed to activate the ouabain-insensitive ATPase. The physiologic significance of Ca^{2+}-activated ATPase is not known.

CALCIUM METABOLISM

The role of calcium in normal lens metabolism and cataract formation has been studied by Hightower et al. (134–136). They confirmed the finding that calcium levels are approximately 0.2 mM in rat, bovine, dog, and rabbit lenses, a level that is at least an order of magnitude lower than that in aqueous humor. They showed an active efflux of calcium that is not dependent on Ca^{2+}-Na^+ exchange or on the external level of Na^+. Propranolol and lanthanum, both inhibitors of Ca^{2+}-ATPase, were able to block $^{45}Ca^{2+}$ efflux. Earlier data by Iwata (137) suggested a relation between Ca^{2+}-ATPase and active Ca^{2+} extrusion. This enzyme is present in lens epithelium, cortex, and nucleus, but in the nucleus the enzyme was inactive at a Ca^{2+} concentration of 10^{-5} M and very active at 10^{-7} M calcium. The specific activity of this enzyme is approximately 30 times higher in the epithelium than in the cortex nucleus.

Calcium transport can be supported by glycolysis alone if exogenous glucose is present (138,139); respiration does not contribute to Ca^{2+} efflux. Both anterior and posterior fibers are involved, as was shown by the accumulation of more Ca^{2+} in the de-epithelialized lens treated with iodoacetate than the de-epithelialized lens treated with buffer. Also, more Ca^{2+} accumulation occurred in a lens whose posterior aspect was immersed in medium when glycolysis was inhibited than when it was not.

If the calcium level is elevated to at least 1.8 mM in a cultured rabbit lens, a superficial anterior and posterior opacity develops. There is no associated increase in lens Na^+, and the calcium appears to be solely responsible for the opacity. There is considerable binding of calcium to fiber cell membranes, and the bound form resists release through dialysis. However, over 95% of total lens Ca^{2+} is easily removed by dialysis from the water-soluble protein fraction. These studies are interesting in light of recent work showing a correlation between calcium content and severity of human cataracts classified by the Cooperative Cataract Research Group Guidelines (136,140). This study showed bound Ca^{2+} in senile cataracts with varying degrees of brunescence to be as follows: 30 ng/mg (immature cataract) and 200 ng/mg (mature cataract); 100 ng/mg (pale yellow nuclear color) and 300 ng/mg (brown nuclear color). The unpigmented, freshly excised rabbit lens contains 16 ng/mg. Of equal interest is their observation that the increased protein-bound Ca^{2+} is not derived from a fixed pool of free or dialyzable Ca^{2+}. Rather, as total lens Ca^{2+} increases, a constant proportion of the total becomes bound to membranes and proteins. They speculate that the elevation of lens Ca^{2+} may result from the gradual deactivation or inhibition of Ca^{2+}-ATPase. Jedziniak et al. (141) have shown that the high-molecular-weight fraction (first peak on an agarose calcium) of soluble lens proteins contains three times as much calcium as the second, lower molecular weight peak. The calcium densities of water-insoluble and water-soluble lens proteins were the same. The correlation between total Ca^{2+}, free/bound Ca^{2+}, and cataract type is beclouded by the failure of most scientists to use a reliable system of classifying human cataracts. Such studies as Ref. 142 are useful in that they provide a comparison of normal and three groups of colored lenses (Pirie classification groups I–III), but it is incorrect to use this scheme to quantitate or classify cataractous change for reasons recently published (143).

Fagerholm (144) has found a differential effect of Ca^{2+} on subcapsular and deeper cortical lens fibers; the former swell and rupture in media containing Ca^{2+} in a concentration equal to that in aqueous humor; the latter were hardly affected. Fagerholm postulates a role for Ca^{2+} in subcapsular cataract formation, one that may be related to a modification of the lens membrane's permeability to Na^+ and K^+.

The valinomycin cataract (145) may be an exception to this proposal: lenses exposed to 10^{-5} M valinomycin and low concentrations of Ca^{2+} do not accumulate Ca^{2+} and do not become opaque in spite of the increase in lens sodium. In the presence of higher Ca^{2+} concentrations, K^+ decreases, Na^+ increases, Ca^{2+} increases, and a superficial cortical opacity develops. The authors believe that Ca^{2+} accumulation rather than a disturbance in Na^+,K^+ accounts for the opacification. That there are differences in the response of cortical fibers to different Ca^{2+} concentrations is suggested by Clark et al. (146), who studied calf lens fibers. They showed that the calf lens is transparent and normal fiber morphology is preserved in medium containing calcium between 0.5 and 1.3 mM, the concentration in the normal lens. At concentrations either below or above this concentration range, cortical fibers swell and rupture.

In looking at the effect of Ca^{2+} deficiency on lens metabolism, it has been known for many years that calcium-deficient incubation media induce cataracts in rabbit lenses (147,148), presumably as a result of the lens's inability to maintain normal Na^+ and K^+ gradients. Potassium is lost and Na^+ is gained as the lens opacifies. Until recently, this effect was attributed to an increase in the lens membrane's permeability to Na^+ and K^+. In 1978, Delamere and Paterson (149) showed in frog lens that Ca^{2+} deficiency selectively increases permeability to Na^{2+} alone. In a similar study (150) using rabbit lenses, they showed a similar effect; namely, calcium deficiency leads to a loss in K^+, in an amount due to membrane depolarization alone and a gain in Na^+ due to a marked increase in the membrane's passive permeability to Na^+ alone.

MYOINOSITOL TRANSPORT IN THE LENS

In 1964, Pirie and van Heyningen (151) reported that inositol levels decreased in the human diabetic lenses and that there was an inverse relation between glucose and inositol contents in the senile cataracts. They suggested that the decrease of inositol might be due to membrane dysfunction in both the diabetic lenses and senile cataracts.

Stewart et al. (152) investigated inositol levels in nerve and lens of galactose-fed rats and found that low inositol seemed to be associated with the increased galactitol and water contents of the lens; the data again suggested membrane damage in sugar-induced osmotic cataracts.

The transport mechanism for myoinositol in the lens was studied by Cotlier (153), Broekhuyse (154), and Varma et al. (155). They have found that

1. Accumulation of myoinositol required ATP.
2. Myoinositol transport was ouabain sensitive, suggesting coupling to cation pump.
3. Uptake of myoinositol was inhibited by hypertonicity in the lens due to accumulation of sugar alcohols, when incubated in high-aldose (glucose or galactose) medium.
4. The efflux rate of myoinositol increased in lenses incubated in the high-aldose medium.

These results explain the apparent disappearance of inositol from the diabetic lenses.

The physiologic significance of myoinositol in the lens is not clear; however, recent evidence showing myoinositol's effect in restoring motor nerve conduction velocity (156) has revived lens researchers' interest in defining the role of myoinositol in the lens. For example, in a preliminary study, Cheng et al. (157) were able to show that myoinositol affected events that occurred in rabbit lenses incubated in high glucose; that is, it diminished ATP content but decreased α-glycerophosphate levels; these results partially mimic those obtained with aldose reductase inhibitors.

Inositol, in conjunction with ARI, could prove to be an extremely effective therapeutic measure for the treatment of diabetic complications.

AMINO ACID METABOLISM

The lens contains much higher levels of amino acids than those present in the aqueous humor and vitreous body (158,159). Amino acids enter the lens through active transport (16,160); however, the carriers have not been clearly identified (evidence for the role of the γ-glutamyl cycle is still inconclusive; see Ref. 161). Reddy and his coworkers have proposed a "pump-leak" system to describe amino acid and cation transport in the lens (161); that is, potassium and amino acids are actively transported into the epithelium, diffuse toward the posterior pole and exit through the posterior capsule, and the transport of sodium is opposite in direction to that of potassium and amino acids.

Transport of amino acids has been shown to covariate with cation transport. For example, removal of the epithelium substantially reduced the influx of α-amino isobutyric acid as well as that of potassium (162). Further, Reddy (159) showed that ouabain, which inhibited Na^+,K^+-ATPase, also inhibited transport of taurine and α-amino isobutyric acid, although inhibition of the latter showed a lag of 2 hr, suggesting an indirect effect. Evidence of a direct involvement of Na^+,K^+-ATPase in amino acid transport, however, has not been clearly established.

Amino acids, once they have entered the lens, are utilized for biosynthesis and energy production.

Trayhurn and van Heyningen (163) and Kern and Ho (164) reported that the influx of glutamine was several times higher than glutamic acid and was probably the major source of glutamic acid for the synthesis of glutathione and crystallins in the lens. Trayhurn and van Heyningen (16) further showed that both glutamine and glutamic acid were oxidized to CO_2 and were involved in the synthesis of proline. They also showed that, in the presence of air, there was an increased oxidation of gluatmic acid (16). This indicates that oxidation of amino acids may be a source of energy for the lens. In a more extensive study (17), they followed the fate of nine L-amino acids in the lens, including alanine, aspartic acid, glutamic acid, leucine, lysine, proline, serine, tryrosine, and tryptophan. All nine amino acids were taken up by the lens, and all except two were further metabolized. The two exceptions were tryptophan and tryrosine, which were incorporated into the lens protein directly. In lenses incubated aerobically, the production of NH_3 increased 46% in the absence of glucose, indicating greater amino acid breakdown. Thus, amino acids can be utilized as endogenous substrates for the tricarboxylic acid cycle to produce ATP during energy deprivation.

INTERACTION OF METABOLIC PATHWAYS

Glucose Metabolism

The overall regulation of glycolytic activity and its interaction with other metabolic pathways can be summarized as follows:

Hexokinase, phosphofructokinase, and pyruvate kinase are the key control points of glycolysis. Both PFK and PK are sensitive to ATP inhibition; they are activated by elevated levels of glucose, fructose-6-phosphate, and fructose-1,6-diphosphate, respectively. This forms the basis of the coordinated regulation of glycolysis. Two such examples are described below.

1. When there is an increased glucose flux due to hyperglycemia, the
   sorbitol pathway is activated. This activation requires the support of the
   hexose monophosphate shunt to provide NADPH for aldose reductase;
   concomitantly, there is an increase of NADH production due to polyol
   dehydrogenase activity. The excess NADH is redirected for α-glycero-
   phosphate synthesis, which is seen to increase gradually with $[^{31}P]$NMR
   spectroscopy (123). NMR also showed that there was no increased ATP
   hydrolysis and, therefore, no release of PFK and PK inhibition; lactate
   production remained stationary.
2. During the induction of the Pasteur effect (e.g., with the addition of CN^-
   in the lens incubating medium or incubation under N_2), there is no
   activation of the hexose monophosphate shunt because there is no
   demand for NADPH. There is, however, a disinhibition of PFK because
   of a decreased energy charge that in turn activates PK. This in turn
   results in 1.5-fold increase in lactate production (165,166). Excess
   NADH, accumulated due to incomplete activation (limited by the extent
   of PFK activation) of PK, is redirected for α-glycerophosphate synthesis,
   and this can be seen clearly with $[^{31}P]$NMR spectroscopy.

Duel Oxidative–Osmotic Stress (167)

Since both AR and GSH reductase require NADPH as cofactor, activation of
either enzyme during a dual oxidative–osmotic stress will depend on the
enzyme's relative affinity for NADPH. Indeed, GSH reductase, with a K_m four
times smaller than that of AR, was found to be preferentially activated.
Although lactate production was not disturbed, in the lens undergoing the
double stress, production of sorbitol and fructose declined 79% and 45%, respectively; further, there is a slight but significant loss of K^+, and morphologically
the posterior pole of the lens showed far more extensive swelling and disruption
than with either stress alone. Thus it appears that the two stresses exert a
synergistic effect on the lens, even though the activation of the sorbitol pathway
is submaximal.

REFERENCES

1.   *Webster's New Collegiate Dictionary*. G. and C. Merriam Co., Springfield,
     Massachusetts, 1980, p. 715.
2.   L. T. Chylack, Jr. Mechanism of "hypoglycemic" cataract formation in the
     rat lens. I. The role of hexokinase instability. Invest. Ophthalmol., *14*: 746
     (1975).
3.   L. T. Chylack, Jr., and F. L. Schaefer. Mechanism of "hypoglycemic"
     cataract formation in the rat lens. II. Further studies on the role of hexo-
     kinase instability. Invest. Ophthalmol., *15*: 519 (1976).

4. H. N. Fukui. The effect of hydrogen peroxide on the rubidium transport of the rat lens. Exp. Eye Res., *23*: 595 (1976).

5. L. J. Takemoto, P. Azari, and W. C. Gorthy. Role of sufhydryl groups in the formation of a hereditary cataract in the rat. Exp. Eye Res., *20*: 12 (1975).

6. N. H. Ansari and S. K. Srivastava. Role of glutathione in the prevention of cataractogenesis in rat lenses. Curr. Eye Res., *2*: 271 (1982/83).

7. S. Iwata and J. H. Kinoshita. Mechanism of development of hereditary cataract in mice. Invest. Ophthalmol. Vis. Sci., *10*: 504 (1971).

8. P. F. Kador, H. N. Fukui, S. Fukushi, and J. H. Kinoshita. Philly mouse: A new model of hereditary cataract. Exp. Eye Res., *30*: 59 (1980).

9. H. M. Cheng, L. T. Chylack, Jr., and I. von Saltza. Supplementing glucose metabolism in human senile cataracts. Invest. Ophthalmol. Vis. Sci., *21*: 812 (1981).

10. T. P. Sippel. Energy metabolism in the lens during aging. Invest. Ophthalmol., *4*: 502, (1965).

11. J. H. Kinoshita, H. L. Kern, and L. O. Merola. Factors affecting the cation transport of calf lens. Biochim. Biophys. Acta, *47*: 458 (1961).

12. J. E. Harris, L. Gruber, E. Talman, and G. Haskmson. The influence of oxygen on the photodynamic action of methylene blue on cation transport in the rabbit lens. Am. J. Ophthalmol., *41*: 528 (1959).

13. II. M. Katzen, and R. T. Schimke. Multiple forms of hexokinase in the rat: Tissue distribution, age, dependency and properties. Proc. Natl. Acad. Sci. USA, *54*: 1218 (1965).

14. V. E. Kinsey and V. N. Reddy. Studies on the crystalline lens. X. Transport of amino acids. Invest. Ophthalmol., *2*: 229 (1963).

15. H. L. Kern. Accumulation of amino acids by calf lens. Invest. Ophthalmol., *1*: 368 (1962).

16. P. Trayhurn, and R. van Heyningen. Aerobic metabolism in the bovine lens. Exp. Eye Res., *12*: 315 (1971).

17. P. Trayhurn and R. van Heyningen. The metabolism of amino acids in the bovine lens. Biochem. J., *136*: 67 (1973).

18. P. Trayhurn. Ph.D. Thesis. Oxford University, 1972.

19. P. Trayhurn and R. van Heyningen. The role of respiration in the energy metabolism of the bovine lens. Biochem. J., *129*: 507 (1972).

20. L. T. Chylack, Jr. Control of glycolysis in the lens. Exp. Eye Res., *11*: 280 (1971).

21. L. T. Chylack, Jr., and H.-M. Cheng. Sugar metabolism in the crystalline lens. Surv. Ophthalmol., *23*: 26 (1978).

22. K. M. Giles and J. E. Harris. The accumulation of ^{14}C from uniformly labeled glucose by the normal and diabetic rabbit lens. Am. J. Ophthalmol., *48*: 508 (1959).

23. J. B. Coulter, III, J. A. Engelke, and D. K. Eator. Insulin concentrations in aqueous humor after paracentesis and feeding of rabbits. Invest. Opthalmol. Vis. Sci., *19*: 1524 (1980).

24. J. R. Reddan, N. J. Unakar, C. V. Harding, M. Bagchi, and G. Saldana. Induction of mitosis in cultured rabbit lens initiated by the addition of insulin to medium KEI-4. Exp. Eye Res., *20*: 45 (1975).
25. C. Orhloff, J. Beusch, M. Jaeger, and O. Hockwin. Immunologic detection of inactive enzyme molecules in the aging lens. Exp. Eye Res., *31*: 573 (1980).
26. O. Hockwin, H.-D. Bergeder, R. Herrmann, and S. Walter. Investigations on the heat lability of the phosphofructokinase (EC 2.7.111) in bovine lenses after X-irradiation and under various incubation conditions. Ophthalmic Res., *11*: 453 (1979).
27. C. Ohrloff. Age changes of enzyme properties in crystalline lens. Interdiscipl. Topics Gerontol., *12*: 158 (1978).
28. P. P. Fagerholm and B. T. Philipson. Human lens epithelium in normal and cataractous lenses. Invest. Ophthalmol. Vis. Sci., *21*: 408 (1981).
29. M. M. Perry, T. Tassin, and Y. Courtois. A comparison of human lens epithelial cells in situ and in vitro in relation to aging: An ultrastructural study. Exp. Eye Res., *28*: 327 (1979).
30. T. K. Kuwabara. The maturation of the lens cell: A morphologic study. Exp. Eye Res., *20*: 427 (1975).
31. N. S. Rafferty and W. Goosens. Cytoplasmic filaments in the crystalline lens of various species: Functional correlations, Exp. Eye Res., *26*: 177 (1978).
32. Y. Kabayashi and T. Suzuki. The aging lens: Ultrastructural changes in cataract. In *Cataract and Abnormalities of the Lens* (J. G. Bellows, ed.), Grune and Stratton, New York, 1975, p. 313.
33. J. V. Greiner, S. J. Kopp, D. R. Sanders, and T. Glonek. Organophosphates of the crystalline lens: A nuclear magnetic resonance spectroscopic study. Invest. Ophthalmol. Vis. Sci., *21*: 700 (1981).
34. L. T. Chylack, Jr. Human lens hexokinase. Exp. Eye Res., *15*: 225 (1973).
35. L. T. Chylack, Jr. Soluble, insoluble and latent hexokinases in the mammalian lens. Ophthalmic Res., *6*: 93 (1974).
36. L. T. Chylack, Jr., J. H. Kinoshita, and R. Kasabian. Nature and distribution of two distinct forms of hexokinase within the mammalian lens. Exp. Eye Res., *10*: 250 (1970).
37. R. van Heyningen. Some glycolytic enzymes and intermediates in the rabbit lens. Exp. Eye Res., *4*: 298 (1965).
38. H. Green, C. A. Bocher, and I. H. Leopold. Anaerobic carbohydrate metabolism of crystalline lens. Am. J. Ophthalmol., *39*: 106 (1959).
39. J. E. Harris, J. D. Hauscheldt, and L. T. Nordquist. Transport of glucose across the lens surfaces. Am. J. Ophthalmol., *39*: 161 (1955).
40. J. W. Patterson and K. W. Bunting. Changes associated with the appearance of mature sugar cataracts. Invest. Ophthalmol., *4*: 167 (1965).
41. A. Pirie. The biochemistry of the eye. Bibl. Ophthalmol., *49*: 287 (1957).
42. A. M. Gonzalez, M. Sochor, J. S. Hothersall, and P. McLean. Effect of experimental diabetes on the activity of hexokinase in rat lens: An example of glucose overutilization in diabetes. Biochem. Biophys. Res. Commun., *84*: 858 (1978).

43. J. B. Typell and J. W. Anderson. Endocrinology, *89*: 1178 (1971).

44. M. Sochor, N. Z. Baquer, and P. McLean. Diabetologia, 1978.

45. J. A. Grunt and R. O. Howard. Eye findings in children with ketotic hypoglycemia. Can. J. Ophthalmol., *7*: 151 (1972).

46. A. J. McKenna. Neonatal hypoglycemia: Some ophthalmic observations. Can. J. Ophthalmol., *1*: 56 (1966).

47. S. Merin and J. S. Crawford. Hypoglycemia and infantile cataract. Arch. Ophthalmol., *86*: 495 (1971).

48. W. A. Wilson. Ocular findings in ketotic hypoglycemia. Trans Am. Ophthalmol. Soc., *67*: 355 (1969).

49. G. W. Barber, S. B. Rosenberg, I. Mikuni, H. Obazawa, and J. H. Kinoshita. Net proteolysis in glucose-deprived rat lenses incubated in amino acid-free medium. Exp. Eye Res., *29*: 663 (1979).

50. O. Hockwin, H. Fink, and C. Orhloff. Carbohydrate metabolism of the lens depending on age. Evaluation of factor analysis. In *Fifth European Symposium on Basic Research in Gerontology* (U. J. Schmidt, G. Bruschhe, E. Lang, A. Viidik, D. Platt, V. V. Frolbis, and F. H. Schulz, eds.), Perimed, p. 632.

51. H. M. Cheng and L. T. Chylack, Jr. Thiol oxidation in the crystalline lens. Invest. Ophthalmol. Vis. Sci., *19*: 522 (1980).

52. H.-M. Cheng and L. T. Chylack, Jr. pH-dependent temperature sensitivity of rat lens phosphofructokinase. Invest. Ophthalmol. Vis. Sci., *15*: 505 (1976).

53. H.-M. Cheng, L. T. Chylack, Jr., J. Chien, and E. C. Baranano. Stability of mammalian lens phosphofructokinase. Invest. Ophthalmol. Vis. Sci., *16*: 126 (1977).

54. H. M. Cheng and L. T. Chylack, Jr. Properties of rat lens phosphofructokinase. Invest. Ophthalmol. Vis. Sci., *15*: 279 (1976).

55. J. F. R. Kuck, Jr. Chemical constituents of the lens. In *Biochemistry of the Eye* (C. N. Graymore, ed.), Academic Press, New York, 1970, p. 183.

56. J. K. Wolfe, B. D. Peczon, and L. T. Chylack, Jr. Kinetics of calf lens phosphofructokinase activity at physiological substrate concentrations. Exp. Eye Res., *34*: 945 (1982).

57. J. K. Wolfe, B. D. Peczon, and L. T. Chylack, Jr. Calf lens phosphofructokinase: Purification by single-step affinity chromatography and partial characterization. Exp. Eye Res., *32*: 553 (1981).

58. A. M. Gonzalez, M. Sochor, and P. McLean. Effect of experimental diabetes on glycolytic intermediates and regulation of phosphofructokinase in rat lens. Biochem. Biophys. Res. Commun., *95*: 1173 (1980).

59. G. D. Reinhart and N. A. Lardy. Rat liver phosphofructokinase kinetic activity under near physiological conditions. Biochemistry, *19*: 1477 (1980).

60. H. W. Hofer. Influence of enzyme concentration on the kinetic behavior of rabbit muscle phosphofructokinase. Hoppe Seylers Z. Physiol. Chem., *352*: 997 (1971).

61. H. M. Cheng and L. T. Chylack, Jr. Control of pyruvate kinase activity in the lens. Exp. Eye Res., *27*: 39 (1978).

62. C. Ohrloff, S. Zierz, and O. Hockwin. Investigation of the enzymes involved in the breakdown of fructose in the cattle lens. Ophthalmic Res., *14*: 221 (1982).

63. J. H. Kinoshita and C. Wachtl. A study of the C^{14}-glucose metabolism of the rabbit lens. J. Biol. Chem., *233*: 5 (1958).

64. H.-M. Cheng and L. T. Chylack, Jr. Regulation and stabilization of lens glucose-6-phosphate dehydrogenase. Exp. Eye Res., *24*: 459 (1977).

65. B. Bullard and A. Pirie. Glucose-6-phosphate dehydrogenase and 6-phosphogluconate dehydrogenase in lens and blood of different species. Exp. Eye Res., *3*: 118 (1964).

66. J. S. Nisselbaum and S. Green. A simple ultramicro method for determination of pyridine nucleotides in tissues. Anal. Biochem., *27*: 212 (1969).

67. F. J. Giblin and V. N. Reddy. Pyridine nucleotides in ocular tissues as determined by the cycling assay. Exp. Eye Res., *31*: 601 (1980).

68. H. M. Cheng, L. T. Chylack, Jr., C. N. Sang, N. Orzalesi, and F. P. Corongiu. GSSG-reducing activity in lenses deficient in glucose-6-phosphate dehydrogenase. Metab. Pediatr. Syst. Ophthalmol., *7*: 53 (1983).

69. F. J. Giblin, D. E. Nies, and V. N. Reddy. Stimulation of the hexosemonophosphate shunt in rabbit lens in response to the oxidation of glutathione. Exp. Eye Res., *33*: 289 (1981).

70. F. J. Giblin, J. P. McCready, and V. N. Reddy. The role of glutathione metabolism in the detoxification of H_2O_2 in rabbit lens. Invest. Ophthalmol. Vis. Sci., *22*: 330 (1982).

71. S. D. Varma. Aldose reductase and the etiology of diabetic cataracts. Curr. Topics Eye Res., *3*: 91 (1980).

72. J. C. Crabbe, S. Wolfe, A. B. Halder, and H. H. Ting. Diabetic cataracts—is aldose reductase important? Metab. Pediatr. Ophthalmol., *5*: 33 (1981).

73. J. W. Patterson. Effect of high fat, fructose and casein diet on diabetic cataracts. Proc. Soc. Exp. Biol. Med., *90*: 706 (1955).

74. J. W. Patterson, M. E. Patterson, E. V. Kinsey, and V. N. Reddy. DVN: Lens assays on diabetic and galactosemic rats receiving diets that modify cataract development. Invest. Ophthalmol., *4*: 98 (1969).

75. J. C. Hutton, P. J. Schofield, J. F. Williams, H. L. Regtop, and F. C. Hollows. The effect of unsaturated-fat diet on cataract formation in diabetic rats. Br. J. Nutr., *36*: 161 (1976).

76. S. D. Varma and J. H. Kinoshita. The absence of cataracts in mice with congenital hyperglycemia. Exp. Eye Res., *19*: 577 (1974).

77. J. Okuda, I. Mua, K. Inagaki, T. Houe, and M. Nakayama. Inhibition of aldose reductase from rat and bovine lenses by flavonoids. Biochem. Pharmacol., *31*:

78. L. T. Chylack, Jr., and J. H. Kinoshita. A biochemical study of a cataract formed in high glucose medium. Invest. Ophthalmol., *9*: 98 (1965).

79. D. Dvornik, N. Summard-Duquesne, M. Krami, K. Sestang, K. H. Gabbay, and J. H. Kinoshita. Polyol accumulation in galactosemic and diabetic rats: Control by an aldose reductase inhibitor. Science, *182*: 1146 (1973).

80. S. Fukushi, L. O. Merola, and J. H. Kinoshita. Altering the course of cataracts in diabetic rats. Invest. Ophthalmol. Vis. Sci., *19*: 313 (1980).

81. T. S. Hu, M. Datiles, and J. H. Kinoshita. Reversal of galactose cataracts with sorbinil in rats. Invest. Ophthalmol. Vis. Sci., *24*: 640 (1983).

82. N. J. Unakar, T. Smart, J. Reddan, and I. Devlin. Regression of cataracts in the offspring of galactose fed rats. Ophthalmic. Res., *11*: 52 (1964).

83. V. N. Reddy, D. Schwass, B. Chakrapani, and C. P. Lim. Biochemical changes associated with the development and reversal of galactose cataracts. Exp. Eye Res., *23*: 483 (1976).

84. N. J. Unakar, C. Genyea, J. R. Reddan, and V. N. Reddy. Ultrastructural changes during the development and reversal of galactose cataracts. Exp. Eye Res., *26*: 123 (1978).

85. M. K. Gillis and L. T. Chylack, Jr. Depletion of sorbitol and fructose from young and older rat lenses. Exp. Eye Res., *25*: 183 (1982).

86. V. J. Stevens, C. A. Ronzer, V. M. Monnier, and A. Cerami. Diabetic cataract formation: Potential role of glycosylation of lens crystallins. Proc. Natl. Acad. Sci. USA, *75*: 2918 (1978).

87. S.-H. Chiou, L. T. Chylack, Jr., H. F. Bunn, and J. H. Kinoshita. Role of nonenzymatic glycosylation in experimental cataract formation. Biochem. Biophys. Res. Commun., *95*: 894 (1980).

88. P. F. Kador, J. H. Kinoshita, W. H. Tung, and L. T. Chylack, Jr. Differences in the susceptibility of various aldose reductases to inhibition. II. Invest. Ophthalmol. Vis. Sci., *19*: 980, 1980.

89. J. A. Jedziniak, L. T. Chylack, Jr., H.-M. Cheng, M. K. Gillis, A. A. Kalustian, and W. H. Tung. The sorbitol pathway in the human lens: Aldose reductase and polyol dehydrogenase. Invest. Ophthalmol. Vis. Sci., *20*: 314 (1981).

90. L. T. Chylack, Jr., H. F. Henriques, III, H.-M. Cheng, and W. H. Tung. Efficacy of alrestatin, an aldose reductase inhibitor in human diabetic and nondiabetic lenses. Ophthalmology, *86*: 1579 (1979).

91. L. T. Chylack, Jr., H. F. Henriques, III, and W. H. Tung. Inhibition of sorbitol production in human lenses by an aldose reductase inhibitor. Doc. Ophthalmol. Proc. Ser., *18*: 65 (1979).

92. A. Beyer-Mears, E. Cruz, J. Nicolas-Alexandre, and E. Varagrammis. Sorbinil protection of lens protein components and cell hydration during diabetic cataract formation. Pharmacology, *24*: 193 (1982).

93. M. Laurent, P. Kern, and F. Regnault. Thickness and collagen metabolism of lens capsule from genetically prediabetic mice. Ophthalmic Res., *13*: 93 (1981).

94. P. F. Kador, J. S. Zigler, and J. H. Kinoshita. Alterations of lens protein synthesis in galactosemic rats. Invest. Ophthalmol. Vis. Sci., *18*: 696 (1979).

95. T. Shinohara and J. Piatigorsky. Regulation of protein synthesis, intracellular electrolytes and cataract formation in vitro. Nature, *270*: 406 (1977).

96. J. Piatigorsky, H. N. Fukui, and J. H. Kinoshita. Differential metabolism and leakage of protein in an inherited cataract and in normal lens cultured with ouabain. Nature, *274*: 558 (1978).

97. J. Liang and B. Chakrabarti. Sugar-induced change in near ultraviolet circular dichroism of α-crystallin. Biochem. Biophys. Res. Commun., *102*: 180 (1981).

98. H. M. Cheng, N. Hutson, and M. Peterson. Effect of sorbinil on the thiol maintenance in the lens, presented at the 1983 Japan–U.S. CCRG Meeting, Hawaii.

99. R. H. Harding, L. T. Chylack, Jr., and W. H. Tung. The sorbinil pathway as protector of the lens against glucose-generated osmotic stress. Invest. Ophthalmol. Vis. Sci. (Suppl.), *20*: 34 (1981).

100. T. J. C. Jacob and G. Duncan. Glucose-induced membrane permeability changes in the lens. Exp. Eye Res., *34*: 445 (1982).

101. J. H. Seland and L. T. Chylack, Jr. Short-term morphological changes in glucose stressed rabbit lenses in vitro. Invest. Ophthalmol. Vis. Sci. (Suppl.), *22*: 199 (1982).

102. F. J. Giblin, B. Chakrapani, and V. N. Reddy. Glutathione and lens epithelial function. Invest. Ophthalmol. Vis. Sci., *15*: 381 (1976).

103. V. N. Reddy, S. D. Varma, and B. Chakrapani. Transport and metabolism of glutathione in the lens. Exp. Eye Res., *16*: 105 (1973).

104. W. B. Rathbun and S. K. Hanson. Glutathione metabolic pathway as a scavenging system in the lens. Ophthalmic Res., *11*: 172 (1979).

105. R. C. Augusteyn. On the possible role of glutathione in maintaining human lens protein sulfhydryls. Exp. Eye Res., *28*: 665 (1979).

106. V. N. Reddy and N. J. Unakar. Localization of gamma-glutamyl transpeptidase in rabbit lens, ciliary process and cornea. Exp. Eye Res., *17*: 405 (1973).

107. W. B. Rathbun and K. Wicker. Bovine lens γ-glutamyl transpeptidase. Exp. Eye Res., *15*: 161 (1973).

108. D. L. Epstein and J. H. Kinoshita. The effect of diamide on lens glutathione and lens membrane function. Invest. Ophthalmol. Vis. Sci., *9*: 629 (1970).

109. D. L. Epstein and J. H. Kinoshita. Effect of methyl phenyldiazenecarboxylate (azoester) on lens membrane function. Exp. Eye Res., *10*: 228 (1970).

110. H. M. Cheng, R. G. Gonzalez, I. von Saltza, N. H. Ansari, and S. K. Srivastava. Effect of GSH deprivation on lens metabolism. Exp. Eye Res., In press.

111. P. L. Bergard, W. B. Rathbun, and W. Linden. Glutathione peroxidase from bovine lens: a selenoenzyme. Exp. Eye Res., *34*: 131 (1982).

112. A. Spector and W. H. Garner. Hydrogen peroxide and human cataracts. Exp. Eye Res., *33*: 673 (1981).

113. L. Spinker, J. Harr, P. Newberne, P. Whanger, and P. Weswig. Selenium deficiency lesions in rats fed vitamin E supplemented rations. Nutr. Rep. Int., *4*: 335 (1971).

114. A. Swanson and A. Truesdale. Elemental analysis in normal and cataractous human lens tissue. Biochem. Biophys. Res. Commun., *45*: 1488 (1971).

115. R. P. Saneto, Y. C. Awasthi, and S. K. Srivastava. Interrelationship between cationic and anionic forms of glutathione S-transferase of bovine ocular lens. Biochem. J., *191*: 11 (1980).

116. Y. C. Awasthi, R. P. Saneto, and S. K. Srivastava. Purification and properties of bovine lens glutathione S-transferase. Exp. Eye Res., *30*: 29 (1980).

117. M. C. Neville, C. A. Paterson, J. C. Rae, and D. E. Woessner. Nuclear magnetic resonance studies and water ordering in the crystalline lens. Science, *184*: 1072 (1974).

118. P. Racz, K. Tompa, and I. Pocsic. The state of water in normal and senile cataractous lenses studied by nuclear magnetic resonance. Exp. Eye Res., *28*: 129 (1979).

119. J. M. Pope, S. Chandra, and J. D. Balfe. Changes in the state of water in senile cataractous lenses as studies by nuclear magnetic resonance. Exp. Eye Res., *34*: 57 (1982).

120. S. Lerman, D. L. Ashley, R. C. Long, Jr., J. H. Goldstein, J. M. Megaw, and K. Gardner. Nuclear magnetic resonance analyses of the cold cataract: Whole lens studies. Invest. Ophthalmol. Vis. Sci., *23*: 218 (1982).

121. S. Kopp, T. Glonek, and J. Greiner. Interspecies variations in mammalian lens metabolites as detected by phsophorus-31 nuclear magnetic resonance. Science, *215*: 1622 (1982).

122. J. V. Greiner, S. J. Kopp, D. R. Sanders, and T. Glonek. Dynamic changes in the organophosphate profile of the experimental galactose-induced cataract. Invest. Ophthalmol. Vis. Sci., *22*: 613 (1982).

123. R. G. Gonzalez, P. Barnett, H. M. Cheng, and L. T. Chylack, Jr. Altered phosphate metabolism in the intact rabbit lens under high glucose conditions and its prevention by an aldose reductase inhibitor. Exp. Eye Res., In press.

124. T. Glonek, J. Greiner, S. J. Kopp, and D. R. Sanders. The intracellular pH during cataractogenesis with intact lens as determined by phosphorus-31 nuclear magnetic resonance spectroscopy. Invest. Ophthalmol. Vis. Sci. (Suppl.), 89 (1981).

125. R. G. Gonzalez, J. Willis, T. Aguayo, P. Campbell, L. T. Chylack, Jr., and T. Schleich. ^{13}C-Nuclear magnetic resonance studies of sugar cataractogenesis in the single intact rabbit lens. Invest. Ophthalmol. Vis. Sci., *22*: 808 (1982).

126. H. M. Jernigan, H. N. Fukui, J. D. Goosey, and J. H. Kinoshita. Photodynamic effects of rose bengal or riboflavin on carrier-mediated transport system in rat lens. Exp. Eye Res., *32*: 461 (1981).

127. S. Kobayashi, D. Roy, and A. Spector. Na^+K^+ATPase in normal and cataractous human lens. Invest. Ophthalmol. Vis. Sci. (Suppl.), 88 (1981).

128. C. A. Paterson, N. A. Delamere, and L. G. Mawhorter. Na-K-ATPase activity in normal and cataractous human lenses. Invest. Ophthalmol. Vis. Sci. (Suppl), 157 (1982).
129. M. Pasino and G. Maraini. Cation pump activity and membrane permeability in human senile cataractous lenses. Exp. Eye Res., *34*: 887 (1982).
130. G. Maraini and M. Pasino. Active and passive rubidium influx in normal human lenses and in senile cataracts. Exp. Eye Res., *34*: 543 (1982).
131. H. Matsuda, F. J. Giblin, and V. N. Reddy. The effect of X-irradiation on Na-K-ATPase and cation distribution in rabbit lens. Invest. Ophthalmol. Vis. Sci., *22*: 180 (1982).
132. A. Bergner and D. Glasser. Demonstration of a magnesium calcium-dependent ATPase on the outer surface of bovine lens epithelial cells. Ophthalmic Res., *11*: 322 (1979).
133. P. M. Hamilton, N. A. Delamere, and C. A. Paterson. The influence of calcium on lens ATPase activity. Invest. Ophthalmol. Vis. Sci., *18*: 434 (1979).
134. K. R. Hightower, V. Leverenz, and V. N. Reddy. Calcium transport in the lens. Invest. Ophthalmol. Vis. Sci., *19*: 1059 (1980).
135. K. R. Hightower and V. N. Reddy. Ca^{++}-induced cataract. Invest. Ophthalmol. Vis. Sci., *22*: 263 (1982).
136. K. R. Hightower and V. N. Reddy. Calcium content and distribution in human cataract. Exp. Eye Res., *34*: 413 (1982).
137. S. Iwata. Process of lens opacification and membrane function. A review. Ophthalmic Res., *6*: 138 (1974).
138. K. R. Hightower and V. N. Reddy. Metabolic studies on calcium transport in mammalian lens. Curr. Eye Res., *1*: 197 (1981).
139. N. A. Delamere and C. A. Paterson. Studies on calcium regulation in relation to sodium-potassium balance in the rabbit lens. Ophthalmic Res., *14*: 230 (1982).
140. L. T. Chylack, Jr. Classification of human cataracts. Arch. Ophthalmol., *96*: 888 (1979).
141. J. A. Jedziniak, D. F. Nicoli, E. M. Yates, and G. B. Benedek. On the calcium concentration of cataractous and normal human lenses and protein fractions of cataractous lenses. Exp. Eye Res., *23*: 325 (1976).
142. G. Duncan and R. van Heyningen. Distribution of nondiffusible calcium and sodium in normal and cataractous human lenses. Exp. Eye Res., *25*: 183 (1977).
143. L. T. Chylack, Jr., B. J. Ransil, and O. White. Classification of human senile cataractous change by the American Cooperative Cataract Research Group (CCRG) Method. III. The association of nuclear color (sclerosis) with extent of cataract formation, age and visual acuity. Invest. Ophthalmol. Vis. Sci., In press.
144. P. P. Fagerholm. The influence of calcium on lens fibers. Exp. Eye Res., *28*: 211 (1979).
145. R. H. Hightower and S. E. Harrison. Valinomycin cataract: The relative role of calcium and sodium accumulation. Exp. Eye Res., *34*: 941 (1982).

146. J. I. Clark, L. Menzel, A. Bagg, and G. B. Benedek. Cortical opacity, calcium concentration and fiber membrane structure in the calf lens. Exp. Eye Res., *31*: 399 (1980).

147. J. E. Harris and L. B. Gehrsitz. Significance of changes in potassium and sodium content of the lens. Am. J. Ophthalmol., *34*: 131 (1951).

148. L. O. Merola, H. L. Kern, and J. H. Kinoshita. The effect of calcium on the cations of calf lens. Arch. Ophthalmol., *63*: 830 (1960).

149. N. A. Delamere and C. A. Paterson. The influence of calcium-free EGTA solution upon membrane permeability in the crystalline lens of the frog. J. Gen. Physiol., *71*: 581 (1978).

150. N. A. Delamere and C. A. Paterson. The influence of calcium-free solutions upon permeability characteristics of the rabbit lens. Exp. Eye Res., *28*: 45 (1979).

151. A Pirie and R. van Heyningen. The effect of diabetes on the content of sorbitol, glucose, fructose and inositol in the human lens. Exp. Eye Res., *3*: 124 (1964).

152. M. A. Stewart, M. M. Kurien, W. R. Sherman, and E. V. Cotlier. Inositol changes in nerve and lens of galacose fed rats. J. Neurochem., *15*: 941 (1968).

153. E. Cotlier. Myo-inositol permeability in the lens due to cataractous conditions. Biochim. Biophys. Acta, *163*: 269 (1968).

154. R. M. Broekhuyse. Changes in myo-inositol permeability in the lens due to cataractous conditions. Biochim. Biophys. Acta, *163*: 269 (1968).

155. S. D. Varma, B. Chakrapani, and V. N. Reddy. Intraocular transport of myoinositol. II. Accumulation in the rabbit lens in vitro. Invest. Ophthalmol. Vis. Sci., *9*: 795 (1970).

156. D. A. Green, R. A. Lewis, S. A. Lattimer, and M. J. Brown. Selective effects of myo-inositol administration on sciatic and tibial motor nerve conduction parameters in the streptozocin-diabetic rat. Diabetes, *31*: 573 (1982).

157. H. M. Cheng, R. G. Gonzalez, and T. Kawaba. The effect of myo-inositol on the metabolic status of the lens incubated in high glucose. (Submitted for publication).

158. V. N. Reddy. Distribution of free amino acids and related compounds in ocular fluids, lens, and plasma of various mammalian species. Invest. Ophthalmol. Vis. Sci., *6*: 478 (1967).

159. V. N. Reddy. Transport of organic molecules in the lens. Exp. Eye Res., *15*: 731 (1973).

160. V. N. Reddy. Studies on intraocular transport of taurine. II. Accumulation in the rabbit eye. Invest. Ophthalmol., *9*: 206 (1970).

161. V. N. Reddy. Dynamics of transport systems in the eye. Friedenwald lecture. Invest. Ophthalmol. Vis. Sci., *18*: 1000 (1979).

162. V. E. Kinsey and D. V. N. Reddy. Studies on the crystalline lens. XI. The relative role of the epithelium and capsule in transport. Invest. Ophthalmol. *4*: 104 (1965).

163. P. Trayhurn and R. van Heyningen. The metabolism of glutamine in the bovine lens: glutamine as a source of glutamate. Exp. Eye Res., *17*: 149 (1973).
164. H. L. Kern and C. K. Ho. Transport of L-glutamic acid and L-glutamine and their incorporation into lenticular glutathione. Exp. Eye Res., *17*: 455 (1973).
165. R. van Heyningen. The metabolism of the bovine lens in air and nitrogen. Exp. Eye Res., *20*: 393 (1975).
166. M. K. Gillis, L. T. Chylack, Jr., and H.-M. Cheng. Age and the control of glycolysis in the rat lens. Invest. Ophthalmol. Vis. Sci., *20*: 457 (1981).
167. H.-M. Cheng, P. P. Fagerholm, and L. T. Chylack, Jr. Response of the lens to oxidative-osmotic stress. Exp. Eye Res., *37*: 11 (1983).

7
Physical Basis of Lens Transparency

FREDERICK A. BETTELHEIM / Adelphi University, Garden City, N.Y.

The performance of the lens, that is, the proper focusing of images on the retina, depends on the transparency of the lens. In a perfectly transparent body the light intensity exiting the body is the same as that of incident beam, and furthermore there is no change in the direction of the light beam. Obviously, the lens is not perfectly transparent. The diminution of the intensity of the exiting light beam may be due to absorption, refraction-reflection, and scattering of light. The absorption of light is due to molecular species that have energy requirements for transition between ground and excited states that lie within the visible range of the electromagnetic radiation spectrum. Since these processes are described by Zigman (Chap. 8) and Dillon (Chap. 9), we do not deal with the basic aspects of absorption in this chapter.

The interaction of light with a particle induces polarization in the particle, and the degree of polarization is related to the refractive index. Furthermore, since the induced dipole of the particle oscillates, this gives rise to light scattering. Therefore, our first task in understanding transparency is to establish the relation between polarizability and light scattering.

Beyond that the practical aspect of back-reflection of light when the incident angle reaches a critical angle is considered, but the main thrust is the discussion of light scattering. This particular phenomenon has dual aspects: phenomeno-logic and analytical. Phenomenologic investigations try to answer such questions as: How much light reaches the retina? How much is scattered and absorbed? What is the spectrum of light reaching the fovea? Analytical studies try to answer such questions as: What are the structural or optical units that cause scattering? What are the dimensions, shape, and so on, of these structural units?

PRINCIPLES OF TRANSPARENCY

Polarization and Refractive Index

When a molecule is placed in an electrical field between two condenser plates, the applied electrical field is reduced. The reason is that the field polarizes the molecule in such a fashion that the positive end of the molecule aligns itself with the negative plate, and vice versa. This total polarization of the molecule can be measured by the dielectric constant ϵ, which is the ratio of the capacitances of the dielectric compound in question to that of vacuum at the same geometric setting of the condenser.

$$\frac{C}{C_0} = \epsilon \tag{1}$$

The total molar polarization of the compound is given by Eq. (2).

$$P_T = P_I + P_p = \frac{4\pi N\alpha}{3} + \frac{4\pi N\mu^2}{9kT} = \frac{\epsilon - 1}{\epsilon + 2} \frac{M}{\rho} \tag{2}$$

where P_T is the total molar polarization; P_I is the induced molar polarization; P_p is the molar polarization due to the permanent dipole moment μ; M is the molecular weight; and ρ is the density of the compound in question. The α is the polarizability of the molecule (in cubic centimeters), which indicates how much distortion is induced by the field in the nuclear and electronic charge distribution. The N is Avogadro's number, k is the Boltzmann constant, and T is the absolute temperature (1).

The dielectric dispersion, that is, the change of the dielectric constant with the frequency of the field, is given schematically in Fig. 1. There are three plateaus in this dispersion curve. At very low frequencies (radio frequencies) both the induced and the permanent polarization contribute to the total polarization. At higher frequencies the orientation of the permanent dipole cannot follow the field; hence its contribution decreases until the second plateau is reached. At this point only the induced polarization is present. This induced polarization is the contribution of two kinds of polarization: P_A, the atomic polarization due to the displacement of the nuclei under the influence of the field, and P_E, the electronic polarization that represents the distortion of the electronic cloud induced by the field. At higher frequencies (beyond the infrared region) and contribution of P_A diminishes to zero, and the dielectric constant measured at such high frequencies is due to electronic polarization alone. When the refractive index n of a compound is measured in the visible region, the light can be considered a field of high frequency, and the Maxwell relation holds.

$$n^2 = \epsilon_\infty \tag{3}$$

where n is the refractive index measured and ϵ_∞ is the dielectric constant at high frequencies.

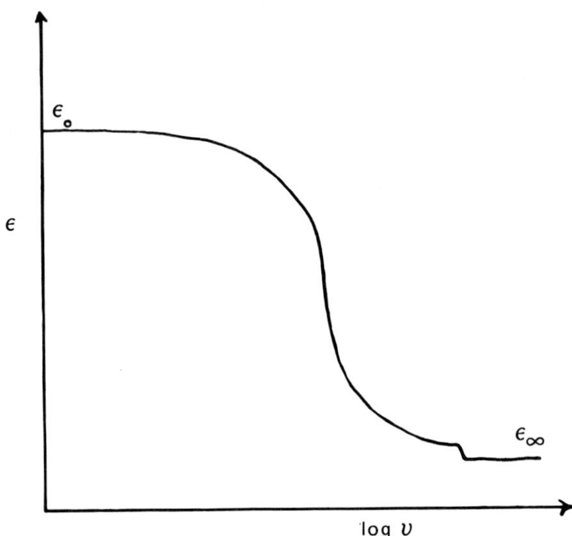

Figure 1 Variation of the dielectric constant (ϵ) with the frequency (ν) of the applied field.

Hence, in the region of interest of lens transparency,

$$P_E = \frac{4\pi N\alpha}{3} = \frac{n^2 - 1}{n^2 + 2} \frac{M}{\rho} = R \tag{4}$$

The second part of this equality is called the Lorentz-Lorenz equation, and R is the Lorentz-Lorenz molar refraction. The molar refractivity is a characteristic of the molecular structure, and it was long thought to be an additive property of its parts. However, the increments are dependent not only on the atoms but also on the types of chemical bonds between the atoms; hence the property is both additive and constitutive.

The lens is a multicomponent system, and its refractivity is the result of the contribution of its components. Upon mixing two components one would expect the additivity to be operative: hence,

$$R_{add} = X_1 \frac{n_1^2 - 1}{n_1^2 + 2} \frac{M_1}{\rho_1} + (1 - X_1)\frac{n_2^2 - 1}{n_2^2 + 2} \frac{M_2}{\rho_2} \tag{5}$$

The experimental molar refractivity has the following formula:

$$R_{exp} = \frac{n^2 - 1}{n^2 + 2} \frac{X_1 M_1 + (1 - X_1)M_2}{\rho} \tag{6}$$

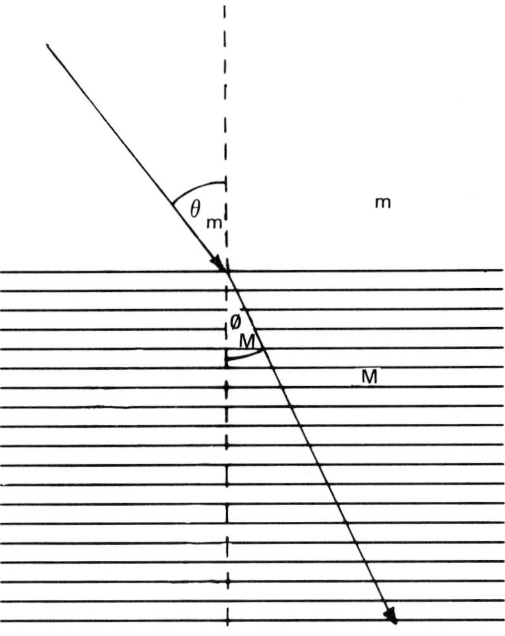

Figure 2   Schematic representation of Snell's law.

In Eqs. (5) and (6) the subscripts 1 and 2 refer to components 1 and 2 and the quantity without subscript is the property of the mixture measured. The X_1 is the mole fraction of compound 1.

The difference between the experimental and the predicted values demonstrates the deviation from additivity.

$$\Delta R = R_{exp} - R_{add} \tag{7}$$

When ΔR exceeds 0.1 ml, one may assume that strong interaction forces between components 1 and 2 cause electronic deformations.

In using Eq. (4) in attempting to obtain a polarizability value α for either of the pure components or for a mixture, one should keep in mind that the single value of α obtainable from Eq. (4) represents merely an average of the polarizabilities in various directions.

When a light passes from one isotropic medium to another (m→M) both its direction and its velocity will change unless it is normal to the boundary separating m from M (Fig. 2).

The relation between the angle of incidence θ_m and the angle of refraction ϕ_M is given by Snell's law, Eq. (8).

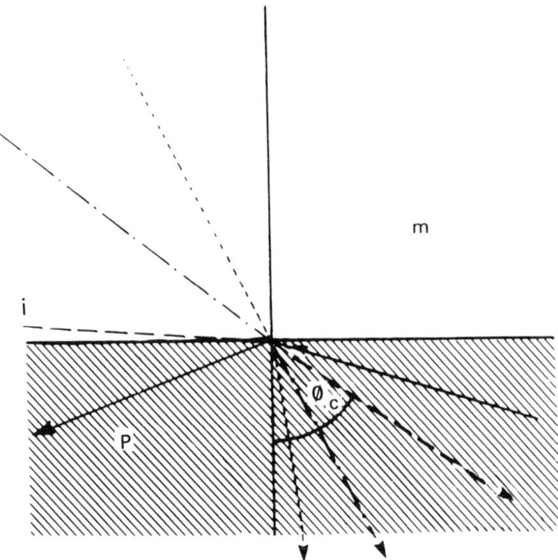

Figure 3 Schematic representation of the critical angle of refraction ϕ_c that occurs when the angle of incidence is glancing incidence (i). Any angle of refraction ϕ higher than the critical angle ϕ_c will result in reflection (solid line).

$$n = \frac{\sin \theta_m}{\sin \phi_M} \tag{8}$$

In this case n is the refractive index of M relative to m. Usually the reference substance m is air or, more properly, vacuum, which has a refractive index of 1.00. In measuring critical angles of refractions one must consider that a sample m is in contact with a body M, which has a greater refractive index than m.

Rays entering medium m at different angles have corresponding angles of refraction according to Snell's law (Eq. (8)]. If we follow different angles of incidence θ_m from the normal to the glancing incidence i, we find that the angle of refraction increases (Fig. 3). At the glancing incidence we obtain the critical angle of refraction, which is the last angle that lets light through in the body P. Any higher angle of refraction will be reflected back.

Considering the lens as a whole, reflection of the light will occur if the angle of incidence exceeds a critical angle.

The primary role of the lens in the visual process is focusing the image on the retina. This is achieved by refraction. The refractive power is due to (1) the difference between the refractive indices at the interface and (2) the curvature of

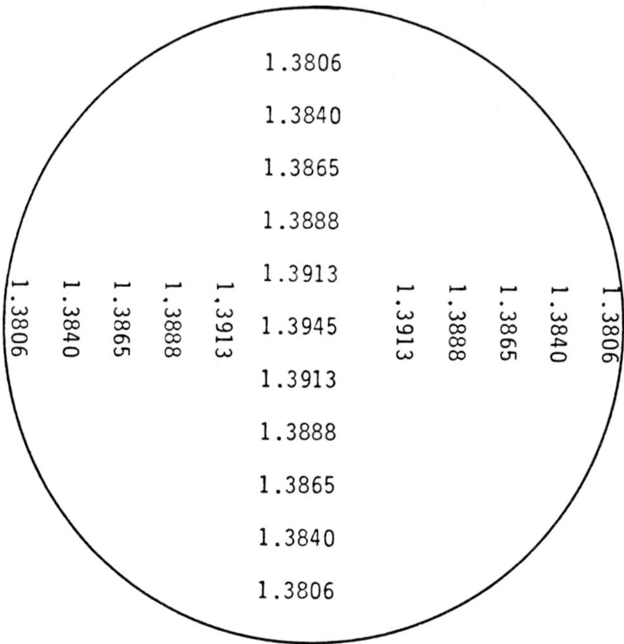

Figure 4   Topographic distribution of refractive indices in bovine lenses (frontal view).

the lens. The front of the lens is embedded in the aqueous humor and the posterior in the vitreous. Both have refractive indices of 1.336 (2). The average refractive index of the lens is given as 1.420 (3).

The lens capsule almost matches the refractive index of its naturally embedding media, the vitreous and the aqueous (4). The mapping of protein concentrations of the lens along the optic axis by microradiography showed a gradual increase in protein content from cortex to the center of the nucleus in all but the youngest (few days old) lenses (5). This is the reason the lens has its highest refractive index in the nucleus and there is a refractive index gradient from cortex to nucleus.

Bettelheim and Wang (4) also found a refractive index gradient along the surface of the lens; the highest refractive index was on the central anterior part of the lens (Fig. 4). This means that the highest surface refractive index is at the position where there is the least curvature in the lens, and vice versa. The difference in refractive index of several layers of cortical fibers provide a zone of discontinuity observed in clinical examinations that are due to light reflection from these fiber cells (3).

Light Scattering

When a light interacts with a particle, be it an atom, a molecule, or an aggregate of molecules, a dipole moment is induced in the particle. This is due to the fact that the particle in the electric field is subjected to polarization; that is, the nuclei and the electrons move in opposite directions. Since electromagnetic radiation, such as light, carries an oscillating field, the induced dipole in the particle will also oscillate. An oscillating dipole, however, is the source of electromagnetic radiation, that is, the scattered light is propagated in all directions.

There are two kinds of light scattering: elastic and nonelastic (quasi-elastic). We concern ourselves mainly with the first kind. If we are dealing with elastic scattering (i.e., none of the energy of the incident radiation is used to put the atom or molecule in an excited—translation, rotational, vibrational, or electronic—state), the frequency of the scattered light will be the same as that of the incident light. We call this Rayleigh scattering. It was Lord Rayleigh who first derived the equation for the intensity of scattered radiation in 1871:

$$\frac{I_\theta}{I_0} = 16\pi^4 \alpha^2 \frac{\sin^2\theta}{\lambda^4 r^2} \tag{9}$$

where I_θ is the intensity of the scattered light at any angle θ, θ is the scattering angle (i.e., the angle between the incident beam and the scattered beam), I_0 is the intensity of the incident beam, λ is the wavelength of the incident beam in vacuum, r is the distance of the observer from the scattering center, and α is the polarizability, a molecular parameter stating how much dipole moment is induced by a unit electric field. Combination of the intensities with the distance of the observer from the scattering center,

$$\frac{I_\theta}{I_0} \frac{r^2}{V} = R_\theta \tag{10}$$

is known as the Rayleigh ratio, which describes the intensity of the scattered light per unit volume V.

Equation (9) predicts that the intensity of the scattered light will depend on the inverse of the fourth power of the wavelength that is scattered. Blue light will have a greater intensity than scattered red light, which accounts for the blue sky.

The polarizability α is related to the refractive index n [Eq. (4)]. In dilute gases the refractive index in turn can be expressed as a result of the fluctuation in the refractive index as a result of local concentration fluctuations (dn/dc).

Equations (9) and (10) combined can be rewritten as

$$R_\theta = \frac{[4\pi^2 \sin^2\theta (dn/dc)^2 Mc]}{N\lambda^4} \tag{11}$$

where the new quantities are M, the molecular weight of the particles, and c, the concentration of the particles (g/cm^3).

In contrast to the random position of gas molecules within the scattering volume, the atoms in a crystal are rigidly fixed in a geometric array. Since the wavelength of the light is much larger than the individual scatterers (atoms), we always can find a pair of atoms (equal scatters) that scatters out of phase from the point of view of the observer (or the detector) at any particular scattering angle. Thus, the destructive interference of a pair of scatterers is complete, and no scattered light will be observed.

The expression "crystal clear," the historical name "crystalline lens," and the isolated protein fraction designated "crystallins" have more to do with the transparency of the lens than with a crystalline organization of molecules within the lens.

Scattering from pure liquids is intermediate between that from crystals and gases. We could select pairs of small-volume elements (much smaller than the wavelength of light) that would be independent scatterers that are out of phase, and destructive interference from these pairs would follow. However, since liquids are not orderly, the packing of particles in each member of the pair of volume elements may not be the same; therefore, the intensity of the scattered light from volume element 1 may differ from that from volume element 2. As a consequence, the destructive interference is not complete, and some scattering will occur.

The scattering from a two-component solution can be understood on a similar basis. The solution is construed as a system of small-volume elements, each of which acts as a single scattering source. Each of these volume elements contains a large number of solvent molecules and a few solute molecules.

Based on the theories of Einstein (6) and Debye (7) the whole system can be looked upon as a fluctuation in concentration; that is, each small volume element has a different concentration, $\bar{c}$ + dc, where $\bar{c}$ is the average concentration of the solution and dc is the deviation from this average. The deviation can be both positive and negative. This fluctuation of concentration is completely random. The concentration fluctuation corresponds to a refractive index fluctuation, and the intensity of the scattered light or the corresponding Rayleigh ratio will be given by the equation

$$\frac{I_\theta}{I_0} \frac{r^2}{V} \frac{1}{1 + \cos^2\theta} = R'_\theta = \frac{2\pi^2 n_0 \, (dn/dc)^2 \, c}{N\lambda^4 (1/M) + 2Bc + 3Cc^2 + \cdots)} \tag{12}$$

where R'_θ is the Rayleigh ratio corrected for unpolarized light, c is the concentration of the solute (g/cm^3), n_0 is the refractive index of the solvent, and B and C are the second and third virial coefficients representing thermodynamic interaction parameters between solute and solvent and solute and solute, respectively.

If all the constants are lumped together in an optical constant K,

$$K = 2\pi^2 n_0^2 \left(\frac{dn}{dc}\right)^2 N\lambda^4 \tag{13}$$

then Eq. (5) can be rewritten in the familiar form

$$\frac{Kc}{R_\theta'} = \frac{1}{M} + 2Bc + 3Cc^2 \tag{14}$$

This equation shows that the intensity of the scattered light increases with concentration and molecular weight. When such an equation is applied to very dilute (infinitely dilute) solutions, the second and third terms on the right-hand side approach zero, and thus Kc/R_θ' will be equal to the reciprocal of the molecular weight (8).

If the solutions contain macromolecules with at least one dimension comparable to the wavelength of the light, that is, greater than $\lambda/20$, one macromolecule cannot be contained in the small-volume element but will be part of a number of volume elements. This means in essence that the macromolecule will contain more than one scattering source. The light scatterers within the macromolecules will be somewhat out of phase with each other and will cause destructive interference. How much interference occurs depends on the scattering angle and the size and shape of the particle. The effect of the large size of the macromolecule is expressed by the particle-scattering function $P(\theta)$:

$$P(\theta) = \frac{\text{scattered intensity for macromolecule}}{\text{scattered intensity without interference}} \tag{15}$$

Since the interference reduces the scattered light intensity, $P(\theta)$ will always be less than 1, but it approaches 1 at zero scattering angle where the scattered light from the different parts of the macromolecule will be in phase and no destructive interference will occur.

The analytical expression of $P(\theta)$ for differently shaped molecules has been worked out by a number of authors and can be found in textbooks on light scattering, such as those by Stacey (9), Kerker (10), and Tanford (11). However, in concentrated solutions or gels, such as the lens, the interaction between solute and solute particles become important. The third virial coefficient C in Eq. (14) may be sufficiently large and negative so that the light-scattering intensity will not increase with concentration but decrease. This has been shown to be true by using random fluctuation theory applied to a model system of randomly dispersed hard spheres (12).

A small isolated spherical scatterer provides a scattered light intensity I_θ at angle θ:

$$I_\theta = I_0 \frac{N}{V} V\sigma_{i(\theta)} S(q) \tag{16}$$

where I_0 is the intensity of the incident beam, $\sigma_{i(\theta)}$ is the scattering cross section of the sphere, and $S(q)$ is the structure factor. In the Born approximation (10), $\sigma_{i(\theta)}$ is given as

$$\sigma_{i(\theta)} = \frac{8\pi^4 a^6 n_2^4}{r^2 \lambda_0^4} \left(\frac{m^2 - 1}{m^2 + 2}\right)^2 (1 + \cos^2\theta) \tag{17}$$

where m is the relative refractive index $m = n_1/n_2$ of the refractive index of the sphere n_1 and that of the surrounding n_2, a is the radius of the sphere, r is the distance between the scatterer and the detector photo multiplier tube, and λ_0 is the wavelength of the light in vacuum.

For N particles comparable to the size of the wavelength, in volume V, the Rayleigh ratio [Eq. (10)] becomes

$$R_\phi = \frac{6\pi^3 a^3 n_2^4}{\lambda_0^4} \left(\frac{m^2 - 1}{m^2 + 2}\right)^2 (1 + \cos^2\theta) \, S(q)P(\theta)\phi \tag{18}$$

where the new quantities are the particle-scattering function or form factor given previously in Eq. (15); and ϕ the concentration in terms of volume fraction of the spheres. The structure factor itself is related to the concentration ϕ, radius of the spheres a, and scattering angle θ and can be obtained by the Percus-Yevick approximation (12,13).

Calculations using the hard sphere model yielded concentration-dependent light-scattering curves similar to that presented in Fig. 5. These curves show that, although the light-scattering intensity increases with concentration in dilute solution, it decreases in concentrated solutions or gels, such as the lens, largely due to destructive interference. This is true also when the scattering particle can be of any shape, not just a hard sphere. This modeling (12) proved that a random distribution of scattering particle can explain turbidity as caused by (1) size of the scattering particle, (2) refractive index difference between particle and environment, and (3) concentration effects. The last one in the lens will cause an increased turbidity with dilution, which is the cause of many metabolic cataracts due to osmotic pressure-generated water influx.

One can also obtain analytical data from the random fluctuation theory. Those are: the size and concentration of the scattering centers and their relative refractive indices (compared to the surrounding). By using polarized light, the random fluctuation theory of Debye-Bueche (14) can be extended (15). There are two kinds of fluctuations that can cause light scattering: density fluctuations and fluctuations in the orientation of optically anisotropic material. Both will result in variation in the local refractive index as the photon passes through the material (Fig. 6). In Fig. 6, the solid straight line represents the average polarizability (refractive index) of the medium. Here η shows the local deviation from this average, $\bar{\eta}^2$ is the average (mean squared) deviation from the average refractive index, and a is the size of the region correlated.

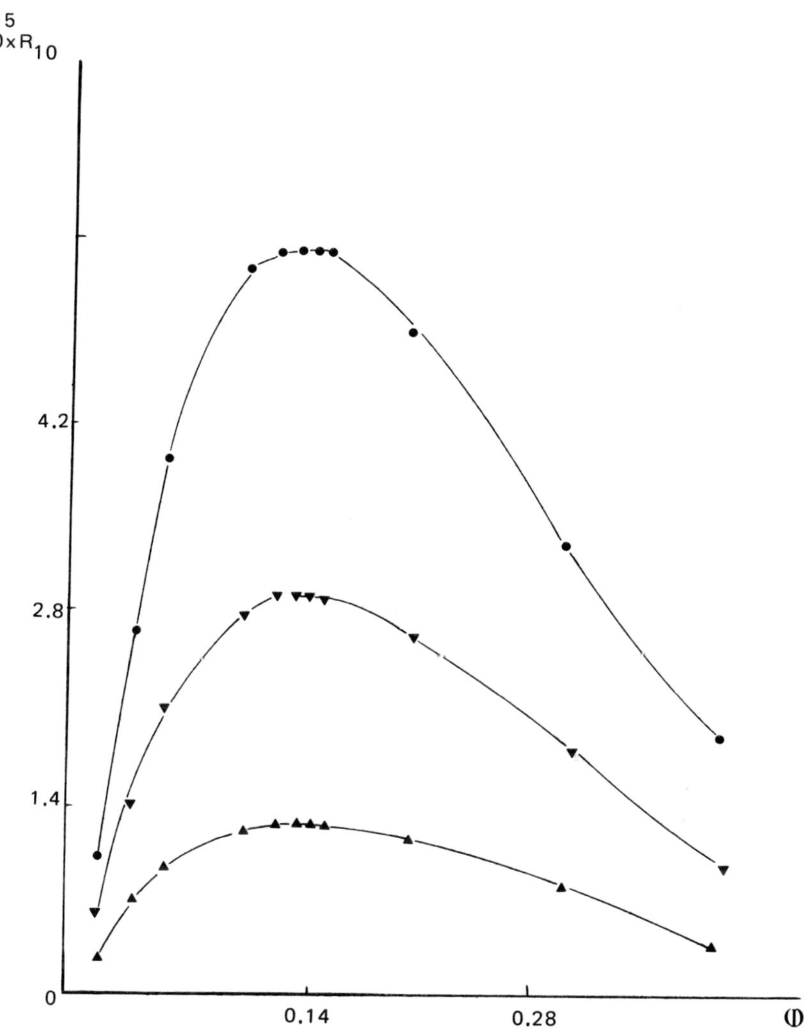

Figure 5 Calculated scattering intensities (Rayleigh ratios at $10°$ scattering angle; nm^{-1}) as a function of volume fraction of randomly dispersed spheres of 300 nm diameter; ($\bullet$——$\bullet$), $n_1 = 1.58$, $n_2 = 1.43$; ($\blacktriangledown$——$\blacktriangledown$), $n_1 = 1.48$, $n_2 = 1.38$; ($\blacktriangle$——$\blacktriangle$), $n_1 = 1.46$, $n_2 = 1.43$.

The above symbols are used in the designation of density fluctuations. Similar symbols ($\bar{\delta}^2$ and b) are used in designating the orientation fluctuations. The reason for using two different designations for the same kind of refractive index fluctuation is that in most cases one can separate the contributions from the

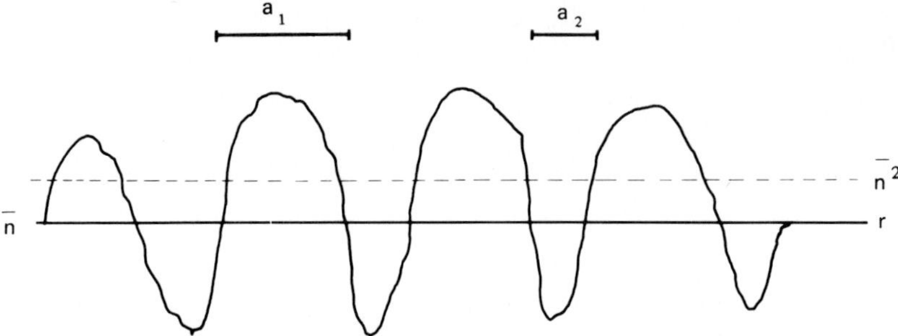

Figure 6 Random fluctuation in refractive indices as perceived by a photon traveling r distance in a medium, where $\bar{n}$ is the average refractive index of the medium, $\bar{\eta}^2$ is the amplitude factor, the mean squared deviation from the average refractive index, a_1 is the correlation length representing the length of the scattering center (high refractive index domains), and a_2 is the correlation length representing the separation of the scattering centers.

two fluctuations. This is done by the use of plane-polarized light as a source. If the scattering sample is placed between two polarizers whose axes of polarization are parallel, the measured scattered light intensity is designated $I_{\parallel}$. On the other hand, if the polarizer and analyzer are set 90° apart, the scattered light intensity is measured in the I_+ mode.

The theory of light scattering from random density orientation fluctuation (15) predicts that

$$I_{\parallel} = K\left\{\bar{\eta}^2 \int\gamma(r)\frac{\sin hr}{hr} r^2\,dr + \frac{4}{45}\bar{\delta}^2\int f(r)\left[\frac{\bar{\eta}^2}{\alpha^2}\gamma(r) + 1\right]\frac{\sin hr}{hr} r^2\,dr\right\} \quad (19)$$

$$I_+ = K\left\{\frac{1}{15}\bar{\delta}^2\int f(r)\left[\frac{\bar{\eta}^2}{\alpha^2}\gamma(r) + 1\right]\frac{\sin hr}{hr} r^2\,dr\right\} \quad (20)$$

where $\bar{\eta}^2$ and $\bar{\delta}^2$ are the average deviations in density and orientation fluctuations, respectively, as explained before, $h = (4\pi/\lambda)\sin(\theta/2)$ is a function of the scattering angle θ, where λ is the wavelength of the light in the medium, and r is the distance of penetration of light in the medium. The correlation function of the density function $\gamma(r)$ is set up in such a manner that, when two volume elements are the same (i.e., their separation is zero), $r = 0$ and $\gamma(0) = 1$. This is a perfect correlation. The form of the correlation function is

$$\gamma(r) = <\eta_1, \eta_2>_r/\bar{\eta}^2 \quad (21)$$

For all pairs of volume elements separated by a distance r, $<\eta_1\,\eta^2>$ is the average of η_1 and η_2. On the other hand, η_1 and η_2 are the local deviations (of refractive indexes from the average) of two volume elements separated by a distance r.

At the other extreme, when the two volume elements are separated by $r = \infty$, no correlation exists and $\gamma(\infty) = 0$.

The experimental $\gamma(r)$ must be fitted to an analytical function obeying the boundary conditions. In some cases (14,16) a single exponential function,

$$\gamma(r) = e^{-r/a} \tag{22}$$

fits the experimental density correlation function, where a is the correlation length of the density fluctuation.

In other experiments it was found (17,18) that the scattering over the entire angular range could be described by a sum of Gaussians:

$$\gamma(r) = \Sigma X_i \exp\left(\frac{-r^2}{a_i^2}\right) \tag{23}$$

where $X_i = 1$.

To interpret such an experimental correlation function, one can consider the inhomogeneous system a quasi–two-phase system (dispersed and dispersant; $i = 2$). Thus, two different correlation lengths can measure the size of the inhomogeneities under these conditions: a_1 and a_2 are the average sizes of the dispersed and the dispersant phases, respectively (19–21).

When a proper analytical form of the experimental $\gamma(r)$ is obtained, the $\bar{\eta}^2$ values can be evaluated from Eqs. (19) and (13) by using the Rayleigh ratio of intensities at a set angle.

Once this has been accomplished, the nature of the orientation correlation function f(r) can be predicted and thus the experimental orientation correlation function f(r) can be obtained. The experimental orientation correlation function can be expressed as the result of two kinds of inhomogeneities, and similar to Eq. (23), the corresponding form of f(r) will be

$$f(r) = \Sigma X_i \exp\left(\frac{-r^2}{b_i^2}\right) \tag{24}$$

where $i = 2$.

With the aid of the analytic form of f(r), Eq. (20) is rearranged and solved for $\bar{\delta}^2$ at a set scattering angle. In this manner eight structural parameters can be obtained from the light-scattering measurements. In relation to lens transparency and cataract formation, these parameters are size, volume fraction of the aggregates within the fiber cell, average distance of their separation, and refractive index difference between the scattering units and their environment. Similarly, four parameters are obtained from the depolarized light-scattering envelope, giving information on the size, volume fraction, separation, and refractive index difference of the optically anisotropic units (cytoskeletal filaments, microtubules, and so on) within the lens (22–24).

NORMAL LENSES

Light Scattering from Membranes

In the normal lens most of the light scattering comes from fiber cell membranes, which have higher refractive index than the surrounding cytoplasm. Since fiber cells are regularly packed, this light scattering gives rise to a diffraction pattern. Ocular halos that surround bright objects are due to this diffraction effect (25).

When lasers became available, we were the first to produce sharp diffraction patterns from thin sections of bovine lenses that showed even second- and third-order diffractions (26,27). The repeating distances calculated from the laser diffraction patterns corresponded well to the membrane-to-membrane distances in fiber cells observable in electron micrographs. These repeating distances changed as one proceeded from the cortical region to the nucleus, becoming progressively smaller. The diffraction patterns could be observed only with unpolarized light or polarized light in the $I_{\parallel}$ mode. This indicates that the membranes causing the diffraction have only periodic density fluctuations and no optical anisotropy. Similar diffraction patterns were found in thin sections of human lenses by Philipson (28), and the calculated spacings from such patterns corresponded well to the thickness of cortical fiber cells, 7 μm.

Using the laser diffraction patterns from thin lens sections, we studied (29) the behavior of the fiber cells and their orientation under stress. When the lens sections were submitted to periodic vibrations the diffraction spots showed periodic changes in their intensities. This could be interpreted as a change in the refractive index difference between membrane and surrounding due to the migration of water. Cortical fiber cells had a capacity to store water, possibly between two adjacent fiber cells, when subjected to stress. In the nucleus there was less barrier to water migration under stress than in the cortex.

Studies by Benedek et al. (30) and Clark et al. (31) have shown that the intensity of the diffraction pattern also increases upon cold cataract and salt cataract formation in calf lenses, although the positions of the diffraction spots do not change. Scanning electron microscopic observation showed that, in these cataracts, patches occurred on the surface of the cell membrane, which increased the refractive index fluctuation and therefore could account for the increase in intensity of the diffracted laser beam. These observations further reinforce our belief that diffraction in the lens is a membrane-caused phenomenon.

Light Scattering from Cytoplasm

The amount of light scattered by normal human lens is less than 5%. More than half of this comes from the membrane scattering described above. This happens in spite of the fact that the volume fraction of membrane in the lens is about

0.05. This means that the cytoplasm is largely transparent. Trokel (32) thought that an even distribution of lens protein within the fiber cells may account for the transparency. Benedek (33) accounted for the transparency by proposing that in normal lens the sizes of the spatial Fourier components are smaller than the wavelength of the light. Thus, either the absence of large fluctuations in refractive index (28) or the absence of scattering units comparable to the wavelength of the light provide a clear (non-turbid) appearance.

The cytoplasm with its 35% solid content, most of which is macromolecular compounds, is an unlikely candidate for a transparent tissue. It is the close packing of these components with little refractive index fluctuation between them that provides transparency. A model of uniform scatterers randomly distributed (12) showed that calculated light-scattering intensities are greatly reduced by destructive interference when interactions among all scatterers are considered, as compared with calculations when only interactions between nearest neighbors are taken into account. The same is true if the scatterers are not randomly aligned but are distributed in a certain lattice (34). This concentration effect accounts for the phenomenon that, if lens proteins are extracted and a more dilute solution than the lens concentration is prepared, the protein solution becomes turbid. In this case both the separation of the scattering particles is increased, hence the correlation between pairs of destructive scatterers decreases, and also the amplitude of the refractive index fluctuation increases.

Delaye and Gromiec (35) found that crystallines in the calf cortex behave as "hard spheres," the radius of which does not depend on concentrations. The intensity of the scattered light increases in dilute solutions with increases in concentration up to about 14% concentration. At that point a decrease in scattering intensities ensues as the matrix becomes more and more homogeneous.

These results are in excellent agreement with the predictions of the random fluctuation theory of Bettelheim and Siew (12), which also indicated maximum scattering at 14% volume fraction.

Further confirmation was also provided by Delaye and Tardieu (36) in studying the small-angle x-ray scattering of calf cortex cytoplasm. The normalized intensity curve versus concentration had the same shape as in light scattering (35). This, therefore, reinforces the arguments that the transparency of the normal lens is accounted for by the "interference effects due to correlations in the particles' position" (12) or in other words due to the short-range order of the crystallin proteins in the lens (36).

Clark et al. (37) obtained undiluted cytoplasm both from the cortex and nucleus. They separated the two by forming cold cataract at 4°C in the nucleus and dissecting it. After homogenization they centrifuged the sample to obtain membrane-free fractions. The distribution and the average size of the cytoplasmic bodies seem to be the same both in the cortex and nucleus: 12.5 nm in diameter as calculated from light-scattering spectroscopy (quasi-elastic light scattering). This is in contrast to our findings that the size of the scatterers in normal human lens is at least a magnitude larger, 100–300 nm (22,23,38).

One reason for the disagreement is the different techniques of investigations. While the quasi-elastic light scattering measures the diffusion coefficients, and using certain assumptions, one can calculate an average diameter size, the static light scattering using the random distribution theory of Debye and Bueche (14) yields correlation lengths that are the size of the *significant* scatterers, although their concentration may be small. For example, the size of the scatterers in a 16-year-old human lens cortex was about 100 nm; this was confirmed by electron microscopy (38), but it constituted only 1% of the total volume. Thus, the very small scatterers, 17.5 nm, do not show up in our calculations because they do not contribute significantly to the light scattering. In a similar manner we find that the nucleus, even in normal human lens, contains larger scatterers (300–500 nm) than the cortex. Furthermore, the volume of these large scatterers is between 10 and 25%. Again, this difference between Clark et al.'s (37) and our observations is explainable by the difference in technique, that in our measurement we derive size from the elastic scatterer and not from the energy absorbed by the particle to enhance translational motion. That means that we pick up only those scattering centers that contribute significantly to turbidity.

Besides the density fluctuation contributed by the size and the refractive index difference between scatterers and environment, one must deal with orientation fluctuation in the cytoplasm. This is due to the fluctuation in the orientation of optically anisotropic particles. Such particles exhibit intrinsic birefringence when aligned; that is, their refractive index along the long geometric axis is different than perpendicular to it.

In the normal lens, optical anisotropy is provided by the alignment of supramolecular structures, such as cytoskeletal thin and intermediate filaments (39–43) made of optically anisotropic vimentin and actin molecules or microtubules made of tubulin (44–46). Other supramolecular structures that could provide intrinsic birefringence may be the α- and β-crystallins aligned in semi-crystalline matrices by low-molecular-weight phosphate-containing peptides (47–49). Regarding the distribution of different cytoskeletal components, it was mainly observed that microfilaments are localized along the cell boundaries containing actin. Because of the presence of high ATP concentrations in this region, Ramaekers et al. (50) suggest that these microfilaments are contracted, providing an optimal curvature of the lens and thus playing a role in accommodation. These authors also find that the intermediate filaments, composed mainly of vimentin, are distributed throughout the cytosol of the fiber cell, thus forming a flexible skeleton. Lieska et al. (51) have shown that the intermediate filament of fiber cells form a triple-helix superstructure from three strands of α-helices. These are the protofilaments of 1.9 nm diameter. The protofilament twist into double helices of 5 nm diameter, and these form a superhelix of 8–14.8 nm diameter. These structures can be obtained both with keratin and vimentin. Lieska et al. (51) find the positive fluorescence for vimentin in bovine lenses surprising since lens has an epithelial origin and vimentin is thought to be of mesenchymal origin.

In spite of these highly organized optically anisotropic elements in the normal lens fiber, there is very little light scattering coming from orientation fluctuations. This can be seen from the small contribution of the I_+ mode of scattering compared to that of $I_{\parallel}$ of normal lenses. Also, the total birefringence of normal lens is small, of the order 10^{-5}–10^{-7} (52–54). For lens cytoplasmic elements of the proper size to be optically anisotropic, they must possess two properties: the molecules and molecular segments must be optically anisotropic, and their orientation in space must not be random. These two properties combined will give rise to intrinsic birefringence, that is, optical anisotropy due to molecular structure. Another kind of birefringence, form birefringence, develops when geometrically anisotropic bodies, such as needles or plates, are embedded in a matrix with a different refractive index. The form birefringence increases with the preferentially parallel alignments of the needles or thin plates.

The sum total of form and intrinsic birefringence will give the total birefringence that will measure the optical anisotropy of the sample.

Because of the particular morphology of the lens—thin membranes (plates or tubes) surrounded by a cytoplasmic medium of different refractive index—the lens must possess form birefringence. The estimated form birefringence is of the order of 10^{-4} (52). That the total birefringence is 100 or 1000 times less implies that an intrinsic birefringence of the opposite sign must cancel out the total birefringence. This theory (53) has been proved also by light scattering.

The theory of minimal total birefringence of the lens (53) through cancellation of the intrinsic birefringence by the form birefringence has a certain teleologic significance. The lens is a cellular body composed of onionlike layers of fiber cells. To minimize the light scattering that is inevitable from such a morphology (i.e., resulting from form birefringence), some supramolecular organization within the fiber cells must provide intrinsic birefringence to minimize the optical anisotropy (55). A corollary statement to this is that any disturbance that perturbs the delicate balance between form and intrinsic birefringence will inevitably lead to turbidity, be it the disintegration of membranes or the disorganization of cytoplasmic supramolecular structures.

The role the form birefringence plays in counteracting the effect of (molecular organization) instrinsic birefringence is illustrated convincingly in fish lenses. It has been known since the work of Brewster (56) that fish lenses have three distinct optical areas: the cortex and the nucleus have positive birefringence and the intermediate transition zone negative birefringence. Burke et al. (57)

have shown that in the transition zone the lateral interdigitations of the fiber cells have increased greatly compared with that exhibited by the cortex and nucleus. This means that the surface area of membranes is much greater in the transition zone than elsewhere. Membrane of fiber cells gives rise to negative birefringence. If the cytoskeletal elements are aligned along the long axis of the fiber cell, they give rise to positive intrinsic birefringence. Where the surface area of the membranes is relatively small (cortex and nucleus), the positive intrinsic birefringence overcomes the effect of form birefringence and the total birefringence is positive. In the transition zone, however, where the surface area of membranes is large, the form birefringence dominates and the total birefringence is negative.

CATARACTOUS LENSES

In contrast to the normal lens, cataractous lenses scatter significant amounts of light. The result is twofold: less light will reach the retina, especially the fovea, and the image will not be formed. The net effect is a glare at the beginning and partial or complete blockage of the passage of light, which manifests in impaired vision or complete blindness. The glare produced with different light sources seem to be independent of the wavelength λ of the light. Therefore, Moon and Spencer (58) concluded that the scattering is not a Rayleigh type; that is, it does not have λ^4 dependence. Therefore, the scattering particles must be within the order of the wavelength $\lambda/20$ or larger. Using a reduced eye, in which all the components are reduced to a uniformly scattering field, Fry (59) calculated the amount of light scattered reaching the fovea. He used Mie-like (60) scatterers to predict the relation between transmitted and scattered light. However, the eye is not a uniform scatterer. The lens scatters much more light than the aqueous humor or even the vitreous. The scattering from the lens is the main contributing factor to glare, especially where cataract formation occurs. The most important glare causing cataract is the posterior cataract, which rescatters the already focused (by the cornea and the anterior part of the lens) light rays.

There are a number of mechanisms by which cataract may form. These are aggregation processes, syneresis, microphase separation, membrane disintegration, and change in the orientation of cytoskeletal elements. These mechanisms can act independently in individual cases as a sole cause, or two or three mechanisms may act simultaneously, providing a synergetic effect enhancing the development of cataracts.

Aggregation

Proteins and other macromolecules in the normal cytoplasm are highly hydrated. They are in constant thermal motion. The effective number of collisions between these molecules are modulated by the viscosity of the medium, the hydration layer, and the surface charge density on the proteins. Changes in any of these properties may bring the molecules in close contact, resulting in aggregations. The importance of the aggregation process lies in the fact that domains that were previously too small to be perceived by the light as fluctuations will grow in size when they aggregate. They become domains $\lambda/20$ or larger and will scatter light.

Mach (61) proposed that the formation of water-insoluble aggregates from water-soluble proteins caused the turbidity of lenses. The light scattering of human lens was known to increase with age (62). There is also a concomitant increase in the water-insoluble fraction of lens protein with age (63,64). Thus, the increased turbidity of human lens and increased aggregate formation in the form of insoluble proteins were linked in the aging process. Benedek (33) suggested that the aggregates must reach a molecular weight of more than 1×10^6 to cause significant scattering. Such high-molecular-weight proteins were found mainly in the nucleus of lenses (65–70). The proportion of such high-molecular-weight soluble protein increases with aging (66–68,70,71) and in human senile cataract as well as in x-ray–caused cataracts (72,73). It was suggested that high-molecular-weight proteins are the precursors of insoluble protein aggregates. [See, for review, Harding and Dilley (74).] The size of such large irregularly shaped particles was found to be variable, up to 500 nm in diameter, by electron microscopy (75).

The cause for aggregation can be a multitude of changes in the secondary and tertiary structures of proteins. Harding and Dilley (74) reviewed these proposed schemes: disulfide bridge formations, Ca^{2+} salt bridges, and nonsulfide covalent cross-links. Among the latter, transamination is the latest mechanism proposed for aggregate formation, Lorand et al. (76) not only showed the presence of transglutaminase in lenses (especially in the cortex) but they also isolated the γ-glutaminyl-ϵ-lysin isopeptide residues from cataractous human lenses, which are the cross-link products of the enzyme. Further evidence that transamination may be a significant contributor to the aggregation process comes from the work of Azari et al. (77). Hereditary cataractous rat lenses showed significantly higher transglutaminase activity than normal lenses of comparable age. The rat enzyme isolated could use crystallins as substrates; the γ-crystallin was the most effective substrate. This is in contrast to the findings of Lorand et al. (76), who found β-crystallin is the most suitable among the crystallins.

Other types of aggregation products have also been found in which only certain crystallin subunits participate. Roy and Spector (78) found that the high-molecular-weight aggregates contain only α and β subunits. Furthermore, γ-crystallin is excluded from the crystallin aggregates when such are prepared from the subunits by in vitro reaggregation (79). Molecular self-assemblies using crystallin subunits were also void of γ-crystallins (80,81).

On the other hand, Siezen and Owen (82) found that calf lens nuclear α-crystallin preparations all contain minute amounts of β-crystallin and also a 43 kd component. The highest molecular weight subpopulation of calf lens nuclear α-crystallin also contained some γ-crystallins.

On the basis of water vapor sorption studies on α-crystallin subunits, Finkel and Bettelheim (83) found that the aggregation of the subunits into α-crystallin assemblies occurs through hydrophobic interactions. The water-sorbing capacity increases with the complexity of the aggregates. This means that an aggregate of αA_1, αA_2, αB_1, and αB_2 can sorb more water than the aggregates of αA_1 or the other subunits or in any binary or ternary combination. This implies that thermodynamic stability of α-crystallin is achieved by hydrophobic interactions in the interior of the assemblies, turning hydrophylic polar groups to the outer surface of the assemblies. The Mörner α-crystallin that contains α and β subunits had even higher water sorptive capacity than α or β alone, again indicating a thermodynamically favorable interaction between these chains.

We also have shown that α- and β-crystallin chains can be organized into a crystalline matrix by phosphate-containing small peptides (47,49). In these crystalline organizations, similar to the molecular self-assemblies of Bloemendal et al. (80,81), the γ-crystallin was not present. Furthermore, the addition of γ-crystallin to a mixture of α- and β-crystallin and phosphate-containing peptide prevented the organization of these polymers into a crystalline lattice that gives x-ray diffraction pattern.

On the basis of ultrafiltration experiments on concentrated calf lens extracts, Malinowski and Manski (84,85) claimed to detect $\alpha\beta\gamma$, $\alpha\beta$, and $\beta\gamma$ complex formations. However, Siezen has shown recently that ultrafiltration is not a suitable technique to differentiate between mixed associations and self-associations (86).

Thus it appears that there are a number of ways high-molecular-weight aggregates can form, which may eventually end up in water-insoluble protein formation as one of the mechanisms of cataract formation. However, a cataractous process may not be a simple, one-track process. As the Marlat et al. (87) study indicates, different human cataracts have different and possible multi-channel pathways. Unselected human senile cataract has increased Na^+ and Ca^{2+} and decreased K^+ and increased hydration and insoluble protein. Senile cortical cataract have increased hydration and Na^+ content; nuclear cataracts have increased insoluble protein as well as increased brown pigmentation. The two processes are not correlated. Na^+ content is correlated with the overall decrease of soluble protein content but not with the insoluble protein content. The decrease in soluble protein is not due simply to reduced protein synthesis but also degradation and leakage; thus not all soluble protein that is lost appears as insoluble protein (87).

Syneresis

The factors influencing the increase in density and orientation fluctuations have been enumerated (p. 274–277). One way the intensity of the scattered light can be increased is through aggregation, that is, increasing the size of the scattering units, as discussed. Another way to increase turbidity is to increase the amplitude of the density fluctuations.

An increase in the amplitude of fluctuations means an increase in the refractive index difference between the scattering unit and its surroundings. The process by which such change in the refractive index difference can be brought about is called the syneresis (88). It must be emphasized that this syneretic process is independent of aggregation. Under certain cataract formation one finds the aggregation and syneresis occur simultaneously (24,89), although under other conditions an increase in turbidity occurs due to syneresis alone or even in the presence of decrease in the size of the scattering units (23).

We shall call syneresis any molecular or supramolecular process that enhances the difference in the refractive index between the scattering unit and its environment. A collapse of the protein network in which the water of hydration is lost and ends up as bulk water is the primary mechanism for syneresis (88). The water of hydration is often referred to as bound water; the bulk water is called free water. Other names used in this context are freezable (bulk) and nonfreezable water (water of hydration). Some or all of the water of hydration may have icelike structure, and thus it is unfreezable. Neville et al. (90) have shown that, in the lens, part of the water is bound or is in an ordered state. Rink (91) divides the lens water content into extracellular (10%) and intracellular (90%) water. Of the 90% intracellular water, according to Rink, 54% is bound and 36% is free. Further classification indicates that, of the 54% bound water, only 3–4% is tightly bound. The others are loosely bound and therefore interchangeable with bulk water. Bettelheim and Christian (92) have calculated that the amount of water of hydration converted to bulk water in a cataractous process may amount to as much as a third of the total intracellular water. Using the minimal hydrogen bond energy of 3 kcal/mol, they calculated that a syneretic process may require a minimum of 33 cal/g lens tissue. The process itself is endergonic (endothermic).

Evidence that the movement of water in and out of the lens results in cataract formation was accumulated early. Cataracts can be caused by external conditions, leading to dehydration. It was observed early (93) that the rat lens developed cataracts when eyelid closure was prevented. Similar data are available on guinea pig lens (94). In both cases the dehydration was caused by exposure of the cornea. Other experiments (95,96) showed that drug-induced cataract formation due to analgesics, morphine, or epinephrine could be prevented if the eyelids were kept closed. In each of these cases the dehydration of the cornea set up a water gradient, which lead to the loss, probably first, of extracellular water from the lens. Later this was replaced by the migration of intracellular water into the extracellular place, and the loss of intracellular free water lead to the dehydration of the protein network. Nuclear magnetic resonance (NMR) studies have often been used to estimate the amount of bound and free water from relaxation times. Rácz et al. (97) measured the spin-spin and spin-lattice relaxation times in normal and senile cataractous human lenses as a function of temperature. They found that below $-9°C$ there was no difference between the relaxation times of normal and cataractous lenses. Above $-9°C$ both the spin-spin and the spin-lattice relaxation times were higher in the cataractous lenses than in the normal lenses. This was interpreted as an increase in the free water versus the bound water content upon cataractogenesis. Although the overall hydration of the lens was the same in normal and cataractous lenses, only 76% of the total water was free in normal lenses but 87% was free in cataractous lenses.

This strongly supports a syneretic process, that is, the release of strongly bound water from protein aggregates into the surroundings.

A similar syneretic exudation of water from protein aggregates was found in ultraviolet (UV)-induced cataracts by studying the Raman spectra of lenses (98). Pope et al. (99) also measured the NMR spectra of normal and cataractous lenses. To obtain the spin-lattice relaxation time two modes were employed: (1) inversion-recovery and (2) locking. The total water content was measured by vacuum drying at $50°C$ to constant mass, which may not have removed all the bound water. These authors find that, although spin-lattice relaxation times above the freezing point are somewhat different for cataractous than normal, this is not sufficiently large to change the bound to free water proportions in nuclear cataract, only in swollen completely opaque cortical cataract. The discrepancy between the Rácz et al. (97) and the Pope et al. (99) findings may not simply be due to the different experimental techniques employed. Rácz et al. (97) observed both the spin-spin and spin-lattice relaxations, Pope et al. (99) only the latter. It has been shown in other tissues, for example in skeletal muscles (100), that although spin-lattice relaxation does not show any significant anisotropy, one can observe significant anisotropy in the spin-spin relaxation times due to the ordering of water chains (water of hydration) along the muscle fibers. Although the NMR data interpretation may not firmly establish the syneretic process as a primary cause of senile nuclear cataract formation, other evidence points in this direction.

Bettelheim et al. (101) investigated the status of different water in the lens with a more direct technique: differential scanning calorimetry (DSC). This measures the amount of freezable ("free") water. A separate measurement by vacuum dehydration measures the "total" water content. The difference is the bound water. In different parts of two representative human lenses (one normal and one cataractous), the finding is as follows: 39, 40, and 46% freezable water of the total water content in normal and 74, 64, and 51% in cataractous lenses in respective areas of cortex, intermediate, and nuclear zones. Although the trend found by DSC measurement is the same as interpreted from NMR measurements by Rácz et al. (102), the absolute values are quite different. The DSC measurements are direct and not prone to different theoretical interpretations, which is the case with converting relaxation times to water contents. On the other hand, Bettelheim et al. (101) values are preliminary and only on two representative lenses. The Rácz et al. (102) values are average values of whole lenses, but they do not give the number of samples investigated or the standard deviation. Further studies are needed to confirm which range of free/bound water is the correct for normal and for cataractous lens.

The analysis of the angular dependence of the intensity of the scattered light from thin lens sections enables us to evaluate independently the size of the scattering units, the concentration, and the amplitude factor reflecting refractive index differences. (See Scattering of Light and Ref. 22.)

The analysis showed that the amplitude factor increased with age in normal human lenses. The rate of increase with age was the same in the different parts of the lenses (23). When one takes into account that in normal human lens the size of the scattering units decreases rather than increases with age (collapse of protein), we can assume that the diminution of transparency with age is strictly due to a syneretic process. A collapse in the protein network would result in a decrease in the light scattering with age, but an increase in refractive index difference counteracts this, and thus the transparency of the lens decreases with age due to syneresis alone.

The same type of analysis applied to nuclear cataracts (24,89) shows that here, too, syneresis is a major pathway to cataractogenesis. Roughly 42% of the opacity found in nuclear cataracts could be attributed to syneresis-caused refractive index changes. Although the aggregation process is evident in nuclear cataract, it is a relatively minor contributor to opacity. Only 14% of the opacity could be accounted for by the aggregation process alone (89). On the other hand, the distance separating the scattering centers correlated much better with opacity, with a gross correlation coefficient of 35%.

It is interesting to note that the morphology of the nuclear cataract is also indicative of the cataractogenetic process. Nuclear cataracts, in which the center of the nucleus is the densest, correlated well with a syneretic process-generated refractive index change (99.6%). However, nuclear cataracts that had a ringlike appearance showed little correlation with the amplitude factor of density fluctuation (89). These cataracts had a better correlation with membrane degradation and cytoskeletal orientation degradation processes, which we shall discuss later.

A syneretic process may also contribute to cortical cataracts. In metabolic cataracts, which show small vacuole and eventually lake formations as the cause of turbidity some part of the process may be of syneretic origin. Kinoshita (103) described a typical sugar cataract, for example, that is attributable to high galactose in the diet, as due to the formation of galactitol that is not metabolized further. Galactitol or other sugar alcohols, such as sorbitol, accumulates in the lens, especially in the cortex. This generates an osmotic pressure imbalance, and water is imbibed into the fiber cells, resulting in turbidity and swelling. An increase in the galactitol or xylitol concentration in the lens fibers results in an effective competition for water between the lens proteins and the smaller molecules. Thus, an increase in sugar alcohol may decrease the hydration of the lens proteins. Li (104) has shown that the gel permeation chromatography of lens proteins is different in water than in sucrose solutions. Part of the difference can be explained by a less hydrated form, hence a smaller hydrodynamic radius of lens proteins in the sucrose medium than in the aqueous medium.

Philipson (105) has proposed that in sugar cataracts the water in the cortex comes from the surrounding fluids and includes water from the nucleus. Bettelheim (53) and Bettelheim and Bettelheim (106) showed that this, in effect, was the case in incipient cataract formation in galactose and xylose cataracts, respectively. Although the cortex showed an increase in form birefringence, hence in the relative contribution of the optical anisotropy as compared with density fluctuations, the nucleus showed a decrease. If such a dehydration process can occur across lens membranes, it can occur with greater ease within the same fiber cell. That migration of water from different parts of the lens can occur as a result of osmotic pressure gradient or other external forces applied was demonstrated dramatically by Chylack et al. (107) and Bettelheim et al. (108). In studying the optical behavior of thin sections of cortical cataracts as well as of whole lenses with cortical cataract, we observed that during a freezing and thawing cycle the turbidity disappears. In essence, during the freeze-thaw cycle the lakes between fiber cells, as well as vacuoles within the fiber cells, disappeared. This finding was substantiated in histologic studies.

In summary we have seen that syneresis can play part in the aging as well as in certain kind of cataract formation of lenses. The question arises as to what kind of molecular processes can contribute to the syneretic behavior. Changes in primary, secondary, tertiary, and even higher structures of the lens proteins can lead to change in the macromolecular suprastructures, resulting in gradual dehydration of proteins and sequestration of water into the surrounding medium. Among the changes in the primary structure one may mention the gradual deamination of glutamine residues with age (109–111). This process progresses differently in different species (74,112,113). Among the changes in the secondary and tertiary protein structure that can lead to syneresis we may include all the cross-linking processes mentioned in aggregation. However, in syneresis the cross-linking is intramolecular rather than intermolecular. Thus, it results in the collapse of network. Disulfide bridges (114–116) and transamination (76,77) can be important contributors. Other conformational changes, such as those due to racemization (117–119), may also lead to a dehydration process. This interrelation is also supported by Raman spectroscopic studies (120,121), where the mouse lens nucleus was rapidly dehydrated for the first 4 months after birth and beyond that only gradual dehydration occurred. The S-H to S-S transition of the lens proteins showed a similar time dependence, thus relating cross-linking to dehydration.

Finally, we should mention in connection with syneresis that this is not necessarily a localized effect. Hydration and dehydration of the whole lens may result from syneretic processes, especially since we have seen that, due to water migration, not only intracellular vacuole formation may occur but also inter-fiber-cell lake formations as well. The extracellular water, on the other hand, is accessible to exchanges between aqueous and lens, and vitreous and lens.

In this connection it is worthwhile to mention that diabetics suffer from myopia when their blood glucose rises and from hyperopia when the blood glucose falls. We have shown that the turbidity of the lens is proportional to the water weight loss and gain (hydration-dehydration) during incubation of lenses in hypertonic media and the reversal into isotonic media (122). Therefore, the fluctuation in blood sugar of diabetes not only causes faulty focusing (dehydration of the lens focuses the light in the vitreous cavity) but also glare due to scattering.

Phase Separations

Young animal lenses have an interesting property. When the temperature is lowered opacity develops, which will disappear upon raising the lens temperature again. This "cold cataract" has been attributed to phase separation by their earliest students (123,124). Lerman and Zigman (125,126) thought that the reversible opacification is due to conformational changes of cryoproteins in the lens, which were identified with the γ-crystallin fractions. However, Horowitz et al. (127) showed that in cold cataract formation there is no apparent conformational change as observed in circular dichroism measurements and suggested that this is simply a precipitation phenomenon occurring with γ-crystallins, which are near their isoelectric point and, therefore, have low solubility.

Tanaka and Benedek (128) were the first who systematically studied the phase separation in lenses. "Phase separation" in this case does not mean the observation of two visible phases; rather it refers to the interpretation of the appearance of opacity. Opacity, the argument goes, is due to the aggregation of protein molecules forming microphases of a size comparable to the wavelength of the light. When one studies the spread of cold cataracts in young rat lenses, one finds that at first an annular cataract is formed approximately 0.5 mm from the cortical surface. When the temperature is lowered, this opacity spreads toward the surface of the lens as well as toward the nucleus. Plotting the positions in the lens where the edge of turbidity becomes a function of temperature, they obtained a bell-shaped coexistence curve that very much resembled the coexistence curves for a protein-water mixture in which the protein concentration was varied and the temperature at which turbidity appeared was plotted against the protein concentration (129).

In calf lenses the protein concentration gradient is different; the critical concentration is in the center and the cold cataract develops in the nucleus but not in the cortex (130). Although in different species the lens cold cataract may have different appearances, what is common in them in most species is their reversibility. Some species, like fishes, have shown that part of the cold cataract, an annular zone where the membrane interdigitation is the greatest in the transition zone between cortex and nucleus, remains irreversibly opaque even after the lens was brought back to room temperature (131). The same is true for pressure-induced cataract formation in fish lenses (132).

In spite of these exceptions, cold cataract formation is accepted and referred to as a reversible phenomenon. Thus the term "phase separation" can be applied to it since typical coexistence curves are associated with its appearance.

The question arises of what makes phase separation a distinct mechanism, apart from the aggregation and syneresis mentioned in the previous sections. The distinction lies exactly in the reversibility of the phenomenon, and, therefore, in the presence of equilibrium conditions. Thus, phase separation is governed by the laws of thermodynamics. On the other hand, both aggregation and syneresis are irreversible phenomenon, occurring far from equilibrium conditions, and therefore they can be understood only by the tenets of irreversible thermodynamics if not too far removed from equilibrium or by phenomenologic laws.

Originally, Tanaka and Benedek (128) proposed that the increase in the relaxation time of the density fluctuation as one approaches the temperature of opacification T_{cat} is due to a growth in the correlation range of concentration fluctuation. Later considerations (133) make this more succinct and propose that the new phase of droplets results from nucleation, a first-order transition. This occurs in isolated cytoplasm (37,130,133), where the protein concentration is less than the critical concentration found in calf nucleus. Under these circumstances there is no increase in the correlation of the concentration (density) fluctuation. The rapid dissolution of the droplets of the new phase when one heats the turbid isolated cytoplasm is consistent with a first-order phase transition. On the other hand, in the intact lens, such as in the nucleus of calf lens where the protein concentration reaches the critical range, only second-order transitions can occur (130). In essence that is what was found by differential scanning calorimetric study of calf lenses. There was no heat of transition (either exotherm or endotherm), only a change in the heat capacity at the "melting" point of the cold cataract, 16°C (92). This confirms that in the intact lens the phase separation is a second-order transition.

Further experiments of Benedek et al. (30) and Clark and Benedek (133) demonstrated that, when bovine lenses were soaked in glycerol and glycols, cold cataract formation at 5°C could be prevented; that is, the critical temperature T_c at which opacity appeared was lowered to below 5°C. Once the glycols and other agents were removed by soaking in a buffered saline solution, the critical temperature returned to 5°C. These authors believe that the reversibility of the coexistence curves by such agents as glycols may have practical use. These same agents not only inhibit cold cataract formation but they have similar effects on pathologic human lens. However, the most important aspect of reversible cold cataract formation is that it may be an early indicator for the irreversible cataract formation (134). Galactosemic cataracts in rat lens are associated with an increase in the temperature of the cold cataract formation (135). This is an interesting association because the two cataracts also occur in different portions of the lens: the galactosemic cataract is cortical and irreversible; the cold cataract is nuclear and reversible. Tanaka et al. (136) ran experiments on cold cataract formation of incubated lenses in low-glucose medium. They found that although the temperature of lenses at which cold cataract occurs was between 5 and 15°C when incubated in isotonic media, the hypotonic incubation showed a temperature range from 15 to 50°C or more, increasing with increasing time in hypoglycemic medium. The temperature dependence of the onset of "cold cataract" occurs prior to the appearance of the actual opacity due to the hypotonic media. The authors conclude that the phase separation of the cold cataract in the nucleus is a preindication of the actual cataract formation that will occur in the cortex due to water influx. Clark et al. (137) investigated the effect of x-irradiation on the temperature at which cold cataract formation occurs in rabbit lenses. Cold cataract formation was studied in x-ray irradiated and control lenses which were incubated up to 10 weeks after irradiation. The control showed a steady decrease in the temperature at which cold cataract formation occurs with incubation time. The x-irradiated lenses, however, showed an increase in cold cataract temperature 6 weeks after the irradiation, and the temperature continued to increase up to 10 weeks. This indicates that phase separation causing cold cataract is enhanced by x-irradiation and this enhancement occurs before the actual irreversible x-ray cataract develops.

Membrane Degeneration

In many cataractous processes morphologic changes occur that are most observable in the degeneration and later in the disintegration of membranes. Since these processes are reviewed more intensely in Chaps. 5 and 10 we shall include only a few typical examples. The importance of membrane degradation and disintegration lies in the multiple effect it has on opacification. The well-aligned geometric packing of normal fiber cells give rise to a diffraction effect (26,27, 29), which in a sense creates transparency by destructive interference over a large angular space section and only provides a certain halo effect, already referred to. Degradation and disintegration diminish the destructive interference, and turbidity will develop due to random fluctuations as opposed to the ordered fluctuations in normal lens.

Furthermore, degradation of membranes may influence another type of cataractogenesis. For example, Garner et al. (138) reported that the aggregation of cytoplasmic component may be initiated at membranes. They isolated membrane preparations containing disulfide-linked cytosol polypeptides in which γ-crystallin was the predominant component. The hypothesis is that the degradation of membrane allows oxidative changes that cross-link cytoplasmic crystallin and membrane proteins, thus allowing the nucleation of large aggregates (139). Morphologic changes that accompany sugar cataracts, in particular galactose cataracts, have been well documented (140–142). These changes include disalignment of lens fibers, membrane-bound vesicles randomly distributed throughout the fibercell, variable degrees of swelling, membrane disruption, enlargement of intercellular spaces, and vacuole and lake formation, electron-dense aggregates, decreased amounts of interdigitation between adjacent fiber cells, and granulation of the surface of fiber cells. Galactose cataracts are reversible, and both the morphologic changes disappear in reverse, paralleling the biochemical changes. The glucose-free media incubation of mouse lenses caused cataract formation similar to hypoglycemic cataracts (143). The membrane proteins are altered, and the appearance of the 23,000 d bound protein is rapid (after 48 hr). Hu et al. (144) believe that these biochemical changes are post-translational modifications of the 26,000 d component because very little synthesis occurs in glucose-deprived mouse lens. In the special case of Nakano mouse cataract, which is a convenient animal model to study osmotic pressure-caused cataracts, it was observed by Tanaka et al. (145) that alterations in the plasmic membrane are involved. There is a marked decrease in gap junction during cataractogenesis. This is also accompanied by the decrease in the main membrane polypeptide (26,000 d) content. Similar morphologic and biochemical changes occurred when rat lenses were incubated in glucose-free medium or in media that inhibitis Na^+,K^+, or Ca^{2+} transport (146). Unakar et al. (147) have shown that the acid phosphatase present in normal rat lenses (148), and senile rat lenses (149) increases its activity as the galactosemia develops in rat lenses. Since phosphatase is present in both extra- and intracellular locations, Gorthy suggested (149) that this enzyme is involved in the degradation of lens membranes. Unakar et al. (147) thought that this enzyme may be involved in the repair of injured tissues. Harding et al. (150) have shown that specific human lens opacities, such as punctate opacity, reveal numerous membraneous folds in scanning electron microscopy. Energy dispersive x-ray analysis (EDXA) of the opacity area yielded high phosphorus and low sulfur content. This was interpreted by the authors as caused by the intrusion of membrane into cytoplasmic space with concurrent degeneration (hence loss of sulfur) of the crystallins. The high phospholipid content of the membrane accounts for the increased P content.

Cataract formation is enhanced by exposure to sunlight. Studies show that people living near the equator have significantly more cataracts than those living in a temperate climate (151-153). Ionizing radiative damage, such as x-ray cataracts, show disalignment of fiber cells, disintegration of membranes, rounding out of swollen fiber cells, intracytoplasmic vacuoles, and lakelike intercellular spaces (70,154,155). Similar vacuolization and intercellular spaces are also found in cataracts due to nonionizing radiation (156).

Disorientation of Cytoskeletal Elements

The presence of cytoskeletal elements in the lens fibers is thought to determine the viscoelastic properties of the lens. This is of utmost importance in the main function of the lens: accommodation. For that reason these elements are reviewed extensively in Chap. 5. However, cytoskeletal elements provide optical anisotropy, that is, intrinsic birefringence, to the lens. Fluctuation in optical anisotropy contributes to refractive index fluctuation in the same way as density (concentration) fluctuations do. It is true that with unpolarized light the density fluctuations are more important than orientation (optical anisotropy) fluctuations, but many light sources are at least partially polarized. Certainly light reflected from shiny surfaces (mirrors) are partially polarized; therefore, the optical anisotropy fluctuations may play also an important role in impairment of vision in cataract formation.

We discussed the role of the orientation of cytoskeletal elements in the lens fiber in contributing to transparency. The basis of this is the compensation positive birefringence plays in counteracting the negative birefringence of membranes and thus providing a minimal total birefringence (52,53). Now if either membrane disintegration or cytoskeletal disorientation occurs, the balance providing minimal optical anisotropy is upset and as a consequence turbidity will develop. The small degradation of membranes will not influence the birefringence much, since the form birefringence is proportional to the volume fraction of membranes, which is small to begin with. However, small disalignment or disorientation of cytoskeletal elements will substantially decrease the intrinsic birefringence and thus upset the normal balance. The normal lens scatters light mainly in the $I_\parallel$ mode. This implies the absence or the relatively small value of optical anisotropy. The first sign of cataract formation is the large increase in scattering in the I_+ mode compared with the small increase in the $I_\parallel$ mode, leading to a drastic increase in the $I_+/I_\parallel$ ratio (106). This is a qualitative indication that in cataract formation something is happening to the orientation of cytoskeletal elements. A more quantitative approach is when one evaluates the actual contribution of cytoskeletal elements in the form of optical anisotropy to the opacification of lenses. In our analysis the depolarized scattered light intensity I_+ is used in conjunction with the Stein et al. (15) formulation to obtain both correlation length and amplitude factors of the orientation fluctuations.

One finds that the correlation length of optical anisotropy fluctuations are higher in normal lens (22,23) than in cataractous lenses (24). If we identify the correlation length with the average dimension of the cytoskeletal bodies, during cataract formation a reduction from about 1.5 nm to one-third or more occurs. This implies disorientation or degradation or both.

A second requirement so that optical anisotropy fluctuations will not generate appreciable light scattering is that the amplitude of such fluctuations be small. This was found to be the case in normal human lenses by Bettelheim and Paunovic (22), where the statistical measure of the refractive index differences due to optical anisotropy fluctuations $\bar{\delta}^2$ was 10^3–10^5 times smaller than that due to density fluctuations $\bar{\eta}^2$. This further demonstrates that intrinsic birefrigence or the alignment of supramolecular structures plays an important role in minimizing light scattering in normal lenses.

In contrast, in cataractous lenses the amplitude of optical anisotropy fluctuation increased tremendously; it became comparable to about one-tenth of the amplitude of density fluctuations (24). No wonder that, in incipient cataract formation, even before turbidity can be perceived, one can detect it by the $I_+/I_\parallel$ ratio (106). The cause of the increase in the amplitude of optical anisotropy fluctuation during cataractogenesis lies in the disorientation and randomization of cytoskeletal elements. This important point is hitherto little appreciated by the researchers in the cytoskeletal field, whose main concern is the chemistry and morphology as related to mechanical properties.

CONCLUSIONS

The transparency of normal lens and in turn the opacification of cataractous lens has a physical basis. It is the random or nonrandom spatial fluctuation in density (concentration) and in optical anisotropy (orientation of) macromolecules. When the size over which such fluctuations occur approaches the wavelength of the light and when the amplitude of the fluctuations are large, significant light scattering will occur, leading to cataract formation. Five processes are enumerated in this chapter that cause such fluctuations: aggregation, syneresis, phase separation, membrane degradation, and disorientation of the cytoskeletal elements. Each process alone or in combination can cause cataract and impaired vision due to lens turbidity.

For the clinician who deals with such symptoms as glare, color vision disturbances, monocular myopia, diabetic myopia and hyperopia, and a host of others, an understanding of the common physical basis of all these pathologic conditions is of utmost importance.

ACKNOWLEDGMENTS

I would like to thank the National Eye Institute for making our research possible throughout the last 14 years and for their continued support with grants EY-02571 and 03407.

I am grateful to Academic Press for allowing me to reproduce parts of my previous article: Biological-physical basis of lens transparency, which appeared in *Cell Biology of the Eye* (ed. D. S. McDevitt) in 1982.

REFERENCES

1. F. A. Bettelheim. *Experimental Physical Chemistry*, Saunders, Philadelphia, 1971, p. 149.
2. J. F. R. Kuck, Jr. In *Biochemistry of the Eye* (C. N. Graymore, ed.), Academic Press, New York, 1970, p. 190.
3. S. Lerman. *Radiant Energy and the Eye*, MacMillan, New York, 1980, p. 33.
4. F. A. Bettelheim and T. J. Y. Wang. Exp. Eye Res., *18*: 351 (1974).
5. B. Philipson. Invest. Ophthalmol., *8*: 258 (1969).
6. P. Debye. J. Appl. Phys., *15*: 338 (1944).
7. A. Einstein. Ann. Phys., *33*: 1275 (1910).
8. P. Debye. J. Phys. Colloid. Chem., *51*: 18 (1947).
9. K. A. Stacey. *Light Scattering in Physical Chemistry*, Academic Press, New York, 1956.
10. M. Kerker. *The Scattering of Light and Other Electromagnetic Radiation*, Academic Press, New York, 1969.
11. C. Tanford. *Physical Chemistry of Macromolecules*, Wiley, New York, 1961.
12. F. A. Bettelheim and E. L. Siew. Biophysical J., *41*: 29 (1983).
13. N. K. Ailawadi. Reports (Review Section of Physics Letters), *57C*(4): 241 (1980).
14. P. Debye and A. M. Bueche. J. Appl. Phys., *20*: 518 (1949).
15. R. S. J. Stein, J. J. Keane, F. H. Norris, F. A. Bettelheim, and P. R. Wilson. Ann. N.Y. Acad. Sci., *83*: 37 (1959).
16. L. Gallagher and F. A. Bettelheim. J. Polym. Sci., *58*: 697 (1962).
17. R. S. Stein. Polym. Lett., *1*: 657 (1969).

18.   K. L. Wun and W. Prins. J. Polym. Sci. Part A-2, *12*: 533 (1974).
19.   L. G. Kahovec, R. Porod, and H. Rick. Kolloid Z., *133*: 16 (1953).
20.   O. Kratky. Pure Appl. Chem., *12*: 483 (1966).
21.   L. E. Alexander. *X-ray Diffraction Methods in Polymer Science*, Wiley, New York, 1969.
22.   F. A. Bettelheim and M. Paunovic. Biophys. J., *26*: 85 (1979).
23.   E. L. Siew, D. Opalecky, and F. A. Bettelheim. Exp. Eye Res., *33*: 603 (1981).
24.   R. L. Siew, F. A. Bettelheim, L. T. Chylack, Jr., and W. H. Tung, Invest. Ophthalmol. Visual Sci., *20*: 334 (1981).
25.   G. C. Simpson. Br. Jr. Ophtlamol., *37*: 450 (1953).
26.   F. A. Bettelheim and M. J. Vinciguerra. Ann. N.Y. Acad. Sci., *172*: 429 (1971).
27.   M. J. Vinciguerra and F. A. Bettelheim. Exp. Eye Res., *11*: 214 (1971).
28.   B. Philipson. Exp. Eye Res., *16*: 29 (1973).
29.   F. A. Bettelheim, M. J. Vinciguerra, and D. Kaplan. Exp. Eye Res., *15*: 149 (1973).
30.   G. B. Benedek, J. I. Clark, E. N. Serralach, C. Y. Young, L. Mengel, T. Sanke, A. Bagg, and K. Benedek. Phil. Trans. R. Soc. Lond. Ser., *A*: 293 (1979).
31.   J. I. Clark, L. Mengel, and G. B. Benedek. Ophthalmic Res., *12*: 16 (1980).
32.   S. L. Trokel. Invest. Ophthalmol., *1*: 493 (1962).
33.   G. G. Benedek. Appl. Optics, *10*: 459 (1971).
34.   F. A. Bettelheim and E. L. Siew. Dev. Biochem., *9*: 443 (1980).
35.   M. Delaye and A. Gromiec. Biopolymers, *22*: 1203 (1983).
36.   M. Delaye and A. Tardieu. Nature, *302*: 415 (1983).
37.   J. I. Clark, M. Delaye, P. Hammer and L. Mengel. Curr. Eye Res., *1*: 645 (1982).
38.   F. A. Bettelheim, E. L. Siew, S. Shyne, P. Farnsworth, and P. Burke, Exp. Eye Res., *32*: 125 (1981).
39.   H. Maisel and M. M. Perry. Exp. Eye Res., *14*: 7 (1972).
40.   H. Maisel, J. Alcala, and N. Lieska. Doc. Ophthalmol., *8*: 121 (1976).
41.   H. Maisel, M. Perry, J. Alcala, and P. Waggoner. Ophthalmic Res., *8*: 55 (1976).
42.   N. S. Rafferty, W. Goossens, and W. F. March. Am. J. Ophthalmol., *78*: 985 (1974).
43.   N. Rafferty and W. Goossens. Exp. Eye Res., *26*: 177 (1978).
44.   T. Kuwabara. Exp. Eye Res., *20*: 427 (1975).
45.   M. D. Longchampt, M. Laurent, Y. Courtos, P. Trenchso, and R. S. Hughes. Exp. Eye Res., *23*: 505 (1976).
46.   P. N. Farnsworth, S. E. Shayne, S. J. Caputo, A. V. Fasano, and A. Spector. Exp. Eye Res., *30*: 611 (1980).
47.   F. A. Bettelheim. Exp. Eye Res., *14*: 251 (1972).
48.   F. A. Bettelheim and K. N. Mehrota. Exp. Eye Res., *14*: 251 (1972).
49.   F. A. Bettelheim and T. J. Y. Wang. Exp. Eye Res., *25*: 613 (1977).

50. F. Ramaekers, P. Jap, G. Mungyer, and H. Bloemendal. Curr. Eye Res.,
     2: 169 (1982).
51. N. Lieska, H. Maisel, and J. Romero-Herrera. Curr. Eye Res., *1*: 339
     (1981).
52. F. A. Bettelheim. Exp. Eye Res., *21*: 231 (1975).
53. F. A. Bettelheim. J. Colloid Interface Sci., *63*: 251 (1978).
54. F. A. Bettelheim. Exp. Eye Res., *31*: 481 (1980).
55. F. A. Bettelheim. In *Cell Biology of the Eye* (D. S. McDevitt, ed.),
     Academic Press, New York, 1982, Ch. 6.
56. O. Brewster. Phil. Trans. R. Soc. Lond., *1816*: 311 (1816).
57. P. A. Burke, P. N. Farnsworth, and F. A. Bettelheim. Curr. Eye Res., *1*:
     689 (1982).
58. P. Moon and D. E. Spencer. J. Opt. Soc. Am., *33*: 44 (1943).
59. G. A. Fry. Illuminating Eng., *49*: 98 (1954).
60. G. Mie. Ann. Phys., *25*: 377 (1908).
61. H. Mach. Klin. Monatsol. Augenheilkd., *143*: 689 (1963).
62. H. Goldman. Am. J. Ophthalmol., *57*: 1 (1964).
63. A. Pirie. Invest. Ophthalmol., *7*: 634 (1968).
64. A. Spector, D. Roy, and J. Stauffer. Exp. Eye Res., *21*: 9 (1975).
65. A. Spector. Isr. J. Med. Sci., *8*: 1577 (1972).
66. A. Spector, T. Freund, L. K. Li, and R. C. Augusteyn. Invest. Ophthalmol.,
     10: 677 (1971).
67. A. Spector, J. Stauffer, and J. Sigelman. Ciba Found. Symp., *19* (New
     Series): 187 (1973).
68. A. Spector, L. K. Li, and J. Sigelman. Invest. Ophthalmol., *13*: 795
     (1974).
69. H. J. Hoenders and G. J. Van Kamp. Acta Morphol. Neerl. Scand., *10*: 215
     (1972).
70. J. A. Jedziniak, J. H. Kinoshita, E. M. Yates, L. O. Hocker, and G. B.
     Benedek, Exp. Eye Res., *15*: 185 (1973).
71. J. Jedziniak, J. H. Kinoshita, E. M. Yates, and G. B. Benedek. Exp. Eye
     Res., *20*: 367 (1975).
72. K. N. Liem-The, A. L. H. Stols, P. H. K. Jap, and H. J. Hoenders. Exp.
     Eye Res., *20*: 317 (1975).
73. F. J. Giblin, B. Chakrapani, and V. N. Reddy. Exp. Eye Res., *26*: 507
     (1978).
74. J. J. Harding and V. J. Dilley. Structural proteins of the mammalian lens:
     A review with emphasis on changes in development, aging and cataract.
     Exp. Eye Res., *22*: 1 (1976).
75. H. A. Kramps, A. L. H. Stols, H. J. Hoenders, and K. deGroot. Eur. J.
     Biochem., *50*: 503 (1975).
76. L. Lorand, L. K. H. Hsu, G. E. Siefring, Jr., and N. S. Rafferty. Proc. Natl.
     Acad. Sci. USA, *78*: 1356 (1981).
77. P. Azari, I. Rahim, and D. P. Clarkson. Curr. Eye Res., *1*: 463 (1981).
78. D. Roy and A. Spector. Exp. Eye Res., *22*: 273 (1976).
79. A. Spector. Invest. Ophthalmol., *1*: 579 (1965).

80. H. Bloemendal, A. Zweers, and H. Walters. Nature (London), *255*: 426 (1975).

81. H. Bloemendal, A. Zweers, and E. L. Benedetti, and H. Walters. Exp. Eye Res., *20*: 463 (1975).

82. R. J. Siezen and E. A. Owen. Biochim. Biophys. Acta, *749*: 227 (1983).

83. J. Finkel and F. A. Bettelheim. Colloid Interface Sci., *5*: 203 (1976).

84. V. Malinowski and W. Manski. Exp. Eye Res., *30*: 527 (1980).

85. V. Malinowski and W. Manski. Exp. Eye Res., *30*: 537 (1980).

86. R. J. Siezen. Biophys. Chem., *19*: 49 (1984).

87. P. Marlat, P. G. Bracchi, and G. Maraini. Ophthalmic. Res., *13*: 293 (1981).

88. F. A. Bettelheim. Exp. Eye Res., *28*: 189 (1979).

89. F. A. Bettelheim, E. L. Siew, and T. L. Chylack, Jr. Invest. Ophthalmol. Vis. Sci., *20*: 348 (1981).

90. M. C. Neville, L. A. Patterson, J. L. Rae, and D. F. Woessner. Science, *184*: 1072 (1974).

91. H. Rink. In *Altern der Linse* (O. Hockwin, ed.), Symposium über die Augenlinse, Strasbourg 1982. Integra GmbH Puchheim, Germany 1982.

92. F. A. Bettelheim and S. Christian. Lens Res., *1*: 147 (1983).

93. H. Goldmann and G. Rabinowitz. Klin. Monatsbl. Augenheilkd., *81*: 771 (1928).

94. F. T. Fraunfelder and R. P. Burns. Proc. Soc. Exp. Biol. Med., *110*: 72 (1962).

95. F. T. Fraunfelder and R. P. Burns. AMA Arch. Ophthalmol., *76*: 599 (1966).

96. M. Weinstock and J. D. Scott. Exp. Eye Res., *6*: 368 (1967).

97. P. Rácz, V. Tompa, and T. Pocsik. Exp. Eye Res., *28*: 129 (1979).

98. D. M. Thomas and K. L. Schepler. Invest. Ophthalmol. Vis. Sci., *19*: 904 (1980).

99. J. M. Pope, S. Chandra, and J. D. Balfe, Exp. Eye Res., *34*: 57 (1982).

100. S. R. Kasturi, D. C. Chang, and C. F. Hazelwood. Biophys. J., *30*: 369 (1980).

101. F. A. Bettelheim, S. Christian, and L. K. Lee. Curr. Eye Res., *2*: 803 (1983).

102. P. Rácz, K. Tompa, I. Pocsik, and P. Banki. Lens Res., *1*: 199 (1983).

103. J. H. Kinoshita. Invest. Ophthalmol., *13*: 713 (1974).

104. L. K. Li. Exp. Eye Res., *27*: 553 (1978).

105. B. Philipson. Invest. Ophthalmol., *8*: 3 (1969).

106. F. A. Bettelheim and A. Bettelheim. Invest. Ophthalmol. Vis. Sci., *17*: 896 (1978).

107. L. T. Chylack, Jr., F. A. Bettelheim, and W. H. Tung. Invest. Ophthalmol. Vis. Sci., *20*: 326 (1981).

108. F. A. Bettelheim, E. L. Siew, L. T. Chylack, Jr., and J. H. Seland. Invest. Ophthalmol., *24*: 403 (1983).

109. J. G. G. Shoenmakers, R. Matze, M. van Poppel, and H. Bloemendal, Int. J. Protein Res., *1*: 19 (1969).

110. J. G. Shoenmakers, J. J. Gerding, and H. Bloemendal. Eur. J. Biochem., *11*: 472 (1969).
111. J. Stauffer, G. Rothschild, T. Wandel, and A. Spector. Invest. Ophthalmol., *13*: 135 (1974).
112. L. K. Li and A. Spector. Exp. Eye Res., *19*: 49 (1974).
113. G. J. Van Kamp and H. J. Hoenders. Exp. Eye Res., *17*: 417 (1973).
114. E. I. Anderson and A. Spector. Exp. Eye Res., *26*: 407 (1978).
115. R. J. W. Truscott and R. C. Augusteyn. Exp. Eye Res., *24*: 159 (1977).
116. R. J. W. Truscott and R. C. Augusteyn. Biochem. Biophys. Acta, *492*: 43 (1977).
117. P. M. Masters, J. L. Bada, and J. S. Zigler, Jr. Nature (London), *268*: 71 (1977).
118. P. M. Masters, J. L. Bada, and J. S. Zigler. Proc. Natl. Acad. Sci. USA, *75*: 1204 (1978).
119. W. H. Garner and A. Spector. Proc. Natl. Acad. Sci. USA, *75*: 3618 (1978).
120. K. Itoh, Y. Ozaki, A. Mizuno, and K. Iriyama. Biochemistry, *22*: 1773 (1983).
121. Y. Ozaki, A. Mizuno, K. Itoh, M. Yoshiura, T. Iwamoto, and K. Iriyama. Biochemistry, *22*: 6254 (1983).
122. E. L. Siew, F. A. Bettelheim, L. T. Chylack, Jr., and W. Tung, Invest. Ophthalmol. Vis. Sci. Lens Res., *1*: 291 (1983).
123. W. F. Bon, Ophthalmology, *138*: 35 (1959).
124. H. Pau and W. Graeber. *Der Augenarzt*, Vol. 2, G. Theime, Leipzig, 1969.
125. S. Zigman and S. Lerman. Nature, London, *203*: 662 (1964).
126. S. Lerman and S. Zigman. Acta Ophthalmol., *45*: 193 (1967).
127. J. Horowitz, N. P. Robertson, M. M. Wong, J. S. Zigler, and J. H. Kinoshita. Exp. Eye Res., *28*: 359 (1979).
128. T. Tanaka and G. B. Benedek. Invest. Ophthalmol., *14*: 449 (1975).
129. T. Tanaka, C. Ishimoto, and L. T. Chylack, Jr. Science, *197*: 1010–1012 (1977).
130. M. Delaye, J. I. Clark, and G. B. Benedek. Biophys. J., *37*: 647 (1982).
131. M. A. Lowenstein and F. A. Bettelheim. Exp. Eye Res., *28*: 651 (1979).
132. M. A. Lowenstein and F. A. Bettelheim. Exp. Eye Res., *30*: 315 (1980).
133. J. I. Clark and G. B. Benedek. Invest. Ophthalmol. Vis. Sci., *19*: 771 (1980).
134. M. D. Delaye, J. I. Clark, and G. B. Benedek. Biochem. Biophys. Res. Commun., *100*: 908 (1981).
135. C. Ishimoto, P. W. Goalwin, S. T. Sun, I. Nishio, and T. Tanaka, Proc. Natl. Acad. Sci. USA, *76*: 4414 (1979).
136. T. Tanaka, S. Rubin, S. T. Sun, I. Nishio, W. Tung, and L. T. Chylack, Jr. Invest. Ophthalmol., *24*: 522 (1983).
137. J. I. Clark, F. J. Giblin, V. N. Reddy, and G. B. Benedek. Invest. Ophthalmol. Vis. Sci., *22*: 186 (1982).
138. W. H. Garner, M. H. Garner, and A. Spector. Biochem. Biophys. Res. Commun., *98*: 439 (1981).

139. A. Spector, M. H. Garner, W. H. Garner, D. Roy, P. Farnsworth, and
     S. Shyne. Science, *204*: 1323 (1979).
140. T. Kuwabara, J. Kinoshita, and D. Cogan. Invest. Ophthalmol., *8*: 133
     (1969).
141. A. Beyer-Maers, P. N. Farnsworth, J. S. C. Fu, and C. K. Yeh. Exp. Eye
     Res., *27*: 627 (1978).
142. N. J. Unakar, C. Genyea, J. R. Reddan, and V. N. Reddy. Exp. Eye Res.,
     26: 123 (1978).
143. L. T. Chylack, Jr. Invest. Ophthalmol., *14*: 746 (1975).
144. T. S. Hu, P. Russell, and J. H. Kinoshita. Exp. Eye Res., *35*: 521 (1982).
145. M. Tanaka, P. Russell, S. Smith, S. Uga, T. Kuwabara, and J. H. Kinoshita.
     Invest. Ophthalmol. Vis. Sci., *19*: 381 (1980).
146. P. Russell, L. Carter-Dawson, and J. H. Kinoshita. Invest. Ophthalmol. Vis.
     Sci. (Suppl.), *20*: 131 (1981).
147. N. J. Unakar, G. Price, and J. Tsui. Ophthalmic Res., *14*: 83 (1982).
148. W. C. Gorthy, M. R. Shavely, and N. D. Berrong. Exp. Eye Res., *12*: 112
     (1971).
149. W. C. Gorthy. Exp. Eye Res., *27*: 301 (1978).
150. C. V. Harding, L. T. Chylack, Jr., S. R. Susan, W. K. Lo, and W. F.
     Bobrowski. Invest. Ophthalmol. Vis. Sci., *23*: 1 (1982).
151. R. Van Heyningen. Sci. Am., *233*: 70 (1976).
152. R. L. Hiller, L. Giacometti, and V. S. Yuen. Am. J. Epidemol., *105*: 450
     (1977).
153. S. Zigman, M. Datiles, and E. Torczynski. Invest. Ophthalmol., *18*: 462
     (1979).
154. B. W. Lambert and J. H. Kinoshita. Invest. Ophthalmol., *6*: 624 (1967).
155. M. Palva and A. Palkama. Acta Ophthalmol., *56*: 587–598 (1978).
156. M. M. Zaret, N. Z. Snyder, and L. Birenbaum. Br. J. Ophthalmol., *60*: 632
     (1976).

8
Photobiology of the Lens

SEYMOUR ZIGMAN / University of Rochester, School of Medicine and
Dentistry, Rochester, New York

INTERACTION OF ENVIRONMENTAL RADIANT ENERGY WITH OCULAR TISSUES

Radiant energy with wavelengths from approximately 295 to 800 nm reach the
surface of the earth in the sunlight. The ozone of the atmosphere absorbs much
of the energy in the 295–310 nm range and all of the energy below 295 nm. For
comparative purposes, if the value of irradiance at 550 nm is considered to be
100, the relative values at 400 and 750 nm would be 40; that at 350 nm would
be 25; at 300 nm nearly zero; and at 770 nm, approximately 10.

The sunlight spectrum provides by far the most intense short-wavelength light
that humans normally encounter. Near-ultraviolet (UV) radiation represents
some 25% of the solar spectrum, so that approximately 10 mW/cm^2 (300–400
nm) can impinge upon the eye from direct sunlight. Light in the near-UV range
at this irradiance level is capable of photochemically damaging mammalian cells.
Fortunately, most of this energy is at the longer UV wavelength range of the
spectrum (350–400 nm) since an equivalent irradiance at the shorter range
(300–400 nm) would be devastating. Much of the short-range UV is removed
from sunlight by the ozone layer. Higher irradiance levels at shorter ultraviolet
wavelengths would reach the earth if the ozone layer were thinner. Direct
viewing of the sun, to have its full irradiance focused into the eye, is painful, and
reflex action does not allow it. Only a very short time of exposure is possible for
direct sunlight exposure of the eye, and the threshold for lens damage requires at
least 15 min of near-UV light at 1 mW/cm^2 focused directly in the eye according
to present standards (1).

The high-UV irradiance of the sun occurs only during limited times of the day (11 a.m. to 2 p.m.) because of the ascending (rising sun) and descending (setting sun) angles of the sunlight with regard to the earth. Another important consideration is that the specific surface reflectance of the UV wavelengths is very strong from such surfaces as metal, sand, snow, and water, but is not great from grass. The time of year and the latitude must also influence the intensity of UV light reaching the eye. Summer is the season when the sunlight reaching the earth is the most intense, but in more southerly areas (e.g., Florida), the UV irradiance of sunlight is greater in the winter than more northerly areas in the summer (e.g., New England). The relation of sunlight UV irradiance to latitude is an important consideration. The lower the latitude, the greater is the shorter wavelength radiant energy reaching the earth (1–3). It has been found that the lowest wavelengths of light reaching the retina are the most damaging to it (4). Such light is also the most energetic. Detailed discussion sof the earth's radiant energy exposure can be found in (5–7).

The above discussion merely indicates the magnitudes of UV radiant energy that can strike the eye. Light must be absorbed to produce photochemical changes. It is necessary to consider the absorption and transmission of UV light by the ocular tissues to know where photochemical damage to the ocular tissues would result from extreme sunlight exposure. The cornea of the eye absorbs virtually all radiant energy with wavelengths below 310 nm, and as a result overexposure to sunlight, which contains only a small percentage of such wavelengths, causes photokeratitis (DNA damage). The aqueous humor absorbs little near-UV energy (5%); the vitreous humor absorbs 10–15% of such energy. Light entering the eye and passing through the cornea is refracted onto the anterior surface of the lens. Between 25 and 50% of the light with shorter wavelengths than 310 nm pass through the cornea and are nearly totally absorbed by the lens. In fact, nearly all light entering the eye in the range between 310 and 400 nm is filtered by the adult lens, which makes it impossible for adults to see UV light when the lens is present.

Repetitive exposure of the lenses of humans to the near-UV light of sunlight seems to induce cataract (8). This results because yellow chromophores, present both in free form (9) and in lens proteins (10), absorb the radiant energy and transfer this energy to the proteins. Such a process often results in extreme conditions of light scattering and absorption in the human lens (or brunescent cataract) due to protein.

In young children, the lens is pale and some long-wavelength UV light and appreciable short-wavelength visible light can reach the retina. The greatest damage to monkey retinas in planned experiments was found to result from exposure to blue light (wavelength of 450 nm). Young children can see slightly into the near-UV region (down to 370 nm) and do see short-wavelength blue visible light vividly. Obviously, some degree of retinal illumination by short-wavelength light results in children that does not occur in adults. The above information suggests that even the retina can be at hazard from the environmental UV and blue light. As a matter of fact, it has been shown that there is a special hazard due to blue light reaching the retina. Photoreceptors that absorb the blue light are particularly sensitive to damage by it (11).

Pupillary response to light also greatly influences the passage of radiant energy to the lens and retina. The reflex responsible for pupillary dilation and constriction results from visible light between 500 and 600 nm, which initiates a retinal reflex response that activates the ciliary muscles to dilate (open) or constrict (close) the pupil. This serves as a diaphragm that enhances the light passage through the pupil when visible light conditions become dimmer and reduces the passage of visible light to the posterior of the eye.

A circumstance is possible in which the visible light irradiance is low in a mixture of wavelengths, while the UV is high. The pupillary reflex to dim or low-intensity visible light is to dilate, thus enhancing additional UV exposure of the lens since there is no pupillary response to UV. The enhanced exposure not only is hazardous because of the additional intensity of UV, when the pupil dilates, but this UV energy can now strike a greater area of the lens epithelial cells. Those cells exposed to UV light due to a dilated pupil are peripherally located in the anterior lens, and the UV light strikes a greater number of mitotically active epithelial cells, with greater potential for abnormalities in fiber cell differentiation, and more potential for cataract stimulation in later life.

Not only are long-wavelength UV and short-wavelength blue light potentially harmful to the lens, vitreous humor, and retina, but if they are unattenuated these wavelengths can greatly disturb the visual process by causing chromatic aberration. Being short in wavelength, blue and UV light are focused slightly short of the photoreceptors of the retina and lead to poor contrast of images. With only a minimum of lens yellow pigment (as in children), visual activity in bright sunlight would admit levels of short wavelength capable of obscuring vision.

LENS BIOMOLECULES ABSORBING NEAR-ULTRAVIOLET LIGHT

The molecular makeup of the lens governs its absorption of near-UV radiation energy and includes mainly peptide-contained tryptophan, tyrosine, and phenylalamime, the same free amino acids and numerous derived pigments and chromophores. Although the proteins represent the greatest bulk of material that absorbs most radiant energy below 300 nm, the most efficient absorption of energy in the 400–300 nm wavelength region is due to other chromophores and pigments.

Nature has provided most diurnal animals with intraocular filters capable of protecting the retina from near-ultraviolet light (300–400 nm). The physiologic significance of these intraocular filters has been underscored with the recognition of the harmful effects of near-ultraviolent light. Walls (12) reported the presence of functional color filters in the eyes of different species of vertebrates, but Merker (13) first observed the yellow pigmentation of the lens while studying lens fluorescence in European squirrels. Pigmentation of the lens was observed by Franz (14) even in lampreys, which are primitive diurnal animals. An extensive subjective study by Walls and Judd (15) showed that the yellow pigment of the lens is most evident in diurnal squirrels and snakes when compared with other vertebrates.

The absorption spectra of aqueous exhibited 370 and 320 nm wavelength maxima, respectively, indicating the presence of a yellow pigment (16). Comparison of the absorption characteristics of extracts from lenses of gray squirrels (17), baboons, and rhesus monkeys (18) showed an absorption maximum between 360 and 370 nm.

Isolated lenses from all these animals exhibit intense absorption in the near-UV region of the light spectrum. Zigman and Gilbert (19), noted that the pigmentation is more prominent in shallow-swimming sharks than deep swimmers. The exact chemical nature of most of these lens pigments remains to be characterized.

Pigmentation in the human lens serves as the main ocular color filter. The physiologic role of yellow pigmented lenses is to reduce chromatic aberration to provide the greatest visual acuity. Furthermore, lens pigments act as ultraviolet filters to protect the retina from the recently discovered hazardous effects of near-UV radiation. Wald (20) has observed that the retina of the aphakic human eye is 100 times more sensitive to radiation of 365 nm light than the retina of the normal eye. Zigman et al. (21) have shown that, in albino mice whose lenses are devoid of normal pigmentation, exposure to near-UV radiation induced degeneration of retinal cells. Ham et al. (22) have shown that aphakic monkey retinas are more sensitive to near-UV damage than are phakic eyes to damage from visible light.

The absorption spectrum of the human lens is characterized by absorption maxima at 370 and 280 nm (18,23). The isolated, intact noncataractous human lens nearly totally transmits light energy with wavelengths greater than 400 nm. Transmission of visible light by brown cataract is markedly reduced due to both scattering and absorption, and the lens nearly totally filters out light with wavelengths greater than 450 nm (18,24,25). Increased extinction at wavelengths greater than 450 nm is also observed with aging, which is, for the most part, a function of light scattering. The aging process in the lens involves increased insolubilization of lens proteins (26), also accompanied by increase in pigmentation. The light extinction above 450 nm of the aged lens is primarily due to increased light scattering attributable to the increase in insoluble proteins (27). However, the extinction of light at less than 400 nm, which is enhanced with

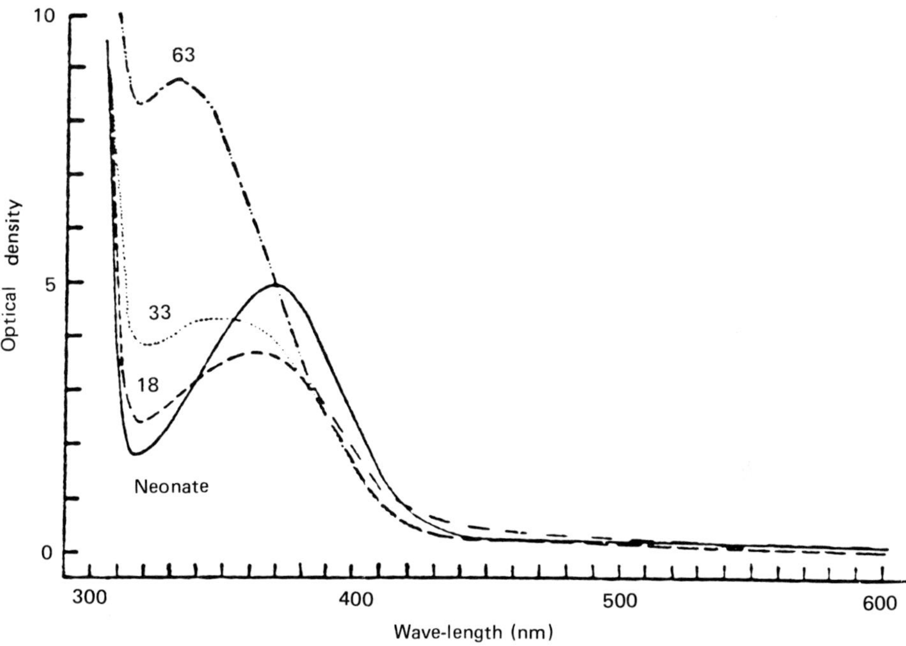

Figure 1   Density spectra of intact human lenses.

aging, is due to an increase in lens pigmentation (18,23,28). A composite of the absorption characteristics of the intact human lens of different ages is shown in Fig. 1 (18).

A low-molecular-weight nonprotein water-soluble pigment is present in the human lens that has absorption maxima at 365 and 260 nm. Normally, the concentration of this pigment is high at birth, decreases in adulthood, and remains fairly constant with advanced age (18). Van Heyningen (28) has identified several kynurenin compounds and their glucosides in the nonprotein water-extractable pigment of the human lens. The pigments isolated from cod (16), gray squirrel (18), and human (17,28) show identical absorption maxima at 360–370 nm and 260 nm.

The water-insoluble fraction of the aged lens is often highly pigmented. This color is tightly bound to the proteins and can only be dissociated by enzymatic or drastic chemical digestion (29–32). Anthranilic acid, the major pigment found in the water-insoluble fraction, appears to be covalently bound to the proteins and to increase with aging (32–34). This chromophore has similar absorption characteristics to the water-insoluble pigments. Anthranilic acid has been hypothesized by many to be an oxidation product of tryptophan, which serves

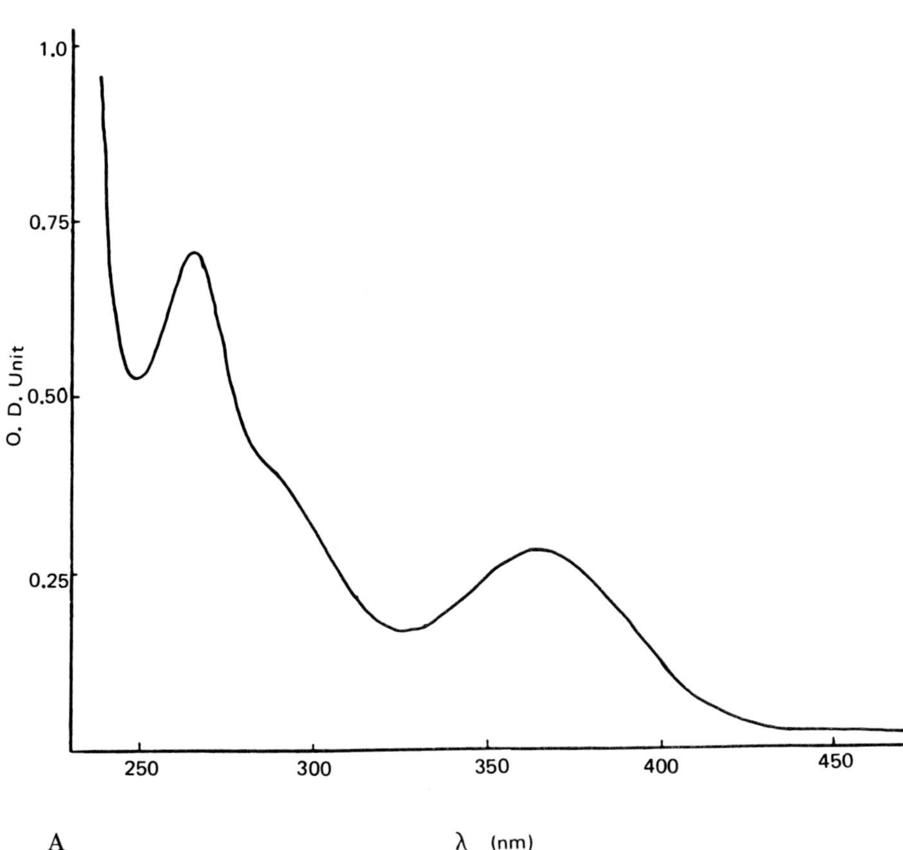

A   λ   (nm)

Figure 2   Optical properties of human lens soluble chromophore. (A) Absorption spectrum. (B) Fluorescence spectrum.

as a cross-link between structural proteins, resulting in increased insolubilization of the proteins with aging.

Most of the chemical and physical changes in the lens are associated with the normal aging process. However, a considerable accumulation of lens pigment is indicative of a pathologic condition. This condition is typified by the nuclear cataract in which the lens becomes more pigmented. The increase in pigmentation is associated with increased insolubilization of the lens proteins (35,36) and with a change in the absorption and fluorescent characteristics of the pigment (37,38).

Van Heyningen (17) and Oguchi et al. (39) independently showed that there is more than one pigment present in the lens. Pirie (40) identified N-formylkynurenine in the lens; 2 years later, van Heyningen (28) characterized 3-hydroxykynurenine, a glucoside of 3-hydroxykynurenine, 3-hydroxyanthranilic acid, and

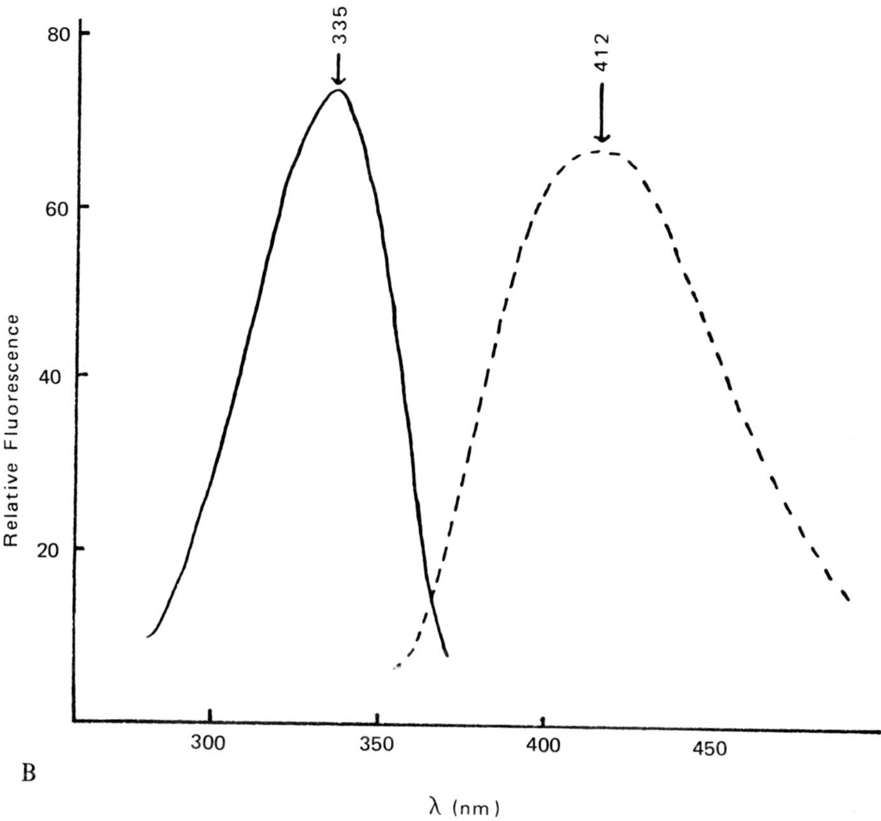

other unidentified fluorescent compounds. These fluorescent chromophores were obtained from the water-soluble extract of the lens homogenate, were of low molecular weight, and had an emission fluorescence of greater than 380 nm. Stein et al. (41) demonstrated that (1) this fluorescence is found in both the water-extractable and guanidine hydrochloride fractions of the lens homogenate, and (2) the guanidine-hydrochloride extract of the nuclear cataractous lens has a higher fluorescence intensity than the normal lens. Satoh et al. (37) found two different kinds of fluorescence in the aging lens, namely, a purple fluorescence (excitation at 290 nm and emission at 340 nm) and a blue fluorescence (excitation at 340 nm and emission at 420 nm). The purple fluorescence is the major fluorescent component of the lens. This fluorescence can be attributed to the protein itself due to the protein-bound tryptophan, and it is relatively constant during the aging process. The blue fluorescence has been observed to be protein linked and appears to increase with aging (30,31,36,41–45).

The blue fluorescence, at present termed nontryptophan fluorescence, appears to be a characteristic fluorescence of several chromophores. Van Haard

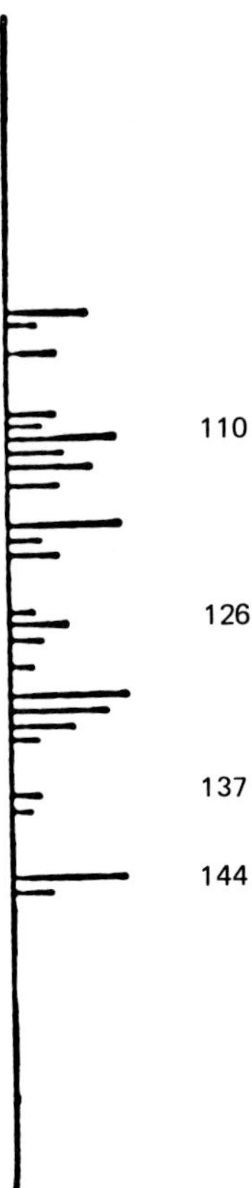

110

126

137

144

Figure 3   Mass spectrum of P2-B.

(32) described a number of low-molecular-weight nontryptophan compounds
that have been isolated from the protease digest of the water-insoluble lens
proteins. Anthranilic acid has been definitely identified as one of the major
fluorescent compounds in the enzymatically digested proteins of the nuclear
cataractous lens (32–34). Bityrosine, an oxidation product of tyrosine, is also
found in the digest (32). The barium hydroxide extraction of the water-insoluble
yellow proteins resulted in the release of several fluorescent compounds; the
principal component is the dicarboxylic acid of the β-carbolines (46).

Another nontryptophan fluorescence was accidentally found by Kuck and Yu
(47) while measuring Raman spectra of the human lenses using a higher excitation wavelength. Further work in the Raman spectroscopy by Yu et al. (48)
demonstrated the presence of a red fluorescence in the nuclear region of the
intact older human lenses. This fluorescent component increases with the

progression of the highly pigmented nuclear cataract, described as a brunescent cataract, and exhibits an emission maximum when excited at 647.1 nm. The fluorescent characteristics of this red fluorophor is different from those of the purple and blue fluorescent chromophores. The identification and characterization of other nontryptophan fluorescent compounds in the pigmented lenses are hampered by technical problems in the isolation procedure because of their minute amount in the lens. The exact nature of these chromophores remains to be investigated.

Details will now be provided concerning a water-soluble human lens chromophore reported to be present in the human lens by Cooper and Robson (18) and by Zigman and Yulo (49). The latter has been called P2-B (its isolation using Biorad P2 columns in the second peak that appears) and has absorption peaks at 365, 265, and 225 nm. This has been shown to be present in the human lens by Zigman and Yulo (49). P2-B chromophore in solution is labile and develops a deeper yellow when allowed to stand at room temperature for an extended period of time. This behavior was also observed by Garcia-Castineiras et al. (33) and Van Haard (32) for some of the chromophores isolated from the water-insoluble lens components.

The absorption and fluorescence spectra of P2-B are shown in Fig. 2a and b. Comparisons of the physicochemical properties of P2-B with other known lens chromophores (Table 1) shows that P2-B is distinct and different from any other lens chromophores and metabolites. A significant amount of anthranilic acid is extracted from the insoluble proteins of the human lens (32,33). The reported characteristics of anthranilic acid show that its absorption and fluorescence spectra, molecular weight, and R_f value differ from those of P2-B. It fluoresces at 385 nm and has a molecular weight of 137 d. In contrast, P2-B is characterized by a corrected fluorescence spectrum with excitation and emission at 335 and 440 nm, respectively, at neutral pH, absorption maxima at 365, 265, and 225 nm, and in thin-layer chromatography (TLC) runs much slower than anthranilic acid. The reported fluorescence spectra of the β-carbolines (46) show emission at 385 nm when excited at 350 nm with ethanol as solvent; its absorption spectra include a number of maxima covering the range from 380 to 250 nm. These spectral characteristics of the β-carbolines do not resemble those of P2-B. On the other hand, bityrosine with a reported molecular weight of 772 d, is considerably heavier than P2-B with an estimated molecular weight of 144 d (see Fig. 3). Thus, it appears that P2-B is not anthranilic acid, β-carbolines, or bityrosine.

Certain photoproducts of tryptophan irradiated under near-UV light have been characterized by Sun and Zigman (50). HPI (2-carboxy-3a-hydroxy-1,2,3, 3a,8,8a-hexahydropyrrol-(2,3b)-indole), which was generated in vitro from UV-treated tryptophan solution, has characteristics different from those of P2-B. Comparisons of other lens chromophores with P2-B also showed that it is not any of the kynurenine series of tryptophan oxidation (i.e.,

Table 1 TLC Characteristics of P2-B Compared with Other Lens Chromophores[a]

Chromophore	Fluorescence maxima	R_f Value	Long-wave-UV lamp	Ninhydrin, lutidine
P2-B	335, 412	0.52	Blue-white	No color
Anthranilic acid	300, 405	0.82	Blue	Peach
3-OH kynurenine	287, 317; 328, 390	0.71	Bright yellow	Yellow
Tryptophan	287, 348	0.70	No fluorescence	Violet
Tyrosine	280, 310	0.76	No fluorescence	Lavender
Ascorbic acid	350, 430	0.75	Blue-white	Yellow

[a]Fluorescence maxima were obtained at neutral pH at room temperature. TLC was done on silica gel plate with metahnol/water (9:1) as solvent.

N-formylkynurenine and 3-OH-kynurenine ± glucoside). The absorption spectrum of HPI (237 and 293 nm maxima and 262 nm minima) and its fluorescence spectrum differ from P2-B. Based on this comparison P2-B is not any of the chromophores that have already been identified and characterized in the human lens.

The absorption and fluorescence spectra are quite sensitive to changes within the molecules, as that in the event of a chemical reaction such as that induced by a change in pH. A displacement in the absorption and fluorescence maxima occurs. The specific shift of the maxima toward the blue in the fluorescence spectra of P2-B as the pH is increased may be indicative of the ionic changes occurring within the P2-B molecules. There is also a corresponding change in the absorption of P2-B at different pH levels. It appears that ionizable groups contribute to the absorption and fluorescence characteristics of P2-B such that changes occurring in these groups as a result of a change in pH affected the spectral properties of P2-B (see Figs. 4a and 4b).

The rate of decrease of P2-B up to 30 years of age is rapid, followed by a gradual, slower decrease beyond age 30 years (see Fig. 5). Several possibilities exist that could explain the age-related changes in the amount of P2-B: (1) the production of the chromophore is decreased due to loss of activity of the metabolic enzymes as a result of the normal aging process, (2) loss of P2-B is due to diffusion out of the lens capsule as a result of a leaky membrane, and (3) P2-B acts as a cross-linker between proteins, which could account for the decrease in P2-B in the soluble extract and in the increased insolubization of the protein with aging.

In terms of the physiologic role of P2-B, its high level at an early age fulfills, at least in part, the need for protection from near-UV, and its decrease in the soluble phase with aging probably represents either the normal aging process or some form of metabolic compensation for the increased generation and appearance of other chromophores in the water-insoluble phase of the human lens.

A consideration of pigment formation in the human lens follows. Coren and Girgus (51) have advocated the growth hypothesis, which states that the increase in absorbance with aging is due to the increase in lens density. These authors maintain that the lens becomes pigmented as a consequence of the natural growth process without taking into consideration the contribution of external factors. Further, this theory does not account for the fact that not all lenses become darkly pigmented with aging, but others are darker at earlier ages. In contrast, experimental evidence showing near-UV induced enhancement of lens pigmentation with aging support a UV hypothesis. Van Heyningen (23) demonstrated that the metabolic oxidative products of tryptophan (the kynurenines and their glucosides) do accumulate with aging. The accumulation of these pigments is accompanied by an increase in the insolubilization of the lens proteins, and pigmentation is higher in the proteins of the lens nucleus rather than cortical

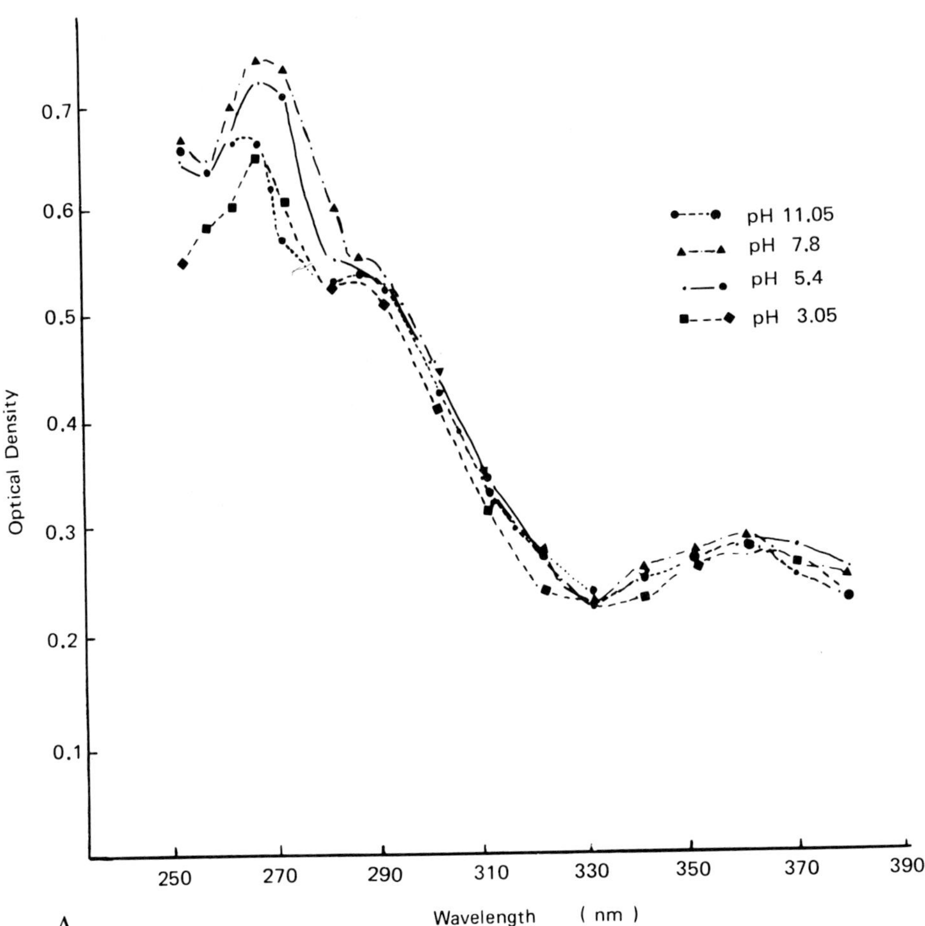

A

Figure 4   Effect of pH on optical properties of human lens soluble chromophore.
(A) Absorption. (B) Fluorescence.

proteins (35,52). These changes in the chemical and physical properties of the
insoluble proteins could not be explained solely by the growth process and
appear to be influenced by near-UV light (53).

   Neither hypothesis can fully explain the phenomenon of increased pigmentation of the lens with aging. As an extension of the UV hypothesis, several
authors have proposed that the increased pigmentation may have resulted from
the generation of free radicals after exposure to near-UV radiation. This free-

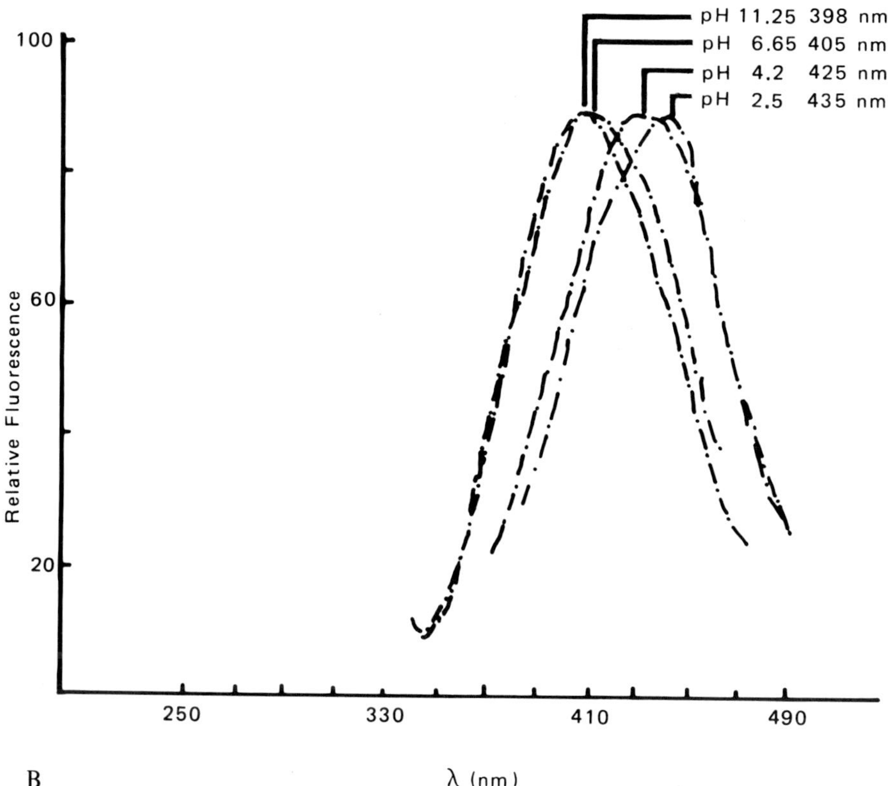

radical hypothesis postulates that tryptophan exposed to near-UV light is involved in the increased pigmentation in a nondestructive fashion by generating free radicals and other oxidants. Both stable and excited species of tryptophan have been observed in the lens (44,54–56). Exposure of the lens to near-UV light in the presence of tryptophan resulted in a higher electron spin resonance (ESR) signal. Inactivation of the enzymes catalase and glutathione reductase would lead to an increase in hydrogen peroxide after exposure of the lens to near-UV radiation (53). Bhuyan and Bhuyan (57) showed that the inactivation of catalase did result in cataracts in animals.

Although these hypotheses, taken separately, explain in part the incidence of increased lens pigmentation with aging, the controversy on the exact mechanism(s) involved cannot be resolved until more information becomes available. For the present, the UV and free-radical hypotheses are more consistent with experimental observations.

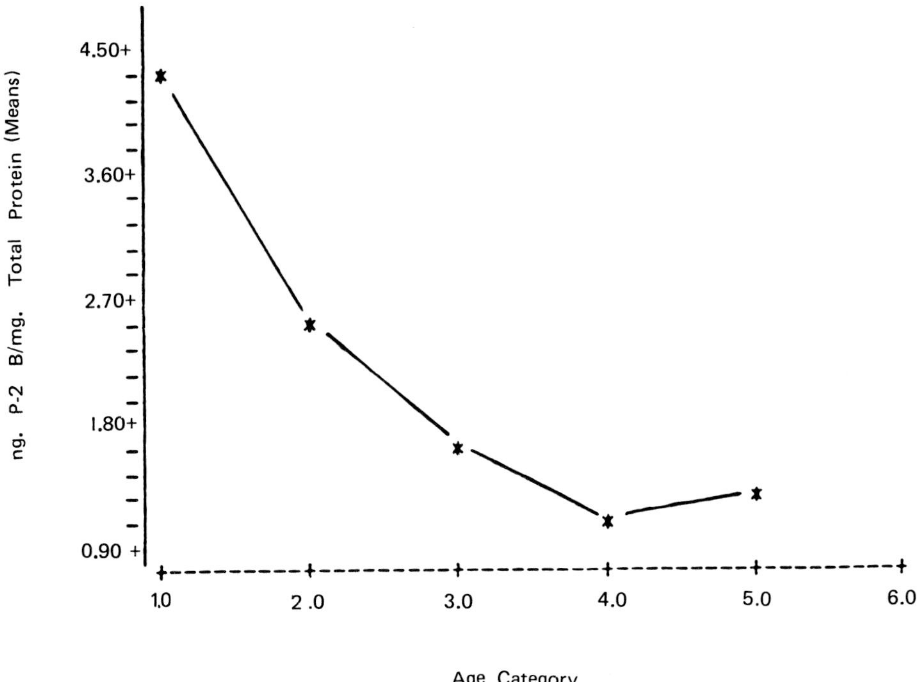

Figure 5   Change of human lens soluble chromophore with age.

PHOTOCHEMICALLY ALTERED LENS CONSTITUENTS

This section is devoted to a discussion of changes induced in the previously referred to lens constituents that are related to near-UV radiation exposure. It includes a consideration of alterations in free radicals, chromophores, tryptophan, protein of both structural and enzymatic types, and lipid or fiber cell membranes.

Tryptophan is the most light sensitive amino acid due to its great extinction coefficient ($E_M^{1\%}$ = 5700). Its presence in the lens in the free and protein-bound state is of major significance to the effect of sunlight UV radiation on the lens. Sun and Zigman (50), proposed a mechanism for the direct photo-oxidation of aqueous tryptophan in the absence of dye sensitizers. Their results paralleled those of Nakagawa et al. (58), who demonstrated the photo-oxidation of tryptophan in the presence of a dye sensitizer, rose-bengal. The major products of tryptophan photo-oxidation by near-UV light at the highly pigmented and fluorescent compounds, *N*-formylkynurenine and kynurenine via the tricyclic intermediates HPI (2-carboxy-3-hydroxy-1,2,3,3a,8,8a-hexahydropyrrolo-(2,3b)-indole) and PPI (a 3a-hydroperoxy intermediate) (50). Pirie (40) had earlier reported that *N*-formylkynurenine is produced as a result of exposure to sunlight of the lens proteins. Supporting evidence was provided by van Heyningen (59), who demonstrated the presence of kynurenine, 3-hydroxykynurenine, and its glucoside in human lenses. Incubation of the human lens with $[^{14}C]$tryptophan resulted in the formation of the labeled metabolites kynurenine and 3-hydroxy-kynurenine glucoside (28). These findings indicate that kynurenine and its derivatives are metabolically produced from tryptophan. The metabolic conversion of tryptophan to kynurenine involves a series of complex enzyme systems, and the enzymatic degradation of tryptophan to its metabolic products is shown in Fig. 6 (28).

Three laboratories have reported the presence of free radicals in human lenses: Weiter and Finch (54), Zigman (55), and Lerman and Borkman (44). intermediates HPI (2-carboxy-3-hydroxy-1,2,3,3a,8,8a-hexahydropyrrolo-(2,3b)-indole) and PPI (a 3a-hydroperoxy intermediate) (50). Pirie (40) had earlier reported that *N*-formylkynurenine is produced as a result of exposure to sunlight of the lens proteins. Supporting evidence was provided by van Heyningen (59), who demonstrated the presence of kynurenine, 3-hydroxykynurenine, and its glucoside in human lenses. Incubation of the human lens with $[^{14}C]$tryptophan resulted in the formation of the labeled metabolites kynurenine and 3-hydroxykynurenine glucoside (28). These findings indicate that kynurenine and its derivatives are metabolically produced from tryptophan. The metabolic conversion of tryptophan to kynurenine involves a series of complex enzyme systems, and the enzymatic degradation of tryptophan to its metabolic products is shown in Fig. 6 (28).

Excited states and stable free radicals of photo-oxidized tryptophan were shown to be present normally in human lenses and especially in brunescent cataracts. These free radicals were reported to be at high levels in cataractous lenses, especially brunescent cataractous lenses, by Weiter and Finch (54). The presence of these free radicals indicates a photochemical change in tryptophan, either in the bound form to protein or in free form, which elevates it into excited-state molecules or free radicals that are very reactive toward proteins. Further reactions of these free radicals with other radicals and proteins would lead to aggregation and an adverse influence on transparency.

When purified lens proteins were irradiated with near-UV light, free radicals were formed, as demonstrated by the presence of ESR signals (60). Zigman (53) has shown that isolated rat lenses incubated in tryptophan solution with exposure to near-UV light exhibited an intense ESR signal characteristic of tryptophan irradiated similarly with UV light. The UV-exposed lenses appeared

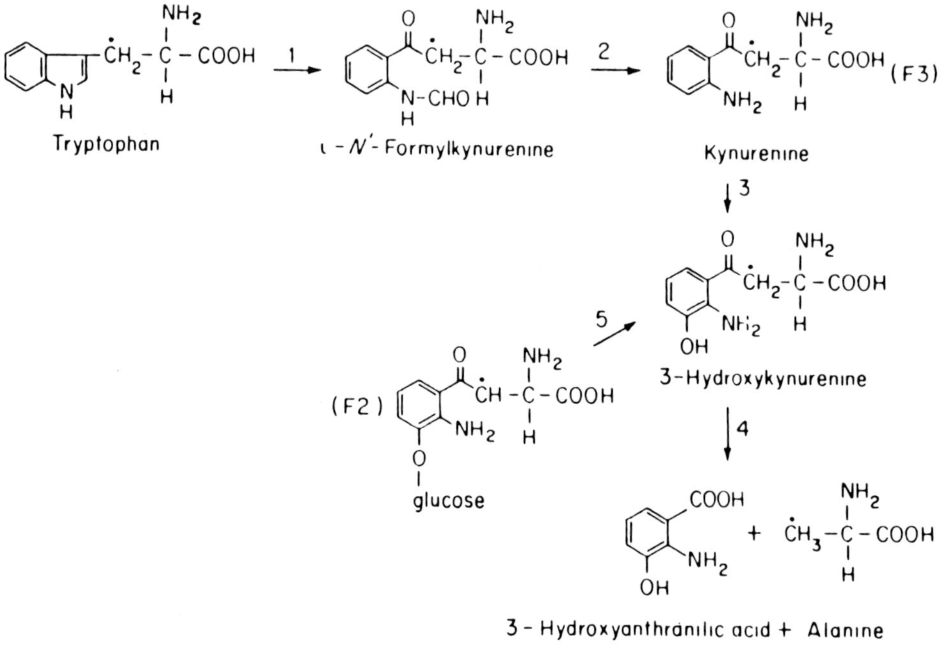

Figure 6 Tryptophan metabolism of human lens.

opalescent and acquired a brown pigmentation. A similar ESR pattern was observed in whole human lenses that had been dried at room temperature for months, suggesting that these excited-state species are stable, as are those derived from similarly irradiated tryptophan (55). This observation concurs with that of Weiter and Finch (54), who reported in the presence of a stable triplet state species of tryptophan in the nuclear region of the human lenses. Thus, the generation of free radicals from near-UV–exposed tryptophan may result in very reactive species capable of binding to and altering the structural proteins and light transmission. An association between increased pigmentation in nuclear cataracts oxidative processes of tryptophan is highly likely on the basis of this information.

The damaging effects of near-UV radiation on ocular tissues and UV-induced accelerated aging of the lens have been the subject of extensive research ever since Duke-Elder postulated in 1926 that exposure to UV radiation from the sun could cause cataract in the human lens. Kurzel et al. (61) and Zigman (53) have written comprehensive reviews focusing on UV radiation-induced damage to the lens. Pirie (35) has reported an increased presence of pigmented, urea-soluble proteins in human lenses with different cataractous states. Pirie (40) then found that solutions of lens proteins and of tryptophan developed brown colorations when exposed to direct sunlight. Near-UV irradiation of the human lens or solutions of partially purified human lens proteins in the presence of fluorescent lens chromophores are exposed to near-UV radiation, and an increase in pigmentation and absorption above 295 nm is observed (59). The evidence confirms that lens chromophores, including tryptophan, act as targets for near-UV–induced changes leading to increased pigmentation.

Several species of stable tryptophan photoproducts were also found in human lenses by the above-named workers, and by Dillon et al. (46) and Garcia-Castineiras et al. (33), who implicated tryptophan as the target for photochemical action by near-UV light in the lens. A constant question raised by the work referred to above was whether the photochemical changes described occurred in the free tryptophan or in the tryptophan intrinsic to the lens proteins. Evidence against such a change occurring within the protein molecules was provided by measurements of the tryptophan levels in the proteins of brunescent cataracts, which indicated no loss of tryptophan. However, in these cataracts there was a loss of free tryptophan along with the loss of many other amino acids, partially due to leakage. The energetics of the changes in tryptophan would also support free tryptophan as the target of near-UV photooxidation that leads secondarily to protein changes. This idea has recently been supported by Dillon (62), who exposed mixtures of tryptophan and other amino acids and peptides to near-UV light. The consensus at present among researchers is that both free tryptophan and that contained within the proteins are susceptible to photochemical changes due to near-UV light. This susceptibility finally renders the state of the lens proteins abnormal, so that they begin to scatter (by aggregation) and absorb light (by pigmentation) rather than transmitting it. Lens protein aggregation due to intermolecular cross-linking appears to be related to the presence of tryptophan oxidation products in the crystallins, which can be produced by photo-oxidation, but can also be formed by metabolism. Such oxidation products of tryptophan were found to be firmly bound to the insoluble proteins of aging normal and especially brunescent human lenses. This cross-linking results in the aggregation of protein to very high molecular weight products, which both scatter and absorb visible light appreciably.

Low-molecular-weight tryptophan oxidation products, such as the kynurenines and others yet to be characterized, are present in the human lens at very low concentrations. Some researchers believe that tryptophan and other easily oxidizable substances present in the lens are artifactually converted into oxidation products during extraction and purification procedures. However, others support their presence as natural.

Zigman and Yulo (49) have partially characterized a low-molecular-weight near-UV light-sensitive water-soluble chromophore present in free form in the human lens that diminishes with aging (see Lens Biomolecules Absorbing

Near-UV Light). The diminution of such an as yet unidentified chromophore is thought to be due to its irreversible covalent binding to lens proteins. A low-molecular-weight yellow lens chromophore present in the squirrel lens serves as a protein sensitizer for near-UV light that enhances lens protein aggregation due to near-UV light exposure in vitro (63). Thus, near-UV–induced alterations in the low-molecular-weight compounds present in the human lens are also important factors involved in the photochemical changes of the lens. The study of Goosey et al. (64) has shown further than lens protein aggregation is stimulated by photodynamic action involving low-molecular-weight sensitizers and short-wavelength (blue) visible light.

Pirie (40) introduced the concept of lens protein alterations by sunlight and reported that exposure of human lens to proteins to sunlight in vitro would cause their browning. Because the irradiation of the proteins with sunlight was through glass-walled vessels, only wavelengths above 320 nm in the ultraviolet region were involved in this photochemical change. Unfortunately, the irradiance levels or the cumulative UV energy infringing upon the protein samples were not reported, and no quantitation can be derived from this study. Besides, in many of these experiments the proteins were not purified, and therefore the initial influence of the sunlight may have been partially due to effects on low-molecular-weight nonprotein, UV-absorbing entities that are also present in the lens. Thus, the protein changes observed could have resulted in photosensitized reactions involving the low-molecular-weight entities, which seemed to indicate direct photochemical changes in the proteins. Again, no quantitation of radiant energy to affect a change was provided.

Grover and Zigman (36) studied the influence of near-UV light in laboratory circumstances on human lenses and extracted human lens proteins, using trypto-phan as a target molecule to stimulate the effects of the UV exposure. The spectral quality of the UV light in these studies was mainly of 365 nm wave-length, with a spread of approximately ±20–30 nm. The near-UV energy photochemically converted tryptophan into molecular material that was pig-mented and fluorescent and that bound to the lens proteins, causing them to become pigmented and altering their fluorescent properties. Changes in the proteins paralleled those of intact human lenses due to similar UV exposure under in vitro conditions. The changes observed were enhanced protein pigmen-tation and aggregation, and the development and enhancement of near-UV excited and blue visible emitted fluorescence.

Lerman and Borkman (44) also examined the nontryptophan fluorescence that is present in the lenses of humans of different ages. They found that this altered fluorescence, which increases with age, can be stimulated by exposing intact lenses of mice and human in vitro to specific irradiances and wavelengths of near-UV light. These researchers indicated that this new type of fluorescence

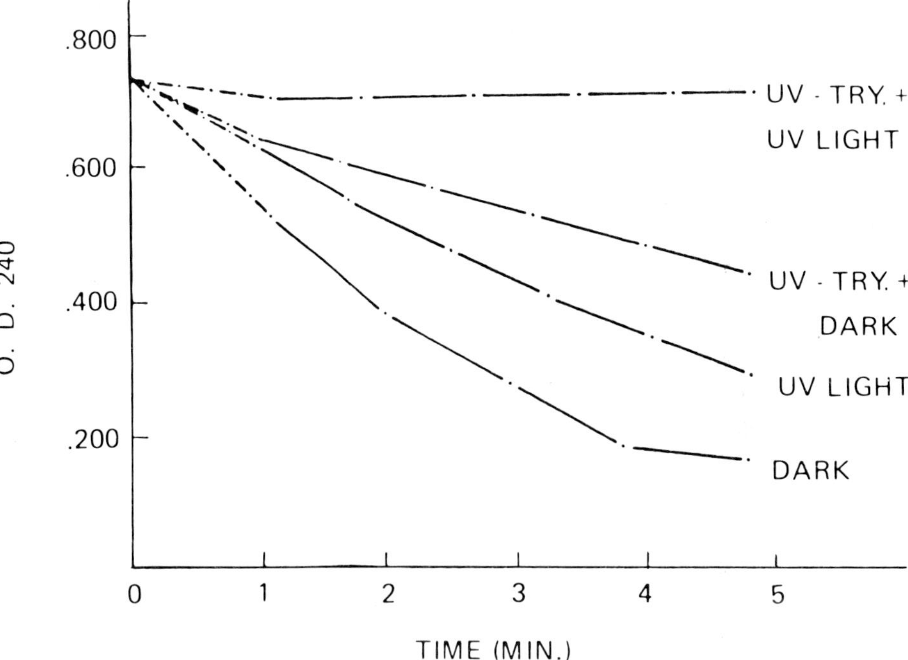

Figure 7 Time course of catalase inhibition by near-UV plus tryptophan.

(excited in the near-UV and emitted in the blue visible spectral region) was produced by the near-UV exposure of the lenses. Enhancement of this fluorescence parallels the changes that occur in human lenses with aging. They implicated altered lens protein side chains due to photo-oxidized tryptophan as the process causing this.

Changes in the oxidative processes in the lens may be involved in nuclear cataract formation following near-UV exposure. Catalase is inactivated when exposed to near-UV light (65). When mouse lenses were incubated in vitro with 3-aminotriazole, a catalase inhibitor, and exposed to near-UV light, the marked depression of amino acid incorporation into proteins indicated an impairment of protein synthesis (66). Lerman and Borkman (44) demonstrated that the colorless lenses of the rat developed a yellow pigmentation after prolonged exposure to near-UV light in the presence of 3-aminotriazole.

Table 2 Quantitative Details of the Influence of Near-UV Light and/or Tryptophan Photoproducts on the Activity of Crystalline Beef Liver Catalase in Decomposing H_2O_2

Time of tryptophan exposure to near-UV light (hr)			Enzyme (units per ml $\times 10^+$) exposed to	
	Darkness	Near-UV light	Dark plus near-UV– exposed TRY	Near-UV light plus near-UV– exposed TRY
3	4.97	4.3	3.8	1.7
24	4.63	3.6	3.2	0.97

*One unit of catalase equals 1 μmol of H_2O_2 decomposition per minute.

Table 3 Influence of Near-UV and Tryptophan Photoproducts on Glutathione Reductase (6R)

Sample	GR activity (μM NADPH oxidized per min per mg protein)		
	Before UV	After UV	After dialysis
GR-no additive-DK	1447	1158	1439
GR-No additive-UV	—	569	603
GR + 20 μM HPI-UV	—	450	419
GR + 10 μM HPI-DK	1190	868	804
GR + 10 μM HPI-UV	—	385	476
GR + 5 μM HPI-DK	1479	933	734
GR + 5 μM HPI-UV	—	405	316

Appreciable effects on enzymes result from their exposure to near-UV light in the presence of tryptophan or some of its oxidation products described above. At present, there is a list of enzymes that have been tested for their sensitivity to near-UV light plus tryptophan photoproducts, either purified or without purification.

The first of this series is catalase (see Fig. 7) and Table 2). As shown by Zigman et al. (65), exposure to near-UV light in the presence of tryptophan leads to an inactivation of this enzyme that occurs because both the prothetic group and the protein chain of the catalase are drastically altered by this treatment. There were also losses of soret and ultraviolent absorption properties, enhancement of nontryptophan fluorescence, and altered electrophoretic mobility of the protein. In this case tryptophan photoproduct was firmly bound to the enzyme protein.

A similar result was obtained for glutathione reductase, whose inactivation by near-UV exposure in the presence of both impure and purified tryptophan photoproducts is documented in Table 3. On the other hand, superoxide dismutase was insensitive to such treatment. Both of these enzymes were inactivated by parachloromercuri benzene sulfonate (PCMBS). A conclusion that can

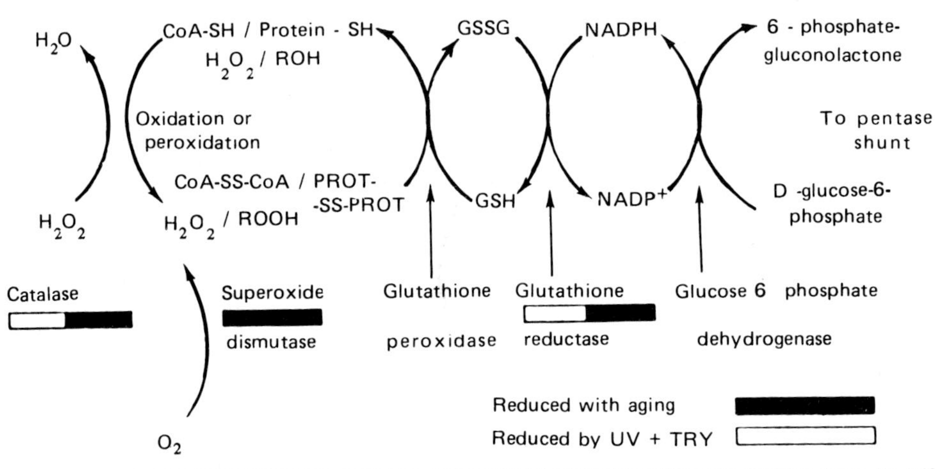

Figure 8   Oxidation-reduction in the lens.

be derived from these findings is that the UV plus photoproducts cause enzyme inactivation not due to –SH group oxidation (67).

It has been shown by Zigman et al. (68) that photoproducts are bound covalently to proteins and polypeptides via amino groups. In the case of glutathione reductase, inactivation by near-UV light exposure in the presence of hydroxypyrrole indole (HPI) was not reversed by extensively dialyzing the enzyme plus bound photoproducts, nor could ^{14}C-labeled HPI be removed from the enzyme protein by dialysis. The conclusion is that HPI (or a photoproduct of it) binds to enzymes irreversibly (unless harsh treatment of the enzymes is carried out), thereby altering chemical and physical properties leading to their inactivation.

The three enzymes discussed above are components of the oxidation reduction system of cells, and their interrelationship is illustrated in Fig. 8. Both glutathione reductase and catalase activity in cells are known to diminish with aging, and both these enzymes are now known to be inhibited by near-UV light in the presence of tryptophan photoproducts. Although superoxide dismutase activity also diminishes with aging, it is not sensitive to near-UV light action. As a result of environmental near-UV exposure, four major defects in cellular function would participate in causing extensive irreversible lens damage. These include inability to destroy superoxide radicals generated in metabolism or by radiation, inability to destroy peroxides, extensive oxidation of both reduced glutathione (GSH) and protein sulfhydril groups, and loss of energy production from oxidative metabolism by inability to regenerate NADP from NADPH.

Other enzymes essential for the normal functioning of many cells and tissues that have shown sensitivity to near-UV light and tryptophan photoproducts include Na^+,K^+- and Mg^+-ATPases, xanthine oxidase, and cytochrome oxidase. Using whole rat lenses in vitro, the Na^+,K^+-ATPase activity was diminished by 20–50% in 17 hr due to exposure to near-UV at lower than sunlight irradiances. Added tryptophan resulted in activity losses of 50–60%. Similar inactivation of Mg^+-ATPase was also observed (69). In vivo losses of ATPase activities would cause osmotic problems of the lens to maintain their integrity. A summary of the photosensitized inhibition of enzymes studied so far is presented in Table 4. This presentation shows that interference with oxidation-reduction reactions of cells results from photosensitized reactions that bind to enzyme proteins to inhibit their catalytic role in the control of cellular oxidation.

Recent findings have indicated that singlet oxygen is the most important oxidant with the potential of damaging the lens. Singlet oxygen may form in the aqueous humor as a result of sensitized photochemistry, but is more likely that singlet oxygen is formed within the lens since there are numerous photosensitizing molecules present that can stimulate its formation. In fact, in some in vitro experiments, it has been shown that aggregation of lens proteins and alterations in their fluorescence emissions result when in vitro preparations are exposed to blue or near-ultraviolet light, with singlet oxygen sensitizers methylene blue, 3-hydroxykynurenine, and pigmented insoluble lens proteins (70). The aggregation and fluorescence change was inhibited by sodium azide and other agents that are known to inhibit the formation of singlet oxygen.

Table 4 Enzymes Tested for Near-UV Radiation Sensitivity[a]

Enzyme	Near-UV Effect
ATPases: Na^+,K^+: Mg^+[b]	Loss of activity
Catalase[b]	Loss of activity
Cytochrome oxidase[b]	Loss of activity
DNA polymerase type I	No effect
Glutathione reductase[b]	Loss of activity
Prostaglandin dehydrogenase[b]	Loss of activity
Superoxide dismutase[b]	No effect
Xanthine oxidase	Loss of activity

[a]Natural sensitizers present (tryptophan photoproducts).

[b]Tested in ocular tissues.

Singlet oxygen would be further stimulated as a result of the increase in tryptophan oxidation products that build up in the lens with aging. Again, reducing agents that are present in the lens, such as glutathione and ascorbic acid, are capable of inhibiting the formation of singlet oxygen and its action.

In contrast to the lens, the retina contains one of the highest oxygen tensions in any living tissue by virtue of the many blood vessels that permeate it. The retina has a very high rate of metabolism and utilizes much oxygen in metabolism. There is quite a good control of the oxygen level in the retina, since high levels of oxygen can destroy the retina, especially in some disease states where the protective function against oxidation (e.g., retrolental fibroplasia) is deficient. There are, however, high levels of superoxide dismutase, glutathione reductase, glutathione peroxidase, and catalase in the retina.

When radiant energy reaches the retina it is absorbed mainly in the outer segments of the photoreceptors and by the melanin granules in the pigment epithelium. In the photoreceptors, vision is the result, whereas in the melanin granules, reactive free radicals are formed. Another photochemical reaction that is known to occur is related to the oxidation by UV light of tryptophan, which yields a number of photoproducts, some of which have found to be toxic to all cells. These products stop the synthesis of macromolecules and the growth of the cell. Normally, UV light does not reach the retina because of ocular lens absorption. However, many humans have no lens as the result of cataract surgery. In this state of the eye, the retina can receive environmental UV light, and therefore in the retinas of the eyes with no lenses or in those of animals that lack lens pigment (e.g., most rodents), near-UV adds to the photo-oxidative capacity of the retina.

The work of Ham et al. (71) has shown that the threshold for retinal damage in monkeys decreases as the wavelengths of visible light to which the eye is exposed are decreased. This also applies to UV light when the ocular lens is absent. The photoreceptors sustain damage initially due to UV light exposure in lensless eyes. Using squirrels, we found that near-ultraviolet light at irradiances in the same range as those of the environment damages the photoreceptors readily, but only in eyes that have had the yellow lens removed (72). The mechanism for this is not clear, but it appears that the damage is related to the initiation of inflammatory processes in the retinal photoreceptor outer segments due to photochemical change, and not heat.

The above discussion raises several interesting points. The first is that there is a delicate balance in the lens and retina between the oxidizing factors and the reducing chemicals and enzymes that protect the tissues from damage. The second point of interest is that there are disease states, such as premature aging and genetic enzyme deficiencies, in which the oxidative reactions outweigh the protective agents. These situations then lead to a loss of function of the tissue. Cataract at an early age may be an example of such a state. The third point of interest is that environmental light in the short wavelength range is capable of overwhelming the balance between oxidative and reductive reactions, thus enhancing damage to the tissues. In this way cataracts can be stimulated and retinal damage can occur in the aphakic eye due to UV light exposure.

Photo-oxidation of amino acids and proteins of the lens has been extensively studied by Zigman (53), Dillon et al. (46), and Borkman and Lerman (73). Metabolic and photo-oxidation have also been proposed as mechanisms leading to the peroxidation of membrane lipids by Bhuyan et al. (74) and by Varma et al. (75), but the findings leave doubts of whether this mechanism has great significance to the physical or functional states of the lens fiber cell membranes. The following attempts to elucidate the potential importance of lipid oxidation in the alterations of human lens fiber cell membranes with aging and cataract formation.

Detailed studies of the lipids present in human lenses have been made by Feldman and Feldman (76), Cotlier et al. (77), Broekhuyse and Daeman (78), and Andrews and Leonard-Martin (79). Lens membranes contain 70-90% of lens lipids, so that altered physicochemical states of these lipids could influence lens transparency.

It has been established by other investigators that a large increase in the membrane-bound (insoluble) protein occurs with aging of human and animal lenses and that the major mechanism to explain this phenomenon is the oxidation of protein –SH groups. This would cause extrinsic proteins to become cross-linked via –SS– and other bonds to membrane intrinsic proteins in cataracts (80,81). Formation of metabolic oxidants, such as H_2O_2 and ultraviolet photo-oxidation, has been suggested. Enhanced light scattering would be a consequence of these events.

A study of human lens lipids has negated the possibility that lipid peroxidation is a major cataract-stimulating factor in the human. Table 5 provides the gross data on the lipid/protein relation in human lenses of both normal and cataractous varieties. Of great interest is the degree of unsaturation present in the normal and brunescent cataractous lens fiber membranes. Mass spectrometry (MS) confirmed that normal lens membranes contain few unsaturated fatty acids (see Fig. 9). By MS, several hydrocarbons not appearing in the gas chromatographic (GC) pattern of human lens fiber lipids were observed. These may be due to low-level sample contaminants. Brunescent cataract lenses, by GC analysis, revealed the presence of several unsaturated fatty acids not seen in the normal lens analysis (see Fig. 10). The significance of these is not known, but they were also shown to be present at appreciable levels by Rosenfeld and Spector (82).

Another point to be derived from these findings is the degree of unsaturation of the fatty acids in the lens membranes (see Fig. 10). In normal lenses, there is

Table 5 Major Lens Membrane Components (μg/Lens)

Lens	Age	Number of lenses	Number of experiments	Fraction	Total Protein[a]	Total lipid[a]	Protein/lipid
Normal	34	4	1	Cortical	440	203	2.2
				Nuclear	280	59	4.7
Normal	56	56	4	Cortical	759.0 ±275.1	338 ±27	2.2
				Nuclear	380 ±152	111 ±23	3.4
Normal	70.9	48	4	Cortical	1,519 ±537	511 ±30	3.0
				Nuclear	649 ±317	129 ±35	5.0 5.0
Brunescent (native)	82	(9)	1	Cortical	11,200	301	29.4
				Nuclear	7,300	112	65.1
Brunescent (Indian)	61	(8)	1	Cortical	3,000	228	13.1
				Nuclear	3,700	131	28.2

[a]Means ± standard error. Total lipid represents the sum of the phospholipid, cholesterol, and triglyceride fractions isolated by TLC and assayed chemically. An average molecular weight of 800 was used for the phospholipids, 387 for cholesterol, and 270 for methylpalmitate, to estimate the total lipid dry weight.

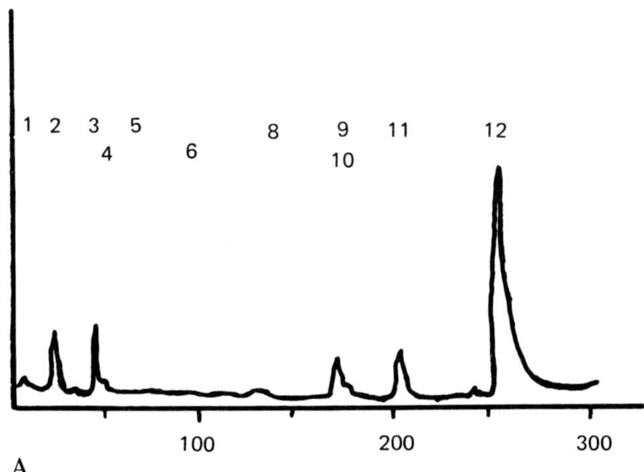

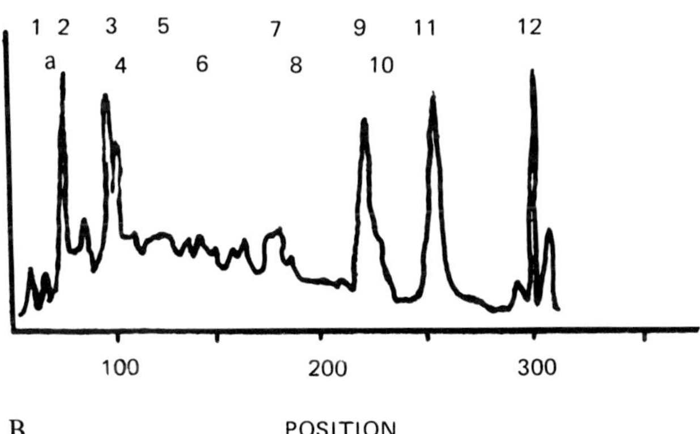

Figure 9 Mass spectral analysis of the GC preparations of normal human lens fiber membrane fatty acids (methyl esters) from cortex (A) and nucleus (B): (1) myristate; (2) palmitate; (3) oleate; (4) stearate; (5) C_{20} saturated hydrocarbon; (6) dioctyl phthalate; (7) C_{22} (single=) hydrocarbon, MW 320 d; (8) C_{22} (saturated) hydrocarbon, MW 354 d; (9) C_{24} (single =) hydrocarbon, MW 382 d; (11) cholestadiene, cholestatriene; (12) cholesterol; (a) C_{18} hydrocarbon.

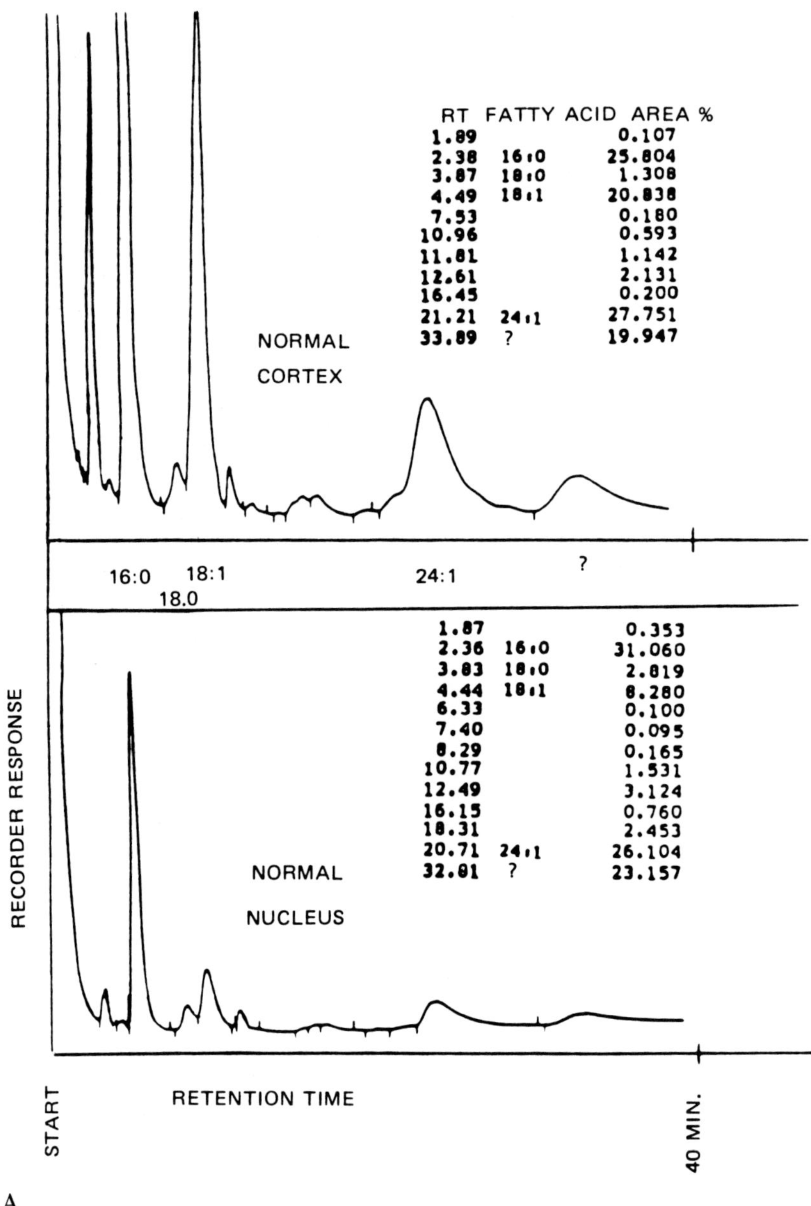

Figure 10 Gas chromatographic analysis of normal (A) and brunescent cataractous (B) human lens fiber cell membranes.

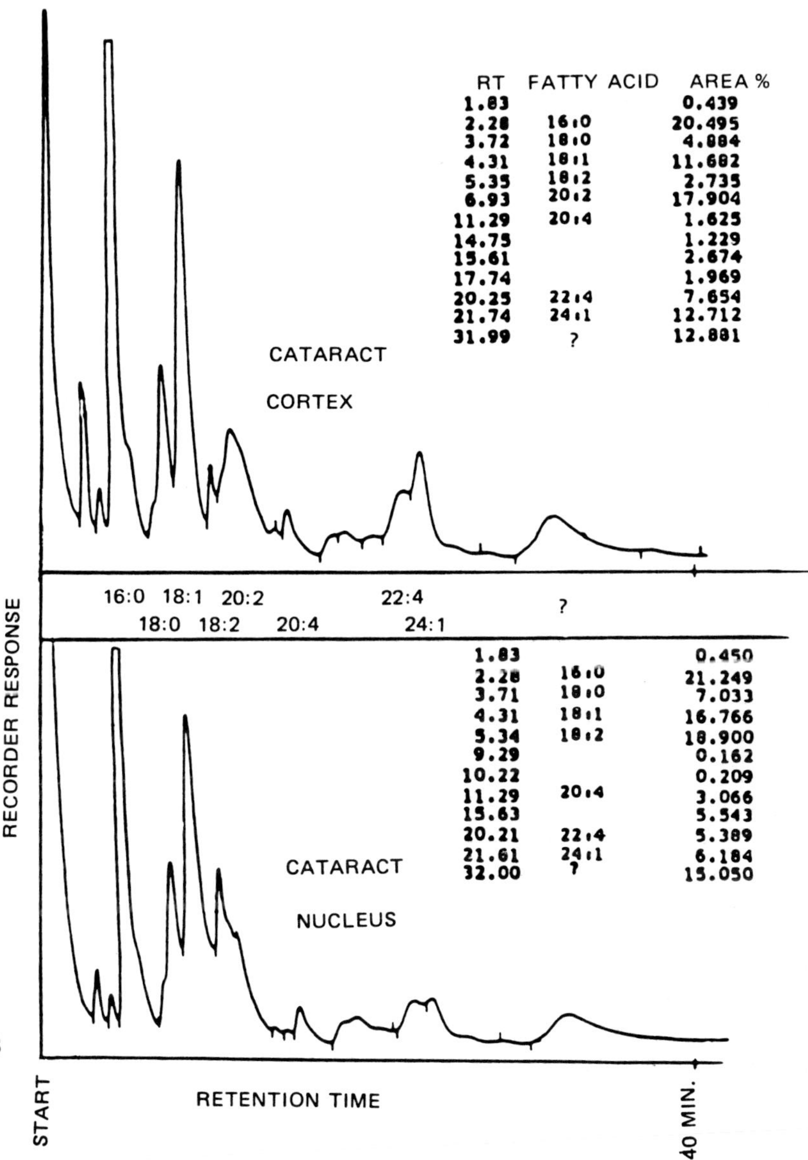
RT FATTY ACID AREA %
1.83 0.439
2.28 16:0 20.495
3.72 18:0 4.884
4.31 18:1 11.682
5.35 18:2 2.735
6.93 20:2 17.904
11.29 20:4 1.625
14.75 1.229
15.61 2.674
17.74 1.969
20.25 22:4 7.654
21.74 24:1 12.712
31.99 ? 12.881
CATARACT
CORTEX
16:0 18:1 20:2 22:4 ?
18:0 18:2 20:4 24:1
1.83 0.450
2.28 16:0 21.249
3.71 18:0 7.033
4.31 18:1 16.766
5.34 18:2 18.900
9.29 0.162
10.22 0.209
11.29 20:4 3.066
15.63 5.543
20.21 22:4 5.389
21.61 24:1 6.184
32.00 ? 15.050
CATARACT
NUCLEUS
RECORDER RESPONSE
B
START
RETENTION TIME
40 MIN.

present a rather low unsaturated fatty acid content. Changes in distribution of
the fatty acids with aging, in cataracts, or by in vitro near-UV exposure did not
diminish the degree of unsaturation. This finding is inconsistent with the idea
that large-scale peroxidation of unsaturated fatty acids is an important change
that occurs with aging, metabolic, or photochemical damage to lens fiber cell
membranes. The importance of peroxidation has been emphasized by several
investigators, including Bhuyan et al. (74) and Varma (75). Our findings,
however, are compatible with the reports of Rosenfeld and Spector (83), who
also found that there was no decrease in unsaturated fatty acid content (and
hence enhancement of saturated fatty acids) in the fiber membrane of humans
in aging or cataractous lenses. An important point to be made here is that the
fatty acids of individual phospholipid classes have not as yet been studied, and
it is still possible that there is an enhancement of saturation in specific phospholipid classes that has not yet been found.

There are natural protective agents within the lens and in the aqueous humor,
such as the reducing agents glutathione and ascorbic acid, and the free-radical
scavenger α-tocopherol, but they are not present at concentrations sufficient to
protect the lens from photo-oxidative damage. Chemical oxidants are also
present that enhance the photo-oxidative damage of UV light on the lens.
Spector and Garner (84) have drawn attention to the hydrogen peroxide, present
in the aqueous humor as a product of metabolism, as an important lens and lens
protein oxidant. Diffusion of H_2O_2 from the aqueous humor could adversely
influence lens state and function. The proper balance of all these factors must be
maintained to protect the lens from loss of transparency due to oxidative
changes. The diagram in Fig. 11 summarizes the central role of the lens in the
oxidation reduction reactions of the eye relative to near-UV radiation.

ANIMAL STUDIES OF LENS-RADIANT ENERGY INTERACTIONS

Modern studies on animal lenses, in which diffuse ambient or focused near-UV
light has produced cataractous changes, have been completed by Pitts et al. (85),
Zigman et al. (86), and Zuclich and Connaly (87). A study on mouse ocular
tissues altered by near-UV exposure is provided in Table 6. For convenience,
near-UV wavelengths are designated as *UV-A* (320–400 nm) and *UV-B* (320 nm).
Generally, UV-B is strongly absorbed by the cornea and produces photokeratitis;
UV-A penetrates the cornea and is nearly totally absorbed by the lens. Corneal
keratitis also results at low exposure thresholds and within several hours after
exposure; cataractous changes in the lens require high thresholds, usually
repetitive exposure, and they develop very slowly unless extremely high
irradiances are used, which can cause heating. Lens damage appears to be
cumulative and results in three major changes: abnormal epithelial cell differentiation into fiber cells, deep pigmentation, aggregations of proteins, and enzyme

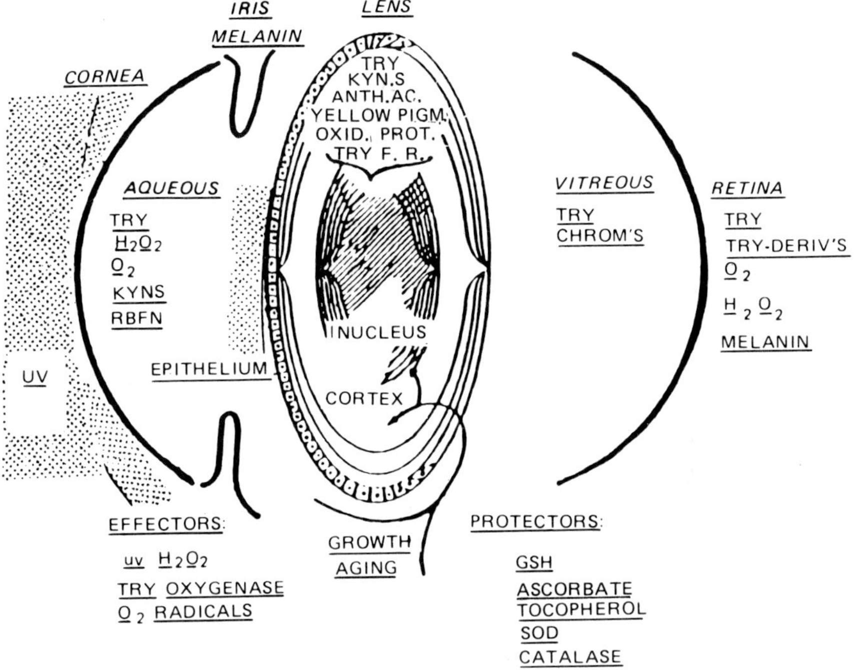

Figure 11 Central role of the lens in ocular oxidation-reduction reactions. Distribution of tryptophan and its oxidation products among the ocular tissues. Also listed are the oxidants and antioxidants present. TRY, tryptophan; KYN, kynurenine; ANTH. AC, anthranilic acid; OXID. PROT., oxidized protein; TRY. FR., tryptophan-free radicals; GSH, reduced glutathione; SOD, superoxide dismutate; RBFN, riboflavin; CHROM'S, chromophores.

inactivation. In summary, animal studies show that the UV-A components of sunlight are those involved in the formation of cataracts. The sunlit environment does provide sufficient irradiances of UV-A to alter proteins to damage the lens.

A search for a suitable animal lens model to mimic the lenses of humans has led to the use of squirrels. There are several species of squirrels, most of which are diurnally active and which have visibly observable yellow pigmented lenses. However, the relation between daytime activity and yellow lens pigment is emphasized more by the night-active squirrel, the flying squirrel, whose lens totally lacks visible yellow pigment. Besides having a yellow lens pigment, the squirrel lens has a size and shape more similar to that of human than other rodents. The oblate lens also can accommodate, but the spherical lenses of most

Table 6 Summary, Near-UV Light Effects on Mouse Eyes

Mouse type	Illumination[a]	Dosages (J/cm^2 day^{-1})	Time (weeks)	Cytotoxic effect on		
				Cornea[b]	Lens[c]	Retina[d]
White albino	40 W black light BLB (UV-A) 12 hr/day	21	40	0	++	++
White albino	40 W black light BLB (UV-A) 24 hr/day	43	19	0	+++	+++
Black pigmented	40 W black light BLB (UV-A) 24 hr/day	43	22	0	0	0
White albino	40 W, BLB (UV-A) 12 hr/day	100	30	0	++	++
White albino	40 W, BLB (UV-A) 12 hr/day	200	30	0	+++	+++
White albino	40 W, FS40 sunlamp (UV-B) 12 hr/day	0.05	30	+++	0	0[e]
White albino	40 W, FS 40 sunlamp (UV-B) 12 hr/day	0.1	30	+++	0	0[e]
White albino	40 W BLB (UV-A) + 40 W FS40 (UV-B) 12 hr/day	200 + 0.01	30	++	++	0

[a]BLB (40 W) black light = 75% emission at 365 nm, spread of ±40 nm, Mylar filter to stop 315 nm. FS 40 (40 W) sunlamp = 65% emission in the 290–320 nm range, spread of –15 and +40 nm.

[b]Keratitis.

[c]Epithelial cell abnormalities.

[d]Photoreceptor damage.

[e]Damage due to spread of inflammation from cornea.

rodents and many fish do not. A comparison of the absorptive properties of the squirrel lens with those of young and old human lenses shows that a strong absorption of light occurs below 450 nm.

The near-UV absorptive chemical species that are present in the squirrel lens include both protein and low-molecular-weight non-protein chromophores. Examination of the proteins of gray squirrel lenses revealed that there are species with similar sizes to those of most other animals, in that there are present α-crystallins (high-molecular-weight species), β-crystallins (heterogeneous population of intermediate-sized crystallins), and γ-crystallins (the lowest molecular weight crystallins). These species make up the total soluble protein that can be extracted from the lens with aqueous solvents, leaving a high-molecular-weight colloidal protein population, which is urea soluble, and an insoluble fraction, which includes mainly the membranes of the fiber cells of the lens, but also aggregated formerly soluble proteins. The proteins of the squirrel lens seem not to have a yellow pigment associated with them in the natural state. There does not appear to be membrane protein-associated nontryptophan fluorescence in the normal squirrel lens. The presence of protein in the lens therefore accounts for the absorption of light by the tryptophan, tyrosine, and phenylaline moieties predominantly. Therefore, lens proteins cause an absorption of light predominantly below 300 nm.

There are interesting low-molecular-weight water-soluble pigments present in the squirrel lens that seem to be similar to pigments that are present only in young human lenses. Passage of a water-soluble extract of squirrel lens homogenates through UM_2 filter membranes (cutoff at 1000 d using pressure dialysis) elutes a pale yellow pigmented solution. The molecular weight of the species in this solution are in the 135–150 d molecular weight range. They have not been totally resolved at this time. The pressure-dialyzed solution, when left standing at room temperature in either light or dark conditions, becomes more intensely yellow, whereas when the same eluent is exposed to near-ultraviolet light, it is bleached to a pale solution.

An investigation of the species present in this eluent reveals the presence of two major components, which may be interrelated. This was discovered by analyzing the material eluted from Biorad P-2 columns in one peak by thin-layer chromatography. Two components were found, one of which was pale yellow in appearance, and the other had a deep yellow color. The absorptive and other properties of the chromophore 1 and chromophore 2 are outlined in Table 7. Changes in the absorptive and fluorescence properties of these two chromophoric substances has revealed a type of equilibrium between the two species based upon their reaction to standing and their reaction to near-UV light exposure. This is illustrated in Fig. 12. Essentially, compound 1, the pale compound most similar to that found in nonprotein extracts of young human lenses, is the material synthesized in the squirrel lens, and upon standing it

Table 7 Some Properties of TLC Components Derived from Biogel P-2 Column Elutant from Squirrel Low-Molecular-Weight Fraction[a]

Property	Spot 1	Spot 2
R_f	0.66 0.52	
Color	Yellow	Yellow-orange
Fluorescent appearance	Dull yellow	Bright orange
Ninhydrin reaction	Dull yellow	Orange
Absorption maxima	225 nm, 265 nm, 365 nm	235 nm, tr 320 nm, 425 nm, 445 nm
Fluorescence measured	285 nm exc., 345 nm em.	

[a]Solvent, MeOH:H_2O (9:1), 3-OH-Kynurenine, R_f = 0.64, yellow color; yellow fluorescence; brown ninhydrin.

becomes bright yellow. The second compound is already bright yellow and appears to be bleached due to exposure to near-ultraviolet radiation. To maintain the yellow pigment in the lens to optimize vision, it seems that a renewal system for darkening the pigment is necessary. A speculative description of how these pigments change in the living lens follows.

Pigment is synthesized and darkens under low-light illumination conditions. During daytime activity, environmental light could bleach the pigment, so that it must be reformed during low-light conditions. This proposed scheme of action must yet be tested by experimental procedures using living animals.

Of great interest is the interactive near-UV light reactivity of squirrel lens chromophore and proteins. When exposed without chromophore, the proteins have little propensity to be aggregated, to have altered electrophoretic mobilities, or to show altered absorptive and fluorescent properties. When the total soluble extract, with both chromophore and protein, is irradiated, the proteins then are altered so that they aggregate and exhibit altered electrophoretic and optical properties. Figure 13 illustrates the aggregation of squirrel total soluble proteins due to near-UV exposure (in vitro) by polyacrylamide gel electrophoresis. Lesser amounts of aggregation occurred in the total soluble protein of squirrel lenses after removal of the low-molecular-weight chromophore. Thus it appears that the presence of the pigment has a role in the action of near-UV light on lens proteins. It may serve as a photosensitizer in this respect.

```
                                  No UV                                  UV-A
SYNTHESIS -CPD 1 (A 365; Pale)  ---------- CPD 2 (A 430; Yellow)  ----------  BLEACH
                                 Room light
                                  or dark
```

Figure 12 Dynamics of squirrel lens pigment alterations.

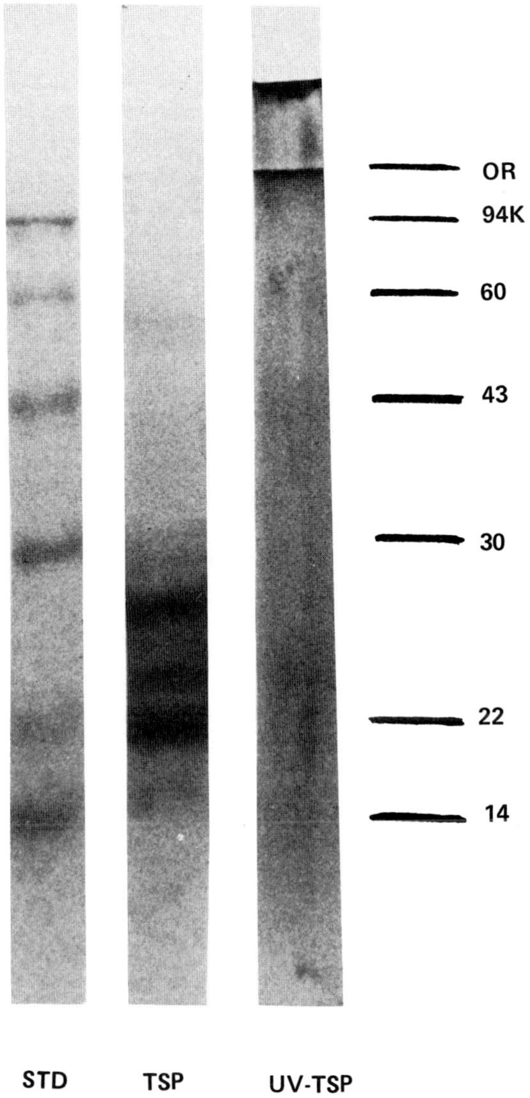

Figure 13 Near-UV induced lens protein aggregation in the squirrel as observed by polyacrylamide gel electrophoresis.

HUMAN CATARACT AND ENVIRONMENTAL NEAR-ULTRAVIOLET RADIATION

Epidemiology

The most noted theoretical suggestions that this relation exists were made by Duke-Elder (88) and by Clark (89). No firm data were collected by them to support these suggestions. Van Heyningen (90) compared the types of cataracts removed in Oxford, England, with those removed in Pakistan. The cataracts with a dark brown appearance were found in much greater abundance in Pakistan. Her findings suggested a link between sunlight exposure and darkly pigmented cataract. The data relating geographic location to numbers of cataracts extracted also indicated a higher extraction rate at earlier ages in Pakistan as compared with Oxford. However, no mention was made of the quantity of exposure that the individuals received or the distribution of irradiances in the near-UV spectral region reaching them. The protein studies were also crude.

Van Heyningen (91) continued the study of human cataract relative to geographic location by comparing the cataract extraction rate in Oxford with that in Israel. The Israeli group was further divided into darker skinned individuals who immigrated to Israel from European countries, and lighter skinned native Israelis. At all ages, the cataract extraction rate (per population) in Israel was greater than in Oxford and occurred at earlier ages. The lighter skinned Israeli population suffered a greater number of extractions at each age than the darker skinned immigrants. Here again, no mention of UV radiation quantitation or exposure levels of the subjects was made, and the types of cataracts were not mentioned.

A study of the relation between cataract extraction rates and the number of sunshine hours occurring in different locations throughout the United States was done retrospectively by Hiller et al. (92) using data from another study (HAHNES). Although there are many questions raised about the validity of the data, there were sufficient statistical data to support the concept that greater numbers of cataracts were extracted in locations where the numbers of yearly sunshine hours were the greatest. These data did not consider types of cataract or subject exposure, and may be influenced by population migration.

Excellent prospective epidemiologic studies of cataracts relative to the environment have been described by Hollows and Moran (93) and by Taylor (94). This work related the number of cataracts observed (percentage of those examined with cataract) to several factors influencing Australian Aborigine populations. When age was considered, over 50% of the cataracts observed were in the age range of 60 years and greater, with only about 15% observed at ages below 60 years. The study relating latitude to the prevalence of cataracts indicated that, at the lower latitudes (20–24°), approximately 54% of the patients had lens opacities observable by slit lamp examination. The non-Aborigines who were not active outdoors showed no such trend. These results were significant at $p < 0.001-0.005$. Again, no classification of cataracts was attempted.

Although the information summarized above supports a general relation between sunlight (UV) exposure and human cataract, none of these studies have considered that there are many different types of cataract, and that some types of cataract are related to sunlight exposure, but many types may not be. Another disadvantage of the studies referred to is that, although very intense or long duration sunlight (UV) may be present in certain geographic locations, the presence of radiation is not equivalent to the amount of exposure of the individuals.

One study (8) has attempted to overcome certain deficiencies of other studies, by relating to the types of cataract to the actual exposure of human to sunlight. These workers examined all cataracts extracted by ophthalmologists at surgical facilities at three geographic locations, differing with regard to latitude. These included: Rochester, New York, at the highest latitude (43° north); Tampa, Florida, at the intermediate latitude (27° north); and Manila in the Philippines at the lowest latitude (14° north latitude). In this work, consideration of the spectral irradiances at various wavelengths of the near-UV radiation reaching the earth at different latitudes was also made. Data were accumulated concerning the nature of the subjects' major activities, to compare the cataracts extracted from outdoor (sunlight-exposed) individuals with those whose major activities were indoors (sunlight-shielded individuals). Cataracts were classified as cortical, nuclear, mixed, and brunescent.

This early classification scheme has been found to be totally compatible with the abbreviated form of the cataract classification scheme used by The Cooperative Cataract Research Group (95). The extracted cataracts were all examined against both white and black backgrounds using a surgical microscope to discern the extent and locations of opacities and pigmentation of the cataracts in this classification scheme. Brunescent cataracts were defined as nuclear cataract with both a high degree of light scattering (opacity) plus dark brown pigmentation. In most lenses examined, a spectrophotometer was used to measure the light extinction directly. It was found that the percentage of brunescent cataracts extracted increased dramatically as the latitude of the subjects' habitat decreased, and that subjects with outdoor occupations in each location presented a severalfold greater percentage of brunescent cataracts compared with those with indoor occupations. These data are presented in Fig. 11 and Table 8.

As in the other studies outlined above, certain features of this geographic study (8) must be examined critically. The individuals included in this study are of somewhat different genetic, ethnic, and economic backgrounds, which could have influenced the findings. The choice of when cataracts should be removed is

Table 8 Geographic Distribution of Human Cataract Types

Location	Number	Type of cataract	Cortical	Nuclear	Mixed	Brunescent
(a)			All cataracts in study (%)			
Rochester	200		20	30	41	9
Tampa	90		6	23	51	20
Manila	133		8	11	38	43
(b)	Occupation		All cataracts from indoor or outdoor individuals (%)			
Rochester	Indoor	154	20	20	53	7
	Outdoor	46	29	14	29	28
Tampa	Indoor	54	8	38	46	8
	Outdoor	36	4	13	55	28
Manila	Indoor	53	11	14	43	32
	Outdoor	80	5	8	38	49

dependent upon the degree of impairment the subjects would accept, and the decision of the ophthalmologist of when surgery should be done. Migration must also be considered, since many individuals of advancing age relocate to milder climates at retirement, and therefore are more inclined to develop age-related cataracts.

In the Zigman et al. (8) study, the three populations examined were of different ethnic and economic backgrounds, and the choice to do surgery could have been made based upon different parameters. The three medical centers that participated in this study were a free clinic (Philippines), a Veterans Administration Hospital (Tampa), and a University Medical Center (Rochester). Migration seemed not to be a problem since the subjects in the Philippines did not migrate, the subjects in Rochester were a stable population, and in Tampa 80% of the subjects were born or had remained in the southeastern United States for more than 20 years. Therefore, the best comparison to test the hypothesis of sunlight-stimulated cataracts is between outdoor- and indoor-active subjects in each area. Examination of these data strongly supports the idea that sunlight exposure does enhance the formation of human cataract. Although all the epidemiologic studies reviewed above suffer certain defects, taken all together, they strongly support a relation between sunlight exposure (UV) and the development of human (brunescent) cataracts.

Enhancement of UV-Stimulated Cataracts

Although sunlight is the greatest source of near-UV radiation normally encountered by humans, artificial light sources used for the illumination of indoor areas of activity or for special occupation or health-related activities also expose humans to near-UV energy. Variations in the quality and quantity of the light energy emitted by these sources are enormous, as is the number of uses to which they are put. A short list of such sources follows: fluorescent lamps (Cool-White, Daylight, and Vitalight); black light lamps (used for recreational, occupational, and health-related purposes); special UV lamps for medical uses (orthopedic cast setting, dermatologic treatment, dental-filling setting, and sterilization); welding arcs; near-UV lasers; and a host of others. Comprehensive coverage of this field is made in the Proceedings of the Symposium on Non-ionizing Radiation (96). However, to date no direct proof that specific human cataracts were caused by these sources or others related to them is available.

Suggestive findings that cataracts in 40- to 50-year-old individuals were caused by exposure to leakage from a dental-filling hardening lamp (NUVA lamp) have been published, but the very small numbers of subjects, the very low irradiances of UV-A energy to which they were exposed, and the lack of uniformity of the clinical findings severely limit confidence in relating the cataracts to this UV exposure (see Refs. 97 and 98). Other suggested findings that cataracts developing in 40- to 50-year-old humans were caused by artificial UV-A radiation from a blueprinting machine suffered the same criticisms as stated above (see Ref. 99).

In the latter case, the subjects were under treatment for gout using allopurinol, and suggestions have been made that drug photosensitization had caused these cataracts. However, as pointed out by Rudy et al. (99), biologic and photochemical studies failed to support this conclusion. Allopurinol alone may produce cataractous changes occasionally without near-UV exposure greater than ambient levels.

Another drug, the known UV-A photosensitizer 8-methoxypsoralen (8-MOP) has been implicated as a cataract-inducing agent in animal studies and by other suggestive human case studies. This drug has been used over many years to treat vitiligo and, more recently, for photochemotherapy for psoriasis. Jose (100) has observed unscheduled DNA synthesis to result in animal lens epithelial cells when the lenses were exposed in vitro to UV-A + 8-MOP. Lerman et al. (101) have used fluorimetric analyses of animal and human lenses to find permanent binding of 8-MOP lens proteins, and have presumed that several cases of human cataract resulted from the 8-MOP + UV-A therapy. Positive proof of the association between cataract formation in humans and the use of photosensitizing 8-MOP has yet to be provided. Furthermore, Parrish et al. (102) have carried out the most extensive modern study of the potential cataract hazard due to 8-MOP feeding plus chronic UV-A exposure of Dutch-belted rabbits, but noted no signs of cataractous changes over a period of 18 months. Only ophthalmic and histologic examinations of the ocular tissues, but no spectroscopic or chemical data were reported, to support the findings of an absence of lens phototoxicity. Still, their conclusion was that, in this species, the dose of 8-MOP + UV-A as used to treat psoriatic lesions was lower than those doses capable of inducing cataracts.

Many other drugs have chemical structures that predict their potential to sensitize ocular lenses to develop cataractous changes due to UV-A exposure. These include other furocoumarins (as may be present in foods), chlorpromazine, eosine, protoporphyrins, and sulfonamides.

Practical Comments

Sufficient factual and suggestive evidence that near-UV radiation has a role in human cataract development exists, so that most research workers in this field have recommended that patients under phototherapy use artificial eye protective lenses that filter out near-UV radiation. For many years, artificial lenses in the form of spectacles have been available to afford eye protection from UV light, particularly as it is present in sunlight. A multitude of products for sun protection is available, some of which are helpful, but others are not. For severe conditions of UV light exposure, special protective devices have been developed, such as the goggles worn by welders and the helmets worn by astronauts traveling in space.

To influence the prevention of human cataracts that may be stimulated by exposure to ultraviolet radiation, eye care specialists can play a major role. The potential to prevent natural or artificially UV-induced cataractous changes can be satisfied by suggesting and providing the proper spectacles or other devices that are designed to reduce or eliminate the level of near-UV light that enters the eye. However, it seems that there is little importance for such eye protective devices to filter infrared radiation (see Ref. 103). There are already many products available for this purpose, but some types of sunglasses may not be beneficial and could even cause more harm than benefit. If the degree of absorption of visible light is greater than the degree of absorption of near-UV light, dilation of the pupil will result. The pupillary reflex is sensitive only to visible and not to UV light. This response could result in the passage of greater levels of near-UV light to the lens than is desirable. Also, the zone of epithelial cells of the lens, where mitosis is maximum, would be more exposed to UV radiation, so that differentiation problems could result. Some sunglasses even have windows in the near-UV region, which would also allow higher levels of near-UV light to reach the lens. The above-stated ideas are detailed and suggested by Anderson and Gebel (104) and Segre et al. (105).

Although this survey recommends no specific products, there are sunglasses available from all of the major optical product manufacturers that have been specifically designed to block near-UV light from entering the eye. Many plastic lenses are available that effectively block the UV portion of sunlight. A relatively new feature is the ability to coat glass lenses with a dielectric reflective coating that has been designed specifically to reflect all ultraviolet light below 400 nm and all infrared light above 700 nm (Human Lens-II). Other plastic lenses have been designed to absorb near-UV light (such as UV-400 lenses). The appropriate eye protective wear for specific individuals should be provided as a function of the device's absorptive properties, the illumination present in the environment of each individual, and the ability to obtain eyewear that has specific absorptive properties to satisfy the particular needs. The cost of the devices do not relate to their effectiveness. Miller (106) has discussed the effects of sunglasses on the visual process.

Another point to be considered is that cataract surgery results in a lensless eye, in which the major filter for near-UV light (300–400 nm) before it reaches the retina has been removed. Recently, Ham et al. (22) have emphasized the importance of the UV protection provided by the natural human lens to prevent retinal photochemical damage due to UV-A. New information based upon measurements of the near-UV light that can damage the monkey retina in aphakic eyes does not indicate that there is sufficient ultraviolet energy present in the environment to markedly damage the retinal photoreceptors, if they are

not prevented from reaching the retina due to the presence of the natural or
artificial filters (see Ref. 22).

REFERENCES

1. D. Pitts. Effects of ultraviolet radiation on the eye in UV-A. In *Biological
    Effects of Ultraviolet Radiation with Emphasis on Human Responses to
    Longwave Ultraviolet* (J. A. Parrish, R. R. Anderson, F. Urbach, and D.
    Pitts, eds.), Plenum, New York, 1978.
2. D. H. Sliney and M. L. Wohlbarsht. *Safety with Lasers and Other Optical
    Sources*, Plenum, New York, 1980.
3. F. S. Johnson, T. Mo, and A. E. S. Green. Average latitudinal variation in UV-
    radiation at the earth's surface. Photochem. Photobiol., *23*: 179–189 (1976).
4. W. T. Ham, H. A. Mueller, and D. Sliney. Nature of retinal radiation
    damage: Dependence on wavelength, power level and exposure time.
    Vision Res., *20*: 1105–1111 (1980).
5. L. R. Killer. *Ultraviolet Radiation*, John Wiley and Sons, New York, 1965.
6. P. Bener. Spectral Intensity of natural ultraviolet radiation and the depen-
    dence on various parameters. In *The Biological Effects of Ultraviolent
    Radiation* (F. Urbach, ed.), Pergammon Press, Oxford, 1969.
7. A. C. Giese. *Living with Our Sun's Rays*, Plenum, New York, 1976.
8. S. Zigman, M. Datiles, and E. Torczynski. Sunlight and human cataracts.
    Invest. Ophthalmol. Vis. Sci., *18*: 462–467 (1979).
9. R. van Heyningen. Fluorescent compounds of the human lens. CIBA
    Symp. 19 (New Series) (1973).
10. J. P. Dillon and A. Spector. Aerobic and anaerobic photochemical modifi-
    cation of lens protein analyzed by synchronous scan fluorescent tech-
    niques. Invest. Ophthalmol. Vis. Sci. (Suppl.), 127 (1979).
11. H. G. Sperling and R. E. Marc. Selective damage to blue sensitive cones by
    blue light. Science *196*: 454–456 (1977).
12. G. Walls. *The Vertebrate Eye and Its Adaptive Radiation*, Haffner, New
    York, 1942 (reprinted 1963).
13. E. Merker. Drei Fälle Vershiedenen Lichtdurch Lässigkeit der augenlinsen
    von Wirbeltiiren. Biol. Zentralbl., *59*: 87–98 (1939).
14. W. Franz. Die Spirale in dem Stäbchen der Wirbeltirenetzhaut und der
    vermeintliche Plattchenzerfall nach verschiedenan untersuchungmethaden.
    Biol. Zentralbl. *54*: 76–84 (1934).
15. G. Walls and H. Judd. The intraocular colour filters of vertebrates. Br. J.
    Opthalmol., *17*: 641–654; 705–720 (1933).
16. D. Kennedy and R. Milkman. Selective light absorption by the lenses of
    lower vertebrates and its influence on spectral sensitivity. Biol. Bull., *111*:
    375–386 (1956).
17. R. van Heyningen. Fluorescent glucosides in the human lens. Nature
    (Lond.), *230*: 393–394 (1970).
18. G. Cooper and J. Robson. The yellow colour of the lens of man and other
    primates. J. Physiol. (Lond.), *203*: 411–417 (1969).

19. S. Zigman and P. Gilbert. Lens color in sharks. Exp. Eye Res., *26*: 227–231 (1978).

20. G. Wald. Alleged effects of the near ultraviolet on human vision. J. Opt. Soc. Am., *42*: 171–177 (1952).

21. S. Zigman, J. Groff, T. Yulo, and T. Vaughan. The response of mouse ocular tissues to continuous near-UV light exposure. Invest. Ophthalmol., *191*: 714–715 (1975).

22. W. T. Ham, H. A. Mueller, J. J. Ruffolo, D. Guerry, and R. K. Guerry. Action spectrum for retinal injury from near-UV in aphakic monkey. Am. J. Ophthalmol., *93*: 299–306 (1982).

23. R. van Heyningen. Fluorescent derivatives of 3-hydroxykynurenine in the lens of man, baboons, and the grey squirrel. Biochem. J., *123*: 30–31 (1971).

24. F. Said and R. Weale. The variation with age of the spectral sensitivity of the living human crystalline lens. Gerontologica, *3*: 213–231 (1959).

25. R. Weale. New light on old eyes. Nature (Lond.) *198*: 944–946 (1963).

26. A. Spector, L. Li, and J. Sigelman. Age-dependent changes in the molecular size of the human lens proteins and their relationship to light scatter. Invest. Ophthalmol., *13*: 795–798 (1974).

27. T. Philipson and P. Fagerholm. Lens changes responsible for increased light scattering in some types of senile cataract. In *The Human Lens in Relation to Cataract*, Ciba Foundation Symposium, *19*, Elsevier, Amsterdam, 1973, pp. 151–168.

28. R. van Heyningen. The glucoside of 3-OH-kynurenine and other fluorescent compound in human lens. In *The Human Lens in Relation to Cataract*, Ciba Foundation Symposium, *19*, Elsevier, Amsterdam, 1973, pp. 151–168.

29. S. Lerman. Lens proteins and fluorescence. Israel J. Med. Sci., *8*: 1583–1589 (1972).

30. A. Bando. The relationship between coloration and fluorescence in the human lens. Acta Soc. Ophthalmol. Japan, *77*: 147–154 (1973).

31. A. Spector, D. Roy, and J. Stauffer. Isolation and characterization of an age-dependent polypeptide from human lens with non-tryptophan fluorescence. Exp. Eye Res., *21*: 9–24 (1975).

32. P. Van Haard. Oxidative challenge to aging human lens leading to cataract. Ph.D. Thesis, Nijmegen, The Netherlands, 1980.

33. S. Garcia-Castineiras, and A. Spector. Non-tryptophan fluorescence associated with human lens protein. Exp. Eye Res., *26*: 461–476 (1978).

34. R. Truscott, K. Faull, and R. Augusteyn. The identification of anthranylic acid in proteolytic digests of cataractous human lens proteins. Ophthalmic Res., *9*: 263–268 (1977).

35. A. Pirie. Color and solubility of the proteins of human cataract. Invest. Ophthalmol., *7*: 634–650 (1968).

36. D. Grover and S. Zigman. Coloration of human lens by near-UV photo-oxidized tryptophan. Exp. Eye Res., *13*: 70–75 (1972).

37. K. Satoh, M. Bando, and A. Nakjima. Fluorescence in human lens. Exp. Eye Res., *16*: 167–172 (1973).
38. K. Yamamoto. Aging of soluble lens proteins. Acta Soc. Ophthalmol. Japan, *77*: 897–906 (1973).
39. M. Oguchi, T. Shimizu, H. Seki, H. Kawase, M. Sakurai, and Y. Uchiyama. On the fluorescent color of the crystalline lens. Acta Soc. Ophthalmol. Japan, *77*: 186–191 (1973).
40. A. Pirie. Formation of *N*-formylkynurenine in proteins from lens and other sources by exposure to sunlight. Biochem. J., *125*: 203–208 (1971).
41. P. Stein, R. Henkens, P. Yamanashi, and M. Wohlbarst. Studies of brunescent cataract. II. Fluorescent studies in normal an brunescent lens proteins. Ophthalmic Res., *8*: 388–394 (1977).
42. R. Kurzel, M. Wohlbarsht, B. Yamanashi, G. Staton, and R. Borkman. Tryptophan excited states in the cataractous human lens. Nature, *241*: 132–133 (1973).
43. R. Jacobs and D. Krohn. Variations in fluorescence characteristics of intact human lens segments as a function of age. J. Gerontol., *31*: 641–647 (1976).
44. S. Lerman and R. Borkman. Spectroscopic evaluation and classification of the normal aging and cataractous lens. Ophthalmol. Res., *8*: 335–353 (1976).
45. R. Augusteyn. Distribution of fluorescence in the human cataractous lens. Ophthalmol. Res., *7*: 217–224 (1975).
46. J. Dillon, A. Spector, and K. Nakanishi. Identification of beta carbolines isolated from fluorescent human lens proteins. Nature, *259*: 422–423 (1976).
47. J. Kuck and N. Yu. Raman and fluorescence emission of human lens new fluorophor. Exp. Eye Res., *27*: 737–741 (1978).
48. N. Yu, J. Kuck, and C. Askren. Red fluorescence in older and brunescent human lenses. Invest. Ophthalmol., *18*: 1278–1280 (1979).
49. S. Zigman and T. Yulo. Properties of a human lens chromophoric component. Invest. Ophthalmol. (Suppl.), *May*: 207 (1980).
50. M. Sun and S. Zigman. Isolation and identification of tryptophan photoproducts from aqueous solutions of tryptophan exposed to near-UV light. Photochem. Photobiol., *29*: 893–897 (1978).
51. S. Coren and J. Girgus. Density of human lens pigmentation: In vivo measurements over an extended age range. Vision Res., *12*: 343–346 (1972).
52. S. Zigman, J. Groff, T. Yulo, and G. Griess. Light extinction and protein in human lens. Exp. Eye Res., *23*: 555–567 (1976).
53. S. Zigman. Photochemical mechanisms in cataract formation. In *Mechanism of Cataract Formation in Human Lens* (G. Duncan, ed.), Academic Press, London, 1981, pp. 117–150.
54. J. Weiter and E. Finch. Parmagnetic species in human lens. Nature, *254*: 536–537 (1975).

55. S. Zigman. Tryptophan excited states in the lens. Doc. Ophthalmol., *8*: 267–274 (1976).
56. J. Weiter and S. Subramian. Free radicals produced in human lenses by a biphotonic process. Invest. Ophthalmol. Vis. Sci., *17*: 869–873 (1978).
57. K. Bhuyan and D. Bhuyan. Cataract induction in rabbits by feeding amino-triozole. Am. J. Ophthalmol., *69*: 147–154 (1973).
58. M. Nakagawa, H. Watanabe, S. Kodato, H. Okajima, T. Hino, J. Flippen, and B. Witkop. Sensitized photoproducts of tryptophan. Proc. Natl. Acad. Sci. USA, *74*: 4130–4133 (1977).
59. R. van Heyningen. Photooxidation of lens proteins by sunlight in the presence of fluorescent glucosides isolated from the human lens. Exp. Eye Res., *17*: 137–147 (1973).
60. S. Lerman, S. Zigman, and W. Forbes. Insoluble protein fraction of the lens. Exp. Eye Res., *7*: 444–449 (1968).
61. R. Kurzel, M. Wohlbarst, and B. S. Yamanashi. Ultraviolet radiation effects on the human eye. Photochem. Photobiol. Rev., *2*: 133–168 (1977).
62. J. Dillon. Photolytic changes in lens proteins. Current Eye Res., *3*: 145–150 (1984).
63. S. Zigman, T. Paxhia, and T. Yulo. Biochemical and photochemical studies of squirrel lenses. Invest. Ophthalmol. Vis. Sci., *22*: 197 (1982) (abstract supplement).
64. J. D. Goosey, J. S. Zigler, I. B. Matheson, and J. Kinoshita. Effects of singlet oxygen on human lens crystallins in vitro. Invest. Ophthalmol. Vis. Sci., *20*: 679–683 (1981).
65. S. Zigman, T. Yulo, and G. Griess. Inactivation of catalase by near-UV light. Mol. Cell. Biochem., *11*: 149–154 (1976).
66. J. F. R. Kuck. Effects of longwave UV on the lens. Doc. Ophthalmol., *8*: 261–266 (1976).
67. A. Kalustian, S. Zigman, and M. Sun. Near-UV light sensitized inactivation of glutathione reductase. Proc. Am. Soc. Photobiol., 112–113 (1978).
68. S. Zigman, J. Schultz, and T. Yulo. Binding of tryptophan photoproducts to calf lens gamma crystallin. Exp. Eye Res., *15*: 209–217 (1973).
69. L. Strong, A. Torriglia, I. Korc, and S. Zigman. Inhibition of lens and retina ATPases by near-UV radiation. Invest. Ophthalmol. Vis. Sci., *25*: 207 (1984) (ARVO abstracts).
70. J. D. Goosey, J. S. Zigler, and J. Kinoshita. Cross-linking of lens crystallins in a photodynamic system. Nature, *208*: 1278–1280 (1980).
71. W. T. Ham, J. J. Ruffolo, H. A. Mueller, and D. Guerry. The nature of retinal radiation damage. Vision Res., *20*: 1105–1113 (1980).
72. R. Collier, W. Waldron, D. Merrill, and S. Zigman. Effects of ambient near-near-UV exposure on the retina. Invest. Ophthalmol., *25*: 18 (1984) (ARVO abstracts).
73. R. F. Borkman and S. Lerman. Evidence of a free-radical mechanisn in aging and UV-irradiated ocular lens. Exp. Eye Res., *25*: 303–309 (1977).
74. K. C. Bhuyan, D. K. Bhuyan, and S. M. Podos. Evidence for increased lipid peroxidation in cataracts. IRCS Med. Sci., *9*: 126–127 (1981).

75. S. D. Varma, V. K. Srivastava, and R. D. Richards. Photoperoxidation in lens and cataract formation. Ophthalmol. Res., *14*: 167–175 (1982).

76. G. L. Feldman and L. S. Feldman. New concepts of human lenticular lipids and their possible role in cataracts. Invest. Ophthalmol., *4*: 162 (1965).

77. E. Cotlier, Y. Obara, and B. Toftness. Cholesterol and phospholipids in fractions of human lens and senile cataract. Biochim. Biophys. Acta, *530*: 267 (1978).

78. R. M. Broekhuyse and F. J. M. Daeman. *The Eye in Lipid Metabolism in Mammals* F. Snyder, ed.), Plenum Press, New York, 1977, p. 145.

79. J. S. Andrews and T. L. Leonard-Martin. Total lipid and membrane lipid analysis of normal animal and human lenses. Invest. Ophthalmol. Vis. Sci., *21*: 40 (1981).

80. A. Spector and D. Roy. Disulphide-linked high molecular weight protein associated with human cataract. Proc. Natl. Acad. Sci. USA, *75*: 3244 (1978).

81. S. Zigman and T. Paxhia, and B. Antonellis. Stability of shark and human lens fiber cell membranes. Comp. Biochem. Physiol., *73B*: 791–796 (1982).

82. L. Rosenfeld and A. Spector. Changes in lipid distribution in the human lens with the development of cataract. Exp. Eye Res., *33*: 641–650 (1981).

83. L. Rosenfeld and A. Spector. Comparison of polyunsaturated fatty acid levels in normal and mature cataractous human lenses. Exp. Eye Res., *35*: 69–75 (1982).

84. A. Spector and W. H. Garner. Hydrogen peroxide and human cataract. Exp. Eye Res., *33*: 673–681 (1981).

85. D. G. Pitts, A. P. Cullen, and P. Hacker. Effects of ultraviolet radiation from 295–365 nm. Invest Ophthalmol. Vis. Sci., *16*: 932–946 (1977).

86. S. Zigman, T. Yulo, and J. Schultz. Cataract induction in mice exposed to near-UV light. Ophthalmol. Res., *6*: 259–270 (1974).

87. J. A. Zuclich and J. S. Connaly. Ocular damage induced by near-UV laser radiation. Invest. Ophthalmol., *15*: 760–776 (1976).

88. W. S. Duke-Elder. The pathological action of light upon the eye. I. Action of the outer eye to photophthalmia. Lancet, *1*: 1127–1141 (1926).

89. J. H. Clark. The effect of ultraviolet radiation on lens proteins in the presence of slats relative to senile cataracts. Am. J. Physiol., *113*: 538–549 (1935).

90. R. van Heyningen. The human lens. I. A comparison of cataracts extracted in Oxford (England) and Shikarpur (W. Pakistan). Exp. Eye Res., *13*: 126–135 (1973).

91. R. van Heyningen. What happens to the human lens in cataract. Sci. Am., *233*: 70–81 (1976).

92. R. Hiller, L. Giacometti, and K. Yuen. Sunlight and cataract. An epidemiological investigation. Am. J. Epidemiol., *105*: 450–459 (1977).

93. F. Hollows and D. Moran. Cataract—the ultraviolet risk factor. Lancet, *2*: 1249–1251 (1981).
94. H. Taylor. The environment and the lens. Br. J. Ophthalmol., *64*: 303–310 (1980).
95. L. T. Chylack. Classification of human cataracts. Arch. Ophthalmol., *96*: 888–892 (1978).
96. Non-ionizing radiation. Proceedings of a Topical Symposium by The American Conference of Governmental and Industrial Hygienists. Cincinnati, Ohio, November 26–28, 1980.
97. S. Lerman. Human ultraviolet radiation cataracts. Ophthalmic Res., *12*: 303–314 (1980).
98. S. Lerman and S. Zigman. Letters to the Editor. Ophthalmic Res., *14*: 70–72 (1982).
99. M. A. Rudy, S. Zigman, S. Girsch, and E. Schenk. Lack of photosensitization of ocular tissues by allopurinol. Arch. Ophthalmol., *99*: 2030–2033 (1981).
100. J. G. Jose, and K. L. Yielding. Photosensitive cataractogens, chlorpromazine and methoxypsorael, cause DNA repair synthesis in lens epithelial cells. Invest. Ophthalmol. Vis. Sci., *17*: 687–691 (1978).
101. S. Lerman, J. M. Megaw, K. Gardner, Y. Takei, and I. Willis. Localization of 8-methoxypsoralen in ocular tissues. Ophthalmol. Res., *13*: 106–116 (1981).
102. J. A. Parrish, L. T. Chylack, M. E. Woehler, H. M. Cheng, M. A. Pathak, W. L. Morison, J. Krugler, and W. F. Nelson. Classification of human cataracts. J. Invest. Dermatol., *73*: 256–258 (1979).
103. D. H. Sliney. The ambient light environment and ocular hazards. In *Retinitis Pigmentosa* (M. Landers, ed.)., Plenum Press, New York, 1977, pp. 211–221.
104. W. J. Anderson and R. K. H. Gebel. Ultraviolet windows in commercial sunglasses. Appl. Optics, *16*: 515–517 (1977).
105. G. Segre, R. Reccia, B. Pignalosa, and G. Pappalando. The efficiency of ordinary sunglasses as a protection from UV radiation. Ophthalmol. Res. *13*: 180–187 (1981).
106. D. Miller. The effect of sunglasses on the visual mechanism. Surv. Ophthalmol., *19*: 38–44 (1974).

9

Photochemical Mechanisms in the Lens

JAMES DILLON / Columbia University, New York, New York

Besides the skin, the eye is the only part of the body that is continuously subjected to ambient radiation. In the eye the main function of the anterior ocular tissue, the cornea and the lens, is to transmit and focus light on the retina undistorted. They perform a further function, though, which is to filter out short ultraviolet wavelengths of light from the retina. In the case of the human eye, this means that wavelengths below 400 nm do not reach the retina. The wavelengths of light filtered out, however, are absorbed by the cornea and lens and can produce deleterious effects. These effects may occur at ambient radiation over an extended period (1–3) or at high doses over a short period of time (4). This chapter examines the possible implications of ambient radiation impinging on the human lens and its possible role in aging and senile cataractous processes in that organ.

OVERVIEW

There are four major parameters that are important when considering the role of some photochemical reaction in the aging and cataractous processes in the human lens: (1) the wavelengths of light that are transmitted through the cornea but are absorbed by the lens; this also includes various components in the aqueous; (2) the level of oxygen in the aqueous and lens; (3) the presence of various quenchers, which may interrupt or negate various photochemical processes, and (4) the site at which any photo-oxidative damage takes place.

All light below about 295 nm is cut off by the cornea and does not reach the lens (5). On the other hand, the human lens absorbs much of the light between 295 and 400 nm (6). The major constituents that absorb this light in the lens are

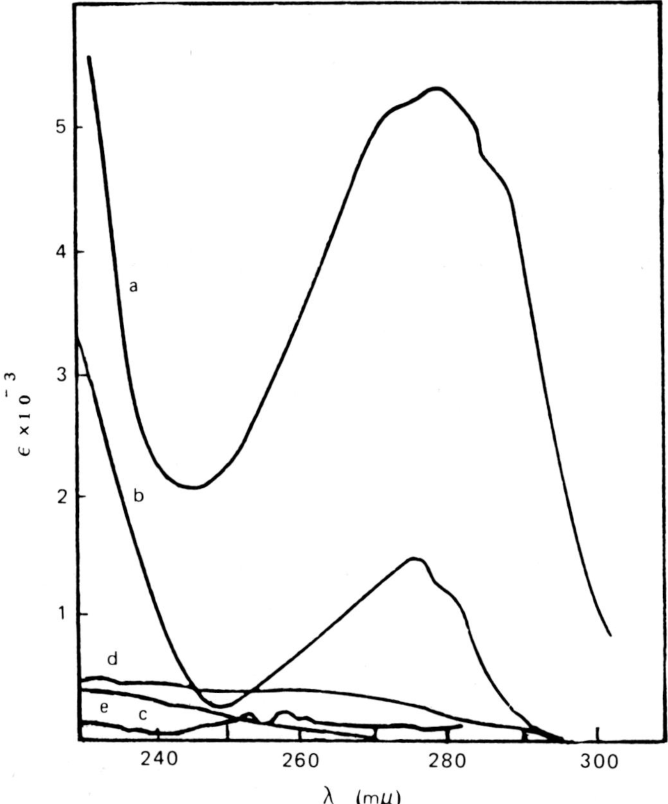

Figure 1 The ultraviolet spectra of the absorbing amino acid residues in lens proteins; (a) Trp, (b) Tyr, (c) Phe, (d) Cys, and (e) cystine.

protein residues (Fig. 1) and 3-hydroxykynurenine glucoside (3-OH Kyn); a metabolite of tryptophan (Trp). Of the amino acid residues in lens protein, Trp absorbs more than 95% of the radiant energy, although Tyr, Phe, and Cys also contribute to some extent. The ultraviolet spectrum of the other major absorbing species in the young lens is given in Fig. 2. It consists of a maximum at 365 nm with end absorption beyond 400 nm (7). Therefore, in all likelihood, any photochemical mechanism operant in the young human lens probably results from the initial absorption of light by lens Trp or the glucoside of 3-OH Kyn.

One of the manifestations of aging in the human lens is the yellowing of the cytosol protein resulting from an apparent modification of that material. This color increases the absorptive characteristics of the lens between 300 and 400 nm (6), adding to the possible initiators of photochemical processes.

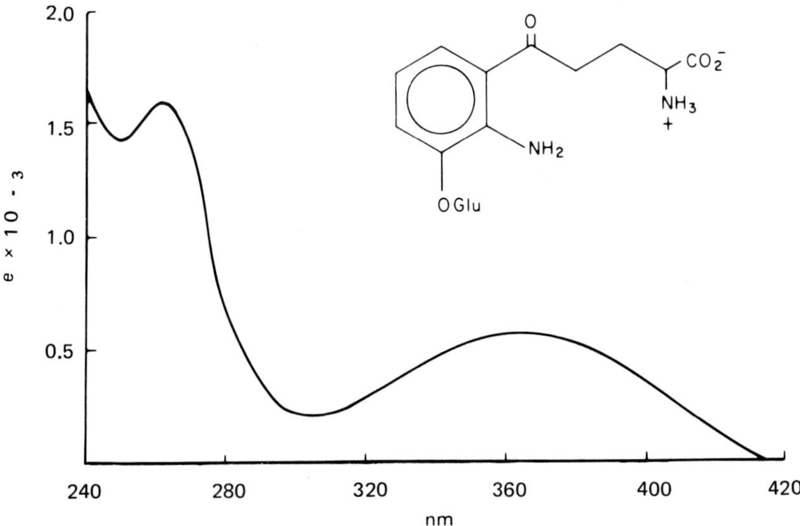

Figure 2 The ultraviolet spectrum of the glucoside of 3-hydroxykynurenine whose structure is given in the figure.

The second parameter important to any photochemical process is the presence of oxygen, since many reactions are slowed down or actually negated at low O_2 tensions. In general oxygen enters the eye primarily through the cornea. From there to the lens a gradient exists with decreasing concentrations of oxygen with a possible range of from 24 to 72 mmHg (8). This is approximately 20% the concentration found in blood.

The third parameter is the presence of quenchers. The lens and the aqueous contain glutathione (9), ascorbic acid (10), and vitamin E (11), all of which quench various photochemical or oxidative reactions.

The final consideration is the actual site of damage. In a cell there are numerous substrates that are susceptible to photolytic change, such as the membrane, proteins, or nucleic acid material. Some of these areas of damage may be more important for the integrity of a living system than others. For example, membrane damage could lead to cell lysis, causing cell death. This and the other parameters are discussed in detail below.

PHOTOCHEMICAL MECHANISMS

Theoretical Considerations

The primary process in any photochemical reaction is the absorption of light by a chromophore (12,13). This causes the formation of an electronically excited

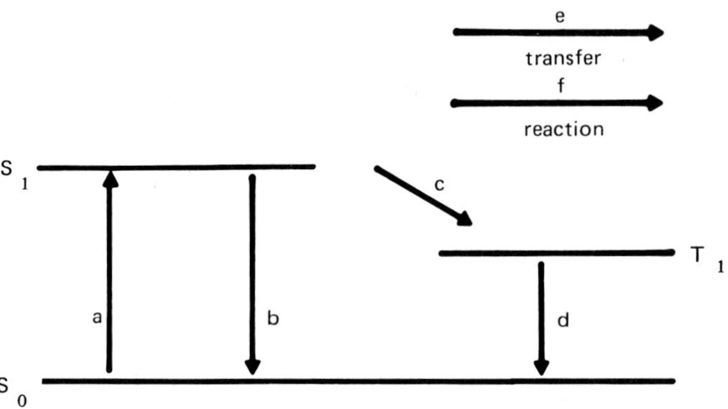

Figure 3 After the initial absorption of light a ground state can make a transition to an excited singlet (a). Once there the electron can go through various processes; (b) return to the ground state, giving off the photon as fluorescence; (c) an intersystem crossing to the triplet state, from which it can also return to the ground state. (d) From both the singlet and triplet state the excited electron has additional processes by which it can become deactivated, which are not shown in the diagram. These are called radiationless transitions, and the excess energy is given off as heat. The excited molecule can also transfer its energy to another chromophore (e) or react (f) to form photoproducts.

state. In most molecules the ground or relaxed state consists of two paired electrons, which means that this state is referred to as a singlet (S_0). Upon absorption of a photon equal to the energy between the ground and excited state (S_1), an electron will make a transition to the latter state (Fig. 3). Once an electron reaches the excited S_1 state there are several processes that can occur. (1) The electron can come down to the ground state, giving off the photon as fluorescence, (2) cross over to the triplet excited state (T_1) via intersystem crossing (here the electrons are unpaired), or (3) once in the triplet it can come back to the ground state, giving off the photon (phosphorescence). In addition, the excited molecule can either react to form photoproducts or transfer its energy to another acceptor molecule. These processes can occur from either the singlet or triplet state.

The extent to which any excited chromophore will go through any of the above processes varies drastically depending on the chromophore involved and the environment it is in. Environment includes functionalities, either in conjugation or near the chromophore, the solvent, and ions in solution. In general, though, intersystem crossing to the triplet state is not a favored process and usually results in low yields. On the other hand, since S_1 to T_1 transition is not favored, neither is the T_1 to S_0 transition. Therefore, once formed the triplet is generally long lived and very reactive. Many of the sensitized reactions discussed below occur via this intermediate.

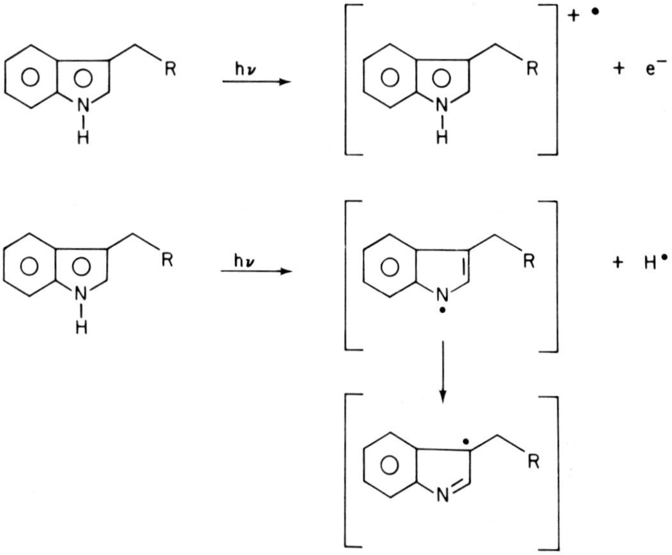

Figure 4 Indoles are thought to react in this manner. The excited singlet can either photoionize or deprotonate, forming the carbon-centered radical depicted.

Besides the photoprocesses outlined above, excited chromophores have additional mechanisms to dissipate its energy, that is, to transfer (14) it to another chromophore and react to form photoproducts. A major criterion for transfer to occur is that the emission energy of the donator molecule must overlap the absorption of the acceptor molecule. For example, in proteins it is very difficult to detect Tyr fluorescence. This is due to the fact that much of its fluorescence is transferred to Trp, which absorbs energy at the same wavelengths that Tyr emits its energy. Numerous photochemical reactions can result from such transfer processes and are referred to as sensitized reactions. Photoproducts can also result from the direct absorption of ambient light by a chromophore. In the lens, therefore, any number of photolytic processes may be operant and may be manifested on molecular, macromolecular, or even the tissue level.

Direct Photolyses

Since, in lens protein, Trp absorbs most of the radiation transmitted by the cornea, the photochemistry of lens protein should be dominated by the initial absorption of light by Trp. It, though, has a rather complex photochemistry. In general, it is dominated by the processes in Fig. 4; either (1) expulsion of an electron, leading to a radical cation (15) or (2) cleavage of the indole N-H, leading to the carbon-centered radical indicated (16). Which of these processes (and possibly others) actually occurs depends on a number of parameters, including (1) wavelength of irradiation (17), (2) the presence of various functionalities (18,19), and (3) the hydrophilicity of the solvent (20). In general, N-H cleavage is thought to dominate at longer wavelengths and in more hydrophobic environments (e.g., protein cores or membranes).

The presence of functionality in the vicinity of the indole core, though, can redirect its photodecomposition. Model experiments on the photolysis of Trp-containing peptides, where an amino function has access to the indole ring, result in deamination and subsequent attachment of the aliphatic portion to the indole moiety (19–21). This process also dominates in air and at longer wavelengths. The tendency of amino functionalities to redirect indole photodecomposition away from N-H cleavage may be due to the apparent formation of an excited state complex between the protonated amine and the indole (18). Other functionality that might be expected to have similar effects are His and disulfides. Such reactions could lead to the modification of protein residues other than Trp, resulting in the destruction of protein function.

The phorolysis of proteins has been studied on both the photophysical (22) and biochemical (23) levels. The photophysical level would include the investigation of various excited states of Trp formed in proteins upon photolysis, the kinetics of those states, and the eventual transfer of this excitation to other residues in the protein. These experiments were performed with lasers, using short pulses of high intensity (22). Biochemical parameters that have been followed would include the loss of enzyme activity and structural modifications, such as cross-linking. These photolyses utilized both lasers and low-intensity lamps.

The excitation of Trp in protein to its singlet state has resulted in the spectroscopic detection of Trp^+ and $Trp^{\bullet}$ from the loss of e^- and possibly $H^{\bullet}$, respectively, from the indole moiety. Besides the destruction of Trp, electronic transfer to a disulfide bridge (24) gives RSSR-, leading to RS- + $RS^{\bullet}$ (the cleavage of the disulfide bond). These experiments were carried out at 265 nm, and it was not clear whether they are valid at 295 nm. In general, though, the direct photolysis of proteins leads to a complex series of reactions manifested by numerous macromolecular and molecular changes, such as cross-linking, yellowing, and the destruction of various amino acid residues (e.g., Trp, Tyr, Cys, and His) (23). Although most of these experiments were performed at relatively short wavelengths (i.e., 265 nm), similar results are obtained at the wavelengths transmitted by the cornea. Since His does not absorb at these wavelengths, its destruction is most probably explained by the conversion of Trp to *N*-formylkynurenine (NFK) (25), which is an efficient singlet oxygen generator (26) and can lead to many of the photosensitized reactions outlined in the next section.

Photosensitized Reactions

There are two biologically important photoxidation mechanisms by which chromophores can act as sensitizers: type I and type II. The main distinction between these two mechanisms is the following: in type I a sensitizer in the triplet state interacts directly with a substrate (protein, DNA, membrane, and so on) by abstracting a proton or transferring an election to the substrate. Thus, a free-radical process is initiated where the substrate can go on to react further with oxygen or other molecules. In a type II mechanism the sensitizer triplet reacts directly with oxygen, either transferring its energy, forming the highly reactive singlet oxygen ($^1O_2$), or (much less likely) transferring an election-forming superoxide (O_2^-) (27).

The photo-oxidation mechanisms mentioned above are extremely important in biologic systems because the substrate that is photochemically altered need not initially absorb radiant energy. This is especially important in the lens, since only light above about 295 nm penetrates to it. There are numerous parameters that can affect photo-oxidation reactions, such as substrate and sensitizer concentration and light intensity. In cells, accessability and/or binding of the sensitizer to the substrate is important (28). In complex biologic systems, photo-oxidation can occur by either a type I or a type II mechanism, or both concurrently.

The substrates affected by these mechanisms are numerous. In proteins, Trp, His, Cys, Met, and Try are most readily photo-oxidized (27). There are several products of Trp; among them is the aforementioned NFK (25). Reactions on model compounds show that the photo-oxidation of His leads to the cleavage of the imidazole ring, which in His would lead to the formation of aspartic acid after acid hydrolysis (29). The other conversions are Met to Met sulfoxide and Cys to cystine and cysteic acid (27). The photoproducts of Tyr are not known, but the initial event is the abstraction of the phenolic proton. In general, Met, His, and Trp are the most susceptible amino acids in proteins via a type II mechanism, but the relative rates in various situations would depend on the access of the sensitizer to the particular residue. The effect of these reactions on proteins is loss of confirmational stability leading to or concurrent with cross-linking.

In membranes, numerous components that contain double bonds have been reported to be susceptible to photo-oxidation. Cholesterol as well as various esters (30) of it have been photo-oxidized, resulting in numerous products, including the 5-α-hydroperoxy derivative. Hydroperoxides are also formed in the case of unsaturated fatty acids (31). Cell lyses can also result from such reactions in both liposomes (32) and ghosts (33), but it is not clear at this time whether cell lyses in intact membranes, that is, those that still contain proteins, is due to primarily to the photo-oxidation of the lipids or proteins. The photosensitized oxidation of red blood cell ghosts leads both to membrane protein photopoly-merization and lipid oxidation (34,35). The net result of this damage is cell membrane disruption. In one study (36) the photosensitized oxidation of only membrane protein was suggested to result in the leakage of cell ions, whereas the destruction of the lipids leads to enzyme leakage. Additionally, the deterioration of cell membrane functions, such as transport, can result from photosensitized oxidation. This effect also seems to be due primarily to changes in the membrane protein (35).

Role of Oxygen

Since the main source of metabolic energy in the lens is via anaerobic glycolysis, the level of O_2 is lower in the lens than in the blood. This may be critical to the relevance of any photochemical process in the lens, since the presence of oxygen can increase the rate at which lens proteins photolyse and is essential for many photosensitized reactions. Thus, any increase of oxygen in the lens will increase the likelihood of photochemical reactions in the lens. The level of oxygen in the human lens and aqueous has not been studied.

Role of Quenchers

With all the possible photochemical processes that conceivably occur in the lens, it is surprising that it can remain intact and functioning for such a long period of time. This may be due, not only to a low level of O_2 but also to the presence of quenchers, which can negate specific reactive intermediates. There two general ways in which a quencher can act to negate the effect of these reactions. They are, first, to interact directly with the excited state, bringing it back down to the ground state. β-Carotene is an example of this type, and quenching by it leaves the system unchanged. The second and most important type of quencher for the lens are those that neutralize or scavenger reactive intermediates, such as free radicals.

The lens aqueous and cytosol contain ascorbic acid, which is both a free radical (37) and superoxide (38) quencher. Studies utilizing whole lenses have shown that photolyses in the presence of riboflavin increased damage to pump function but that the addition of ascorbic acid negated this damage (39). Vitamin E, which is present in the lens membrane, is also a free radical quencher (40), but additionally it is also a singlet oxygen quencher (41). Finally, the lens contains considerable quantities of glutathione (8). Due to the low energy of the SH bond (65 kcal), it is an efficient free-radical scavenger. Reduced glutathione (GSH) (along with penicillamine) has been used by numerous researchers to negate photopolymerization in protein (42) and the production of free radicals in photolysed lenses.

MODEL PHOTOLYSIS OF LENS AND LENS CONSTITUENTS

Whole Lenses

To assess lens damage in the irradiation of whole lenses, investigations have followed a number of different parameters in photolytic studies: (1) opacification, (2) the function of various enzymes, such as transport, (3) protein modification (e.g., cross-linking and yellowing), (4) the production of free radicals, and (5) the production of a fluorescence presumably originating from the photolysis of lens Trp (fluorogens or fluorophors).

The direct irradiation of whole lenses (rat and human) has been shown to produce free radicals, based on electron spin resonance (ESR) studies (43). A marked decrease in the production of this signal was noted upon the addition of penicillamine, a sulfhydryl-containing quencher analogous to GSH. Similar photolyses showed the formation of fluorophors with fluoresecent spectra analogous to those found in aged and cataractous human lenses. The action spectrum for the production of these photoproducts coincided with the absorption spectrum of Trp, suggesting that they originated from that chromophore (44).

It was also demonstrated that the irradiation of whole lenses resulted in decreased incorporation of amino acids (45) and leucine after the preincubation with 3-aminotriazole (46). This latter compound inhibits catalase, leading to decreased levels of GSH. Several investigations have irradiated lenses in the presence of sensitizers. Photolysis with riboflavin, which normally exists in the aqueous, gave reduced rubidium uptake. This damage was retarded by superoxide dismutase (SOD), catalase, and ascorbic acid (47). SOD deactivates superoxide, forming O_2 and H_2O, whereas catalase reduces 2 mol of peroxide to 2 mol of water. That both these enzymes interrupted these photoprocesses suggested to the authors that this reaction was going through a Haber-Weiss process:

$$O_2^- + H_2O_2 \rightarrow OH^- + OH^{\cdot} + O_2$$

producing the highly reactive hydroxide radical. This readily abstracts protons from a wide variety of species (48) initiating polymerization. Ascorbate also blocked these reactions.

Rose bengal, a singlet oxygen generator, has been shown to reduce rubidium, [³H]choline, and [³H]α-aminobutyric acid accumulation in lens water (49). This damage was also retarded with the addition of ascorbate. Rose bengal (and kynurenine) can also cause cross-linking formation in the cytosol protein (50). After preincubation and photolysis in the presence of these sensitizers, rat lenses were separated into water-soluble and insoluble fractions. Cross-linked protein was found primarily in the water-insoluble fractions. Again, ascorbate reduced this damage.

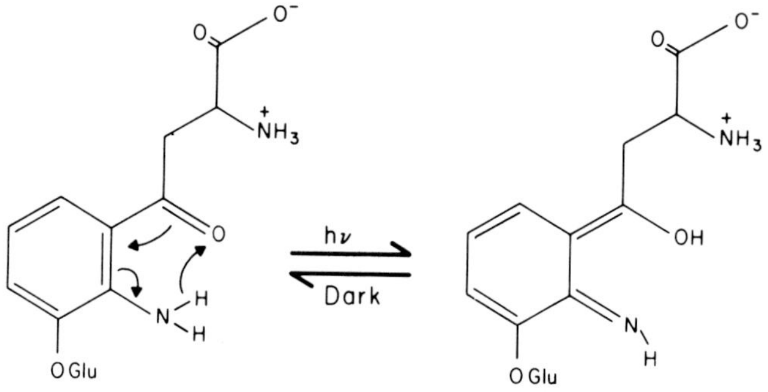

Figure 5 This depicts a possible deactivation path for irradiated 3-OH Kyn glycoside, in which the molecule is left intact.

The irradiation of rat lenses in the presence of 8-methoxypsoralen (8-MOP) leads to increased modification of the lens protein as measured by fluorescence and phosphorescence spectroscopy (51). In addition, it has been shown that this drug can form photoadducts with lens constituents, thus possibly changing their characteristics. The potential phototoxic side effects of various exogenous sensitizers, such as drugs, is an important topic and will be considered in more detail later in the chapter.

Lens Protein, Direct Photolysis

As was stated above there are two chromophores that absorb the majority of light absorbed by the lens, namely, Trp and 3-OH Kyn glucoside. Compounds such as the latter are not readily photolysed. This is due to the mechanism outlined in Fig. 5. Here, abstraction of a proton from the amine in the excited state leads, in the dark, to the original compound. Thus, this molecule has an efficient process by which radiant energy can be dissipated. It has been suggested, though, that this and similar compounds can also act as sensitizers.

Both aerobic and anaerobic photolysis of lens and other proteins have been extensively studied. In general, both photolytic conditions lead to a polymerization (Fig. 6), aggregation, the formation of protein-attached fluorescent components (NT fluorescence), and yellowing (52–55). These changes are the result of a complex series of processes due primarily to the initial absorption of light by Trp. Although this is the initial process, it leads to other reactions, such as the loss of His. In addition, Met is converted to Met sulfoxide and Cys to cysteic acid (56). Since it is not proven that Trp itself can sensitize these reactions, they are best explained by the conversion of Trp to another compound that is itself a sensitizer. Such a compound could be NFK, which is known to an efficient singlet oxygen generator (23) and would bring about these photolytic losses.

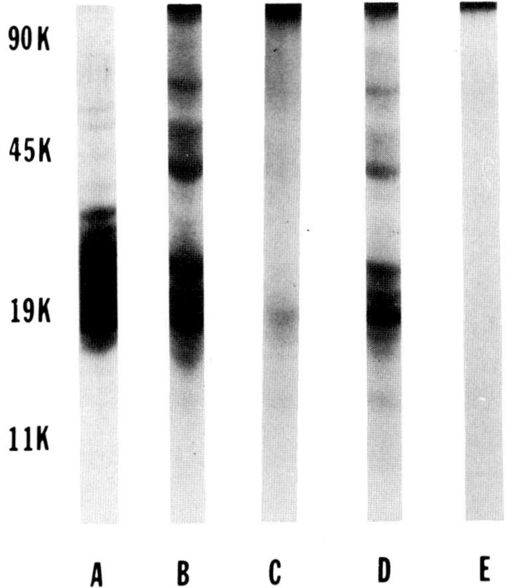

Figure 6   The direct photolysis of lens proteins can lead to a number of modifications, including photopolymerization. This gives the Coomassie blue-stained SDS gels of calf lens protein (A), and after 6 and 16 hr photolysis under vacuum (B) and (C), respectively, and in air (D), and (E), respectively.

The direct photolysis of calf lens membrane leads to similar results (57). The lens membrane consists of a 26 kd polypeptide whose function may be for communication between the fibers (58). With photolysis there is an apparent formation of aggregates that do not enter the gel, similar to that found in the photolysis of the cytosol protein. Therefore, the sites of direct photo-oxidation in human lenses can be both the cytosol and membrane proteins. In addition, other proteins, such as enzymes, could be susceptible to photolytic action.

Sensitized Oxidations

More than a decade ago van Heyningen suggested that some low-molecular-weight components in the lens can act as photosensitizers (59). She demonstrated that the photolysis of lens protein in the presence of human lens dialysates accelerated photoprocesses in terms of amino acid composition and aggregation. A major constituent of the dialysates is the glucoside of 3-OH Kyn, which is in the lens at a level of approximately 0.1 mM (60). Since it absorbs in the 300–400 nm region, its function may be to act as a filter for the retina.

More recently, 3-OH Kyn, a metabolite of the glucoside, has been shown to produce photopolymerization and NT fluorescence in isolated lens proteins (61). In addition, the photolysis of lens cytosol protein in the presence of the glucoside produced increased aggregation in the presence of calcium (62) and increased loss of His (63) when compared with protein photolyzed without the added sensitizers. Similar experiments on calf lens fiber membrane have also demonstrated increased destruction of the main intrinsic protein in the presence of the glucoside of 3-OH Kyn (64).

An additional mechanism has been suggested that involves the direct photolysis of aqueous Trp, which forms photoproducts or reactive intermediates (65). These in turn enter the lens and react with various lens constituents, such as the cytosol proteins, and various protective enzymes, such as catalase (66).

AGED AND CATARACTOUS CHANGES IN THE LENS

With aging and cataractogenesis the structural proteins in the human lens undergo dramatic changes. These are manifested on both the macromolecular and molecular level. The young human lens contains predominantly soluble protein, which can be generally related to the α-, β-, and γ-crystallin fractions delineated in the calf lens; based on elution from DEAE-cellulose, gel filtration, gel electrophoresis, and other chemical and immunological criteria (67). These crystallin polypeptides are generally in the 20,000–30,000 d range but can form aggregates of a molecular weight of up to 1×10^6 d, which are held together by noncovalent forces (68). Older lenses show the appearance of both smaller (11,000 d) (69) and larger (above 40,000 d) (70) polypeptides and a tendency to form water or even urea-insoluble protein (71–73). More recently it has been shown that a major fraction of the 43 kd polypeptide results from the dimerization of the monomeric units (74).

In addition to these macromolecular age-related changes, lens proteins exhibit changes at the molecular level. At birth the human lens is colorless to pale yellow. It progresses to a deeper yellow to brown in some cataractous lenses. There is also a production of fluorescence not originating from Trp or Tyr (75–77). Both these changes suggest an age-related generation of new components attached to cytosol protein. In older normal lenses there is an apparent formation of Met sulfoxide in the intrinsic membrane protein (78). This is the only apparent transformation of the native amino acid residues that occurs as an aging process, other than racemization of Asp (79).

During cataractogenesis a number of changes have been noted. These include (1) an increased NT fluorescence and yellowing (8,80–85), (2) an increased insolubilization of the protein (86–87), (3) the formation of disulfides (88–91), (4) the appearance of Met sulfoxide in all proteins (92), (5) the possible loss of His in some brunescent cataracts (93), and (6) the disruption of the fiber membranes (94–95). It is important to note that there is no apparent loss of Trp (within 5–10%) (96) or unsaturated lipids in the membrane (94). The formation of malonaldehyde has been reported (97) but only in small amounts (about 5% of the total unsaturated lipids).

Numerous attempts have been made to determine the nature of the fluorescent and yellow material attached to the cytosol protein. To date some components have been identified and are depicted in Fig. 7; they are Kyn (81), β-carboline (98), and oxindoylalanine (99). There have also been reports of the presence of anthranilic acid (100–101), but this seems to be an air oxidation product of an in vivo precursor, possibly Kyn. Finally, the detection of bytyrosine has been reported (100), but this is present in only small amounts (102). In addition to the above changes that occur in the lens as aging and cataractous processes, there are other transformations that are not pertinent to the present discussion.

A PHOTOCHEMICAL MODEL FOR LENS AGING AND CATARACTOGENESIS

In summary, then, if photochemical reactions play a role in the lens aging and cataractogensis, the four major factors that must be considered are (1) the chromophores absorbing the ambient radiation, (2) oxygen tension in the lens, (3) the concentration and type of quencher that may be present, and (4) the reaction site. An increase in oxygen or a decrease in the quenchers would accelerate some photochemical processes, but there is no evidence for such changes in the older lens that would conceivably predate the onset of cataractogenesis.

A comparison of the known changes that occur in the human lens with photochemical reactions in model systems lead to the conclusion that many of these in vivo changes are probably due to some photochemical process. Cross-linking can occur by all the photochemical mechanisms outlined in this chapter, as can the formation of NT fluorescence. Most of the compounds isolated from lens digests (Fig. 7) can be explained in terms of the direct photolysis of lens Trp. β-Carbolines form during the anaerobic photolysis of *N*-acetyl Trp (103), whereas Kyn and oxindoyl Ala have been isolated from the aerobic photolysis of Trp peptides (104).

The major oxidative changes that occur in cataractogenesis can, for the most part, be explained by some photosensitized oxidation, possibly due to an increase in the oxygen tension in the lens. Met and His are highly reactive toward singlet oxygen. The preferential oxidation of Met (versus His and Trp) detected in most cataracts has been demonstrated in some model systems. This is due to the Met residues being closer to the surface of the protein than His or Trp, thus have greater access to a sensitizer in the cytosol.

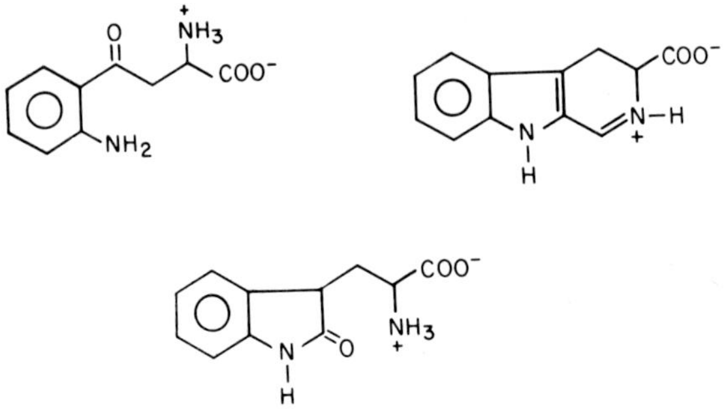

Figure 7 The structures of Kyn (upper left), a β-carboline, and oxindoyl-alanine (below).

Of more importance is the formation of Met sulfoxide in the membrane proteins without apparent oxidation of the lipids. This suggests that whatever reactive intermediate is causing this destruction, it reacts preferentially with the membrane protein. A possible mechanism for this is via some photosensitized oxidation, since similar results were obtained on red cell ghosts.

DRUGS

Since light is continually transmitted through the cornea and lens, the addition of other absorbing chromophores into the eye may put further photochemical stress on this system. Many drugs, such as 8-MOP (105), chlorpromazine (106), and hematoporphryn (107), have the following characteristics: they (1) absorb the wavelengths of light transmitted by the cornea, (2) are known to cause phototoxic side effects in the skin, and (3) get into the aqueous. Thus these drugs could accelerate the aging process in the lens, leading to the early onset of cataractogenesis. The lens is particularly susceptible to such effects, since there is little or no turnover in that tissue.

Although there is no direct proof at this time that some of these compounds cause cataracts, experiments on model systems employing lens or lens proteins show increased photolytic damage in the presence of these drugs. A method that can be used to circumvent these possible phototoxic side effects is to filter out the wavelengths of light absorbed by these drugs (108). For example, chlorpromazine has an absorption maxima at approximately 305 nm. The concomitant use of glasses that absorb these wavelengths should negate these possible side effects.

REFERENCES

1. S. Lerman. In *Interdisciplinary Topics in Gerontology*, Vol. 13 (H. P. von Hahn, ed.), S. Karger, Basel, 1978.
2. R. B. Kurzel, M. L. Wolbarsht, and B. S. Yaminashi. In *Photochemical and Photobiological Reviews*, Vol. 2 (K. Smith, ed.), Plenum Press, New York, 1977.
3. S. Zigman. In *Mechanisms of Cataract Formation in the Human Lens* (G. Duncan, ed.), Academic Press, New York, 1981.
4. S. Lerman. Ophthal. Res., *12*: 303 (1980).
5. A. Bachem. Am. J. Ophthalmol., *41*: 969 (1956).
6. S. Lerman and R. Borkman. Ophthal. Res., *8*: 335 (1976).
7. R. van Heyningen. In Ciba Symposium, *19*, Elsevier, Amsterdam, 1973.
8. M. Kwan, J. Niinikoski, and T. K. Hunt. Invest. Ophthalmol., *11*: 108 (1971).
9. D. V. N. Reddy. Exp. Eye Res., *11*: 310 (1971).
10. H. Heath. Exp. Eye Res., *1*: 362 (1962).
11. J. Dillon, B. Mehlman, L. Ponticorva, and A. Spector, Exp. Eye Res., *37*: 91 (1983).
12. N. Turro. In *Molecular Photochemistry*, W. A. Benjamin, New York, 1965.
13. R. B. Cundall and A. Gilbert. In *Photochemistry*, Appleton-Century-Crofts, New York, 1970.
14. T. Cassen and D. R. Kearns. Biochim. Biophys. Acta, *194*: 203 (1969).
15. J. F. Baugher and L. I. Grossweiner. J. Phys. Chem., *81*: 1349 (1977).
16. M. T. Pailthorpe and C. H. Nicholls. Photochem. Photobiol., *14*: 135 (1971).
17. A. Amouryal, A. Bernas, and D. Grand. Photochem. Photobiol., *29*: 107 (1979).
18. J. Dillon. Photochem. Photobiol., *33*: 137 (1981).
19. J. D. Tassin and R. F. Borkman. Photochem. Photobiol., *32*: 577 (1980).
20. C. Pernot and L. Lindguist. J. Photochem., *6*: 215 (1976).
21. J. Dillon. Photochem. Photobiol., *32*: 37 (1980).
22. L. I. Grossweiner. In *Current Topics in Radiation Research Quarterly*, Vol. 2 (M. Ebert and A. Howard, eds.), North Holland, Amsterdam, 1976.
23. A. D. McLaren and D. Shugar. In *Photochemistry of Proteins and Nucleic Acid*, Vol. 22 (P. Alesander and Z. M. Bacg, eds.), MacMillan, New York, 1964.
24. D. V. Bent and E. Hayon. J. Am. Chem. Soc., *97*: 2612 (1975).
25. A. Pirie. Biochem. J., *125*: 203 (1971).
26. P. Walrant and R. Santus. Photochem. Photobiol., *19*: 411 (1974).
27. C. S. Foote. In *Free Radicals in Biology*, Vol. 2 (W. A. Pryor, ed.), Academic Press, New York, 1976.
28. J. D. Spikes. In *The Science of Photobiology* (K. C. Smith, ed.), Plenum Press, New York, 1977.
29. H. H. Wasserman. Ann. N.Y. Acad. Sci., *171*: 108 (1970).
30. A. A. Lamola, T. Yamone, and A. M. Trozzolo. Science, *179*: 1131 (1973).

31. H. R. Rawls and P. J. Van Santen. J. Am. Oil. Chem. Soc., *47*: 121 (1970).
32. L. I. Grossweiner and J. B. Grossweiner. Photochem. Photobiol., *35*: 583 (1982).
33. M. D. Barratt, J. C. Evans, C. A. Lewis, and C. C. Rowlands, Chem. Biol. Interact., *38*: 215 (1982).
34. T. M. A. R. Dubbelman, A. F. P. M. De Goeij, and J. Van Steveninck. Biochim. Biophys. Acta, *511*: 141 (1978).
35. T. M. A. R. Dubbelman, A. F. P. M. De Goeij, and J. Van Steveninck. Biochim. Biophys. Acta, *595*: 133 (1980).
36. M. R. Deziel and A. W. Girotti. J. Biol. Chem., *255*: 8192 (1980).
37. M. Ballester, J. Riera, J. Castener, and M. Casulleras. Tetrahedron Lett., *643* (1978).
38. M. Nishikimi and K. Yagi. In *Biochemical and Medical Aspects of Active Oxygen* (O. Hayashi and K. Asada, eds.), University Park Press, Baltimore, 1976.
39. S. D. Varma, S. Kumar, and R. D. Richards. Proc. Natl. Acad. Sci., *76*: 3504.
40. L. A. Witting. In *Free Radicals in Biology*, Vol. 4 (W. A. Pryor, ed.), Academic Press, New York, 1980.
41. C. S. Foote, T. Y. Ching, and G. G. Geller. Photochem. Photobiol., *20*: 511 (1974).
42. N. S. Kosower and E. M. Kosower. In *Free Radicals in Biology*, Vol. 2 (W. A. Pryor, ed.), Academic Press, New York, 1976.
43. R. F. Borkman and S. Lerman. Exp. Eye Res., *25*: 303 (1977).
44. R. F. Borkman, A. Darlymple, and S. Lerman. Photochem. Photobiol., *26*: 129 (1977).
45. S. Zigman, J. Schultz, and T. Yulo. Exp. Eye Res., *15*: 201 (1973).
46. J. F. R. Kuck. Invest. Ophthalmol., *15*: 405 (1976).
47. S. D. Varma, V. K. Srivastava, and R. D. Richards. Ophthalmol. Res., *14*: 167 (1982).
48. D. Fredovich. In *Free Radicals in Biology*, Vol. 1 (W. A. Pryor, ed.), Academic Press, New York, 1976.
49. H. M. Jernigan, H. N. Fukui, J. D. Goosey, and J. H. Kinoshita. Exp. Eye Res., *32*: 461 (1981).
50. J. S. Zigler, H. M. Jernigan, N. S. Perlmutter, and J. H. Kinoshita. Exp. Eye Res., *35*: 239 (1982).
51. S. Lerman, M. Jocoy, and R. Borkman. Invest. Ophthalmol., *16*: 1065 (1977).
52. R. H. Buckingham and A. Pirie. Exp. Eye Res., *14*: 297 (1972).
53. S. Zigman, J. Groff, and T. Yulo. Photochem. Photobiol., *26*: 505 (1977).
54. J. Dillon and A. Spector. Exp. Eye Res., *31*: 591 (1980).
55. E. Fujimori. Exp. Eye Res., *34*: 381 (1982).
56. J. Dillon, M. Garner, D. Roy, and A. Spector. Exp. Eye Res., *34*: 651 (1982).
57. J. Dillon, D. Roy, and A. Spector. Exp. Eye Res. (In Press).
58. T. Kuwabara. Exp. Eye Res., *20*: 426 (1975).

59. R. van Heyningen. Exp. Eye Res., *17*: 137 (1973).
60. M. Bando, A. Nakajima, and K. Satoh. J. Biochem., *89*: 103 (1981).
61. J. S. Zigler and J. D. Goosey. Photochem. Photobiol., *33*: 869 (1981).
62. M. Bando, I. Mikumi, and H. Obazawa. Exp. Eye Res., *34*: 953 (1983).
63. J. Dillon. Lens Res. *1(1 and 2)*: 133 (1983)
64. J. Dillon. Curr. Eye Res. *3*: 145 (1984).
65. D. Grover and S. Zigman. Exp. Eye Res., *13*: 70 (1972).
66. S. Zigman, T. Yulo, and G. Griess. Mol. Cell. Biochem., *11*: 144 (1976).
67. K. J. Dilley and J. J. Harding. Biochim. Biophys. Acta, *386*: 391 (1975).
68. A. Spector, J. Stauffer, and J. Sigelman. In Ciba Foundation Symposium, *19*: 185 (1973).
69. D. Roy and A. Spector. Exp. Eye Res., *26*: 429 (1978).
70. A. Spector, D. Roy, and J. Stauffer. Exp. Eye Res., *21*: 9 (1975).
71. S. D. Coghlan and R. C. Augustegn. Exp. Eye Res., *25*: 603 (1975).
72. S. Zigman, J. Groff, T. Yulo, and G. Griess. Exp. Eye Res., *23*: 555 (1976).
73. K. Satoh. Exp. Eye Res., *14*: 53 (1972).
74. J. Jedziniak, J. H. Kinoshita, E. M. Yates, L. O. Hocker, and G. B. Benedek. Exp. Eye Res., *15*: 185 (1973).
75. R. Jacobs and D. L. Krohn. J. Gerontol., *31*: 641 (1976).
76. M. Bando, Y. Ishii, and A. Nakajima. Ophthalmol. Res., *8*: 456 (1976).
77. S. Lerman, O. Hockwin, and U. Dragomirescu. Ophthalmol. Res., *13*: 224 (1981).
78. M. Garner and A. Spector. Proc. Natl. Acad. Sci. USA, *77*: 1274 (1980).
79. R. C. Augusteyn. Ophthalmol. Res., *7*: 217 (1975).
80. W. H. Garner and A. Spector. Proc. Natl. Acad. Sci. USA, *75*: 3618 (1978).
81. K. J. Dilley and A. Pirie. Exp. Eye Res., *19*: 59 (1978).
82. R. B. Kurzel, M. L. Wolbarsht, and B. S. Yamanashi. Exp. Eye Res., *17*: 65 (1973).
83. S. Lerman, B. S. Yaminashi, R. A. Palmer, J. C. Roark, and R. Borkman. Ophthalmol. Res., *10*: 168 (1978).
84. K. Satoh, M. Bando, and A. Nakajima. Exp. Eye Res., *16*: 167 (1973).
85. R. Clark, S. Zigman, and S. Lerman. Exp. Eye Res., *8*: 172 (1969).
86. H. A. Kramps, H. J. Hoenders, and J. Wollensak. Biochim. Biophys. Acta, *434*: 32 (1976).
87. R. H. Buckingham. Exp. Eye Res., *14*: 123 (1972).
88. Z. Dische and H. Zil. Am. J. Ophthalmol., *34*: 104 (1951).
89. E. Anderson and A. Spector. Exp. Eye Res., *26*: 407 (1978).
90. M. H. Garner and A. Spector. Exp. Eye Res., *31*: 361 (1980).
91. E. I. Anderson, D. Wright, and A. Spector. Exp. Eye Res., *29*: 233 (1979).
92. R. J. W. Truscott and R. C. Augusteyn. Biochim. Biophys. Acta, *492*: 43 (1977).
93. J. S. Zigler, J. B. Sidbury, B. S. Yaminashi, and M. Wolbarshi. Ophthalmol. Res., *8*: 379 (1976).
94. T. Matsuto. Acta Soc. Ophthalmol. Japan, *77*: 853 (1973).

95. D. Roy, L. Rosenfeld, and A. Spector. Exp. Eye Res., *35*: 113 (1982).

96. R. Augusteyn. In *Mechanisms of Cataract in the Human Lens* (G. Duncan, ed.), Academic Press, New York, 1981.

97. L. Rosenfeld and A. Spector. Exp. Eye Res., *35*: 69 (1982).

98. J. Dillon, A; Spector, and K. Nakanishi. Nature, *254*: 422 (1976).

99. J. Dillon, S. Garcia-Castineiras, and A. Spector. Exp. Eye Res. *39*: 95 (1984).

100. S. Garcia-Castineiras, J. Dillon, and A. Spector. Exp. Eye Res., *26*: 461 (1978).

101. R. J. W. Truscott, K. Faull, and R. Augusteyn. Ophthalmol. Res., *9*: 263 (1977).

102. M. K. McNamara and R. C. Augusteyn. Exp. Eye Res., *30*: 319 (1980).

103. J. Dillon. Photochem. Photobiol., *33*: 137 (1981).

104. L. A. Holt, B. Milligan, D. Rivett, and F. Stewart. Biochim. Biophys. Acta, *499*: 131 (1977).

105. S. Lerman, J. Megaw, and I. Willis. Photochem. Photobiol., *31*: 235 (1980).

106. J. E. Roberts. Fifth International Congress Eye Research, 1982, p. 61.

107. J. E. Roberts. Am. Soc. Photobiol. (Abstracts), 127 (1982).

108. S. Lerman, K. Gardner, J. Megaw, and R. Borkman. Ophthalmol. Res., *13*: 284 (1981).

10

The Structure of the Human Cataractous Lens

CLIFFORD V. HARDING, JR., STANLEY R. SUSAN, WOO-KUEN LO, and WALTER F. BOBROWSKI / Kresge Eye Institute of Wayne State University, Detroit, Michigan

HARRY MAISEL / Wayne State University, Detroit, Michigan

LEO T. CHYLACK, JR.*/ Harvard Medical School, Massachusetts Eye and Ear Infirmary, Boston, Massachusetts

This chapter concentrates primarily on several classes of opacities that illustrate the basic cellular changes that characterize human cataractogenesis. These cellular changes include alterations in organization (meridional rows), migration (toward posterior pole), swelling, and fragmentation, with the resulting formation of Morgagnian globules and other cellular debris, including plasma membrane fragments. Other changes, such as the formation of crystals and multiple membranes, are also included. In referring to the different classes of human cataracts, the classification system of Chylack (1-3) is used.

Animal models of cataractogenesis provide information basic to an understanding of how cataracts develop in the human (4). In the rat, for example, dietary cataracts, such as those induced by the addition of galactose to the diet, can be reproducibly obtained. This, and a number of other experimentally induced or genetically determined animal cataracts, have been extensively studied in terms of the biochemical and morphologic changes that accompany the development of lenticular opacities. Basic concepts of cataractogenesis, such as the osmotic concept(s) developed by Kinoshita and his colleagues, are based on the study of such animal model systems (5-7). Morphologic studies on reversible dietary cataracts in rats (8-12) have added significantly to our understanding of cataractogenesis.

Because of the difficulties involved, the study of the morphologic and biochemical bases of cataractogenesis in the human lens has lagged behind. A major technical obstacle has been the lack of a generally accepted system of correlating

---

*Brigham and Women's Hospital, Boston, Massachusetts

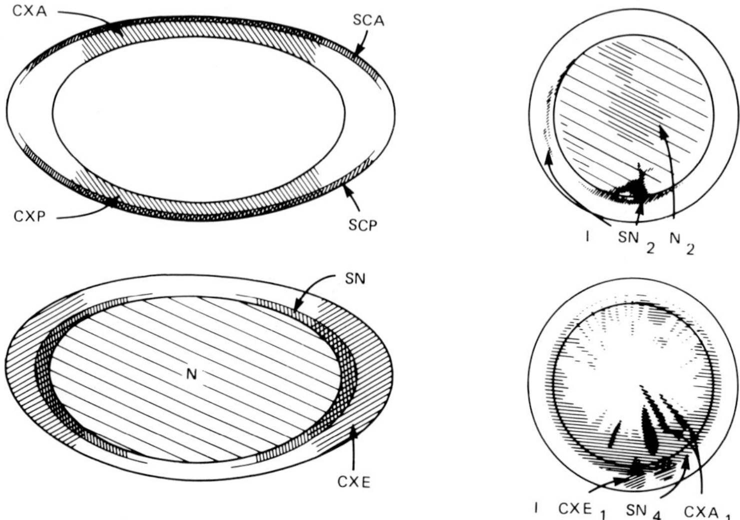

Figure 1 The two diagrams at left show schematically the basic classes of human lens cataracts, modified from Chylack (1). CXA, anterior cortical; SCA, anterior subcapsular; CXP, posterior cortical; SCP, posterior subcapsular; SN, supranuclear; and CXE, equatorial cortical. A more complete explanation of the nomenclature is given in the last section. Two specific examples of the application of this nomenclature are shown in the two diagrams at the right. The top right diagram corresponds to the fresh lens shown in Fig. 9, and the lower right diagram corresponds to the fresh lens in Fig. 17.

the gross morphology of a specific opacity in the fresh human lens with its ultrastructural characteristics.

The in vitro photographic classification procedures described by Chylack and Chylack et al., which have been adopted by the Cooperative Cataract Research Group (CCRG) (3,13), were developed, in part, to enable more accurate correlations between specific kinds of human lens opacities and their structural and chemical characteristics. In our experience, this system has worked very well for the electron microscopic characterization of opacities that have been identified in the fresh lens, and, therefore, we believe, has finally opened the way to a rational study of the structural basis of human cataractogenesis. Figure 1 summarizes in diagrammatic form the basic features of the various classes of human lens opacities.

The nomenclature and symbols used in the Chylack cataract classification system are described in more detail in the last section.

The CCRG photographic procedures referred to in this study were described in detail by Chylack and Chylack et al. The application of these procedures in the ultrastructural analysis of specific opacities has also been reported (14-16). With regard to studies referred to in this chapter from our own laboratories, the following methods are used.

Human lenses are obtained at the time of routine cataract surgery. Immediately following extraction and brief inspection by a pathologist, the lens is brought to the laboratory in normal saline for photography. The standard CCRG stereo camera system is used to record anterior, posterior, and side views of the lens, with and without slit-lamp illumination. One photograph is also made to determine the degree of nuclear sclerosis (yellowing). The lens is then fixed at room temperature in 2.5% glutaraldehyde and buffered with 0.05-0.1 M sodium cacodylate (pH 7.2) for at least 24 hr on a rotator. Dehydration is carried out in 30, 50, 70, and 95% ethanol during a period of 2 hr and then overnight in at least three changes of 100% alcohol. Infiltration of Freon 113 is initiated with a 1:1 Freon/alcohol soak for 1 hr, followed by a minimum of 4 hr in pure Freon 113 with at least three changes of fluid. The lens is dried in a Bomar 800/EX critical point drier after a 2.5 hr infiltration of Freon 13 (at least three changes).

Since maintaining orientation between the location of opacities in the freshly isolated lens and the prepared lens is critical to these studies, a sequence of steps has been established and followed in processing each lens (1,3). Before the lens is photographed it is marked with a spot of ink or dye adjacent to the attachment site of the extraction instrument (cryoprobe or erysiphake). This mark serves as an indicator of the location of possible damage caused by the extraction procedure. More importantly, it becomes a permanent point of reference throughout the entire process. In this way we are able to follow any area of interest in the lens after it has been rendered nontransparent by fixation, dehydration, and drying. In some cases, a small platinum wire is used to replace the dye mark after fixing in glutaraldehyde, and before postfixing in osmium, which obscures the dye mark.

After carefully studying the CCRG stereo slides of a lens, and by reference to the orientation spot, the best plane of fracture to expose internal areas of interest is determined. The lens is fractured (usually in a pole-to-pole cross section) and mounted on scanning electron microscope (SEM) stubs, coated with gold, mapped at low magnification, and studied in a Philips PSEM 500 scanning electron microscope. In some cases the specimens are coated only with carbon, in preparation for subsequent elemental analysis by means of energy dispersive x-ray analysis (EDXA).

In this chapter a major emphasis is placed on several CCRG-identified cataracts: the supranuclear cataract (SN), sutural opacities, and the posterior and anterior subcapsular cataracts. We have chosen this approach because some, but not all, of the cataractogenic changes at the cellular level that occur in these opacities are also found in other human lens opacities. Also, in this way we can analyze these particular classes of opacities in greater depth, from various points of view. Although a variety of human lens opacities and abnormalities are discussed in this chapter, no attempt has been made to review the extensive literature in the field of human cataractogenesis. See, for example, Refs. 17–20. For comparison, reviews of the normal lens morphology should be consulted (see Chap. 1 and Refs. 21–25).

MORPHOLOGY OF CATARACTS

Sutural Cataracts

An example of a sutural cataract as seen in vivo is shown in Fig. 2 (26). Figure 3 shows another sutural cataract photographed in vitro. This CCRG stereo photograph of a lens obtained from a 72-year-old female shows a sutural opacity on

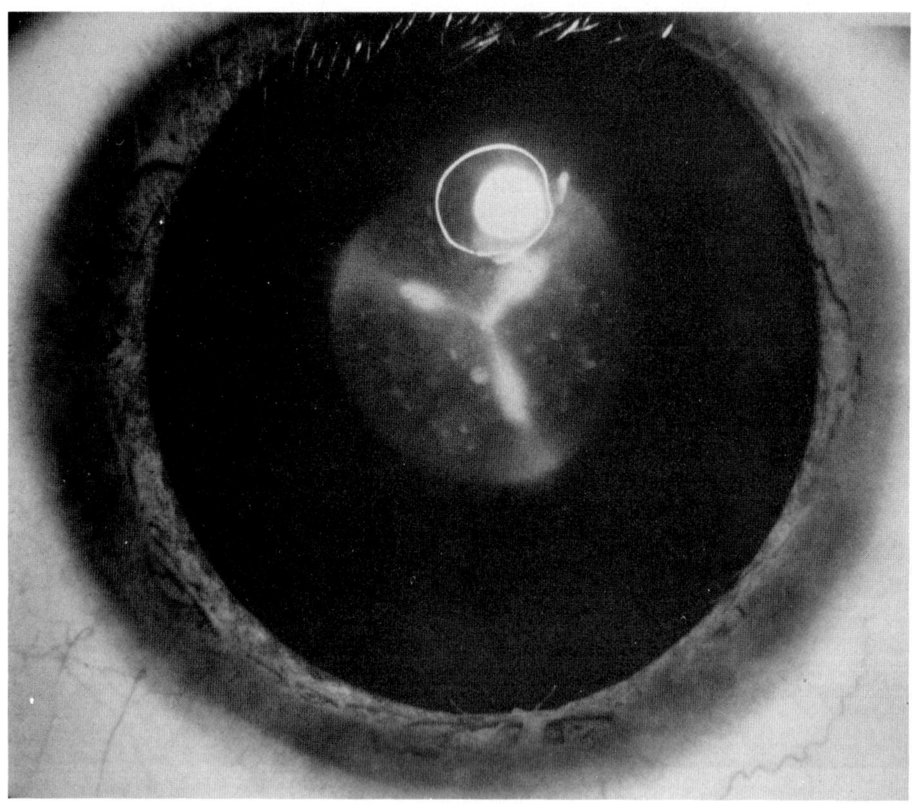

Figure 2 A sutural cataract as seen in vivo. (From Ref. 26; courtesy C. V. Mosby, St. Louis.)

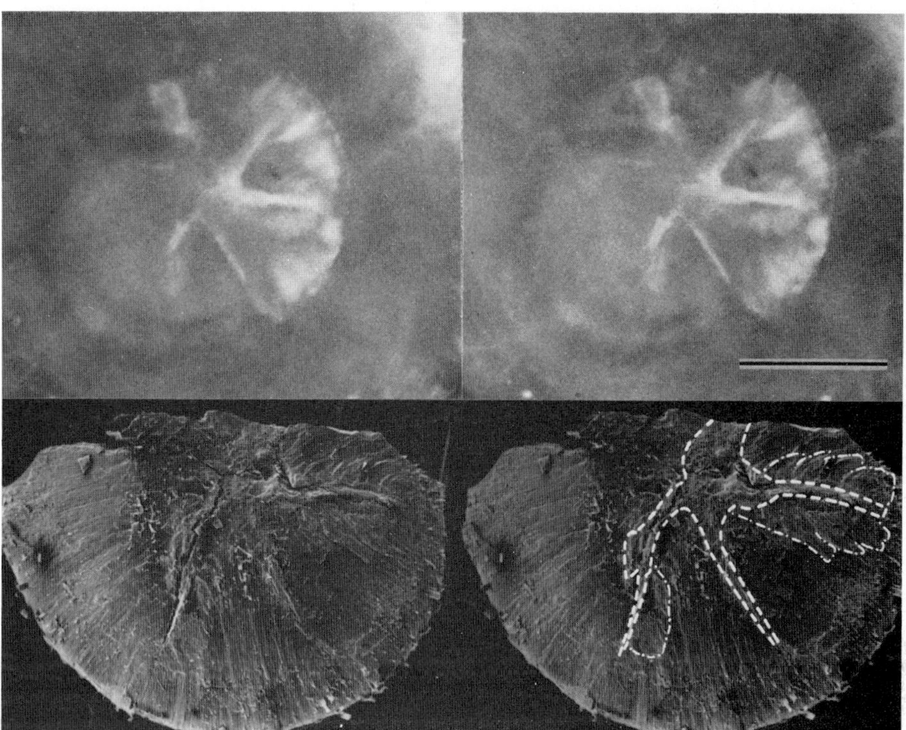

Figure 3 A sutural cataract in the freshly removed lens from a 72-year-old female. The picture is primarily of the nucleus; bar = 2 mm.

Figure 4 An SEM photograph of the isolated lens nucleus showing an area of abnormal structures corresponding to the opacity seen in Fig. 3 and outlined in the right-hand photograph of a stereoscopic pair.

the anterior surface of the fetal nucleus. The nucleus separated from the rest of the lens during the fracturing procedure for electron microscopy, and at the SEM level an identically shaped area of abnormal structures was found on its anterior surface. This area is outlined in the right photograph of an SEM stereoscopic pair (Fig. 4). Stereoscopic examination of Figs. 3 and 4 show that the opacity is associated with the suture system. Even a small fan-shaped area of abnormal fibers on the right arm of the "opacity" corresponds in shape to an opacity visible in the photograph of the fresh lens. At higher magnification (Fig. 5), it can be confirmed that this abnormal star-shaped area is associated with the suture system (compare with normal suture in Fig. 6). At even higher

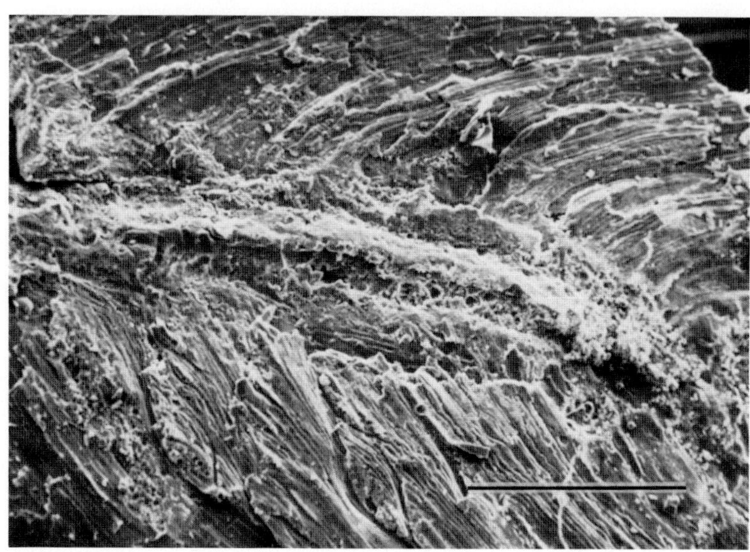

Figure 5 Higher magnification of abnormal sutural system shown in Fig. 4; bar = 250 μm.

Figure 6 Normal structure of the suture system for comparison with Fig. 5; bar = 200 μm.

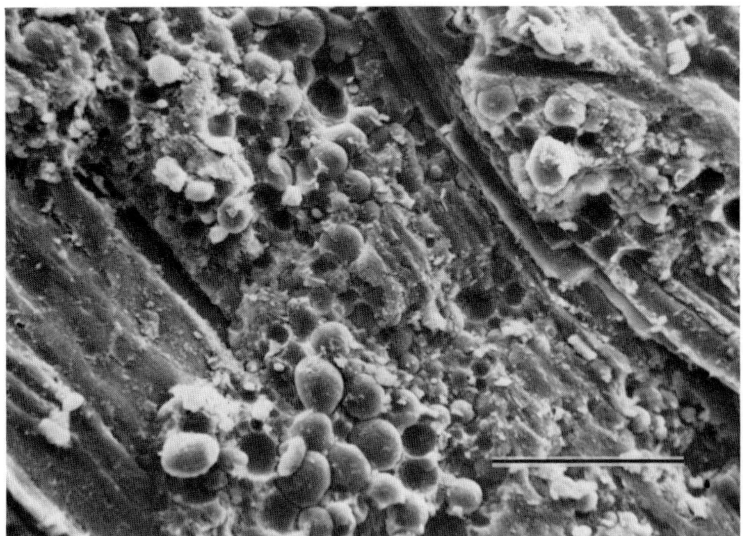

Figure 7 Extensive cellular breakdown at the sutures as seen at higher magnification; bar = 30 μm. (Cf. Fig. 5)

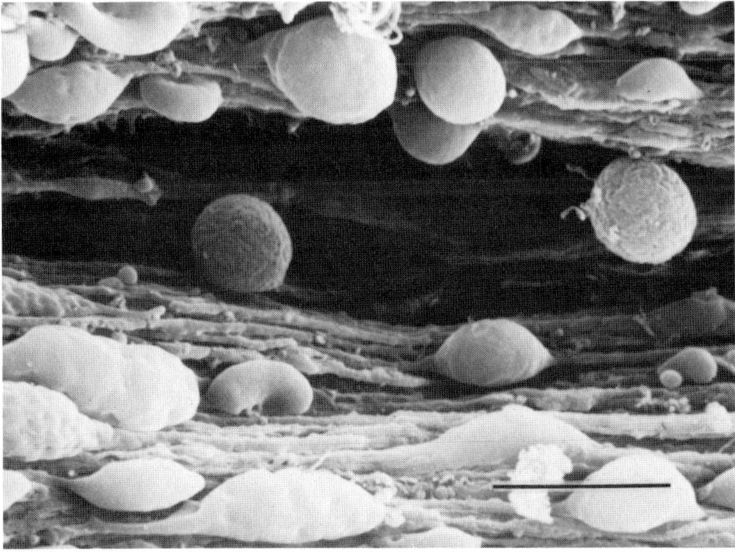

Figure 8 Typical SEM photograph of structures (globules) from fiber cells in human cortical cataracts; bar = 5 μm.

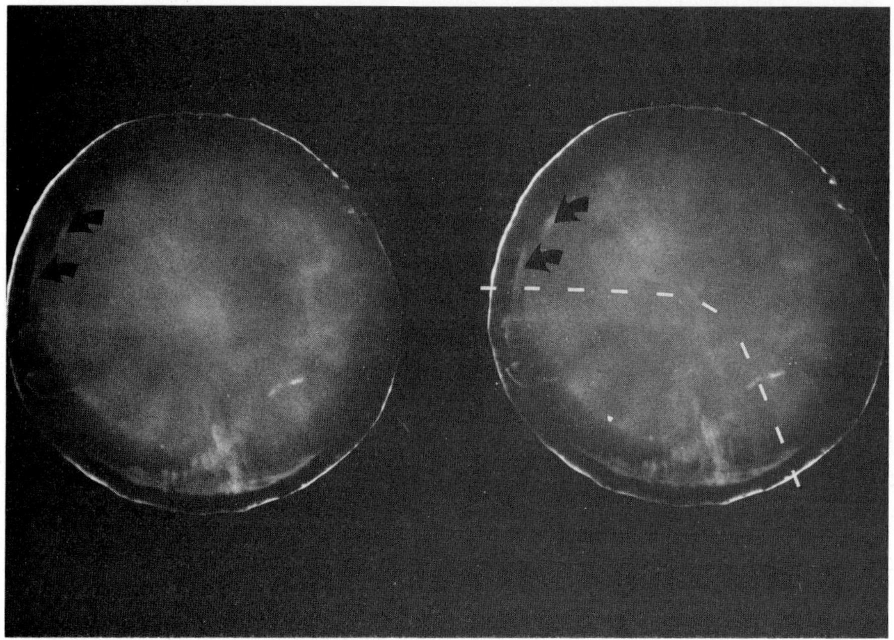

Figure 9 CCRG stereo photograph of a fresh human lens. A portion of the SN opacity is indicated by arrows. This SN opacity as seen in the fresh lens should be compared with the upper right-hand diagram seen in Fig. 1. The line of fracture is indicated by the dotted line.

magnification, extensive cellular breakdown at the sutures can be seen in greater detail (Fig. 7).

This cellular breakdown results in extensive globule formation along the suture lines, where the fiber cell endings are undergoing cytolysis. Spherical cellular breakdown products are not unique to the lens. They occur as the result of cytolysis in a variety of cell types. In the case of the human lens, however, the occurrence of these cellular breakdown products, or globules, has been associated with opacification (see Fig. 8) (27,28).

Supranuclear Cataracts

Although fiber cell breakdown may be initiated at the cellular endings that align the suture system (and cause sutural cataracts), cellular breakdown may also be initiated in the midregion of a bundle of fiber cells. Our studies indicate that this

Figure 10   SEM preparation from lens shown in Fig. 9. The area of localized cellular breakdown indicated by the arrow corresponds to the supranuclear opacity designated by arrows in Fig. 9; bar = 300 μm.

is the basis of the human supranuclear cataract. The supranuclear opacity is characterized in part by its location just exterior to the equator of the lens nucleus. It may extend part way around the nucleus, appearing as a crescent, or it may completely encircle the nucleus. This is designated by SN_1 through SN_4, depending on the number of quadrants involved. Figure 9 shows a CCRG stereo photograph of a fresh human lens with the SN opacity indicated by arrows.

When the lens shown in Fig. 9 is prepared for SEM and fractured after critical point drying, the preparation appears as in Fig. 10. A careful comparison of the stereo photographs of the fresh lens, with the SEM photographs of the same lens, indicates the location of the SN opacity in the SEM preparation (arrow in Fig. 10). In this case, localized cellular breakdown occurs just external to the

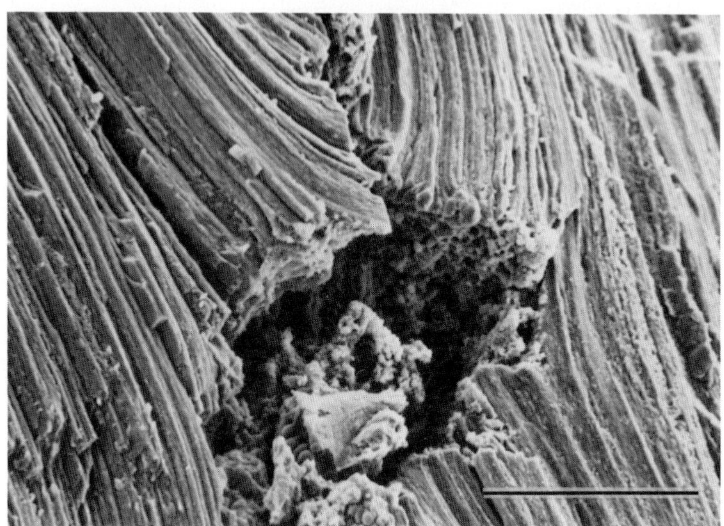

Figure 11   Higher magnification of localized area of cellular breakdown shown in Fig. 10; bar = 100 μm.

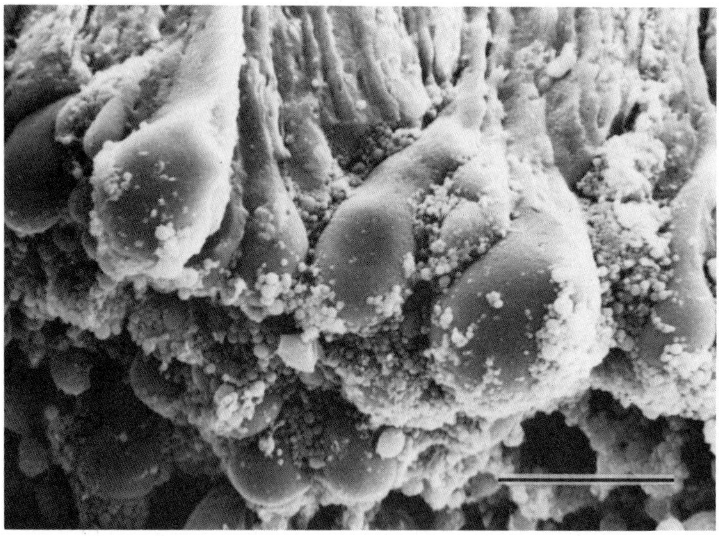

Figure 12   Cellular endings in area of SN opacity; bar = 10 μm.

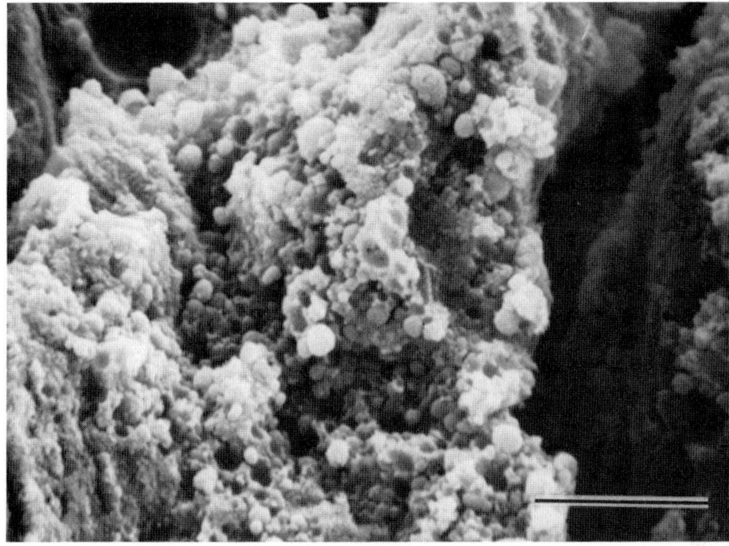

Figure 13 Debris adhering to broken cellular endings at site of SN opacity; bar = 5 μm.

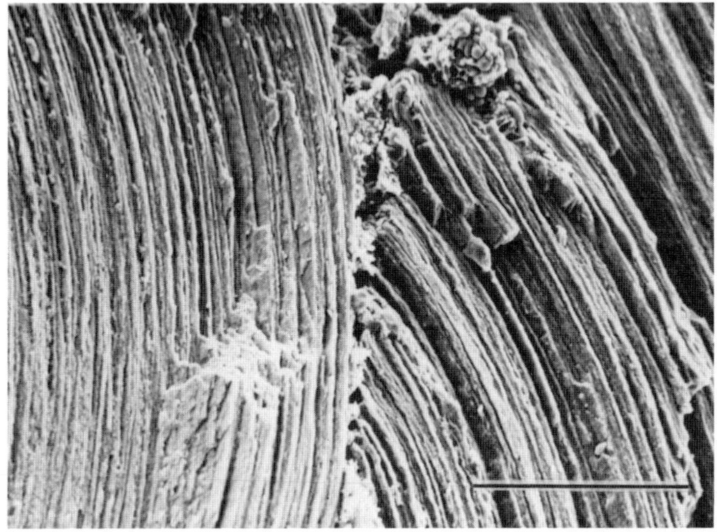

Figure 14 Altered orientation of broken fiber cells from normal concentric arrangement. Bundles of the broken fiber endings abut on the lens nuclear surface; bar = 100 μm.

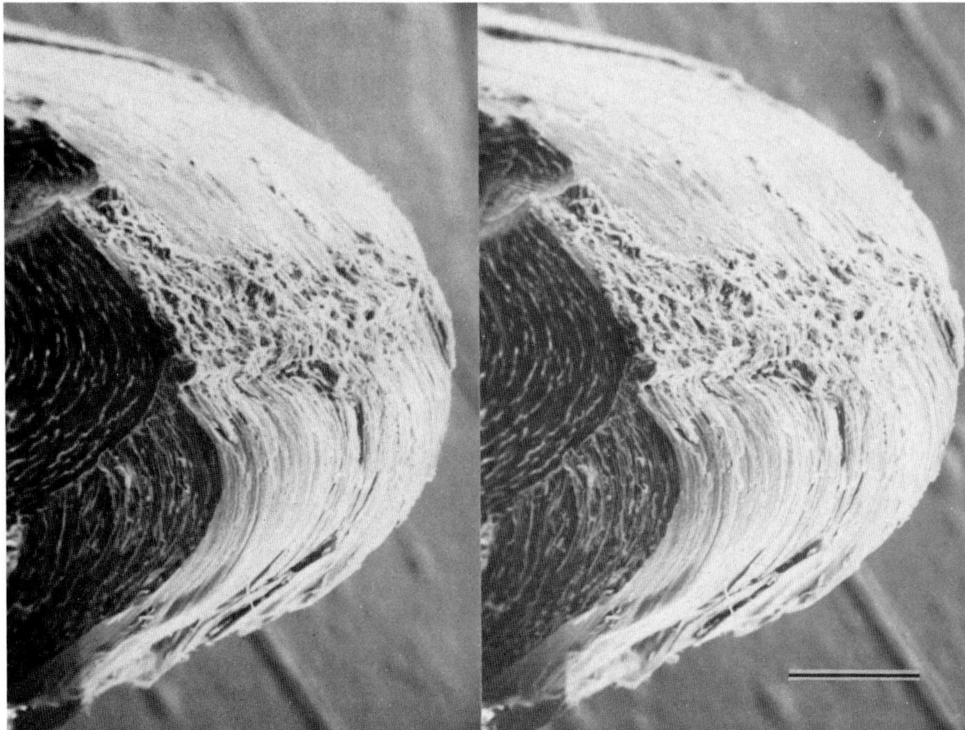

Figure 15 Fiber cell rearrangement at the SN cataract site as seen with stereo SEM. This SEM preparation was made from the fresh lens shown in Fig. 16; bar = 0.25 mm.

equator of the lens nucleus, resulting in a severing of the affected cells into anterior and posterior segments. The broken ends of the fiber cells appear to have sealed off, and in some cases, the endings appear swollen (Figs. 11 and 12) with adhering debris (Figs. 12 and 13). However, the remainder of the fiber cells, both anterior and posterior extensions, can appear perfectly normal. Most of the cellular breakdown products that accumulate between the two groups of newly severed fiber cell endings (Figs. 10 and 11) are in the form of small globules (Fig. 13). Many of the fiber cell endings themselves undergo changes in orientation from the cells' normal concentric arrangement relative to the lens nuclear surface. These changes (Fig. 14) occur at locations that correspond to the SN opacity. This cellular rearrangement at the SEM level is shown stereoscopically in Fig. 15, and the CCRG stereo photograph of the whole lens in Fig. 16.

Among the 14 SN lenses examined, cellular changes at the location corresponding to the SN opacity vary from relatively small sites of cellular degeneration

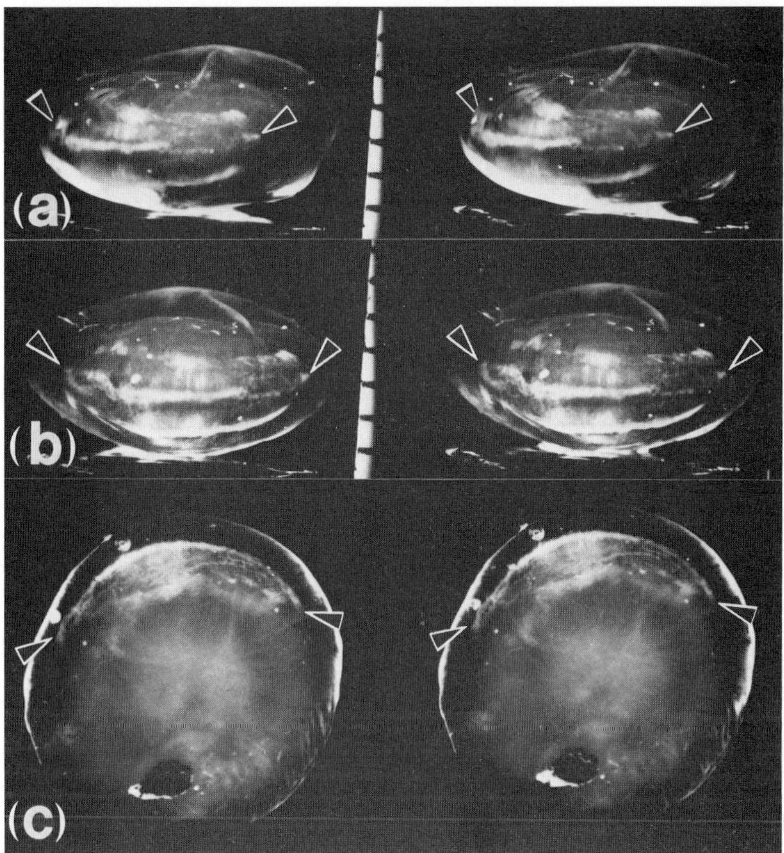

Figure 16 SN opacity as seen stereoscopically in the fresh lens laterally from two direction (a and b) and from the anterior view (c). Arrows indicate the locations of the SN cataract.

to larger, but well-defined areas of cellular breakdown. In the lens (SN_4) shown in Fig. 17, an SEM photograph shows the presence of smooth, glassy fractured surfaces (Fig. 18), which may indicate proteinaceous material resulting from complete cellular dissolution at the SN location. In none of the cases examined, however, is the whole lens cell, or even more than a small fraction of the lens cell, affected. Transmission electron microscopy (TEM) of thin sections of the SN opacity shows that the fiber cell endings have smooth surfaces (Fig. 19) (29), which tend to be in alignment, perhaps accounting for the well-defined outline of many SN opacities. This outline may surround (or envelope) cellular

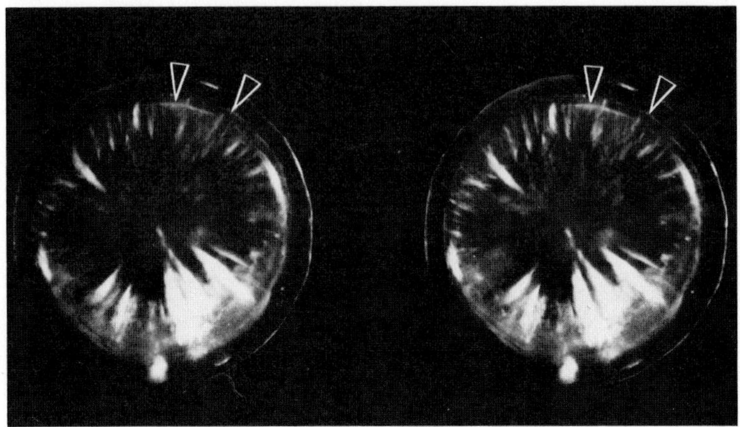

Figure 17 CCRG stereo photograph of advanced SN cataract, indicated by arrows. Spokes are also present. Compare with Fig. 18.

breakdown products, including membrane fragments with gap junctions (Fig. 20) (29).

These results suggest that the SN opacity is a direct result of the breakage of lens fiber cells, which occurs at a location just cortical (external) to the equator of the lens nucleus. The resultant localized cellular degeneration in this region constitutes the SN opacity. The remaining anterior and posterior portions of each affected fiber cell may remain normal in structure and transparency. The basic cellular changes in the supranuclear cataract are shown in a diagram (Fig. 21).

As indicated above, typical sutural and supranuclear opacities cut across bundles of lens fiber cells. Spokelike opacities occur along the length of fiber cells (Fig. 17). At the SEM level corresponding structural abnormalities are clearly evident (Figs. 22 and 23). Extensive cellular breakdown products are seen at higher magnifications (Fig. 24).

Extensive dissolution of the cortical fiber cells in the hypermature cataract (Morgagnian cataract) results in the accumulation of cellular breakdown products, which have been well-described (30). The lens nucleus, however, maintains its physical integrity throughout this process (see Fig. 25).

Anterior Subcapsular Cataracts

A structurally unique opacity falls within the anterior subcapsular cataract class (SCA). This cataract, also designated as anterior polar cataract, has been well

Figure 18 SEM photograph of SN cataract shown in Fig. 17. Complete cellular breakdown is apparent at the site of the SN opacity (arrow); bar = 200 μm.

documented at the light microscope level (19,31), and more recently at the TEM level (32,33). Epithelial cell proliferation at the anterior pole is an early stage in the development of the SCA cataract. The epithelial cells then assume spindle shapes and produce intercellular material, including collagen. The resultant plaque has been described as having the appearance of fibrous connective tissue, and theoretically may involve metaplasia (the transformation of lens epithelial cells to fibroblasts; Fig. 26). Font and Brownstein (34) have raised the question as to whether true metaplasia occurs, based on the morphologic characteristics of the spindle-shaped cells within the plaque: "The formation of basement membrane and of intercellular attachments (desmosomes) and the presence of closely aggregated cells with interdigitations of the plasma membrane are three characteristics of the normal lens epithelium that are not usually shared by fibroblasts." They conclude, therefore, that the morphology of the spindle-shaped

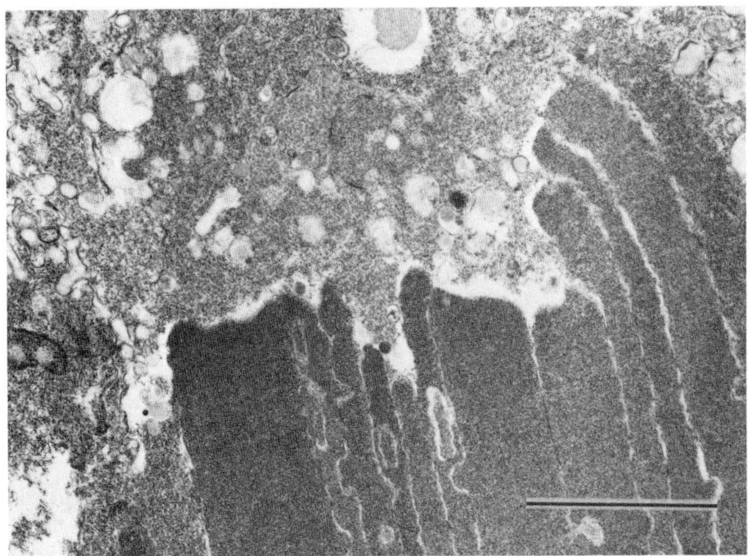

Figure 19   TEM photograph of SN opacity; bar = 2 μm.

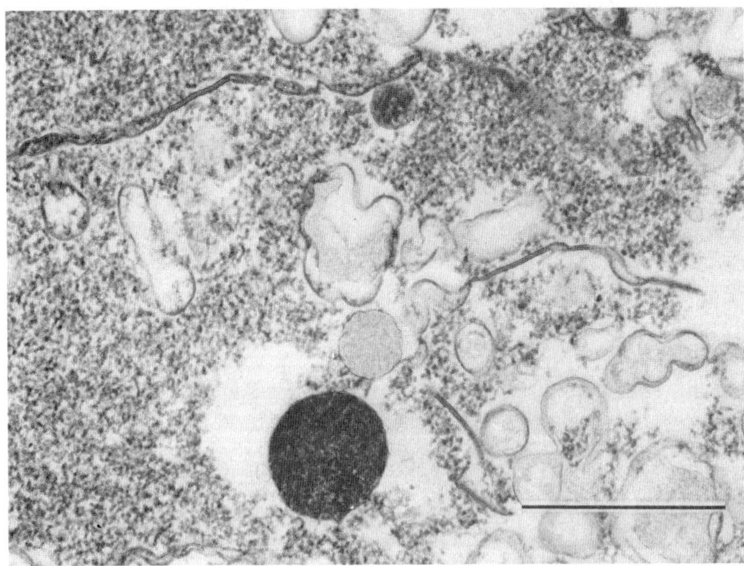

Figure 20   Higher magnification TEM photograph of SN opacity showing membrane fragments, including isolated gap junctions; bar = 0.5 μm.

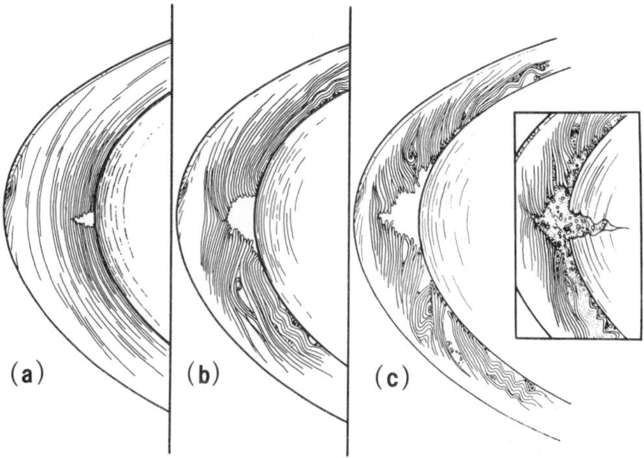

Figure 21 Hypothetical sequence in formation of supranuclear cataract. (a) Initial fiber cell breakage at region just external to lens nuclear equator. (b) and (c) More extensive focal cellular degeneration combined with realign ment of anterior and posterior cell fragments. Inset: Occasionally, a "break" is seen at the equator of the nucleus. Cortical fiber cell breakdown products, which are included in the inset, were not included in the diagrams.

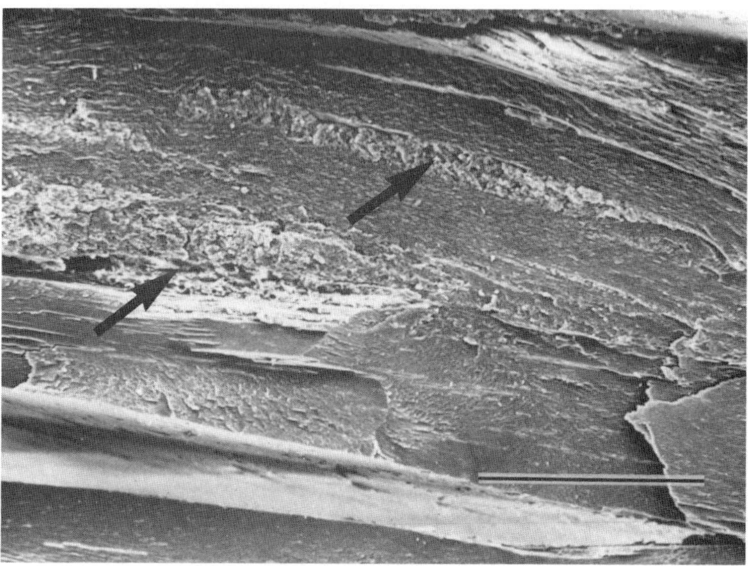

Figure 22 Low magnification of SEM preparation showing abnormal areas corresponding to spoke-shaped cataract in Fig. 17. Arrows indicate location of abnormal areas; bar = 200 μm.

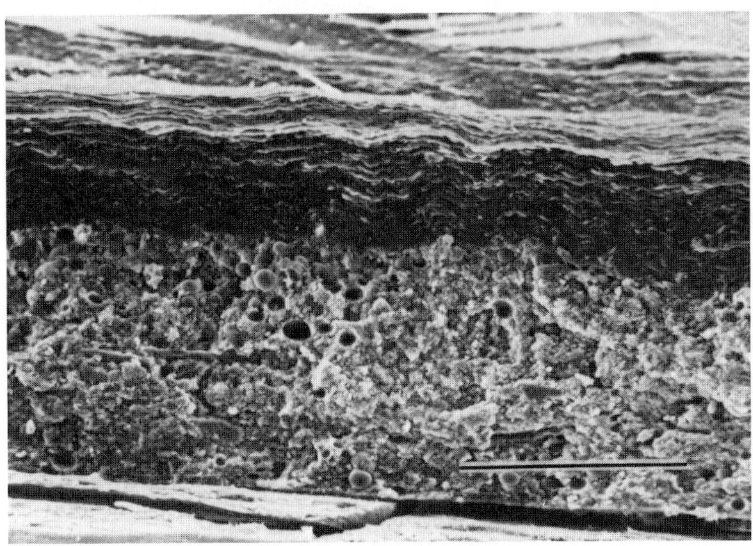

Figure 23 Higher magnification of spoke-shaped opacity. Marker = 100 μm.

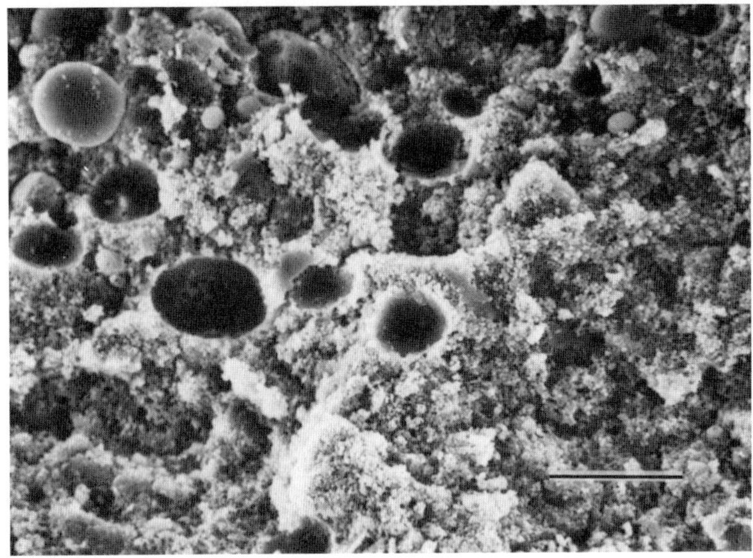

Figure 24 Detail of cellular breakdown products in spoke-shaped cataracts; bar = 15 μm.

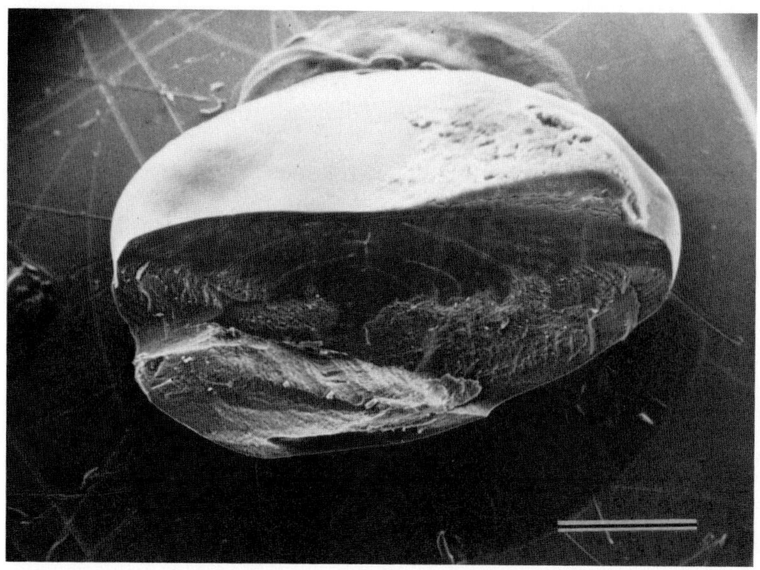

Figure 25 Lens nucleus from a Morgagnian cataract; bar = 1 mm.

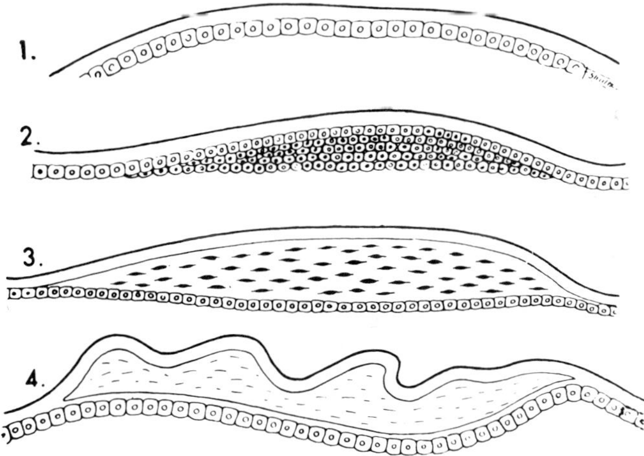

Figure 26 Diagram showing progressive development of anterior subcapsular cataract. (1) Normal capsule and epithelium. (2) Proliferated epithelium. (3) Appearance of spindle-shaped cells (in extracellular matrix containing some collagen, mucopolysaccharides, and calcium). (4) Production of new basement membrane. (From Ref. 34; courtesy of Ophthalmic Publishing Co., Chicago.)

cells is unlike that of fibroblasts, which brings into question the "true metaplasia" theory. They also point out that collagen production by epithelial cells in tissue culture has now been established (35). Pertinent observations were described by Gordon et al. (36), who found that rabbit lenses decapsulated by collagenase were capable of producing new basement membrane in organ culture. Experimentally produced anterior subcapsular cataracts in rabbits have been described by Weinsieder et al. (37).

Posterior Subcapsular Cataracts

Studies on the human posterior subcapsular cataract (SCP) have appeared in the literature for well over a century (38). It is generally agreed that the sequence of events that characterize the development of this cataract include a disorganization of cells at the equator, some of which subsequently migrate to the posterior capsule, where they increase considerably in size. These enlarged nucleated cells are called bladder cells or Wedl cells. The initial stage of cellular disorganization is particularly evident in whole-mount or flat preparations of the capsule with adhering cells (39,40). The normal arrangement of the meridional rows in straight lines (41–47) is altered during cataractogenesis (48). This disorganization and the appearance of the migrating cells on the posterior capsule have been described in experimental radiation cataract in the rabbit (49). Streeten and Eshaghian (40) were the first to describe the morphologic changes in the subcapsular area from the equator to the posterior pole in human SCP. Figure 27 shows the appearance of the equatorial and posterior subcapsular cells in the normal young adult and in cataractous human lenses. Figure 28 is a diagrammatic summary of cellular changes in human SCP, from a mild equatorial cellular disarray in older eyes, to the appearance of the enlarged migrated cells on the posterior capsule. Enlarged cells on the posterior capsule are shown at the SEM level in Figs. 29 and 30. At higher magnification they are seen to have ridged or microvillous surface structures.

Mechanisms other than posterior migration of equatorial cells may account for some kinds of posterior subcapsular cataracts. Osmotic imbalance may be another cause. An experimental posterior subcapsular cataract can be induced in the isolated rabbit lens, maintained in organ culture in a hyperosmotic medium (50).

Other Characteristics of Cortical Cataracts

One specific human lens abnormality resulting in opacification is characterized by the presence of accumulated folded membranes or membrane debris. This concentration of membranes is a general phenomenon, related to aging and cataractogenesis. It is sometimes found in the form of localized "punctate" areas of opacity (or flecks) that are occasionally large enough to be seen in CCRG

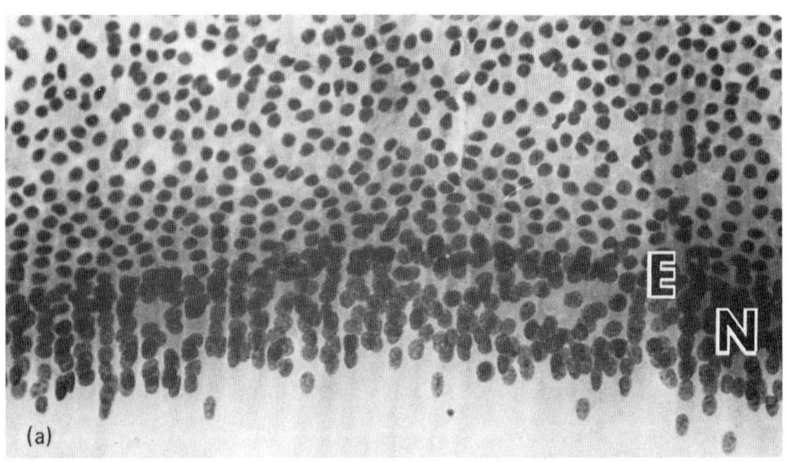

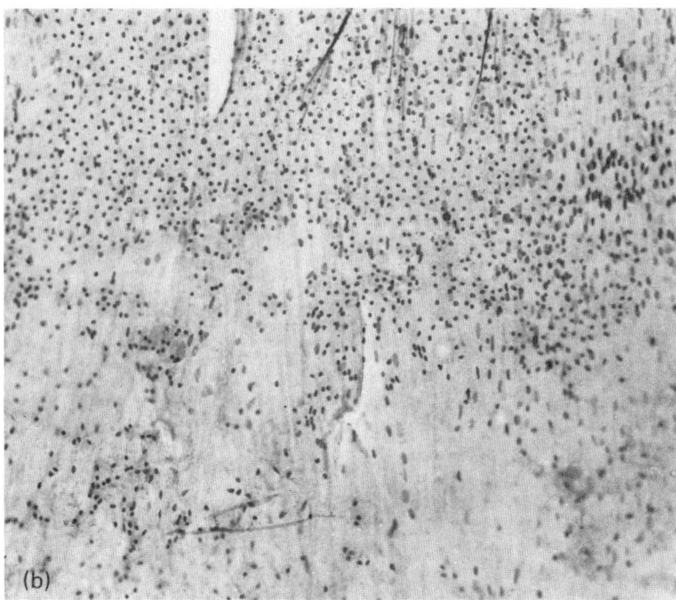

Figure 27 (a) Whole-mount preparation of normal human lens epithelium showing normal alignment of meridional rows. "Compacted equatorial line (E) and several postequatorial nuclear rows (N)" are evident (X 140). (b) Whole-mount preparation of human lens epithelium from lens with steroid-induced posterior subcapsular cataract (SCP), showing disorganization of meridional rows, and migration of cells posteriorly (X 70). (From Ref. 40; courtesy of American Medical Association.)

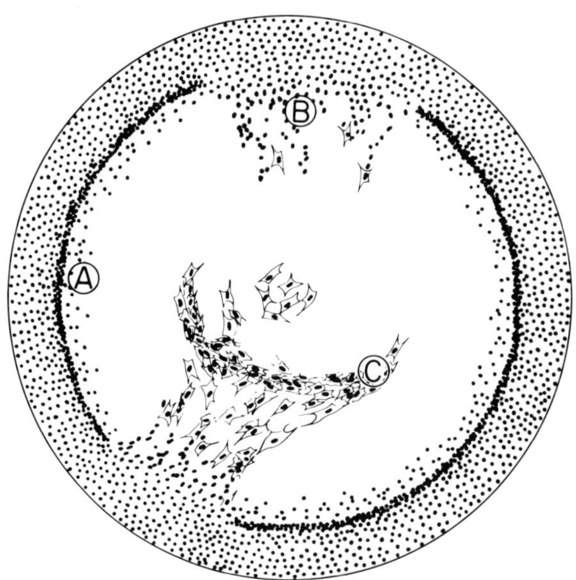

Figure 28 Diagrammatic summary of cellular changes in posterior subcapsular cataracts. (A) Mild equatorial cell disarray in older eyes. (B) Early posterior cell migration. (C) Further migration with cellular enlargement. (From Ref. 40; courtesy of American Medical Association.)

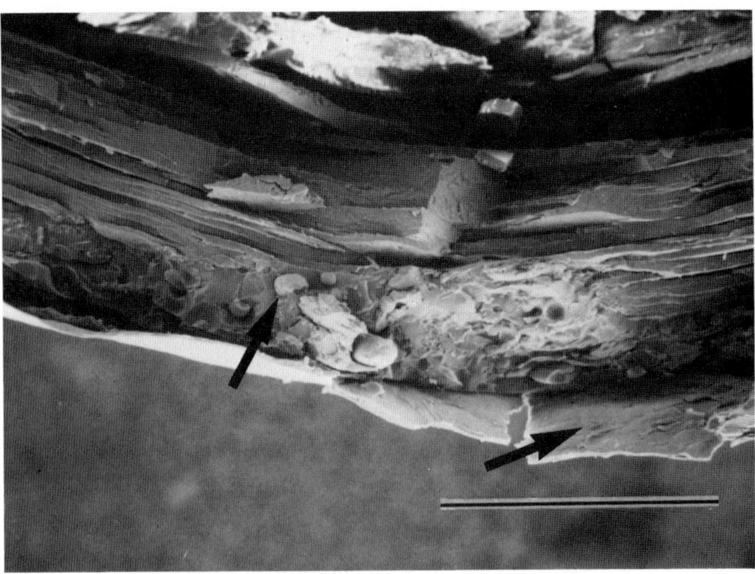

Figure 29 SEM showing enlarged cells (arrow) at the posterior pole. Arrow on right indicates capsule; bar = 0.5 mm.

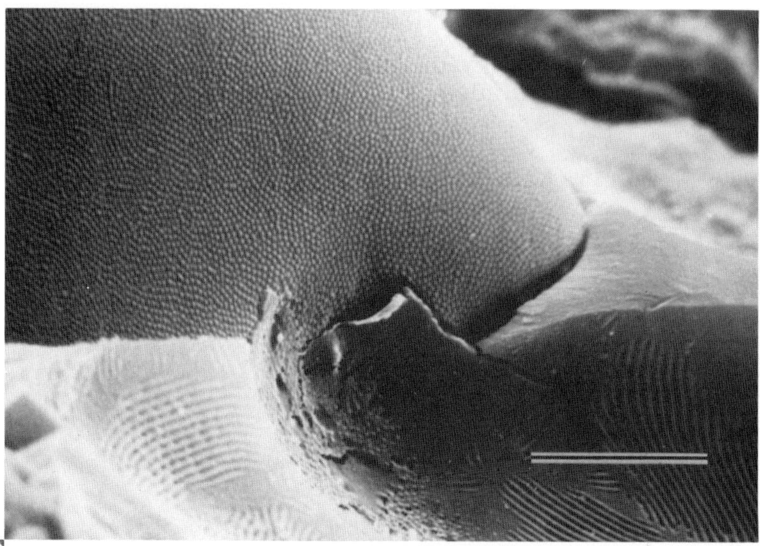

Figure 30 Higher magnification of Fig. 29 showing ridged and microvillous surface patterns of the enlarged cells; bar = 5 μm.

Figure 31 SEM of an individual punctate opacity surrounded by normal-appearing lens fibers; bar = 20 μm. (From Ref. 15; courtesy of C. V. Mosby, St. Louis.)

Figure 32   SEM view of exposed material from a punctate opacity having the appearance of folded membranes; bar = 5 μm. (From Ref. 15; courtesy of C.V. Mosby, St. Louis.)

stereo photographs (15). This cataractous accumulation of membranes is best studied in such punctate opacities. Generally, these distinct sites of degeneration which are immediately surrounded by quite normal-appearing lens fibers (Fig. 31), contain folded membrane debris (Fig. 32).

Elemental analysis (energy dispersive x-ray analysis) of these opacities reveals a high phosphorus-low sulfur composition (Fig. 33), in contrast to the surrounding lens fibers, which are typically low in phosphorus and high in sulfur (Fig. 34). Line scan analyses for phosphorus and sulfur through an individual opacity are shown in Fig. 35. Elemental mappings consistently show the high phosphorus-low sulfur content in the opacities, with the reverse found in the surrounding normal-appearing tissue (Figs. 36 to 38) (15,51).

A source for the abnormally high phosphorus in this type of opacity may be the phospholipids in the accumulated membranes or membrane debris. TEM of thin sections from these lenses indeed reveals such localized sites of concentrated membrane (Fig. 39).

Another type of localized opacity, which has been identified in approximately 3% of human lenses examined, is characteristically spheroidal and dense and has a sharp outline (16). These opacities range in size up to 300 μm and

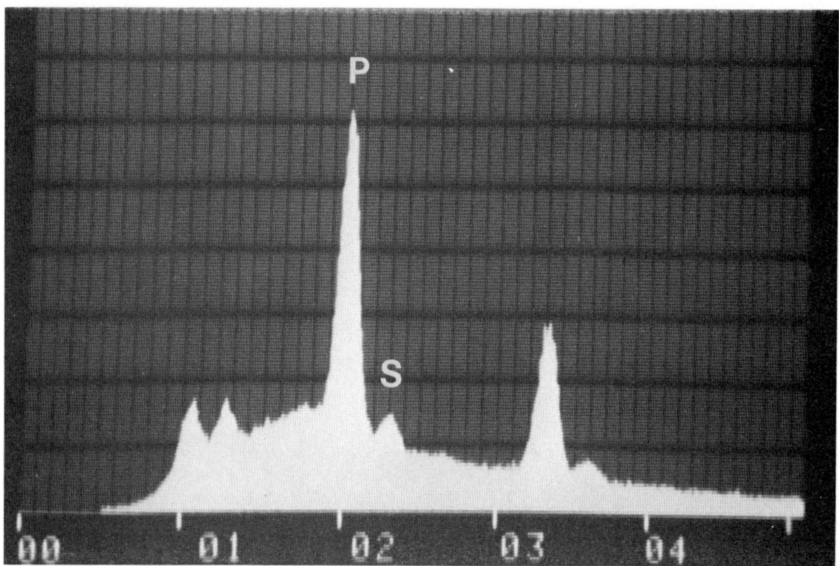

Figure 33 EDXA spectrum of a point in a punctate opacity showing a typically high-phosphorus peak and low sulfur. (From Ref. 15; courtesy of C.V. Mosby, St. Louis.)

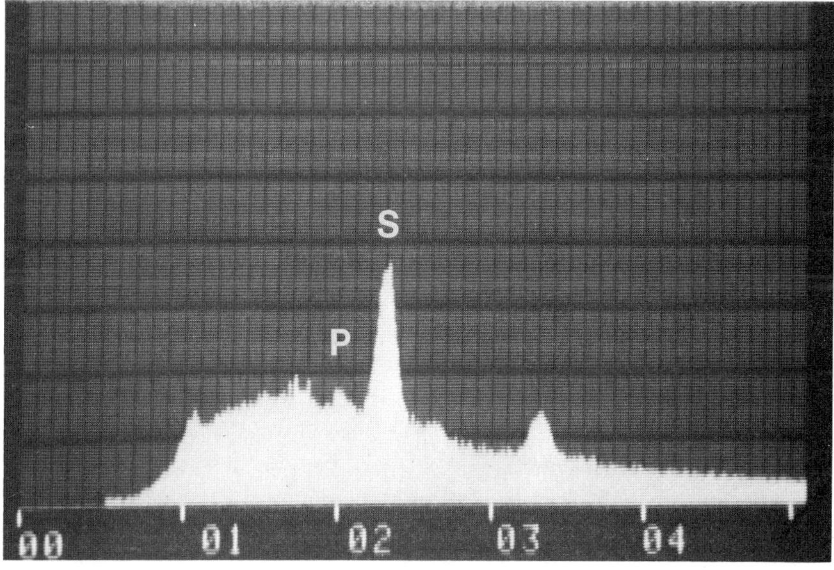

Figure 34 EDXA spectrum for a point on a normal-appearing lens fiber adjacent to the opacity analyzed in Fig. 33. There is a typical high-sulfur peak. (From Ref. 15; courtesy of C.V. Mosby, St. Louis.)

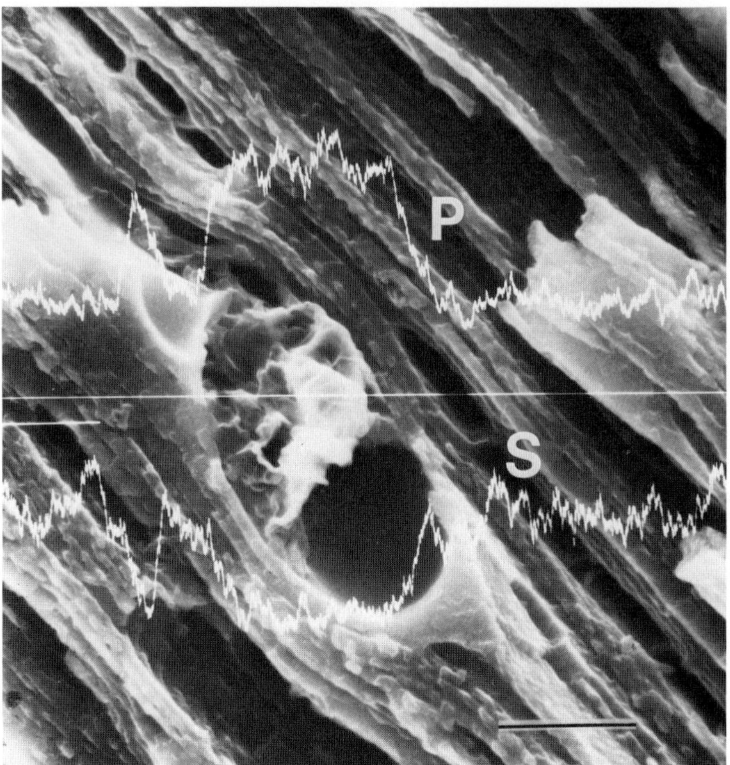

Figure 35   An individual punctate opacity. The horizontal line passing through the opacity material is the line scanned for phosphorus and for sulfur content as determined by EDXA. The upper curve represents the relative quantity of phosphorus in the line scanned, and the lower curve is that of sulfur; bar = 10 μm. (From Ref. 15; courtesy of C.V. Mosby, St. Louis.)

therefore are visible in the CCRG photographs. They are generally found in the deeper layers of the lens, with greater numbers sometimes occurring in a posterior supranuclear region. These opacities are seen to have a definite boundary (Fig. 40), and structures are revealed radiating from the center. Ultra-thin sections (30) reveal a crystal pattern. The surface of these opacities is distinguished by a regular striated pattern (Fig. 41). This opacity is extremely rich in calcium, which accounts for the high portion of the calcium line scan through the middle of the opacity (Fig. 40). Figure 42 also indicates this concentration with a calcium map constructed from the same region.

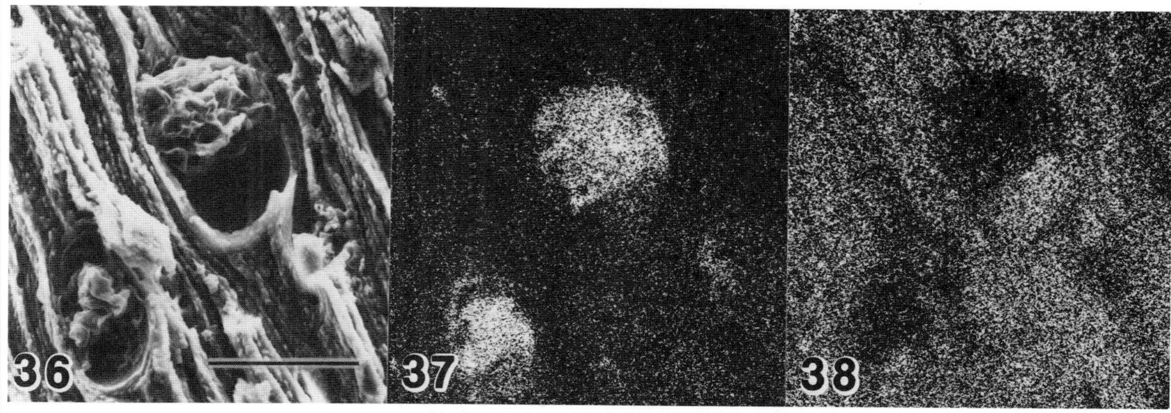

Figure 36   Two punctate opacities with exposed membrane debris; bar = 20 μm.

Figure 37   EDXA phosphorus map of the area in Fig. 36 demonstrating the higher phosphorus counts in the opacity material.

Figure 38   EDXA sulfur map of the area in Fig. 36.

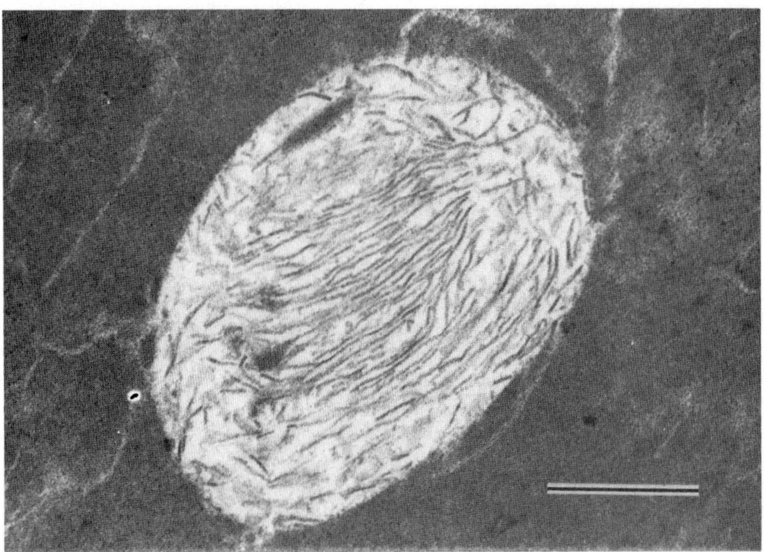

Figure 39 Transmission electron micrograph of a thin section through an opacity revealing multiple unit membranes and a low-density, rather transparent matrix compared to the surrounding lens fibers; bar = 0.5 μm.

Solubility and incineration tests have determined that the major component of these inclusions is calcium oxalate. That these opacities exhibit birefringence also implies that calcium oxalate is the major component. Although the cause for the calcium oxalate opacities is at present open to debate, there is reason to suspect an abnormality in calcium metabolism. This may give rise to related findings elsewhere in the body, such as oxalate crystals in the urine and calcium deposits in blood vessels (as found in two patients with this opacity). Since these calcium opacities can be seen and identified in the fresh lens and possibly even in vivo, their presence in the lens may be of general diagnostic value.

Nuclear Cataract and Other Age-Related Nuclear Changes

A number of independent changes commonly occur in the lens nucleus during aging. Nuclear sclerosis is a process in which the lens nucleus becomes hardened, and its index of refraction increases. Also, the formation of compounds that accumulate give a gradually deepening color to the older lens. Nuclear opacities may also occur.

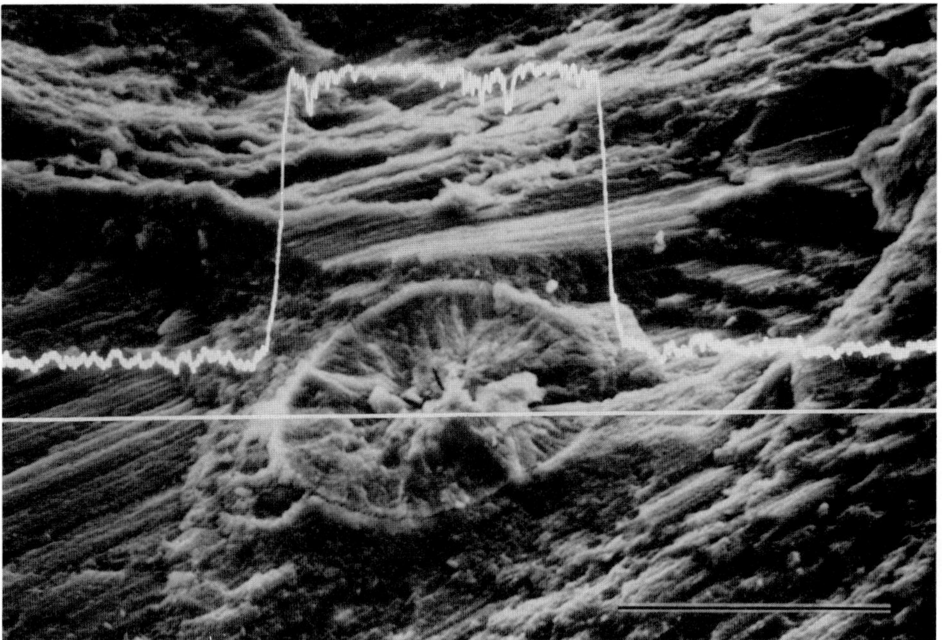

Figure 40 A fractured spheroidal opacity. The curve shows the relative Ca content as determined by EDXA along the line passing through the center of the opacity; bar = 100 μm.

During sclerosis, or induration of the nucleus, the fibers become condensed and very closely packed. The resultant hardening of the nucleus is accompanied by an increase in refractive index. In some reported cases, the refractive index of a central portion of the nucleus increases to such an extent that significant changes in vision occur [i.e., it may result in a "lens with a double focus" (17)]. At one time it was thought that the changes in refractive index were the physical basis for the observed changes in color. However, the results of Rollet and Bussy (52), as pointed out by Bellows (17), show that "many weakly colored lenses are as highly sclerotic as those with intense brown or black pigmentation."

The formation of compounds that give a characteristic color to the lens is a gradual process. The color varies from light yellow through darker shades of yellow and even in some cases to dark brown and, rarely, black [cataracta brunescens (53); cataracta nigra (54)]. The early stages of this process do not obstruct vision, and in fact, the yellow-colored lens nucleus has the beneficial effect of filtering out (absorbing) ultraviolet light and short-wavelength visible

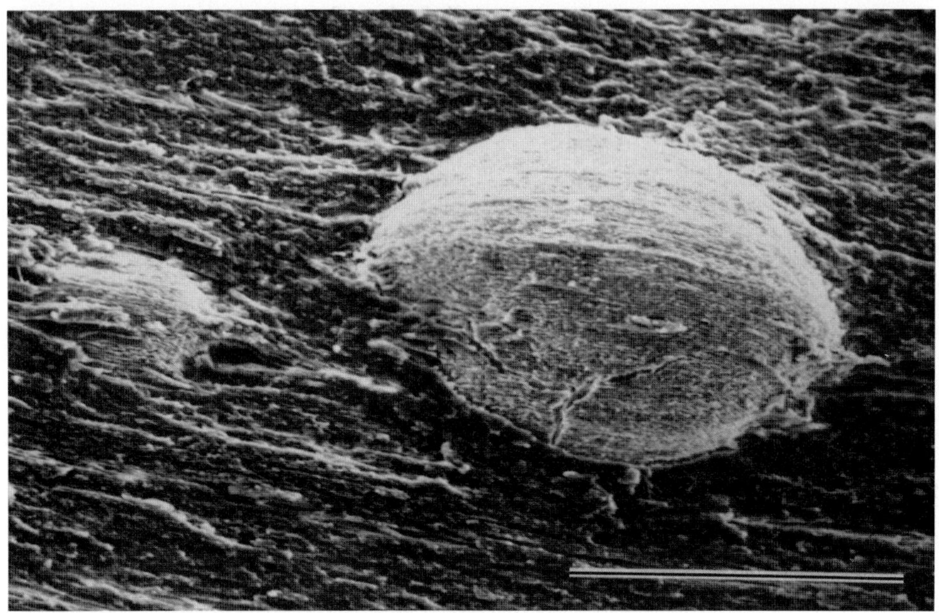

Figure 41   SEM picture of two intact opacities that show the distinct boundary these inclusions typically possess. A characteristic surface striation is also visible in this view; bar = 100 μm.

light, which may be harmful to the retina (Ref. 55; Chap. 8). The later stages of nuclear coloring (dark brown to black) do, of course, obstruct vision.

A considerable effort has gone into studies on the formation of the compounds responsible for color formation during sclerosis, and the causative environmental factors (i.e., radiation: see Chap. 8). In contrast, relatively little has been reported on the structural characteristics of nuclear cataracts at the electron microscope (EM) level. This is due in great measure to the physical properties of the hardened sclerotic nucleus, which make it difficult to obtain ultrathin sections.

In spite of the technical difficulties, several investigators have reported excellent EM descriptions of normal and cataractous human lens nuclei (23, 24,56).

Kuwabara (24) has shown that fiber cells in the normal lenticular nucleus are relatively small and irregular in shape. The cytoplasm is uniform and finely granular in appearance. Numerous infoldings and ridges of the cell membranes are evident, and the whole nucleus "appears to be a solid conglomerate without

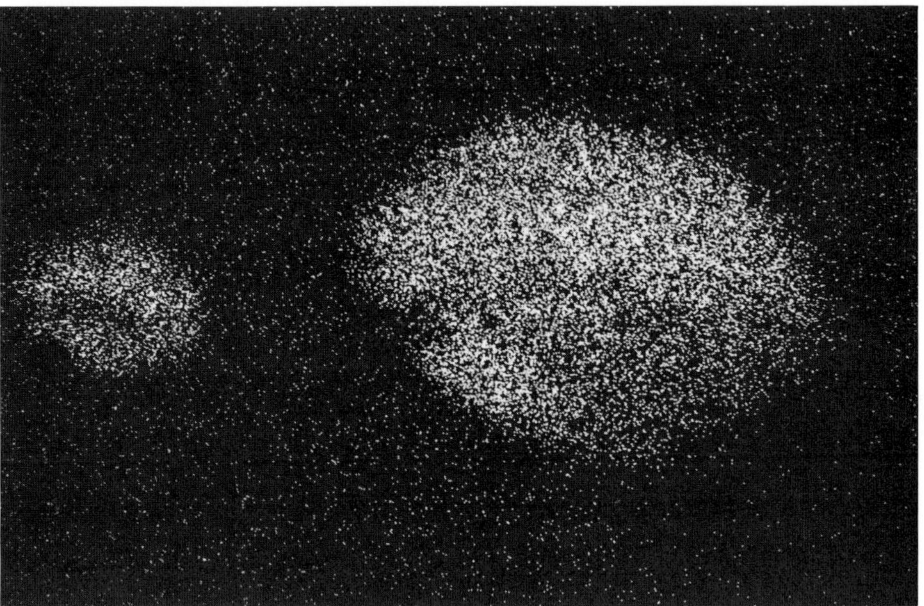

Figure 42   An EDXA calcium map of Fig. 41.

any movement." Although the cell borders are well outlined by their membranes, there are occasional breaks in the membranes that could, according to Kuwabara (24), be due to processing artifacts. Lenses with nuclear cataract have a very different appearance in the nuclear region from the normal. The nuclear fiber cells in nuclear cataract "show an absence of the cell borders and the lens is opaque" (24). It appears, therefore, that nuclear cataractogenesis is accompanied by marked changes in the fiber cell membranes.

RELATION OF STRUCTURE TO CATARACT FORMATION

Morphologic studies on human lenses have revealed a great variety of structural changes associated with aging and cataractogenesis. Some of the structural changes at the cellular level have been summarized by Yanoff and Fine (57), as the diagram in Fig. 43 indicates. At the SEM level, Clark et al. (58) have shown reversible structural changes in lens fibers in "cold" cataracts. This may represent one of the earliest, but reversible, structural changes in cataractogenesis.

The most widely accepted theory of cataractogenesis, which involves the osmotic swelling of the fiber cells, has been tested in dietary and hereditary

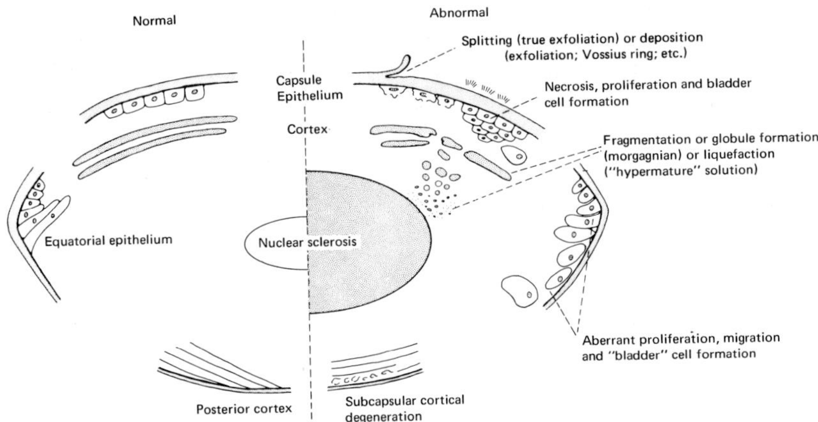

Figure 43 A diagrammatic summary of some of the structural changes associated with human cataractogenesis. (From Ref. 57; courtesy of Harper and Row, Hagerstown, Maryland.)

animal cataracts (5,6). The histologic data on galactose cataract (8) show varying degrees of swelling (i.e., some cells swell more than others). This differential swelling of the fiber cells could cause physical strains within the lens that may lead to localized regions of cell breakdown, with accompanying opacification. At the same time, hardening of the lens nucleus may exert increasing pressures on the overlying cortical fibers and thus indirectly play some role in the development of the pattern of cortical opacities.

Perhaps the most clear-cut example of this interrelation is the supranuclear (SN) class of cataracts. The human SN opacities occur just over (or external to) the equator of the lens nucleus, where the nucleus has its sharpest curvature. Compression of the deep cortical fiber cells against the nucleus at this point may be a factor leading to the focal degeneration, and consequent severing of the bundles of cortical fiber cells, so characteristic of the SN cataract.

Figures 16 and 17 show SN opacities in relation to the lens nucleus in the fresh lens. The SEM photograph in Fig. 12 shows that the severed fiber cell endings may have smooth surfaces. They may appear bulbous or swollen, with debris, presumably derived from cellular breakdown products at the SN site (Fig. 13). The shape of these severed fiber cell endings must be determined in part by the fiber cell cytoskeleton (25). Cell membranes may have formed at the severed fiber cell endings, sealing off the remainder of the fiber cell fragments, perhaps accounting for the retention of transparency by these fragments, as confirmed by the CCRG photography. The sealing of cell membranes, following a limited injury to the cell surface, has been the subject of considerable investigation in other cell types. If the newly formed membrane at the severed fiber cell endings plays a role in preserving the normal composition of the cell interior, it may play a role in maintaining the normal transparency, characteristic of the cell fragments. Nothing is known of the permeability and/or transport characteristics of these fiber cell fragments. Tracer experiments, utilizing horseradish peroxidase (HRP) as a probe (59), may indicate whether a permeability barrier is established at the severed fiber cell endings at the SN site. In conclusion, the SN cataract offers a well-defined system for studies on the cell biology of human cataractogenesis.

In the sutural cataract the interplay of osmotic forces and mechanical pressures might also be occurring, causing the separation of fiber cell endings from the suture system and leading to localized damage and sutural cataract.

Localized injury to the lens fiber cells leads to the formation of globules or spherical structures, some of which may be membrane enclosed. Such breakdown products occur in other cell types and are removed in the extracellular medium. Because of the unique structure of the lens, however, lens fiber cell breakdown products are retained within the tissue for relatively long periods of time. Therefore, globular structures, resulting from localized fiber cell breakdown, have become recognized as a characteristic of many lens opacities. It has been suggested that digestion and removal of the crystallins from such cellular debris would concentrate the remaining cell membrane fragments, accounting for those opacities characterized by dense accumulations of unit membranes (15). Such loss of the crystallins and concentration of unit membrane fragments may also account for the low sulfur-high phosphorus contents of these opacities.

Other opacities, which have an entirely different mechanism of formation, are those consisting of deposits of crystals. For example, numerous small opacities, consisting of deposits of calcium oxalate, may occur in approximately 4% of human cataractous lenses. Such changes may reflect changes in nutrition or metabolism of the individual.

The eventual understanding of cataractogenesis must be based on a firm foundation of physicochemical and morphologic information. The many and varied types of human cataracts make the study of human cataractogenesis especially difficult. The only workable approach is to relate all the findings to specific classes of human cataracts, defined by a mutually agreed upon system. In recent years, this approach has been used more and more frequently. The system of Chylack, which is based on a photographic analysis of the fresh human lens, serves this purpose very well. The details of this system are summarized in the next section.

CCRG CLASSIFICATION

The CCRG classification system employs a Zeiss OpMi-1 microscope with a Zeiss-Urban stereo attachment. Slit-lamp and fluorescent ring light illumination is used. Photographs are taken of anterior and posterior aspects of the fresh lens with a black background, with and without slit illumination. An anterior view against a white background is also taken to determine the intensity of nuclear color. The resulting 35 mm color stereo slides are used to determine the zone, extent, and depth of opacities. The opacities in a cataract are coded into a formula that can have up to nine statements relating to zone and extent of opacification. The complete general classification formula is I, M, H, $CXA_{(\ )}$, $CXE_{(\ )}$, $CXP_{(\ )}$, $SCA_{(\ )}$, $SCP_{(\ )}$, $SN_{(\ )}$, $N_{(\ )}$, and $NS_{(\ )}$. Only those categories pertaining to a given lens are used in writing its classification formula (see Fig. 1).

I, M, or H: This first notation, immature, mature, or hypermature, is an indication of the overall state of the cataract. An immature cataract has areas not obscured by opacities; mature and hypermature lenses are completely opaque. Both M and H lenses have extensive degeneration and most likely some liquefaction of lens fibers. In addition, the hypermature lens is swollen beyond the typical dimensions.

CXA, CXE, and CXP: Cortical opacities are subdivided into anterior, equatorial, and posterior, and extend from the supranuclear region outward to the subcapsular (SC) level. The CXA and CXP zones occupy the center of the anterior and posterior cortex, respectively (see diagram). The subscript in the $CX_{(\ )}$ notation indicates the number of quadrants containing that type of opacity.

SCA and SCP: Subcapsular opacities are confined to a thin layer just beneath the capsule on the anterior or posterior surface. The subscripted number is an estimate of the percentage of the total surface area affected (3–100%).

SN: The supranuclear zone is found just external to the adult nucleus. The number of quadrants containing the SN opacities is indicated with a subscript number (1–4).

N: The nuclear cataract is characterized by its density and not extent. The subscript 4 symbolizes a completely opaque nucleus; 3 and 2 are less opacified, and number 1 is a faint nuclear opacity.

NS: The nuclear sclerosis notation is a determination of the color of the nucleus and is rated with a subscript rating the degree. This indication can be: c (clear), vpy (very pale yellow), py (pale yellow), y (yellow), dy (dark yellow), vdy (very dark yellow), br (brown or brunescent), or bl (black).

ACKNOWLEDGMENTS

Supported in part by NIH Grants EYO-1874, EYO-7034, EYO-1276, EYO-3247, and Research To Prevent Blindness, Inc.
We wish to thank Dr. Drusilla Harding for much important help and advice. We thank Miss Georgean Grubb for her expert help in preparing the manuscript.

REFERENCES

1. L. T. Chylack, Jr. Classification of human cataracts. Arch. Ophthalmol., *96*: 888–892 (1978).
2. L. T. Chylack, Jr. CCRG Newsletter (description of supranuclear cataract) (1980).
3. L. T. Chylack, Jr., M. R. Lee, W. H. Tung, and H. M. Cheng. Classification of human senile cataractous change by the American Cooperative Cataract Research Group (CCRG). Method I. Instrumentation and Technique. Invest. Ophthalmol. Vis. Sci., in press. (1983).
4. W. C. Gorthy. Cataracts in the aging rat lens. Their morphological characterization and evaluation as a model for human senile cataracts. Ophthalmic Res., *9*: 329–342 (1977).
5. J. H. Kinoshita. Cataracts in galactosemia. The Jonas Friedenwald Memorial Lecture. Invest. Ophthalmol., *4*: 786 (1965).
6. J. H. Kinoshita. Mechanisms initiating cataract formation (Proctor Lecture). Invest. Ophthalmol. Vis. Sci., *13*: 713–724 (1974).
7. V. N. Reddy, D. Schwass, B. Chakrapani, C. P. Lim, et al. Biochemical changes associated with the development and reversal of galactose cataracts. Exp. Eye Res., *23*: 483–493 (1976).
8. M. Datiles, H. Fukui, T. Kuwabara, and J. H. Kinoshita. Galactose cataract prevention with sorbinil, an aldose reductase inhibitor: A light microscopic study. Invest. Ophthalmol. Vis. Sci., *22*: 174–179 (1982).
9. T. Kuwabara, J. H. Kinoshita, and D. Cogan. Election microscopic study of galactose-induced cataract. Invest. Ophthalmol., *8*: 133–149 (1969).
10. N. J. Unakar, C. Genyea, J. R. Reddan, and V. N. Reddy. Ultrastructural changes during the development and reversal of galactose cataracts. Exp. Eye Res., *26*: 123–133 (1978).
11. N. Unakar, J. Tsui, and C. V. Harding. Scanning electron microscopy. Development and reversal of galactose cataracts. Ophthalmic Res., *13*: 20–35 (1981).
12. N. J. Unakar, J. Y. Tsui, and C. V. Harding. Scanning electron microscopy. II. Development and reversal of in utero galactose-induced cataracts. Ophthalmic Res., *13*: 165–179 (1981).
13. L. T. Chylack, Jr., and J. H. Kinoshita. The Cooperative Cataract Research Group. Invest. Ophthalmol. Vis. Sci., *17*: 1131–1134 (1978).
14. C. V. Harding, L. T. Chylack, Jr., S. Susan, J. G. Decker, and W. -K. Lo Morphological changes in the cataract: The ultrastructure of human lens opacities, localized by Cooperative Cataract Research Group procedures. In *Red Blood Cell and Lens Metabolism* (S. K. Srivastava, ed.), Elsevier-North Holland, New York, 1980, pp. 27–39.
15. C. V. Harding, L. T. Chylack, Jr., S. R. Susan, W.-K. Lo, and W. F. Bobrowski. Elemental and ultrastructural analysis of specific human lens opacities. Invest. Ophthalmol. Vis. Sci., *23*: 1–13 (1982).
16. C. V. Harding, L. T. Chylack, Jr., S. R. Susan, W.-K. Lo, and W. F. Bobrowski. Calcium-containing opacities in the human lens. Invest. Ophthalmol. Vis. Sci., *24*: 1194–1202 (1983).

17. J. G. Bellows. In *Cataract and Anomalies of the Lens* (J. G. Bellows, ed.), C.V. Mosby, St. Louis, 1944.
18. J. G. Bellows. In *Cataract and Abnormalities of the Lens* (J. G. Bellows, ed.), Grune and Stratton, New York, 1975.
19. S. Duke-Elder. Congenital deformities. In *System of Ophthalmology*, Vol. 3, Part 2, *Normal and Abnormal Development*. C.V. Mosby, St. Louis, 1963.
20. A. Vogt. *Lehrbuch und atlas der spaltlampenmikroskopie des lebenden auges, zweiter teil; linse und zonula*, Julius Springer, Berlin, 1931.
21. A. I. Cohen. The electron microscopy of the normal human lens. Invest. Ophthalmol., *4*: 433–446 (1965).
22. P. N. Farnsworth, S. E. Shyne, P. A. Burke et al. Maturation of the human lens fiber. In *The Red Blood Cell and The Lens* (S. Srivastava, ed.), Elsevier, New York, 1980, pp. 44–48.
23. M. J. Hogan, J. A. Alvarado, and J. E. Weddell. *Histology of the Human Eye*, Saunders, Philadelphia, 1971.
24. T. Kuwabara. The maturation of the lens cell: A morphologic study. Exp. Eye Res., *20*: 424–443 (1975).
25. H. Maisel, C. V. Harding, J. R. Alcala, J. Kuszak, and R. Bradley. The morphology of the lens. In *Molecular and Cellular Biology of the Eye Lens* (H. Bloemendal, ed.), John Wiley and Sons, New York, 1981, pp. 49–84.
26. D. Donaldson. The crystalline lens. In *Atlas of Diseases of the Anterior Segment of the Eye*, Vol. 5, C.V. Mosby, St. Louis, 1976.
27. M. O. Creighton, J. R. Trevithick, G. Y. Mousa, D. H. Percy, A. J. McKinna, C. Dyson, H. Maisel, and R. Bradley. Globular bodies: A primary cause of the opacity in senile and diabetic posterior cortical subcapsular cataracts? Can. J. Ophthalmol., *13*: 166–181 (1978).
28. T. Matsuto. Scanning electron microscopic studies on the normal and senile cataractous human lenses. Acta Soc. Ophthalmol. Japan, *77*: 155–174 (1973).
29. W.-K. Lo, C. V. Harding, and H. Maisel. Gap junctions in the human cataractous lens. J. Cell Biol., *87*: 56a (1980).
30. A. Bron. Morgagnian cataract. Trans. Ophthalmol. Soc. UK, *96*: 265–277 (1976).
31. H. P. Lamb. Anterior subcapsular cataracts: An example of true metaplasia. Arch. Ophthalmol., *17*: 877 (1937).
32. M. Ghosh and C. McCullock. Anterior subcapsular cataract: An electron microscopic study. Can. J. Ophthalmol., *10*: 502–510 (1975).
33. P. Henkind and R. Prose. Anterior polar cataract. Am. J. Ophthalmol., *63*: 768–771 (1967).
34. R. I. Font and S. Brownstein. A light and electron microscopic study on anterior subcapsular cataracts. Am. J. Ophthalmol., *78*: 972–984 (1974).
35. J. W. Dodson and E. D. Hay. Secretion of collagenase stroma by isolated epithelium grown in vitro. Exp. Cell Res., *65*: 215–220 (1971).
36. S. Gordon, M. Bagchi, C. V. Harding, and J. Reddan. Cellular proliferation in the cultured lens, decapsulated by collagenase. J. Cell Biol., *55*: 92A (1972).

37. A. Weinsieder, R. Briggs, J. Reddan, H. Rothstein, D. Wilson, and C. V. Harding. Induction of mitosis in ocular tissue by chemotoxic agents. Exp. Eye Res., *20*: 33–44 (1975).

38. C. Wedl. *Atlas der Pathologischen Histologie des Auges*, Georg Wigan Verlag, Leipzig, 1860.

39. J. Eshaghian and B. W. Streeten. Human posterior subcapsular cataract. An ultrastructural study of the posteriorly migrating cells. Arch. Ophthalmol., *98*: 134–143 (1980).

40. B. W. Streeten and J. Eshaghian. Human posterior subcapsular cataract. A gross and flat preparation study. Arch. Ophthalmol., *96*: 1653–1658 (1978).

41. L. Bito and C. V. Harding. Patterns of cellular organization and cell division in the epithelium of the cultured lens. Exp. Eye Res., *4*: 146–161 (1965).

42. C. V. Harding, J. R. Reddan, N. J. Unakar, and M. Bagchi. The control of cell division in the ocular lens. Int. Rev. Cytol., *31*: 215–300 (1971).

43. A. Howard. Whole-mounts of rabbit lens epithelium for cytological study. Stain Technol., *27*: 313–315 (1952).

44. N. S. Rafferty. Proliferative response in experimentally injured frog lens epithelium: Autoradiographic evidence for movement of DNA synthesis toward injury. J. Morphol., *121*: 295–310 (1967).

45. J. R. Reddan. Development and structure of the lens. In *Cataract and Abnormalities of the Lens* (J. Bellows, ed.), Grune and Stratton, New York, 1975, pp. 29–42.

46. H. Rothstein. Experimental techniques for the investigation of the amphibian lens epithelium. In *Methods in Cell Physiology*, Vol. III (D. Prescott, ed.), Academic Press, New York, 1968, pp. 45–74.

47. B. Srinivasan and C. V. Harding. Cellular proliferation in the lens. Invest. Ophthalmol., *4*: 452–470 (1965).

48. B. V. Worgul. Lens. In *Biomedical Foundations of Ophthalmology* (D. T. Dwayne, ed.), Harper and Row, Philadelphia, 1982, pp. 1–32.

49. D. Cogan and D. Donaldson. Experimental radiation cataracts. I. Cataracts in the rabbit following single x-ray exposure. Arch. Ophthalmol., *45*: 508–522 (1951).

50. M. Bagchi. Effect of hyperosmotic medium on the cultured ocular lenses. Fifth International Congress of Eye Research. Eindhoven, Holland, 1982.

51. C. V. Harding, S. Susan, B. Worgul, A. Solway, and M. Maser. Energy dispersive X-ray analysis of human lens cataracts. Ophthalmic Res., *11*: 159–163 (1979).

52. Rollet and Bussy. La cataracte noire. Arch. Ophtal., *38*: 65–82 (1921).

53. O. Becker. *Zur Anatomie der gesunden und kranken Linse*, J. F. Bergmann, Wiesbaden, 1883.

54. M. J. B. de Wenzel. *Traité de la cataracte*, Paris, 1786.

55. S. Lerman. *Radiant Energy and the Eye*, MacMillan, New York, 1980.

56. H. Obazawa. Electron microscopic studies on the lens nucleus of cataract. Acta Soc. Ophthalmol. Japan, *71*: 1019–1028 (1967).

57. M. Yanoff and B. S. Fine. *Ocular Pathology: An Outline and Atlas*, Harper and Row, Hagerstown, Maryland, 1975.
58. J. L. Clark, L. Mengel, and G. B. Benedek. Scanning electron microscopy of opaque and transparent states in reversible calf lens cataracts. Ophthalmic Res., *12*: 16–33 (1980).
59. W.-K. Lo and C. V. Harding. Tight junctions in the lens epithelia of human and frog: Freeze-fracture and protein tracer studies. Invest. Ophthalmol. Vis. Sci., *24*: 396–402 (1983).

11
Aspects of the Biochemistry of Cataract

ABRAHAM SPECTOR / College of Physicians and Surgeons, Columbia University, New York, New York

The biochemistry of cataract covers a broad spectrum of observations encompassing many aspects of metabolism, protein, and lipid structure, membrane and matrix integrity, terminal differentiation, and molecular biology. Because of space restrictions, the discussion of many subjects will be cursory. A number of texts and reviews have appeared in the last few years that consider topics that are either omitted or only briefly mentioned in this discussion (1–5). This chapter is not intended to be a compilation of all available facts. Rather, selected information is presented from the perspective of the writer's prejudice. An attempt has been made in certain situations to develop a rationale for the sequence of events observed in cataract development. Human cataract is the major concern, and animal systems are discussed primarily in terms of their contribution to elucidating human cataractogenesis.

THE NORMAL LENS

The lens presents a deceiving view to the investigator. The normal young lens is exquisitely transparent, appearing homogeneous and uniform in structure, but the lens is a deceivingly complex tissue. It is avascular, enclosed by a metabolically inert basement membrane composed of protein resembling collagen, as well as glycoproteins and perhaps other components. The lens contains a single layer of epithelial cells that are only present on the anterior side of the organ. In the equatorial region, terminal differentiation into fiber cells occurs with a dramatic increase in cell volume and the production of new membrane and cytosol constituents. The cell extends to both the anterior and posterior sides of the tissue displacing already formed cells in toward the center of the organ.

Thus, the earliest formed lens fibers are in the center of the organ; the nucleus and the most recently formed fibers are at the periphery of the tissue. A gradient of the youngest to the oldest fibers is therefore established within the tissue. Pyknosis of the cell nucleus occurs with degradation of DNA and RNA. As a result of this process, the fiber cell gradually loses its ability to synthesize protein (6,7). More than 90% of the protein synthesis of the lens occurs in the epithelium and the outer superficial fibers of the tissue. Thus, there is preservation of both cellular and macromolecular structure, making the lens a highly attractive tissue for the study of aging. This peculiar biology of the tissue also makes it difficult in some cases to distinguish between normal aging and events leading to cataract in older individuals.

The loss of the ability to replace denatured protein leads to a decrease in general metabolic activity in the inner regions of the tissue (8,9). Although, surprisingly, some enzyme activities can still be detected in the nucleus of the old lens, in general, multienzyme systems appear to be inoperative. There is only a small peripheral band of tissue that maintains the homeostasis of the entire tissue. It should also be noted that, in the transformation from epithelial to fiber cell, dramatic changes in structure and composition occur that are reflected in both metabolic activity and chemical composition. Certain biologic activities required for the viability of the tissue may be predominantly present in the epithelial cell layer.

The structure of the cell is believed to be maintained by an intracellular matrix. Such a matrix in the lens has been visualized and isolated (10–14). However, it has only been found in the peripheral region of the tissue. Whether it collapses or is degraded in the older regions of the tissue is not clear at this time. Cell constituents, such as mitochondria and microsomes, although in abundance in the epithelial cell layer and outer fiber cells, decrease rapidly as one proceeds into the lens and are generally absent from the nuclear region (11).

From this brief discussion, it is clear that the constituents of the lens can be assigned to one of the following cell regions: capsule, cell membrane, nucleus, matrix, subcellular components, and cytoplasm. Furthermore, it should be recognized that the constituents of the epithelial cell differ markedly from those of the fiber cell. And, as will be discussed later on, with aging there appears to be the synthesis of new protein not found in younger lenses, as well as posttranslational changes of old protein.

NORMAL LENS PROTEIN

The human lens is remarkable in that it has a very high protein content of approximately 35% on a wet weight basis in the periphery, which increases to approximately 40% in the central region of the tissue (15). In some other animals, such as the cow, the lens protein concentration in the central region can be greater than 65% on a wet weight basis (16,17). Such protein must be organized in a highly sophisticated manner to maintain the uniform refractive index required for transparency (18,19). Most of the protein in the lens is represented by the "crystallins," the so-called structural proteins. They are described in this manner because no biologic activity has been clearly associated with them, they are believed to contribute significantly to the refractive index and transparency of the tissue, and they are isolated from an organ that in an earlier period was called the crystalline lens.

Until recently almost all studies on lens proteins have involved the crystallins. However, it is clear from high-resolution two-dimensional fractionation of the proteins of the lens that the tissues contain a multitude of protein species, reflecting the varied metabolic activities of the organ. Although the general composition of the lens varies, depending on the section of the tissue being studied, most studies consider either the whole lens or the outer region, the cortex, and the inner region, the nucleus. These regions are not well defined. The cortex may vary from the outer 20% to 60% of the lens. Recently, there has begun an effort to develop techniques for removing small sections of lens tissue for analysis (20). With such techniques it is possible to consider opaque and clear regions in close proximity, allowing for detection of subtle differences that may be obscured by analyses of larger sections of material. Individual lens variation is also minimized by this approach.

To understand the changes in lens chemistry that occur with the development of cataract, it is necessary to briefly consider certain aspects of the normal lens. A consideration of the crystallins, membrane proteins, and proteins of the matrix will be presented. This will be followed by a brief discussion of normal metabolism of the lens.

The major structural lens proteins are designated as α-, β-, and γ-crystallins (1–4). Alpha-crystallin is the largest of the crystallins. It is physically homogeneous only at the time of synthesis (21). Newly synthesized bovine α-crystallin has a molecular weight of approximately 7×10^5 (21). Within a short time physical heterogeneity is observed, with the population of macromolecules having a weight average molecular weight of approximately 1×10^6 (22). With further aging, particularly in the nuclear region of the tissue, the preparation becomes grossly heterogeneous, and species greater than 50×10^6 d can be observed (23,24). Similar results are obtained with bovine and human lenses (25,26). All α-crystallin macromolecules in the normal lens are composed of polypeptide chains held together by noncovalent interaction. The newly produced bovine α-crystallin aggregate is composed of two directly synthesized polypeptides, designated the A_2 and B_2 chains (21,27,28). Based on sequence work the two chains have approximately 55% homology (29). The polypeptides are primarily in a β-chain conformation, with some random coil and almost no helical content (30–32). In α-crystallin isolated from 6-week-old human lenses the A_2 and B_2 chains are also found, but there is another polypeptide that, on the basis of gel electrophoresis and two-dimensional analysis of tryptic peptides, appears to be closely related to the A_2 chain (26). In bovine lenses it has been reported that the ratio of A to B chains in α-crystallin varies, being 1:3 in the epithelial cells in the central region where only small amounts of the protein are synthesized, changing to 3:2 before differentiation, and finally to 3:1 in the fibers (33). Such observations suggest a varying control of gene expression at either the DNA or RNA level.

The B chain is more alkaline than the A chain, and the chains are easily separated (25). A number of posttranslational changes involving deamidation and specific peptide bond cleavage at a limited number of sites near the COOH-terminal end of the chains have been observed (34,35). In all, five B chains and six A chains have been reported in both bovine and human lenses (25). Most of these chains are not directly synthesized but arise as a result of posttranslational modification. With the exception of γ-crystallin all the crystallin polypeptides appear to be blocked at their NH_2-terminal group by an acetyl group, in most cases blocking an amino terminal methionine. Work from Bloemendal's laboratory suggests that the acetylation occurs during synthesis of the polypeptide and involves acetyl coenzyme A (36,37).

Both A and B chains of α-crystallin are capable of forming aggregates independently (38). These aggregates are, like their parent macromolecules (39), held together noncovalently and are physically heterogeneous, having a size range approximately half that of the native protein. It is interesting that, when bovine α-crystallin is deaggregated by 7 M urea and then allowed to reaggregate by removal of the urea, the polymers are similar in size to those formed with either A or B chains. These results suggest that there may be a stable subunit of half the observed aggregate size of α-crystallin. Such a viewpoint is supported by studies involving modification of the thiol groups of the protein, as well as recent observations of Augusteyn (40,41).

It is apparent that the addition of Ca^{2+} during reaggregation will affect the size of the reconstituted aggregate (42). The origin of the polypeptides involved in the reaggregation process are important. If low-molecular-weight bovine α-crystallin is used (i.e., α-crystallin in the $1-2 \times 10^6$ d range), then no Ca^{2+} effect is observed, but if high-molecular-weight α-crystallin ($>50 \times 10^6$ d) is utilized, then a large proportion of the polypeptide chains reaggregate to high-molecular-weight species. These results suggest that specific modification of the α-crystallin polypeptides are associated with the loss of physical homogeneity and the shift to higher molecular weight. Jedziniak et al. (43) have shown that calcium can cause an aggregation of the α-crystallin native polymer, and it has also been demonstrated that, although calcium may be required for the formation of high-molecular-weight protein, it is not required to stabilize the already formed macromolecule (44).

The most abundant crystallin group is the β-crystallin group. It contains aggregates ranging from approximately 4×10^4 d to 2.5×10^5. There appear to be four major size populations of β-crystallin, with the larger species called β high (H) and smaller components β low (L) (45,46). The subunit composition of these aggregates appears to be rather similar. These aggregates, like those of α-crystallin, are bound together noncovalently. The major β-crystallin subunit, termed βBp, is approximately 24×10^3 d. Its amino acid sequence has been determined (47). The βBp polypeptide shows some homology between NH_2-terminal and COOH-terminal halves of the peptide chain, as well as approximately 25% homology with γ-crystallin. Such homology suggests divergence from a parent gene at some early point in evolutionary development. A number of other β subunits in the $20–30 \times 10^3$ d range have been reported. In this group is a polypeptide delineated as β_S, which is believed to exist as a monomer and is usually isolated from the γ-crystallin fraction (48). Analysis of the β-crystallin polypeptide population, as with α-crystallin, suggests posttranslational changes, which are reflected by a large group of polypeptides with differing isoelectric points and molecular weights (49–51). Similar to α-crystallin, the β-crystallin group is primarily in antiparallel β-pleated sheet configuration (52,53). All β-crystallin polypeptides appear to be blocked at their NH_2 terminal position. β_S is claimed to have an unusual N-acetyl Trp amino terminal (54).

The γ-crystallins are unique in being the only major group of lens proteins found entirely in the monomeric form (55). γ-Crystallin is expressed primarily at the time of terminal differentiation of an epithelial cell to a fiber cell (56). The γ-crystallins have a free NH_2-terminal glycine and, as discussed elsewhere, have an unusual two-domain Greek key motif structure with a remarkable degree of homology between the NH_2-terminal and COOH-terminal domains (57,58). The β- and γ-crystallins are relatively rich in thiol groups. γ-crystallin has five cysteines in the NH_2-terminal domain and one in the COOH-terminal domain. The γ-crystallins are the most basic of the crystallins and the α-crystallins, the most acidic. The γ-crystallins are characterized by unusually high levels of tyrosine and arginine. The γ-crystallins generally have a molecular weight of approximately 20,000. Based on sequencing of a calf γ-crystallin a (γ_{II}) molecular weight of 19,871 was calculated (57). It is believed that a number of genes code for γ-crystallin (59). In the adult bovine lens a γ-crystallin of approximately 24,000 d has been reported based on analyses of sodium dodecyl sulfate gels (60).

Besides the crystallins there are two other groups of structural proteins that should be briefly mentioned: the proteins of the membrane and of the intracellular matrix. A major problem in studying membrane polypeptides is the difficulty in obtaining pure materials. Since with aging and cataract formation there is an increase in the amount of insoluble protein derived primarily from the crystallins, the problem is clearly greater in older or diseased lenses. There are two types of protein associated with the membrane, the intrinsic polypeptides imbedded in the lipid bilayer and the extrinsic components bound to the surface of the membrane. The latter polypeptides are held to the membrane relatively weakly and are usually removed in the process of obtaining pure membrane components. This situation has led to considerable confusion with respect to which polypeptides are extrinsically bound to the membrane. Claims have been made by a number of laboratories suggesting that one or more of the crystallins are extrinsically bound to the membrane (61–63). To eliminate the problem of contamination, such procedures as acylation and alkaline treatment have been developed (64,65). Such procedures dramatically reduce the number of components obtained by the more conventional urea extraction and buoyant density techniques. In the young human lens fiber membrane a major intrinsic 26,000 d polypeptide has been found (64–67). With aging, increasing amounts of 22,000 d polypeptides are observed. The composition of the 22,000 d population is not well defined, but the available evidence suggests that the fraction is composed of degraded 26,000 d material as well as a polypeptide fraction that is directly synthesized (68–71). The membrane also contains other polypeptides, including components functioning as receptors and pumps, such as Na^+/K^+-ATPase, which are present in very small amounts and are difficult to detect with the usual Coomassie blue-staining techniques.

An extrinsic membrane polypeptide population of particular interest is the 43,000 d species (72,73). This fraction appears to be bound specifically to the membrane of the fiber cells. Addition of EGTA, which is relatively specific for Ca^{2+}, causes the release of the polypeptide. Initially it was concluded that the population did not contain an actin. This was based on immunochemical analysis utilizing an anti-43,000 d preparation and muscle actin and a dissimilarity in amino acid composition. However, recent experiments suggest that the 43,000 d species may contain small amounts of actin. Most of the work on this polypeptide fraction has been carried out with material released into the soluble phase by either physical insult (homogenization) or chelation of Ca^{2+}. Based on radial immunodiffusion studies there appears to be an additional 43,000 d component present in the water-insoluble fraction that does not react with the anti-43,000 d antibody generated from the soluble fraction 43,000 population (74). There also appears to be heterogeneity in the soluble 43,000 d species, based on isoelectric focusing experiments suggesting greater complexity than has as yet been delineated. Very recent results from our laboratory suggest that this fraction contains nondisulfide-linked dimers of β and γ polypeptide chains. Undoubtedly, there are other human lens extrinsic membrane polypeptides, but they have not as yet been characterized because they are either present in small amounts or do not manifest unique size or charge properties.

The lens appears to contain a cytoskeleton consisting of approximately 7 nm (thin) and 10 nm (intermediate) filaments and possibly larger tubular and filamentous structures as well (75). The microfilaments contain actin. Observations suggest the presence of both β- and γ-actins in both the filamentous (F) and the globular (G) forms (76–78). With older lenses and in the inner parts of younger lenses it has been difficult to detect actin. What has happened to the actin in the inner regions of the lens is not understood. It is possible that the actin fibers collapse along the inner side of the fiber cell plasma membrane and are no longer detected.

The intermediate filaments contain vimentin, a polypeptide of approximately 57,000 d (4). Similar to the thin filaments, the intermediate filaments disappear with aging. However, it is not known if the disappearance of all matrix structures, except for the tubules, is indicative of a disappearance of the constituents of the cytoskeleton or a change in the structure of these components so that they are no longer detected by electron microscopy. It should also be noted that a number of other components are associated with the matrix. However, they are present in smaller concentrations and have not as yet been definitely delineated.

A number of reactions appear to be initiated after the synthesis of lens protein. Some of these reactions have briefly been alluded to in the previous discussion. They are all grouped together under the term "posttranslational change." One of the major problems facing the investigator of pathologic alteration in the lens is that of ascertaining whether a biochemical change is related to normal aging or is directly induced by pathologic insult, which causes the loss of normal function. It can be argued that the loss of biochemical activity makes the tissue susceptible to insult that can cause cataract. It can be tentatively assumed that changes in protein structure, which occur decades before the onset of cataract, which are found in essentially all lenses examined, and which are progressive, are age related and not necessarily related to pathology.

Visual inspection of the human lens as a function of aging immediately indicates a number of alterations (79). The lens grows larger and heavier with aging, it becomes more translucent, and it appears to yellow. On the molecular level a number of changes are observed, including an increase in insoluble protein and increased aggregation to high-molecular-weight species. Although most investigators find increases in water-insoluble protein with aging, values vary. This problem is undoubtedly due to the differences in methodology of isolation. Obviously, the concentration of the protein is critical in determining solubility. It is apparent that the solubility of the protein in the cortex is greater than that in the nuclear region. Values have been reported from this laboratory (80) indicating only small quantities of water-insoluble protein in the very young human lens, which increase to approximately 15% by the age of 20. With further aging the values increase so that lenses in their sixties manifested approximately 50% water-insoluble protein (80,81). Kramps et al. (82) have reported similar values; an earlier study by Mach reported somewhat less water-insoluble protein (83).

A number of tentative conclusions can be drawn from the studies of lens protein solubility. The water-insoluble fraction in very young lenses represents the membrane and matrix components. With aging the major contribution to the water-insoluble fraction must come primarily from the crystallins, since no other group of proteins is large enough to account for the substantial increase in this fraction. The generation of water-insoluble protein represents a real change in the chemistry of the protein and not an artifact due to the isolation procedure. It does not seem likely that there are present in a relatively transparent older normal lens the levels of water-insoluble protein that are observed. It is probable that the protein is in a soluble gelatinous state in the lens, but that with the addition of fluid and the breakdown of native protein interactions, insolubilization occurs. It is not possible to delineate at this time with precision which of the posttranslational changes observed with the lens protein contributes to the loss of solubility.

Unlike the bovine lens, where studies generally span lens from 3 months to 8 years, aggregation to high-molecular-weight (HMW) aggregates in aging human lenses involves not only α-crystallin but the other crystallins as well (84–87). The size spectrum included in this fraction varies from a few million to more than 50×10^6 d. There are a number of unfortunate aspects to the use of the term "HMW aggregates." Obviously, large protein aggregates have limited solubility and can be removed from a preparation by centrifugation. Thus, it is necessary to impose an operational definition for this fraction. Initially, HMW aggregates referred to the noncovalent interactions of α-crystallin and then the description was extended to HMW aggregates of the normal human lens, which are primarily noncovalent and contain all the crystallins. However, as will be discussed in a later section, covalent HMW aggregates are found in cataract (72,88–90). It is thus important to designate the aggregates as covalent or noncovalent. In our laboratory the water-insoluble fraction is defined as the pellet obtained when 0.7–1.0 g (wet weight) of lens, homogenized in 10 ml of 0.1 M KCl, 0.01 M Tris, pH 7.6, is centrifuged at $9,900 \times g$ for 15 min at 4°. The HMW aggregate fraction is then obtained by an additional centrifugation of the supernatant at $59,400 \times g$ for 15 min (87). By resuspending this precipitate and running it through gel filtration columns, it can be shown that the $59,400 \times g$ pellet is composed of material greater than 50×10^6 d. It has been suggested that the HMW species is a precursor to the water-insoluble fraction. This viewpoint is supported by the observation that in bovine as well as normal human lenses the amino acid composition and polypeptide profiles are qualitatively similar. Of course, the membrane and matrix components, which are always a part of the water-insoluble fraction in normal lenses, distort the results but with increasing age represent an ever-smaller part of the material.

Aside from the more apparent changes in protein solubility and generation of the HMW aggregates, there are a number of other posttranslational changes that should be considered in the normal aging lens before discussion of the cataractous condition.

Although early studies of the oxidation of thiol groups in young normal human lens protein indicated little disulfide formation in either water-soluble or water-insoluble protein, with aging appreciable amounts of thiol have been reported to be oxidized (91,92). It is now apparent that much of the oxidation observed in these studies was a result of the isolation procedure (93). If a thiol-reacting reagent, such as iodoacetic acid, is employed, then almost no disulfide is observed (81,94). Since water-insoluble protein continues to be observed, it can be concluded that the formation of this material in the normal human lens is not associated with disulfide formation. Investigation of methionine oxidation to methionine sulfoxide also suggests a similar pattern (94): no oxidation in young normal human lenses and only slight oxidation in protein isolated from older lenses. If the protein from older normal lenses is fractionated, a most interesting result is observed. Only in the intrinsic membrane fraction is there disulfide formation, with 60% of the thiol in this form. Furthermore, there has been a change in protein conformation, so that whereas in protein isolated from the membrane of young lenses all the thiol is buried, in older lenses the nonoxidized thiol is exposed. In a similar manner approximately 20% of the methionine is oxidized to the sulfoxide. These studies would suggest that the oxidation of normal lens proteins occurs only at the membranes of the cells and that it is an age-related phenomenon that is probably accompanied by considerable unfolding of the protein.

It is now rather well established that the racemization of protein amino acids can be used to measure the aging of the protein, and where little protein turnover occurs precursor relations and the age of the tissue can be established (95,96). Masters et al. (97,98) and Garner and Spector (99) found that only protein aspartic acid in the lens racemizes to a significant extent and that racemization proceeds at a constant rate of 0.14%/year in the inner region in both normal and cataractous lens. Garner and Spector (99) found that the D/L Asp ratio of the water-soluble fraction of 0.036 remained constant with age but the ratio in the water-insoluble protein fraction increased at a linear rate of approximately 0.17%/year. Since the level of racemization of the water-soluble fraction does not change and the D/L ratio of the water-insoluble protein increases at a constant rate, the results suggest a constant rate of insolubilization of lens protein, which indeed has been observed. It is possible that in normal lenses the racemization process may contribute to the transformation of native protein to high-molecular-weight and water-insoluble components. The unfolding of the native structure, exposing thiol and other groups, may in part be due to perturbations caused by L-to-D transformation at critical sites.

Studies with the red cell have established that nonenzymatic glycosylation of protein is dependent on glucose concentration (100). It has been shown with hemoglobin that the NH_2-terminal residue of the B chain of the protein reacts with glucose to form glucosylated hemoglobin A_{1C}. The higher concentration of this form of hemoglobin in diabetes is caused by the elevation of plasma glucose levels. The low pK of the terminal amino group makes this a preferred site for condensation with glucose via a Schiff base and an Amadori rearrangement forming the *N*-deoxyfructosylvaline derivative. In the lens most proteins have a blocked N-terminal. The ϵ-NH_2 group of lysine can also participate in such a reaction, but because of its higher pK is not as favorable a site. The extent to which this reaction occurs is usually determined by utilizing tritiated borohydride, reducing the deoxyfructosyl to the alcohol with the incorporation of one nonexchangeable tritium atom. Cerami's laboratory has suggested that nonenzymatic glycosylation is a factor in both aging and cataract formation (101). A problem in evaluating results obtained with tritiated borohydride is that the reduction is not specific: any group susceptible to borohydride reduction may incorporate tritium. Analysis of tritiated borohydride-reduced protein of human lens suggests that approximately 60% of the incorporated tritium represents true glucosylated products (102). The results indicate that from 0.1 to 0.2 mol of tritium are incorporated per mole of protein monomer of 2×10^4 d, representing modification of up to 2% of the lysine groups. It is interesting that no differences were obtained between normal and diabetic subjects. It is difficult to evaluate the impact of this low level of glycosylation upon the protein structure. An average of 1 macromolecule in 10 may be modified at a single site. However, it is conceivable that modification may occur preferentially with macromolecules that are unfolded where ϵ-NH_2 groups are more readily available, leading to a higher level of glycosylation. Examination of the population of polypeptides that might be considered to represent degraded material, such as the water-insoluble, HNW, and the 1×10^4 d fraction, failed to indicate any preferential glycosylation.

Although the level of glycosylation is low, the question has been raised of whether there is increasing glycosylation with aging. Racemization studies with lenses in the 65–72-year-old range indicate that the oldest lens polypeptide fraction is the 10,000 d fraction isolated from the water-insoluble fraction, which has a D/L ratio of 0.24 compared with the relatively young water-soluble fraction, where the D/L ratio is approximately 0.04. Based on such observations, it can be tentatively concluded that the average age of the macromolecules in the soluble fraction may be as much as 40–50 years younger than the 10,000 d fraction. Nevertheless, the level of glycosylation does not differ significantly between the water-soluble fraction, 0.18 mol of ^3H per mole protein, and the 10,000 d fraction, 0.12 mol of ^3H per mole protein.

These results differ somewhat from those of Chiou et al., who examined bovine lens preparations (103). They found that there was an approximately 30% greater incorporation in protein fractions from 6-year-old than from

2-year-old lenses. The reason for this difference is not immediately apparent. However, it has been reported that γ-crystallin isolated from 4-month-old calves already has one-half the level of glycosylation observed in 50- to 70-year-old human lens protein (104). This observation raises a number of questions. Is there glycosylation occurring during the synthesis of the protein? Is the keto sugar an intermediate state, and does further reaction of the Maillard type occur leading to nonreducible products? Is there a species-dependent rate of glycosylation?

It is now apparent that significant proteolysis occurs in the lens (105,106). This has been shown in the human lens by the accumulation of low-molecular-weight polypeptides (the approximately 10,000 d range proteins) with aging (80). This fraction is not found in very young lenses. Although little work has been conducted with the human lenses, it is apparent from work in other species, particularly cow, that a variety of proteolytic activities are present. Hoenders and his group have extensively studied posttranslational modification of the α-crystallin polypeptides and have shown a number of specific cleavages of both the A and B chains (34,35). The degradation in most cases occurs in the C-terminal region, usually at a Ser-Ala linkage, as well as Ser-Ser, Thr-Ala, and Asn-Glu bonds. It has been also suggested that some of these cleavages may occur by nonenzymatic reactions.

The lens contains two types of proteolytic activity, an exopeptidase activity resulting in the cleaving of one residue at a time from the N- or C-terminal and an endopeptidase activity, which cleaves internal bonds. It is well established that the lens contains a remarkably high concentration of an aminopeptidase that requires a free N-terminal (107,108). There had been some question concerning the presence of the enzyme in human lenses, since little activity had been observed. However, immunochemical studies by Taylor and Daims suggest that the enzyme is present at levels similar to that observed in the bovine lens but is not active (109). It is not known whether the enzyme can be activated or is in an irreversibly denatured state.

Endopeptidase activity has been reported by a number of investigators (105, 110–112). Although partial purification has been reported it has not as yet been possible to completely characterize the endopeptidase activity. The enzyme studied first by van Heyningen and Waley (105) is metal ion dependent and has its maximum activity at neutral pH. Wagner et al. report that the lens β_2-crystallin fraction is the best substrate and γ-crystallin is not hydrolyzed (111). Since γ-crystallin has been shown to be extensively degraded in human lens preparation, it is clear that other endopeptidases must be present. Thus, it is not surprising that Tse and Ortwerth have found a number of proteases that can be activated by ultraviolet (UV) light or by a preincubation period (113). In this respect, it is interesting to note that a lens protease inhibitor has been observed that apparently is released during the activation incubation (112,113). In the normal lens the activity of proteases is very low. The situation reflects the low level of protein metabolism in most of the lens—hardly any synthesis and negligible degradation.

As with other aspects of lens biochemistry, the lipid story is also somewhat unusual (133-135). Approximately half the lens lipid is cholesterol, and spingomylein is the major phospholipid. The fatty acids are predominantly saturated (136,137), with only approximately 2% polyunsaturated (138). The lens membrane has been found to be very rigid, probably because of its high cholesterol content and the low concentration of unsaturated fatty acids (139). The membranes of the inner regions of the tissue have less fluidity than the cortical region, reflecting the increase in cholesterol content found in this part of the lens. The literature suggests that the membranes of the fiber cells are not homogeneous, varying in both chemical and physical properties with their location in the tissue.

METABOLISM AND OXIDATION-REDUCTION REACTIONS

Lens sugar metabolism has been extensively studied, and a number of reviews summarizing the work have appeared (114,115). As noted earlier, almost all coordinated metabolic activity is located in the periphery of the tissue. This is apparent on a morphologic level, where intact cell nuclei are found only in the epithelial layer and mitochondria and microsomes are abundant only in the outer region of the tissue. As might be anticipated from such observation, as well as the low oxygen tension in the region of the tissue, the citric acid cycle has a minor role in sugar metabolism. It is estimated that approximately 5% of the glucose metabolized goes through this pathway. Because of its much greater efficiency in producing ATP, the citric acid cycle does result in the production of an appreciable amount of ATP, yet addition of fluoroacetate causes little effect and incubation of the lens in N_2 causes no change in amino acid incorporation, ATP + ADP levels, or alteration in Na^+/K^+ ratios. The major route for the metabolism of glucose is anaerobic glycolysis, with the end product, lactate, eliminated into the aqueous fluid. Approximately 80% of the glucose is metabolized by this route. The remaining 10–15% of the glucose is metabolized by the hexose monophosphate shunt, making this cycle remarkably active in comparison with other tissues. The hexose monophosphate shunt does not appear to be an ATP generator but produces new metabolites, particularly pentoses, which are utilized in biosynthetic reactions. The shunt also contributes a reducing capability through the reduction of NADP to NADPH.

The reducing capacity of the lens as expressed through reduction of NADP may be of importance in protecting the tissue from oxidative insult. Studies by Giblin and Reddy show that the levels of pyridine nucleotides vary somewhat from species to species (116). In the human lens, the values for NAD + NADH of 260 nmol/g wet weight are about half of the levels found for rat, 566 nmol/g wet weight. Human values for NADP + NADPH of 21 nmol/g wet weight are approximately equal to rat. The predominant human pyridine nucleotide is NAD, equal to 208 nmol/g wet weight. In the human the ratio of NAD to NADH is about 4; NADP has a value of 15 nmol/g wet weight and a ratio of NADP to NADPH of 1.4. Studies on the distribution of the pyridine nucleotides in rabbit lenses indicates that the concentration of NADP + NADPH is approximately 25-fold greater in the epithelium than in the cortex, with little change in the ratio of the oxidized to the reduced form, but NAD + NADH are approximately 5-fold greater in the epithelium. Surprisingly high values similar to those in the cortex were found in the nucleus, although more reduced NADH was observed.

Perhaps more important than the steady-state concentration of the reduced pyridine nucleotides is the capacity of the lens to generate a reducing potential. It can be assumed that the lens utilizes approximately 5 μmol of glucose per hour per gram wet weight (based on studies with the rat lens) and that NAD and NADH are predominantly involved in glycolysis and the reduction of pyruvate to lactate (115,117). The reducing capacity available would then come primarily from the glucose metabolized via the hexose monophosphate shunt, generating 2 mol of NADPH for every glucose metabolized. If 10% of the glucose was metabolized by this route, then 1 μmol/hr of reducing capacity would be available. Since only approximately 25 nmol of NADP(H) is present, this would suggest that the lens utilizes approximately 1000 nmol of reducing capacity per hour per gram wet weight. It does not store its reducing capacity in the form of NADPH. Obviously, some of the reducing capacity is used for the synthesis of new compounds, but much of it is utilized to maintain compounds in a reduced state.

The lens contains a remarkably high level of glutathione (118–120). This thiol tripeptide is synthesized by the lens and is found almost entirely in the reduced form. It is present in dramatically high levels in the epithelium, where a concentration of 60 μmol/g wet weight has been reported (120). In the epithelium there is more glutathione thiol than protein thiol. The concentration decreases in the cortex and is much lower in the nucleus, resulting in average values of approximately 10 μmol/g wet weight for the total lens. In the entire lens, glutathione represents approximately one-fifth as much thiol as the protein. It is interesting that the systems protecting the lens from oxidation are so efficient that essentially no protein or glutathione disulfide is observed in the normal lens.

How then does the lens succeed in maintaining its protein and glutathione in a reduced state? There are two approaches available to the tissue. The insulting agents can be detoxified, and if oxidation occurs repair can be initiated by reductive systems. The lens has the usual group of enzymes designed to detoxify deleterious oxidizing agents. Catalase is present in the lens and will detoxify H_2O_2, forming H_2O and O_2 (121). This enzyme is present in relatively low levels and is not efficient in removing small amounts of H_2O_2 because of its relatively high K_m. Another enzyme present in the lens, which will detoxify H_2O_2, is glutathione peroxidase (122-125).

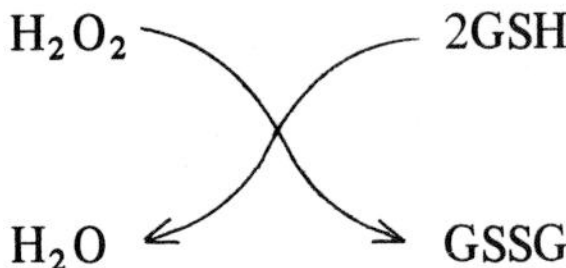

In this reaction glutathione is a cofactor. This selenium-containing enzyme has a low K_m for H_2O_2 and is also capable of hydrolyzing hydroperoxides.

Superoxide can be detoxified by superoxide dismutase, which has been shown to be present in the lens. For other potent oxidizing compounds, such as singlet oxygen and the hydroxyl radical, no enzymatic defenses are known. Glutathione is believed to be a key compound involved in the oxidative defense of the lens (120). The oxidized glutathione formed in such reactions as those involving glutathione peroxidase can be reduced by glutathione reductase (126), an enzyme

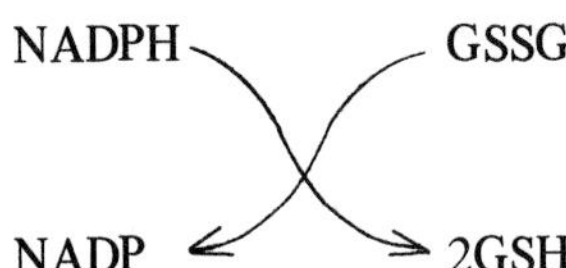

found in the lens that utilizes NADPH as a cofactor. Thus, the reducing capability of the hexose monophosphate shunt can be linked to the maintenance of reduced glutathione. Although the role of glutathione with respect to glutathione reductase is clearly delineated, its relation to reduction of disulfides is ambiguous. Reactions involved with detoxification of a potent oxidant are more easily defined than reactions involved with repair of oxidative damage, such as disulfide formation. It is possible that glutathione is capable of spontaneously reacting with protein disulfides in the following manner:

$$PSSP + GSH \longrightarrow PSSG + PSH$$

and that the mixed disulfide could then be reduced with a second molecule of glutathione with glutathione reductase and NADPH driving

$$PSSG + GSH \longrightarrow PSH + GSSG$$

the reaction to the right. It would also appear that glutathione reductase may react directly with the mixed disulfide (127). The observation that glutathione reductase is still active in cataract and that the hexose monophosphate shunt is operational but mixed disulfides are present has caused some investigators to conclude that this mechanism may not be of primary importance for the reduction of lens protein disulfides.

Are there other mechanisms by which protein disulfides can be reduced? Recently, attention has been drawn to other types of reactions that may be involved in reduction of disulfides. An enzyme system of particular interest with respect to disulfide reduction is that of thioredoxin (128,129). The system consists of thioredoxin, a dithiol protein of 12,000 d, thioredoxin reductase, a flavin adenine dinucleotide containing enzyme, and NADPH. The system is highly effective in reducing protein disulfide bonds. It operates in the following manner:

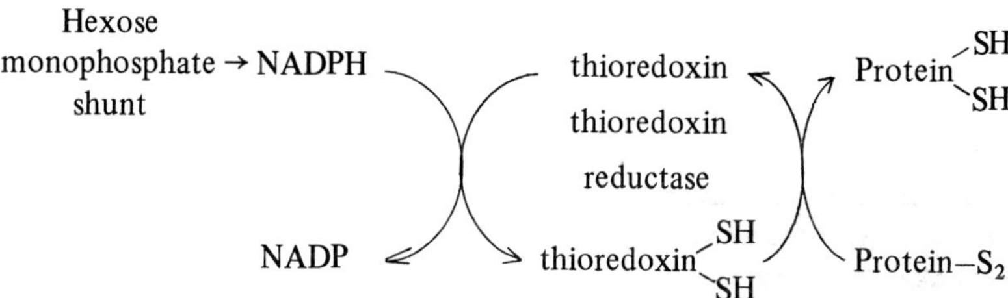

Thioredoxin is a relatively powerful reductant (130) with an $E_0 = -0.26$ V at pH 7.0 for the polypeptide isolated from *Escherichia coli*, in comparison with -0.33 V for dithiothreitol and -0.25 V for glutathione. Thioredoxin has been shown to effectively reduce disulfides in a number of proteins, such as insulin, choriogonadotropin, and fibrin. The polypeptide can act as an enzyme under appropriate conditions, catalyzing, for example, the reduction of insulin by dithiothreitol. Thioredoxin requires dithiols for reduction. It is not reduced by monothiols, such as glutathione. This laboratory has recently obtained evidence for the presence of both thioredoxin and thioredoxin reductase in calf lenses, particularly in the epithelial and the outer cortical region.

The actual function of this oxidation-reduction system is confused, since it appears to be involved in a number of apparently unrelated reactions, such as the oxidation of ribonucleotides to form deoxyribonucleotides, and is a required subunit of phage T7-induced DNA polymerase. Thioredoxin has been shown to be involved with the regulation of fructose-1,6-diphosphatase in photosynthesis, and it has been suggested that insulin activity is regulated by thioredoxin. It is present in most tissues, and in calf liver its concentration has been estimated to be of the order of 15 μM. It is attractive to consider that thioredoxin may be a key system for regulating protein disulfide redox reactions and protecting protein thiol from oxidative insult.

Another set of enzymes involved in thiol redox reactions are the thiol transferases (131). These enzymes, like glutathione reductase, catalyze thiol-disulfide exchange and also utilize glutathione.

$$PSSP + GSH \longrightarrow PSSG + PSH$$

$$\text{thiol}$$
$$\text{transferase}$$

$$PSSG + GSH \longrightarrow GSSG + PSH$$

The oxidized glutathione would then be reduced by glutathione reductase and NADPH coupling the hexose monophosphate shunt to reduction of the protein disulfide. One of the thiol transferases has been isolated from rat liver. The enzyme has a molecular weight of 11,000 d and a pH optimum of 7.5 and has been shown to reduce disulfide bonds of a number of proteins. However, disulfides in native proteins are relatively resistant to reduction. Thus, it is conceivable that only disulfides that have been formed by oxidative insult will be accessible to these enzymes. It is known that the liver glutathione concentration varies in a diurnal manner that involves equilibrium with mixed glutathione protein disulfides, and it is conceivable that thiol transferases are involved in regulating such reactions. It should be noted that there appear to be a number of thiol transferases, some located in the cytosol, some membrane bound. Preliminary experiments with the lens from this laboratory suggests that, possibly, a low level of thiol transferase activity is present in this tissue.

Not only thiol but other groups, such as the S-CH$_3$ in methionine, may also be subject to oxidative damage, forming methionine sulfoxide. Recently it has been reported that an enzyme capable of reducing protein methionine sulfoxide is present in the human lens (132). The greatest activity is present in the epithelium. The enzyme appears to be partially bound to the water-insoluble fraction but can be released by repeated washing. The cofactor requirements for the enzyme relate it to the disulfide redox reactions discussed above. Monothiols, such as mercaptoethanol and glutathione, are ineffective cofactors. However, dithiothreitol and the thioredoxin system are highly effective. Thus, the overall reaction may be described as follows:

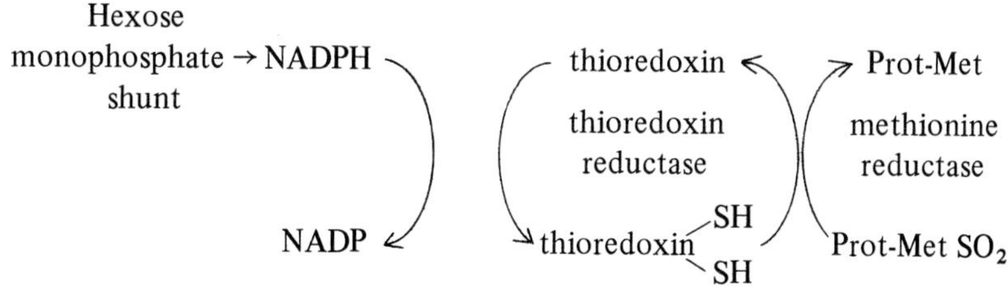

A simple assay utilizing labeled *N*-acetylmethionine sulfoxide has been developed that facilitates monitoring of the enzyme.

CATARACT

Cataract is usually defined as an opacification of the lens. This simple definition obscures the numerous problems involved with the delineation of cataract for experimental purposes. First, it is rare that the entire lens becomes opaque. With human cataract one occasionally finds a lens that has been extracted because of a small opacity in the posterior subcapsular region along the optical axis. It is possible that such a lens may have a normal chemistry, with pathologic changes limited to a small region of the tissue. In some cases there is the appearance of total opacity of a particular region, whereas it is in fact only a shell of opacity surrounding a clear interior. In most cases, even in regions of opacity, there are transparent areas. Thus, examining the entire lens or even a generally opaque region may give results that do not accurately reflect the chemistry of the opacity. The problem is further compounded by questions concerning the diversity of factors that cause loss of transparency. Indeed, it can be argued that opacity is an end result, the last step of a complex process. A transparent lens is not necessarily a healthy lens. It is probable that significant change in lens chemistry must occur before loss of transparency is observed. Indeed, since the lens depends on an actively metabolizing outer shell for homeostatic maintenance, it is conceivable that an insult to this region may cause opacity elsewhere in the tissue. Still another problem involved with cataract classification is that of color. The lens not only yellows with age but in some cases becomes a dark brown. The pigmentation decreases the amount of light passing through the tissue, specifically absorbing certain wavelengths. What is the relation of pigmentation to opacity? It is possible to have a darkly colored lens that is clear and an opaque lens that is hardly colored. Thus, at present the relation between color and transparency is confused. This has led to some disagreement with respect to classification. In the United States most investigators use the Chylack system for describing cataract based on a specific localization of the opacities (140). This system appears to be working reasonably well and allows investigators to describe their material in sufficient detail to allow others to use similar material. Another approach used primarily in Europe is based on color and sodium content, with higher sodium levels observed with increasing color (141,142).

Still another problem encountered by the investigator considering human cataract is the freshness of the material. At some research centers cataract surgery is performed at nearby hospitals and thus fresh material may be available for study. With normal lenses a delay is almost always encountered, and caution must be exercised in evaluating results from postmortem eyes. Furthermore, so-called normal lenses require careful inspection, since unreported opacities may be present and patients treated for diseases of other parts of the organism may have been subjected to medications that may affect lens biology.

CHANGES IN PROTEIN STRUCTURE ASSOCIATED WITH CATARACT

One approach to the study of cataract may be directed to the examination of the HMW protein population of macromolecules in the tissue. Theoretical work had suggested that HMW aggregates with molecular weights of 50×10^6 are capable of scattering light and causing cataract (18). In the normal lens the gradual decrease in transparency with aging (80–82) can be attributed to the accumulation of such protein polymers held together by noncovalent forces, but it does not appear likely that such aggregation can explain cataract. What other changes occur with protein in the cataractous lens? Although the literature is somewhat confused with respect to actual amounts, ther is no question concerning an increase in water-insoluble protein with cataract development (80). In mature cataracts there is almost no water-soluble protein. Initially, it was found that in normal lenses the formation of HMW aggregates and water-insoluble protein occurred predominantly in the nuclear region and did not involve covalent linkage of the polypeptide monomers. However, in cataracts, the formation of water-insoluble protein is accelerated and, as will be discussed below, involves covalent linkage of monomers and is not limited to the nuclear region. It would appear that most structural protein components of the lens are involved in the transformation. The discrepancies between different laboratories concerning the amounts of material involved in this process are due to a number of differences in experimental conditions. In the early work, absorbance at 280 nm was used to estimate the concentration of HMW aggregates. However, because of the light scattering caused by these giant aggregates, the initial estimates were approximately twofold too high. Possibly the most accurate determinations are now based upon amino acid analysis.

Another factor that has contributed to differences in the estimate of water-insoluble and HMW protein is related to the method of isolation (143). Solubility is, of course, a function of concentration. It has been shown that repeated washing of the water-insoluble fraction from normal lenses will solubilize a large fraction of the material. Insoluble protein can be defined as protein which is no longer soluble under conditions where it had previously been in solution. Thus, in the isolation of the water-insoluble material from lens homogenates, attention to pH, volume, and salt concentration is important. HMW protein is found in both the soluble and insoluble fraction. In mature cataracts it is predominantly covalently linked and is to a large extent located in the water-insoluble fraction. There are two types of linkages involved in the covalently bound HMW protein fractions. One involves disulfides and appears to be the predominant class. The second type contains other kinds of covalent linkages, most of which have not as yet been defined.

It can be argued that the quantity of water-insoluble and HMW protein in cataract is not a sufficient condition to cause opacity, since some older normal lenses have a greater amount of water-insoluble protein than severe cataracts obtained from younger individuals (80). This observation would suggest that if the insoluble protein is contributing to the opacification of the tissue, it involves not simply the quantity but a change in the quality of the protein, perhaps related to refractive index. The proteins become more effective light-scattering units. This observation has stimulated an investigation of the nature of the water-insoluble and HMW protein of human cataracts.

If the polypeptide composition of the water-insoluble protein fraction from normal and cataractous lenses is examined, similar profiles are obtained on sodium dodecyl sulfate (SDS) polyacrylamide gels (90). Furthermore, reduction does not appear to alter the pattern significantly. Similar patterns are found with the total water-insoluble material or fractions of the total solubilized with solvents, such as urea and guanidine. If a different type of support is used for electrophoresis, one containing agarose as well as polyacrylamide, a dramatic difference is observed (90). The normal protein now does not appear to contain a HMW species: all the material goes into the gel. Thus, the bonds holding the large protein aggregates formed in the normal aging lenses are weak, noncovalent linkages disrupted by such agents as detergents, urea, and guanidine. It is probably these aggregates that contribute to the normal age-dependent changes in transparency. The binding of the polypeptides of such aggregates to poly-acrylamide but not to agarose plus polyacrylamide is not understood but is indicative of a modification in protein structure. Now, with the protein from cataractous lenses, the band at the top of the gel remains; the protein polymers present in this region of the gel are not dissociated by detergent. Only upon reduction with such compounds as mercaptoethanol or dithiothreitol does this band disappear. Thus, this material is held together by bonds that are broken upon reduction. When the bonds are cleaved the released polypeptides are cap-able of moving into agarose-polyacrylamide but not simply polyacrylamide gels.

Utilizing 7 M urea as a dissociating agent, it is possible to dissolve a consider-able proportion of the water-insoluble protein (90). The protein thiol groups can be alkylated to prevent disulfide formation. This material then can be fractionated on the basis of size, utilizing gel filtration columns, with the largest protein passing through the column first. The normal lenses contain much less HMW peak 1 material. Isolation of the fractions and analyses by SDS agarose-polyacrylamide gel electrophoresis with and without a reducing agent indicates that the peak 1 fraction from cataract contains bonds that upon reduction release polypeptide chains. Upon reduction, the peak 1 material from normal lenses has approximately the same pattern as before reduction, but the cataract HMW peak releases material in the 20 and 43 kd range, some low-molecular-weight material, and a streaky component. It is most likely because of the characteristics of the reducing agents used to break up the HMW aggregate that disulfide bonds stabilize the aggregates. Comparable isolated fractions from normal and cataractous lenses were therefore analyzed for disulfide content. The large aggregates isolated from the cataract preparations were found to be rich in disulfides, but there are essentially no disulfides in the fractions from the normal lenses. There is also an increase in thiol, detected after reduction, in the material isolated from cataract, suggesting disulfide formation with other constituents, such as glutathione.

Thus, these experiments strongly suggest that a new type of aggregate is formed in cataract, one stabilized by covalent disulfide bonds. With increasing opacity, there appears to be increased accumulation of this type of giant macromolecule. These studies confirm and extend previous observations indicating that oxidation of thiol groups accompanies cataract formation (91,144).

On the basis of size, the aggregates are composed of three major populations of polypeptides of approximately 43, 20, and 10 kd and the nondescript streaky material (90). The 20 kd range fraction would be expected to contain the polypeptides of the crystallins. Preliminary evidence suggests that the 10 kd fraction is heterogeneous in terms of size and composition and contains breakdown products of the crystallins. The 43 kd fraction is puzzling. Studies indicated that, when the lens was homogenized in the usual manner by grinding the tissue, this polypeptide fraction was present in both the water-soluble and insoluble material, but if the lens protein was obtained by hypotonic shock, then it remained primarily with the water-insoluble fraction (72,145). Treatment with EGTA, a calcium chelator, also released this material. Such results suggest that this fraction is normally associated with the insoluble components and may involve Ca^{2+} as a stabilizing factor. Since isolation of a component from the water-insoluble fraction is difficult, the 43 kd polypeptide was released into the water-soluble fraction and then purified by a number of gel filtration columns (72). The 43 kd fraction was then injected into rabbits to prepare an anti-43 kd polypeptide preparation. As expected, the antibody appeared to react quantitatively with the 43 kd polypeptide isolated from the urea-solubilized water-insoluble material and the HMW disulfide-linked aggregates. Furthermore, when the anti-43 kd polypeptide was chemically bound to a gel matrix (Sepharose-4B) and the water-insoluble HMW disulfide-linked fraction (solubilized by modification with citarconic anhydride) passed through such a column, almost all the HMW species were bound to the column. A similar column without the antibody did not bind the aggregates. Thus, this observation would suggest that essentially all disulfide-linked aggregates contain the 43 kd polypeptide fraction.

To get an idea of the distribution of this polypeptide species in the lens, indirect staining immunofluorescent techniques were employed utilizing the anti-43 kd preparation and fluorescein-conjugated goat antiserum to rabbit immunoglobulin (73). In both longitudinal and cross sections the fluorescence was found primarily in the vicinity of the lens fiber membrane. Antibody to total water-soluble protein gave fluorescence throughout the cell.

Thus, it appeared from this work that the 43 kd polypeptide fraction is present in all S-S–linked aggregates, can be released by reduction, and appears to be an extrinsic membrane fraction. Such a component is relatively weakly bound to the surface of the cell membrane, in contrast to an intrinsic membrane component, which is embedded in the membrane and strongly linked to it. The work suggests that the formation of the disulfide aggregates may begin at the membrane.

In cataract it is difficult to isolate a pure membrane fraction because of the large amount of water-insoluble protein that remains with the membrane. However, if the preparation is citraconylated, a reaction where the amino groups are modified, becoming carboxyl residues, then all nonmembrane-associated components are solubilized and can be separated from the membrane (64). Another approach utilizing weak alkali appears effective with animal lenses (65).

When this citraconylation process was used to examine membrane fractions from normal and cataractous lenses, a surprising result was obtained (146,147). In nuclear cataracts, approximately 50% of the fiber cell protein remained bound to membrane isolated from the nuclear region, but in normal or transparent regions and in cortical cataracts this did not occur. On reduction an interesting effect was observed. The major intrinsic polypeptides usually associated with the membrane, the 26 and 22 kd polypeptides, are released and can be detected by conventional SDS polyacrylamide gels. However, to visualize the nonmembrane components, the membrane was citraconylated, reduced, and removed by centrifugation. When the supernatant was examined upon agarose-polyacrylamide gels, major bands were detected of approximately 24 kd and in the 10 kd range, as well as a broad minor band of approximately 43 kd. Excellent agreement in amino acid composition was found between the released polypeptides and one of the lens cytosol proteins, γ_H-crystallin (148). That the major polypeptides linked to the membrane are indeed γ-crystallins or derived from γ-crystallin was confirmed by sequencing the amino-terminal region of γ-crystallin and the polypeptides released by reduction (148). Immunochemical data utilizing Ouchterlony immunodiffusion with anti-γ also supported this conclusion and also indicated that γ-crystallin was in HMW disulfide-linked aggregates isolated from the other types of cataract. It can be concluded that γ-crystallin and its breakdown products are generally present in disulfide-linked aggregates.

Using an antibody to the major intrinsic polypeptides of the membrane, it can be shown that HMW disulfide-linked aggregates also contain intrinsic membrane polypeptides (149). Furthermore, analysis of the distribution of lipid indicated that there was a clear loss of lipid from the membrane fraction but not from the whole lens. Thus, the finding of intrinsic membrane polypeptides associated with HMW disulfide-linked aggregates that can be separated from the intact membrane and the implied presence of lipid in such fractions suggests a general breakdown of membrane in certain types of cataract.

There appear to be two types of aggregates. In nuclear cataracts γ-crystallin and perhaps other cytosol polypeptides remain disulfide linked to the membrane; in cortical cataract complete rupture of the polypeptide membrane complex occurs, and segments of the membrane break off and remain with the disulfide-linked aggregates in the cytoplasm. In both cases it seems that oxidation is a required reaction for the formation of the aggregates.

Analyses of the oxidation of cysteine and methionine in lens protein with aging and cataract have also been undertaken (94). To prevent oxidation from occurring during the isolation of the lens proteins, the tissue was homogenized under N_2 in the presence of iodoacetamide, a reagent that will modify thiol groups. Thus, all exposed thiol groups would react with iodoacetamide, forming carboxyamidomethylcysteine before isolation of the protein. The buried cysteine groups were then exposed by denaturing with an appropriate solvent and modified with methyliodide to form methylcysteine. Differentiation could, therefore, be made between exposed and buried cysteines. By reducing the protein and then alkylating the protein thiol groups, disulfides could be determined by noting the increase in alkylated cysteine groups. Classic techniques were used to determine the state of oxidation of methionine. When normal human lenses 60–65 years of age were examined, no oxidation was found in the soluble lens proteins, as exemplified by α- and β-crystallins and the low-molecular-weight fraction containing γ-crystallin and smaller polypeptides. Water-insoluble proteins including the 10 and 20 kd populations, as well as the 43 kd polypeptide, also did not show any oxidation. Much of the thiol was found to be buried. Only the intrinsic membrane fraction showed some oxidation of its methionines and cysteines. Some oxidation of the methionine in the heterogeneous high-molecular-weight fraction was also observed. Examination of younger normal lenses revealed no oxidation in any fraction including the intrinsic membrane protein.

Thus, in young normal lenses there is no protein oxidation, and in older lenses, oxidation is found to a slight extent but is confined to the membrane or membrane-associated components. In contrast, in cataract, extensive oxidation is found in all fractions. In most cases, more than 50% of the methionine and as much as 75–100% of the cysteine was found to be oxidized, the most extensive oxidation occurring in the membrane. In one membrane sample cysteic acid was observed. The overall data suggest that oxidation commences at the membrane in apparently normal lenses and extends to cytosol components in cataract. There is generally somewhat more oxidation in the membrane region in cataract. Furthermore, although in normal lenses much of the thiol is buried, in cataract, these groups become exposed, suggesting a marked change in protein conformation. The data indicate that, prior to extensive oxidation, including disulfide formation, the native structure is perturbed, causing an unfolding of the macromolecule and exposure of oxidizable groups. Although the cause of the unfolding is not clear, it is conceivable that oxidation of accessible groups starts the process.

X-ray diffraction analyses of γ_{II} from the bovine lens by Blundell and his group, discussed elsewhere in this text, have shown that this protein consists of two tightly packed domains with a total of approximately 174 amino acid residues (57). Each domain contains two motifs of approximately 40 residues. The polypeptide chain in each domain is twisted in two right-handed β-pleated sheets. The polypeptide is remarkably rich in thiol groups containing six cysteines. There is also an unexpected distribution of the thiols, namely, one in the C-terminal domain and five in the N-terminal domain. Utilizing the x-ray coordinates obtained from Blundell's laboratory, the static accessibility (i.e., the exposure) of the different functional groups have been examined. It is interesting to note that three of the cysteines are essentially inaccessible in the native form, and the tryptophans are all buried (150). The most accessible groups susceptible to oxidation in γ-crystallin are methionine-65, the C-terminal tyrosine, tyrosine 174, and histidine-113. It is possible that oxidation to one or more of the accessible groups will cause an unfolding of the molecule. Further oxidation may then lead to protein-protein disulfide formation and aggregation resullting in opacification.

Utilizing the atomic coordinates, it has also become possible to examine the electrostatic free-energy stabilization of the N- and C-terminal domains (150). Of particular interest is the finding that the N-terminal domain, which contains most of the cysteines, is much less stable than the C-terminal domain. Thus, slight perturbation of this region of the molecule could cause unfolding and further reaction of such groups. The accessibility of certain groups in the protein, which may destabilize the domains, has recently been examined by photochemically induced nuclear polarization (photo-CIDNP) utilizing a lumiflavin dye (150). Such methodology allows for analysis of the degree of exposure of His, Tyr, and Trp to the dye. The results indicate that, as shown previously, all Trp are buried. His_{113} and $Tyr174$ were completely exposed, and

His$_{14}$ and Tyr$_{62}$ partially exposed. It would, therefore, appear that susceptible groups are available to inciting agents that may destabilize the domains of the protein.

The hypothesis of destabilization and unfolding caused by oxidation of critical groups presumes that oxidation precedes opacification and that disulfide-linked aggregates would be found primarily in opaque regions. If an examination is made of cataracts that are either essentially nuclear or cortical, extensive oxidation is found in both clear and opaque regions (147). But, in nuclear cataracts, only in the nucleus are large protein-membrane aggregates observed, and in cortical cataracts, primarily in the cortex are the HMW disulfide-linked aggregates found. Thus, these observations support the concept that oxidation precedes the aggregation process.

H$_2$O$_2$ AND OTHER OXIDIZING AGENTS: THEIR EFFECTS ON THE LENS

What are the oxidizing agents involved in this process? The observation that the membrane appears to be most susceptible to oxidation suggested that possibly the oxidizing agent(s) might come from outside the lens. To test this hypothesis, an examination was undertaken of the aqueous fluid that surrounds the tissue. Limited results (151) suggest that in some cataract patients there is a remarkable elevation in the concentration of H$_2$O$_2$, an agent capable of oxidizing both methionine and cysteine. Both normal and diabetic human aqueous were used as a control. Because of the very limited material available, diabetics were included, since the initiating cause of this type of cataract is believed to be different from that involved in most senile types of cataract. Normal primate aqueous was also examined. Both populations appear to have normal distributions, with a mean of approximately 26 μM. When 17 patients with differing degrees of opacification were examined, elevated H$_2$O$_2$ levels were observed in 7 of 17 cases, with one sample having a concentration of 663 μM H$_2$O$_2$. Elevated aqueous H$_2$O$_2$ appears to be related to elevated lens H$_2$O$_2$, with the higher value usually found in the aqueous. In some lens samples for which a corresponding aqueous was not available, elevated H$_2$O$_2$ values were also observed. No relation between aqueous H$_2$O$_2$ and ascorbate was observed. The inconsistency in the level of H$_2$O$_2$ in the aqueous from cataract patients may be due either to a fluctuation in H$_2$O$_2$ concentration due to unknown causes or that H$_2$O$_2$ is simply not elevated in certain cataracts. It is not possible to chose between such alternatives.

Is H$_2$O$_2$ in the observed range capable of causing cataract? To answer this question, bovine lenses were incubated in the presence of 1 mM H$_2$O$_2$ (152). Opacification of the superficial cortex became apparent after 18 hr. After 48 hr, an obvious superficial cortical opacity was observed. With a concentration of 0.5 mM H$_2$O$_2$, a somewhat longer period is required to produce an opacity.

Of course, even if one accepts the hypothesis that oxidation is a major factor in cataract development, there is less certainty of what the inciting factors may be. Although H_2O_2 may contribute to the oxidation of the tissue, a number of other candidates should be considered.

Oxygen itself may be an oxidant even though the oxygen tension in the vicinity of the lens is relatively low (153,154). Oxygen has a somewhat unusual chemistry (127). It is a relatively stable compound able to slowly oxidize thiols. However, upon reduction it becomes highly reactive. The oxygen molecule is reduced by a series of univalent reactions producing highly toxic intermediates, such as superoxide, hydrogen peroxide, and the hydroxyl radical, before finally being reduced to $H_2O\cdot$ In biologic systems, reduction of O_2 occurs rapidly on the surface of enzymes and is thus innocuous. However, in other situations, the reduction of oxygen may result in the release of potent damaging radicals. It is ironic that under certain circumstances, reduction of oxygen by reducing compounds, such as glutathione, can produce toxic oxygen species. Thus, it is not surprising that Schocket et al. (155) and others have shown that an increase in the partial pressure of O_2 in the environment of the lens may cause cataract.

Ascorbic acid, which is present in the lens and its environment, may act as both a reductant and as a free-radical scavenger generating H_2O_2 in the process (156). It is also known that ascorbic acid in the presence of light and metal ion or riboflavin reacts spontaneously, forming H_2O_2. Pirie (122) suggests that H_2O_2 detected in the aqueous could be formed from ascorbate. There are also a number of normal metabolic reactions, such as those utilizing oxidases, that give rise to peroxide.

Some of the potent oxidants of oxygen may react with biologic components to form new radicals; thus, superoxide can react with certain polyunsaturated fatty acids, forming lipid peroxides that may then break down to form malondialdehydes. Such compounds are reactive and may react further forming Schiff bases with amino groups of proteins or phospholipids, producing pigmented and fluorescent components. The lens, as indicated previously, has a very low level of polyunsaturated fatty acids that do not appear to be oxidized in cataract, in contrast to protein (138). However, it has been suggested that there are small amounts of malondialdehyde in the lens and increases in this material in cataract (157). It can be argued that high levels of malondialdehyde would not be found since the compound is very reactive. However, if substantial formation of malondialdehyde occurs in cataract, it should be reflected in a significant change in the fatty acid composition of the tissue. This does not appear to occur.

Cataracts have been induced in rabbits and rats utilizing either 3-aminotriazole, an effective catalase inhibitor, ultraviolet light, or x-irradiation (158). It was found that all these treatments produced increased levels of H_2O_2, the only oxidant measured. Some of the oxidative defenses of the lens, such as superoxide dismutase, catalase, and glutathione peroxidase, have been measured after rabbits were treated with 3-aminotriazole, as well as in certain hereditary mouse cataracts and in human cataract expressed in old age. In all cases, a dramatic drop in catalase, superoxide dismutase, and glutathione peroxidases was found. The data would support the viewpoint of decreased enzymatic defenses to oxidative insult and increased levels of oxidants in these cataract. The mechanism by which ultraviolet light and x-ray produce deleterious effects is not well understood. X-ray causes effects similar to photochemical insult, resulting in disulfide formation, insolubilization, and possibly tyrosine destruction, as well as other changes (120). Ultraviolet and other types of photo-oxidation also may be important factors in cataract formation and are discussed in detail elsewhere in this text.

The mechanism by which cataract is produced is not understood. This is true even with diabetic cataracts, where the general outline of the development of the cataract has been known for some time. Although the activities of aldose reductase and the formation of sugar-alcohol initiates the process, the precise manner by which the osmotic insult produces the cataract is still unclear.

The examination of human cataract indicates that the initial insult may occur at or near the membrane. Studies with bovine lenses that have been incubated with H_2O_2 suggest that Na^+,K^+-ATPase is affected very early in the process (159). Active $^{86}Rb^+$ influx can be completely inhibited with little change in membrane permeability. However, the enzyme continues to hydrolyze ATP. Thus, energy is utilized nonproductively. Initially, anaerobic glycolysis appears to increase markedly, and the hexose monophosphate shunt is also stimulated. Yet, in spite of this increased metabolic activity, ATP levels decrease slightly, probably due to the increased hydrolysis of ATP by the Na^+,K^+-ATPase. Although definitive data have not as yet been reported, it would be expected that loss of pump function would lead to significant changes in cation concentration. Increased Na^+ levels have been found in cataract, and Marcantonio et al. have proposed use of Na^+ levels as one of the indicators of the severity of cataract (142). The initial effect of H_2O_2 would appear to be rather specific. Examination of the kinetics of Na^+,K^+-ATPase indicates that only when the enzyme is in the conformation where all ATP binding sites are in a low-affinity form (K+ form) is it possible to obtain a H_2O_2 modification similar to that found with H_2O_2-exposed lenses.

Very recent studies indicate that in human cataract a variety of oxidizing agents may act together. As previously noted, a small but significant concentration of covalent nondisulfide bonds are found in cataracts. Such bonds are not likely to be produced by H_2O_2. However, it can be shown that photochemical insult with light above 295 nm (the cutoff point for the cornea) will produce nondisulfide polymers with γ- or β-crystallins (160). Based on immunochemical analysis, it would appear that some of these aggregates (dimers) are present in the 43 kd population isolated from cataracts.

There is also a possibility that in some cases the initiating pathology that causes cataract may arise in tissues other than the lens. This is suggested by work involving retinal dystrophies. It is well known that the outer segment of the retina is high in polyunsaturated fatty acids, particularly in 22:6 unsaturated fatty acid. Work of Zigler and Hess (161) suggests that oxidative breakdown of such compounds produces active aldehydes, which move through the vitreous and react at the posterior side of the lens, eventually causing cataract.

SOME OTHER FACTORS INVOLVED IN CATARACT

It has been suggested that another transport system that may be of importance in cataract formation is that involved with calcium transport (162). An active pump maintains low Ca levels in the lens. This system also appears to be sensitive to oxidative insult, and increased Ca levels have been reported in human cataract (163). It is possible that the increased levels of Ca may activate transglutaminase, an enzyme Lorand has shown causes changes in the red cell membrane and that may be involved in protein polymerization (164). The enzyme catalyzes an acyl transfer reaction in which the amide of a protein glutamine is removed as NH_3 and primary amines, such as the E-NH_2 group of protein-bound lysine, are acyl acceptors. The polyamines putrescine, spermidine, and spermine are known to be substrates for transglutaminase and could possibly act as cross-linking agents. Recently, it has been shown that these polyamines are present in the lens and are covalently bound to lens protein (165). Whether they act as cross-linkers remains to be determined, but the presence of bis-γ-glutamyl derivatives is suggested by the stoichiometry of isolated polyamine-γ-glutamyl fragments. Whether polyamines are involved or the reaction is directly between lysine and glutamine groups, the potential for protein polymerization is apparent. It is also of interest to note that transglutaminase appears to be a membrane-bound enzyme, possibly explaining the involvement of membrane polypeptides. Thus, oxidative insult to Ca^{2+}-ATPase may lead to a series of deleterious reactions that do not necessarily involve oxidation.

It is apparent from the above discussion that perturbation of a protein, either an enzyme or a structural protein, may lead to further problems. Not only aggregation must be considered but proteolysis as well. Once the protein has been denatured by any given agent, it becomes susceptible to proteolysis because previously buried areas containing sites for proteolytic attack are now exposed. Thus, there are numerous reports of decreases in protein and the appearance of low-molecular-weight degradation products. Of course, some of the loss of protein can be attributed to migration through leaky membrane. This is particularly true of the low-molecular-weight γ-crystallin fraction that has been detected in the aqueous fluid.

The protein breakdown products are excellent candidates for further reactions. It is not surprising to find that a significant fraction of the high-molecular-weight disulfide-linked aggregates are low-molecular-weight polypeptide fragments. In nuclear cataracts, it would appear that much of the 10,000 d polypeptide material linked to the membrane arises from the γ-crystallin fraction (148).

CONCLUSIONS

The argument has been made that oxidation is an initiating factor in the development of cataract. Cataract is primarily a disease of older people. What has changed to make the lens more susceptible to cataract development? It has been noted that metabolism decreases with aging. Thus, the ability to detoxify insulting agents, such as H_2O_2, diminishes. Second, there appear to be a group of spontaneous reactions that make the protein more susceptible to damage. Racemization of aspartic acid can possibly expose groups susceptible to oxidation, making such regions targets for oxidizing agents that are less effectively metabolized. The same argument may apply to nonenzymatic glycosylation, which undoubtedly perturbs local regions of protein structure. Experiments from a number of laboratories suggest that, with aging, deamidation of asparagine and/or glutamine occurs. This conclusion is based upon sequence studies, changes in isoelectric point, and, most recently, specific determination of amide and primary amine groups. Significant changes in charge can also cause localized disruption in structure. It is possible that native low-molecular-weight metabolites, such as tryptophan and its metabolites N-formylkynurenine, 3-hydroxykynurenine, and 3-hydroxykynurenine glucoside and their photoproducts, may bind to protein, accumulating with age. Some of these compounds might then act as photosensitizers, absorbing energy and transmitting it to other regions of the protein with resulting irreversible damage.

Finally, there is the question of the contribution to cataract formation of compounds foreign to the organism. A remarkable variety of drugs is administered to alleviate a broad spectrum of diseases. Some of these drugs are potential photosensitizers and have solubility characteristics that would concentrate them in the membrane fraction. In experiments with human lenses, it is necessary to obtain as complete a medical history of the donors as possible.

It would appear that oxidation in all cataracts is an early event in the formation of cataract. Of course, in certain cataracts, such as of the genetic and diabetic types, it is not the initiating reaction. However, with cataracts found in older age, oxidation may in many cases initiate the overall process. Based upon our present information then it could be suggested that the sequence of events might be as follows: first, oxidation at the cell membrane; next, loss of ion transport immediately resulting in stimulating glucose metabolism, increasing lactate production, and slightly decreasing ATP formation with concomitant changes in ion concentrations resulting in a decrease in protein synthesis. Oxidation of glutathione and initial oxidation of cytosol protein would also be expected. This stage would then be followed by a general decrease in metabolic activity, ATP formation, reducing potential as reflected in NADPH and gluta-thione, and marked increases in membrane permeability, Na^+ and Ca^{2+}, proteo-lytic and transglutaminase activity, and formation of disulfide-linked protein aggregates. Finally, further development of the above trends would lead to cell membrane rupture, loss of low-molecular-weight protein, significant increases in water content, and opacification. The validity of aspects of this concept of cataract development will be more critically examined in the near future with development of new technology to examine the intact lens and the continuation of present-day experimentation.

ACKNOWLEDGMENTS

The writing of this chapter and work from the author's laboratory was sup-ported by grants from the National Eye Institute, National Institutes of Health.

REFERENCES

1. J. J. Harding and K. J. Dilley. Exp. Eye Res., *22*: 1 (1976).
2. S. K. Srivastava (ed.). *Red Blood Cell and Lens Metabolism*, Elsevier/ North Holland, New York, 1979.
3. G. Duncan (ed.). *Mechanisms of Cataract Formation in the Human Lens*, Academic Press, New York, 1981.
4. H. Bloemendal (ed.). *Molecular and Cellular Biology of the Eye Lens*, John Wiley and Sons, New York, 1981.
5. D. S. McDevitt (ed.). *Cell Biology of the Eye*, Academic Press, New York, 1982.
6. C. F. Wannemacher and A. Spector. Exp. Eye Res., *7*: 623 (1968).
7. K. J. Dilley and R. van Heyningen. Doc. Ophthal. (Proc. Ser.), *8*: 171 (1976).
8. O. Hockwin and C. Ohrloff. In *Molecular and Cellular Biology of the Eye Lens* (H. Bloemendal, ed.), John Wiley and Sons, New York, 1981, p. 367.
9. C. Ohrloff. In *Interdisciplinary Topics in Gerontology*, Vol. 12 (H. P. von Hahn, ed.), Karger, Basel, 1978, p. 158.
10. N. S. Rafferty and W. Goossens. Exp. Eye Res., *26*: 177 (1978).
11. T. Kuwubara. Exp. Eye Res., *20*: 427 (1975).
12. R. H. Bradley, M. Ireland, and H. Maisel. Exp. Eye Res., *28*: 441 (1979).
13. Z. Dische. Invest. Ophthalmol., *4*: 759 (1965).
14. H. Maisel, M. Perry, J. Alcala, and P. K. Waggoner. Ophthalmol. Res., *8*: 55 (1976).
15. R. van Heyningen. Exp. Eye Res., *13*: 155 (1972).
16. B. Philipson. Invest. Ophthalmol., *8*: 271 (1969).
17. B. Philipson. Invest. Ophthalmol., *8*: 281 (1969).
18. G. G. Benedek. Appl. Optics, *10*: 459 (1971).

19. F. A. Bettelheim and E. L. Siew. In *Cell Biology of the Eye*, D. S. McDevitt, ed.), Academic Press, New York, 1982, p. 243.

20. J. Horwitz, R. Neuhaus, and J. Dockstader. Invest. Ophthalmol. Vis. Sci., *21*: 616 (1981).

21. Stauffer, C. Rothschild, T. Wandel, and A. Spector. Invest. Ophthalmol., *13*: 135 (1974).

22. A. Spector, L. K. Li, R. C. Augusteyn, R. C. Schneider, and T. Freund. Biochem. J., *124*: 337 (1971).

23. A. Spector. Isr. J. Med. Sci., *8*: 1577 (1972).

24. H. Kramps, A. Stols, H. Hoenders, and K. deGroot. Eur. J. Biochem., *50*: 503 (1975).

25. D. Roy and A. Spector. Biochemistry, *15*: 1180 (1976).

26. D. Roy and A. Spector. Invest. Ophthalmol., *15*: 394 (1976).

27. J. Delcour and J. Papaconstantinou. Biochem. Biophys. Res. Commun., *41*: 401 (1970).

28. J. Delcour and J. Papaconstantinou. J. Biol. Chem., *247*: 3289 (1972).

29. H. Bloemendal. In *Molecular and Cellular Biology of the Eye* (H. Bloemendal, ed.), John Wiley and Sons, New York, 1981, p. 13.

30. L. K. Li and A. Spector. Exp. Eye Res., *19*: 49 (1974).

31. N. T. Yu and E. J. East. J. Biol. Chem., *250*: 2196 (1975).

32. L. H. Cohen, R. A. Schachar, and S. A. Solin. Invest. Ophthalmol., *14*: 380 (1975).

33. A. J. M. Vermorken and H. Bloemendal. Nature, *271*: 779 (1978).

34. F. S. M. van Kleef, W. W. Thijssen, and H. J. Hoenders. Eur. J. Biochem., *66*: 47 (1976).

35. F. S. M. van Kleef, W. W. de Jong, and H. J. Hoenders. Nature, *258*: 264 (1975).

36. G. J. A. M. Strous, A. J. M. Berns, and H. Bloemendal. Biochem. Biophys. Res. Commun., *58*: 876 (1974).

37. H. Bloemendal and G. J. A. M. Strous. In *Lipmann Symposium, Energy, Biosynthesis and Regulation in Molecular Biology*, Walter deGruyter, New York, 1974, p. 90.

38. L. K. Li and A. Spector. Exp. Eye Res., *15*: 179 (1973).

39. A. Spector and E. Katz. J. Biol. Chem., *240*: 1979 (1965).

40. L. K. Li and A. Spector. Exp. Eye Res., *26*: 419 (1978).

41. R. Augusteyn. Unpublished results,

42. A. Spector and C. Rothschild. Invest. Ophthalmol., *12*: 255 (1973).

43. J. A. Jedziniak, D. F. Nicoli, E. M. Yates, and G. B. Benedek. Exp. Eye Res., *23*: 325 (1976).

44. M. Fein, A. Pande, and A. Spector. Invest. Ophthalmol., *18*: 761 (1979).

45. J. Björk. Exp. Eye Res., *3*: 248 (1964).

46. M. K. Mostafapour and V. N. Reddy. Invest. Ophthalmol., *7*: 660 (1978).

47. H. P. C. Driessen, P. Herbrink, H. Bloemendal, and W. W. de Jong. Exp. Eye Res., *31*: 243 (1980).

48. A. F. van Dam. Exp. Eye Res., *5*: 255 (1966).

49. J. Bours and O. Hockwin. Klin. Monatsbl. Augenheilkd., *170*: 51 (1977).

50. J. S. Zigler and J. B. Sidbury. Invest. Ophthalmol., *15*: 673 (1976).
51. F. C. S. Ramaekers, P. L. E. van Kan, and H. Bloemendal. Ophthalmol. Res., *11*: 143 (1979).
52. G. Armand, E. A. Balazs, and M. Testa. Exp. Eye Res., *10*: 143 (1970).
53. J. Horwitz, I. Kubasawa, and J. H. Kinoshita. Exp. Eye Res., *25*: 199 (1977).
54. L. R. Croft. In Ciba Foundation Symposium, Vol. 19, 1973, p. 207.
55. I. Björk. Exp. Eye Res., *9*: 152 (1970).
56. J. W. McAvoy. In *Mechanisms of Cataract Formation in the Human Lens* (G. Duncan, ed.), Academic Press, New York, 1981, p. 27.
57. L. R. Croft. Biochem. J., *128*: 961 (1972).
58. T. Blundell, P. Lindley, L. Miller, D. Moss, C. Slingsby, I. Tickle, B. Turnell, and G. Wistow. Nature, *289*: 771 (1981).
59. C. Slingsby and L. R. Croft. Exp. Eye Res., *17*: 369 (1973).
60. I. Kabasawa, Y. Tsunematsu, G. W. Barber, and J. H. Kinoshita. Exp. Eye Res., *24*: 437 (1977).
61. R. M. Broekhuyse and E. D. Kuhlman. Exp. Cell Res., *19*: 297 (1974).
62. E. L. Benedetti, I. Dunia, C. J. Bentzel, A. J. M. Vermorken, M. Kibbelaar, and H. Bloemendal. Biochim. Biophys. Acta, *457*: 353 (1976).
63. J. Alcala and H. Maisel. Exp. Eye Res., *26*: 219 (1978).
64. D. Roy, A. Spector, and P. Farnsworth. Exp. Eye Res., *28*: 353 (1979).
65. P. Russell, W. G. Robison, and J. H. Kinoshita. Exp. Eye Res., *32*: 511 (1981).
66. J. Horwitz, N. P. Robertson, M. M. Wong, J. S. Zigler, and J. H. Kinoshita. Exp. Eye Res., *28*: 359 (1979).
67. J. Alcala, J. Valentine, and H. Maisel. Exp. Eye Res., *30*: 659 (1980).
68. J. Horwitz and M. M. Wong. Biochim. Biophys. Acta, *622*: 134 (1980).
69. D. Roy. Biochem. Biophys. Res. Commun., *88*: 30 (1979).
70. W. H. Garner, S. R. Leff, and A. Spector. In *Developments in Biochemistry, Red Blood Cell and Lens Metabolism*, Vol. 9 (S. K. Srivastava, ed.), Elsevier-North Holland, New York, 1980, p. 81.
71. P. N. Farnsworth, S. E. Shyne, M. Garner, D. Roy, and A. Spector. Curr. Eye Res., *2*: 81 (1982/1983).
72. A. Spector, M. H. Garner, W. H. Garner, D. Roy, P. N. Farnsworth, and S. Shyne. Science, *204*: 1323 (1979).
73. P. N. Farnsworth, A. Spector, J. R. Lozier, S. E. Shyne, M. H. Garner, and W. H. Garner. Exp. Eye Res., *32*: 257 (1981).
74. W. H. Garner, M. H. Garner, and A. Spector. Exp. Eye Res., *29*: 257 (1979).
75. H. Maisel, C. V. Harding, J. R. Alcala, J. Kuszak, and R. Bradley. In *Molecular and Cellular Biology of the Eye Lens* (H. Bloemendal, ed.), John Wiley and Sons, New York, 1981, p. 63.
76. M. Kibbelaar, A. E. Selten-Versteegen, H. Bloemendal, I. Dunia, and E. L. Benedetti. Eur. J. Biochem., *95*: 543 (1979).
77. G. Y. Mousa and J. R. Trevithick. Exp. Eye Res., *29*: 71 (1979).
78. F. C. S. Ramaekers, M. Osborn, E. Schmid, K. Weber, H. Bloemendal, and W. W. Franke. Exp. Cell Res., *127*: 309 (1980).

79. A. Spector. In *Aging and Human Visual Function*, Alan L. Liss, New York, 1982, p. 27.
80. W. H. Garner and A. Spector. Doc. Ophthalmol. Proc. Ser., *18*: 91 (1979).
81. E. I. Anderson and A. Spector. Exp. Eye Res., *26*: 407 (1978).
82. H. A. Kramps, H. J. Hoenders, and J. Wollensak. Biochim. Biophys. Acta, *434*: 32 (1976).
83. H. Mach. Klin. Monatsbl. Augenheilkd., *143*: 689 (1963).
84. A. Spector, J. Stauffer, and J. Sigelman, Ciba Foundation Symposium, *19*, (New Series), 187 (1973).
85. J. A. Jedziniak, J. H. Kinoshita, E. M. Yates, L. O. Hocker, and G. B. Benedek. Exp. Eye Res., *15*: 185 (1973).
86. J. A. Jedziniak, J. H. Kinoshita, J. H. Yates, and G. B. Benedek. Exp. Eye Res., *20*: 367 (1975).
87. D. Roy and A. Spector. Exp. Eye Res., *22*: 273 (1976).
88. J. J. Harding. Exp. Eye Res., *17*: 377 (1973).
89. R. H. Buckingham. Exp. Eye Res., *14*: 123 (1972).
90. A. Spector and D. Roy. Proc. Natl. Acad. Sci. USA, *75*: 3244 (1978).
91. Z. Dische and H. Zil. Am. J. Ophthalmol., *34*: 104 (1951).
92. M. Testa, M. Fiore, C. Bocci, and S. Calabro. Exp. Eye Res., *7*: 276 (1968).
93. J. J. Harding. Biochem. J., *129*: 97 (1972).
94. M. Garner and A. Spector. Proc. Natl. Acad. Sci. USA, *77*: 1274 (1980).
95. P. E. Hare and P. H. Abelson. Carnegie Inst. Yearbook, *66*: 526 (1968).
96. H. M. Williams and G. G. Smith. Origins Life, *8*: 91 (1977).
97. P. M. Masters, J. L. Bada, J. S. Zigler, Jr. Nature, *268*: 71 (1977).
98. P. M. Masters, J. L. Bada, and J. S. Zigler, Jr. Proc. Natl. Acad. Sci. USA, *75*: 1204 (1978).
99. W. H. Garner and A. Spector. Proc. Natl. Acad. Sci. USA, *75*: 3618 (1978).
100. H. F. Bunn, K. H. Gabbay, and P. M. Gallop. Science, *200*: 21 (1978).
101. V. J. Stevens, C. A. Rouzer, V. M. Monnier, and A. Cerami. Proc. Natl. Acad. Sci. USA, *75*: 2918 (1978).
102. A. Pande, W. H. Garner, A. Spector. Biochem. Biophys. Res. Commun., *89*: 1260 (1979).
103. S.-H. Chiou, L. T. Chylack, Jr., W. A. Tung, and H. F. Bunn. J. Biol. Chem., *256*: 5176 (1981).
104. A. Spector and W. H. Garner. In *Red Blood Cell and Lens Metabolism* (S. K. Srivastava, ed.), Elsevier-North Holland, New York, 1980, p. 475.
105. R. van Heyningen and S. G. Waley. Exp. Eye Res., *1*: 336 (1962).
106. A. M. J. Blow, R. van Heyningen, and A. J. Barrett. Biochem. J., *145*: 591 (1975).
107. A. Spector. J. Biol. Chem., *238*: 1353 (1963).
108. H. Hansen and F. Frohne. Meth. Enzymol., *45*: 504 (1976).
109. A. Taylor and M. Daims. Invest. Ophthal. Vis. Sci. (Suppl.), *20*: 148 (1982).
110. S. S. Tse and B. J. Ortwerth. Exp. Eye Res., *32*: 605 (1981).

111. B. J. Wagner, J. W. Margolis, P. N. Farnsworth, and S.-C. J. Fu. Exp. Eye Res., *35*: 293 (1982).
112. S. S. Tse and B. J. Ortwerth. Exp. Eye Res., *34*: 659 (1982).
113. S. S. Tse and B. J. Ortwerth. Exp. Eye Res., *31*: 313 (1980).
114. R. van Heyningen. In *The Eye*, Vol. I (H. Davison, ed.), Academic Press, New York, 1969.
115. J. F. R. Kuck. In *Biochemistry of the Eye* (C. N. Graymore, ed)., Academic Press, New York, 1970.
116. F. J. Giblin and V. N. Reddy. Exp. Eye Res., *31*: 601 (1980).
117. J. Kuck. Invest. Ophthal., *1*: 390 (1962).
118. J. H. Kinoshita and L. O. Merola. Arch. Ophthal., *56*: 36 (1957).
119. V. N. Reddy. Exp. Eye Res., *11*: 310 (1971).
120. V. N. Reddy, F. J. Giblin, and H. Matsuda. In *Red Blood Cell and Lens Metabolism* (S. K. Srivastava, ed.), Elsevier-North Holland, New York, 1980, p. 139.
121. K. C. Bhuyan and D. K. Bhuyan. Am. J. Ophthalmol., *69*: 147 (1970).
122. A. Pirie. Biochem. J., *96*: 244 (1965).
123. K. C. Bhuyan and D. K. Bhuyan. IRCS Med. Sci., *5*: 279 (1977).
124. S. D. Varma, T. K. Ets, and R. D. Richards. Ophthal. Res., *9*: 421 (1977).
125. K. C. Bhuyan and D. K. Bhuyan. Biochim. Biophys. Acta, *4521*: 28 (1978).
126. R. van Heyningen and A. Pirie. Biochem. J., *53*: 436 (1953).
127. R. C. Augusteyn. In *Mechanisms of Cataract Formation in the Human Lens* (G. Duncan, ed.), Academic Press, New York, 1981, p. 104.
128. T. C. Laurent, E. C. Morre, and P. Reichard. J. Biol. Chem., *239*: 2426 (1964).
129. E. C. Moore, P. Reichard, and L. Thelander. J. Biol. Chem., *239*: 3445 (1964).
130. A. Holmgren. J. Biol. Chem., *27*: 10 (1979).
131. B. Mannervik, K. Axelsson, and K. Larson. Meth. Enzymol., *77*: 281 (1981).
132. A. Spector, R. Scotto, H. Weissbach, and N. Brot. Biochem. Biophys. Res. Commun., *108*: 429 (1982).
133. G. L. Feldman. In *Biochemistry of the Eye* (M. U. Dardenne and J. Nordmann, eds.), Karger, New York, 1968, p. 348.
134. R. M. Broekhuyse. In *The Human Lens in Relation to Cataract*, Ciba Foundation Symposium, (K. Elliot and D. K. Fitzsimmons, eds.), Elsevier, New York, 1973, p. 135.
135. J. S. Andrews. Doc. Ophthalmol. Proc. Ser., *18*: 19 (1979).
136. G. L. Feldman, T. W. Culp, L. S. Feldman, C. K. Grantham, and H. T. Jonsson. Invest. Ophthalmol., *3*: 194 (1964).
137. E. Cotlier, Y. Obara, and B. Toftness. Biochim. Biophys. Acta, *530*: 267 (1978).
138. L. Rosenfeld and A. Spector. Exp. Eye Res., *35*: 69 (1982).
139. L. K. Li, A. Spector, U. Cogan, and B. Schacter. Exp. Eye Res., *34*: 145 (1982).

140. L. T. Chylack. Arch. Ophthal., *96*: 888 (1978).

141. A. Pirie. Invest. Ophthalmol., *7*: 634 (1968).

142. J. M. Marcantonio, G. Duncan, A. R. Bushell, and P. D. Davies. Exp. Eye Res., *31*: 227 (1980).

143. J. J. Harding. In *Molecular and Cellular Biology of The Eye Lens*, (H. Bloemendal, ed.), John Wiley and Sons, New York, 1981, p. 332.

144. J. J. Harding. Exp. Eye Res., *13*: 33 (1972).

145. M. K. Mostafapour and V. N. Reddy. Doc. Ophthalmol. (Proc. Ser.), *18*: 193 (1978).

146. A. Spector and M. H. Garner. In *Red Blood Cell and Lens Metabolism* (S. K. Srivastiva, ed.), Elsevier-North Holland, New York, 1980, p. 233.

147. M. H. Garner and A. Spector. Exp. Eye Res., *31*: 361 (1980).

148. W. H. Garner, M. H. Garner, and A. Spector. Biochem. Biophys. Res. Commun., *98*: 439 (1981).

149. M. H. Garner, D. Roy, L. Rosenfeld, W. H. Garner, and A. Spector. Proc. Natl. Acad. Sci. USA, *78*: 1892 (1981).

150. W. H. Garner, A. Spector, T. Schleich, and R. Kaptein. Curr. Eye Res., *3*: 127 (1983).

151. A. Spector and W. Garner. Exp. Eye Res., *33*: 673 (1981).

152. M. H. Garner, W. H. Garner, and A. Spector. Invest. Ophthal. Vis. Sci. (Suppl.), *22*: 34 (1982).

153. M. Kwan, J. Ninikoski, and T. K. Hunt. Invest. Ophthalmol., *11*: 108 (1972).

154. R. E. Barr and E. L. Roetman. Invest. Ophthalmol., *13*: 386 (1974).

155. S. S. Schocket, J. Esterson, B. Bradford, M. Michaelis, and S. Richards. Isr. J. Med. Sci., *8*: 1596 (1972).

156. S. D. Varma, T. K. Ets, and R. D. Richards. Ophthalmol. Res., *9*: 421 (1977).

157. K. C. Bhuyan and D. K. Bhuyan. In *Oxy Radicals and Their Scavenger Systems. III. Cellular and Medical Aspects* (R. Greenwald and G. Cohen, eds.), Elsevier, New York, 1983, p. 349.

158. K. C. Bhuyan and D. K. Bhuyan. In *Oxy Radicals and Their Scavenger Systems. III. Cellular and Medical Aspects* (R. Greenwald and G. Cohen, eds.), Elsevier, New York, 1983, p. 347.

159. M. W. Garner, W. H. Garner, and A. Spector. Proc. Natl. Acad. Sci. USA, *80*: 2044 (1983).

160. J. Dillon, M. Garner, D. Roy, and A. Spector. Exp. Eye Res., *34*: 651 (1982).

161. S. Zigler and H. H. Hess. Invest. Ophthal. Vis. Sci. (Suppl.), *24*: 75 (1983).

162. K. R. Hightower, V. Leverenz, and V. N. Reddy. Invest. Ophthalmol. Vis. Sci., *19*: 1059 (1980).

163. K. R. Hightower and V. N. Reddy. Invest. Ophthalmol. Vis. Sci., *22*: 263 (1982).

164. L. Lorand, L. K. H. Hsu, G. E. Siefring, Jr., and N. S. Rafferty. Proc. Natl. Acad. Sci. USA, *78*: 1356 (1981).

165. L. T. Kremzner, D. Roy, and A. Spector. Exp. Eye Res., *37*: 649 (1983).

12

Clinical Implications of Research on Lens and Cataract

LEO T. CHYLACK, JR.*/ Harvard Medical School, Massachusetts Eye and Ear Infirmary, Boston, Massachusetts

HONG-MING CHENG / Harvard Medical School, Boston, Massachusetts

The definition of "cataract" as an opacitication of the crystalline lens causing some degree of visual impairment is generally useful but not altogether satisfactory. It is useful in that it separates lens opacities into two main groups: those with and those without associated visual loss. It is common knowledge among ophthalmologists that many lens opacities occur far from the visual axis and cause no visual difficulty. Some cataracts (e.g., anterior cortical) can occur in the visual axis and still not affect one's visual acuity. However, the term "cataract" usually implies lens opacification with visual loss when used by the medical profession as well as the lay public. The reasons the definition is unsatisfactory are manifold, but revolve about the central issue of the lack of the precision of the term "cataract." There are so many types of cataracts that increased precision is gained only by adding qualifying terms to the term "cataract." Therefore, age-related lens opacification is called congenital, hereditary, or senile cataract. If an etiology is known or suspected, this is also added (i.e., x-ray cataract, steroid-cataract, traumatic cataract, and so on). The type of lens opacification may be specified further in anatomic terms describing the location of the opacity within the lens (i.e., subcapsular, cortical, supranuclear, and/or nuclear). Although not wishing to belabor the fact that no one definition of cataract satisfies everyone, it is important to recognize that the term "cataract" is only the essence of a complex brew of age-related, disease-related, toxin-related, and treatment-related pathogenetic factors. As one reads this book, one can appreciate the need for increasingly precise terminology to describe lens opacification.

---

*Brigham and Women's Hospital, Boston, Massachusetts

CATARACT AS A PERSONAL PROBLEM

The occurrence of congenital cataract in an infant can be associated with little or
no visual disability. In spite of a focal, permanent lens opacification, the patient
may develop perfectly normal visual acuity (1). In fact, many congenital
cataracts are not diagnosed until late in life due to the lack of signs or symptoms
pointing to the presence of a visual or systemic problem (2). Congenital cataracts
may be genetic and nongenetic (acquired in utero) and often are associated with
other ocular abnormalities (i.e., retinopathy, lens dislocation, and corneal
clouding) or systemic abnormalities. They may occur unilaterally or bilaterally;
the former usually have a poor prognosis after surgery, and the latter have a
better surgical prognosis. A recent report by Parks (3) emphasizes the need for
early surgical intervention before 2–3 months of age. Posterior lens capsulectomy and anterior vitrectomy in one or both eyes (a few days apart) with intra-
or postoperative soft contact lens fitting were recommended as the best way to
permit optimal development of the fixation reflex in infants. The critical period
for the development of such a reflex is 2–3 months of age (4–8). Congenital
cataracts may cause from 14 to 40% of childhood blindness, depending on the
country surveyed (2). The surgery is frequently the least stressful of the therapeutic interventions, contact lens or aphakic spectacle use proving to be more of
an ordeal for both parents and patients. Intraocular lens implantation in children
is being evaluated in carefully controlled studies, and if these show success, the
morbidity associated with congenital cataract and its treatment may decrease
significantly.

In the adult with an evolving senile cataract, the first symptom of its presence
may be glare during night driving or glare and blur on a bright, sunny day. If
cataract maturity is greater in one eye than another, depth perception may be
compromised. However, regardless of the initial symptoms, eventually visual
blurring becomes significant when the better eye no longer sees well enough to
allow the patient to perform the daily routine easily. Many patients accept the
diagnosis of cataract and the recommendations for surgical treatment with surprising equanimity. Undoubtedly this is due to the widespread publicity of the
successes of modern cataract surgery with contact lens or artificial lens implantation to correct aphakia (9). However, there are a surprisingly large number of
patients who do not seek or accept surgical treatment of their cataract. In fact,
senile cataract is the third leading cause of legal blindness in the United States
(10,11). The reasons a potentially curable disease causes so much blindness are
unknown. Even though the likelihood of 20/40 or better visual acuity following
primary cataract extraction with intraocular lens implantation is 84.2% (for
patients more than 80 years of age) to 95.5% [for patients less than 60 years of
age (12)] , there are sight-threatening complications in up to 8.0% of cases during
the first year (12). Although some legally blind patients with cataract, well

enough informed to know the likelihood of complications following cataract surgery, decide against surgery, the large majority are not so well informed. To account for the latter group's failure to seek or receive surgery for their cataract, one must consider the possibility that cataract surgery is not available to these patients for financial, geographic, or other reasons.

For the patient electing to go ahead with cataract surgery, there is usually a brief hospitalization (0–3 days), a moderately uncomfortable injection of local anesthesia to endure, a 1 hr operation, and then a 3–12 week convalescence during which physical activity is more or less restricted depending on the type of surgery performed. Postoperative topical medications are tolerated as a necessary nuisance, and even if a contact lens is fitted successfully, the nuisance of fussing with one's eyes persists. Not to minimize the successes of modern cataract surgery do we present this scenario, but to emphasize that the success is achieved only after considerable effort by the surgeon, the patient, and the patient's family. Adding this to the rapidly increasing costs of cataract surgery, it is not difficult to see why a successful medical treatment for cataract might enjoy even greater success than surgery now enjoys.

The rapidity with which intraocular lens implantation has been accepted by physicians and patients has been interpreted as a strongly positive endorsement of this new technology by both groups. The surgeons' enthusiasm is appropriately tempered by surgical complications and the statistics on complication rates generated by the FDA study of intraocular lenses (13). However, there has been no published survey of patients' attitudes preoperatively and postoperatively to this new technology. It would be extremely useful to know if their quality of life has improved, whether they would recommend the procedure to a friend or family member, and what visual or other problems and unfulfilled expectations remain after going through with intraocular lens implantation.

CATARACT AS A PUBLIC HEALTH PROBLEM

A recent publication from the National Institutes of Health (11) contains an up-to-date discussion of the epidemiology of cataract. As mentioned above, the nonuniform definition and classification of cataractous change complicate the interpretation of epidemiologic data. The only incidence (new cases in a period of time) data from a United States population are for congenital cataracts (14) and reveal 0.16 diagnosed cataracts per 1000 live births. There are no comparable data for senile cataract.

Cataract prevalence (total number of cases present at a point in time in a specific area) are derived from the Framingham Eye Study (15) and the National Health and Nutrition Examination Survey (NHANES) (11). Only prevalence data for senile cataract are available.

Table 1 Prevalence (%) of Senile Lens Changes and of Senile Cataract in the Framingham Eye Study

	Age (years)			
	52–64	65–74	75–85	Total
Senile lens changes[a]				
Men	37.9	68.1	88.2	54.1
Women	44.7	76.7	93.0	63.0
Total	41.7	73.2	91.1	59.2
Senile cataract[b]				
Men	4.3	16.0	40.9	13.2
Women	4.7	19.3	48.9	17.1
Total	4.5	18.0	45.9	15.5

[a] Aphakia of senile etiology, early senile lens changes (vacuoles, water clefts, spokes, and lamellar separations), late senile lens changes (cortical cuneiform opacities, nuclear sclerosis, posterior sucapsular opacities, and miscellaneous late senile changes).

[b] Aphakia of senile etiology and late lens changes accompanied by visual acuity of 6/9 (20/30) or worse.

Although these data are believed to be the best available, they were derived from white residents in one locale and, therefore, are not applicable to the U.S. population at large.

The major conclusions cited in (11) are as follows:

1. Cataracts are the second most frequent cause of existing and new cases of registered blindness.
2. Cataract blindness increases with age and is higher among females than males.
3. At least 43,000 persons were legally blind because of cataracts, and at least 4,700 became blind from cataracts in 1980.
4. Cataracts are the major cause of self-reported visual impairment. Over one-third of the existing cases of visual impairment of presumed known cause are attributed to cataracts by respondents. For new cases of visual impairment, this fraction is over one half.
5. No studies have directly measured cataract incidence rates.
6. Prevalence data are available for senile cataracts [Table 1]. In one study, senile aphakia and early and late senile lens changes were present in 42 percent, 73 percent, and 91 percent of persons in age groups 52-64, 65-74 and 75-85 years, respectively. Senile aphakia and late lens changes plus visual acuity limitation were present in 5 percent, 18 percent and 46 percent of persons in similar age groups. Lens changes were more frequent among females than males.

7. Provisional estimates of a national study of persons 1–74 years of age indicate that lens opacities are present in approximately 60 percent of persons 65–74 years of age. About one-fourth of individuals in this age group have decreased vision because of lens opacities.

CATARACT AS A SURGICAL PROBLEM

The surgical approach to senile cataract has evolved over the last century from simple couching, in which the lens is dislocated into the vitreous with a needle-knife, through extracapsular cataract (linear) extraction, to intracapsular cataract extraction in the 1930s. In the early years of extracapsular surgery, the lens was opened with a needle-knife, the wound enlarged, and the nucleus expressed. The loose cortical material was washed from the anterior chamber and the residual cortex was left to absorb gradually. Postoperative complication rates were high, usually due to the inflammatory response to retained lens material ("phacotoxic" or "phacoanaphylactic") or related to glaucoma caused by macrophages glutted with lens material obstructing the outflow channels in the trabecular meshwork (phacolytic glaucoma). It was the ability of intracapsular extraction (in which the entire lens is removed) to avoid these postoperative inflammation-related complications that led to the appeal and widespread use of intracapsular techniques from the 1930s to the mid-1960s. With Kelman's pioneering work in planned extracapsular cataract surgery through a small incision, the new era of extracapsular surgery began (16). His technique of phacoemulsification elegantly combines aspiration and sonication to remove cortex and nucleus of all but the most sclerotic cataracts through a 3.5 mm incision. Practically all cortical material can be removed, and the postoperative inflammation has proved to be less than the eminently successful intracapsular extraction. There was considerable reluctance among American ophthalmologists to accept this new technique of extracapsular surgery; it was based largely on the high cost of the instrumentation (>$25,000), the technical difficulty of doing the procedure correctly, and the serious complications (irreversible corneal edema, nuclear-cortical remnants in the vitreous, and others) when done incorrectly. But, those surgeons, adhering to Kelman's admonishment to prepare oneself carefully to do this technique with ample surgery on animal and eyebank eyes, have enjoyed excellent results. Patients can be fitted with a contact lens 3 weeks after surgery and can resume near-normal activities immediately after surgery. Extracapsular surgery has increased from 14% of all lens extractions in 1979 to 30% in 1981 (13,17,18). It is likely that the percentage is even higher at present.

That Kelman's technique is a major advance is indicated by the number of immitations that have appeared since phacoemulsification was first introduced. Not only are there other sonicators, but the technique of "planned"

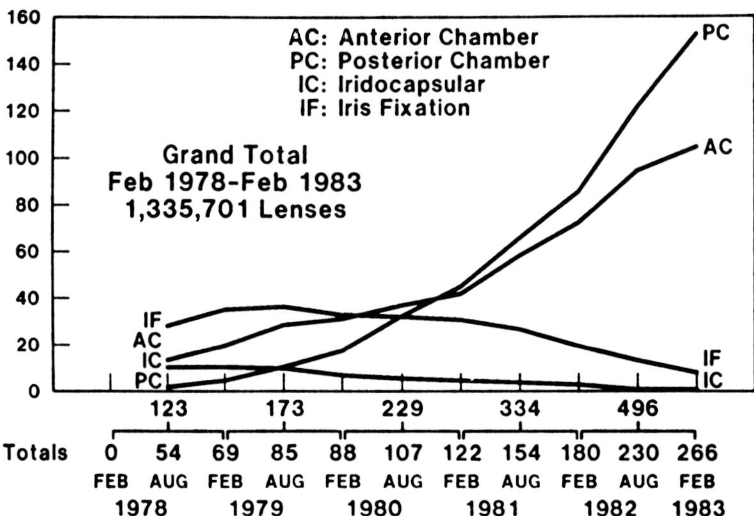

Figure 1 Total number of intraocular lens implants in the United States between February 1978 and February 1983.

extracapsular cataract extraction (PECCE) has regained tremendous popularity in the last few years (19,20). In PECCE the nucleus is expressed, rather than sonicated, and the cortical removal may be done with a series of simple hand-operated aspiration devices (21,22).

The resurgence of extracapsular cataract surgery has to be credited to a number of factors: (1) the reduction in cost and complexity of the instrumentation needed to do the technique, (2) the tremendous success of posterior

Table 2 Final Visual Acuity of 20/40 or Better After IOL Implantation

	Type of IOL[a]							
	AC		IF		IC		PC	
Study	No.	%	No.	%	No.	%	No.	%
Core	4170	83.5	525	80.4	1207	86.5	2524	88.0
Adjunct	18019	83.2	259	83.4	12000	87.7	6839	83.9
Total	22189	83.2	784	81.4	13207	87.6	9361	85.0

[a]Grand total, 45,541 lenses; 84.8%, 20/40 or better. AC, anterior chamber IOL; IF, iris fixation IOL; IC, iridocapsular IOL; and PC, posterior chamber IOL.

Source: From Ref. 12.

Table 3 Incidence of Adverse Reactions: Core Study (8597 Eyes)

Adverse reaction	% Patients
Hypopyon	0.4 (IF, 2.2)[a]
Acute corneal decompensation	0.2
Intraocular infection	0.1
Secondary surgical intervention	
Iridectomy for pupillary block	0.3
Vitreous aspiration for pupillary block	<0.1
Repositioning of lens	0.8 (IF, 2.8)[a]
Lens suturing	0.2
Loop amputation for corneal touch	<0.1
IOL removal for corneal touch	0.1
IOL removal for inflammation	0.1
IOL replacement	0.2
Corneal transplant	0.1

[a]Iris fixation lenses had a significantly greater incidence than other lenses only for hypopyon and repositioning of lens. No other statistically significant difference among different IOL classes was present.

Source: From Ref. 12.

chamber intraocular lens implantation (12), (3) the reduction in postoperative retinal complications (retinal detachment and cystoid macular edema, to name only two) (23), and (4) the development of the YAG (yttrium-aluminum-garnet) laser (24) by which opacified posterior capsules can be opened without anesthesia on an outpatient basis. The development of intraocular lenses has revolutionized treatment of senile cataract. Although this technique was developed in England by Ridley (25) in 1949 and later improved in Holland (26) in the 1950s, it was possible to graduate from one of the finest training programs in the United States in the late 1960s without having performed, observed first-hand, or assisted at a single intraocular lens implantation. Although American ophthalmologists may have been slow off the blocks in the race to implant intraocular lenses, their pace in the later legs of the race has been truly amazing. In 1979, 38.5–39.8% of lens operations involved intraocular lens implantation; in 1981 the figures increased to 81.1–88.3% (13). In 1983, the figure is estimated to be even higher (27).

Originally, lens implants made of polymethyl methacrylate ± platinum were large and heavy (100–200 mg in water); recently, they have been fabricated from polymethyl methacrylate and prolene and are extremely light in weight (10–15 mg in water). Also, the designs have changed significantly; originally rigid anterior chamber lenses or bulky iris clip lenses enjoyed the greatest popularity and success. Now, flexible-loop anterior and posterior chamber lenses are most

Table 4 Core Study Sight-Threatening Complications (%)

	IOL type[a]			
	AC	IF	IC	PC
"Cumulative" (0–12 months)				
Number of eyes	3587	538	1213	2703
Macular edema	8.0%	6.3%	2.8%	3.5%
Secondary glaucoma	5.5	4.3	0.7	1.6
Hyphema	4.9	3.2	2.6	1.0
Lens dislocation	0.2	5.6	1.1	0.4
Pupillary block	0.8	0.6	0.2	0.3
Retinal detachment	0.9	0.4	0.2	0.5
Endophthalmitis	0.1	0.2	0.0	0.0
"Persistent" (at 1 year)				
Number of eyes	4132	538	1213	2465
Macular edema	2.2%	2.4%	0.3%	0.8%
Secondary glaucoma	1.2	0.9	0.1	0.5
Hyphema	0.1	0.0	0.1	0.3
Iritis	1.2	0.9	0.4	1.0
Corneal edema	1.2	1.5	0.6	0.6
Cyclitic membrane	0.1	0.2	0.0	0.0
Vitritis	0.1	0.2	0.1	0.1

[a]IOL, intraocular lens; AC, anterior chamber; IF, iris fixation; IC, iridocapsular; PC, posterior chamber.

Source: From Ref. 12.

popular and successful (12,13). If current trends continue, the posterior chamber intraocular lens will be the most widely used implant in the United States in the next few years (Fig. 1).

The results of intraocular lens implantation have been surveyed by the U.S. Food and Drug Administration and published recently (12). In Table 2 are the numbers of patients and percentages of patients with a final visual acuity of 20/40 or better after anterior chamber (AC), iris fixation (IF), iridocapsular (IC), and posterior chamber (PC) lens implantation. A total of 85.0% (range, 83.9–88.0%) achieved 20/40 or better visual acuity during the 4½ years (February 1978 to August 1982) of the study of 1,088,630 implants. The duration of follow-up (12–14 months) was the same for "core" and "adjunct" patients; the "adjunct" group had three fewer postoperative checks.

The incidence of adverse reactions in core patients is presented in Table 3. The incidence of sight-threatening complications in "core" patients is presented

Table 5 Overall Final Visual Acuity of 20/40 or Better for IOL Patients with Adverse Reactions

Adverse Reaction	No.	%
Hypopyon	27	77.8
Acute corneal decompensation	20	25.0
Intraocular infection	8	37.5
Secondary surgical intervention		
Iridectomy for pupillary block	22	81.8
Vitreous aspiration for pupillary block	4	75.0
Repositioning of lens	55	74.5
Lens suturing	18	88.9
Loop amputation for corneal touch	2	None
IOL removal for corneal touch	6	33.3
IOL removal for inflammation	9	44.4
IOL replacement	4	71.4
Corneal transplant	11	36.4

Source: From Ref. 12.

Table 6 Final Visual Acuity of 20/40 or Better in Group with "Sight-Threatening" Complications

	No.	%
None	2253	88.1
At any time (0–12 months)		
Hyphema	258	82.9
Macular edema	498	65.7
Secondary glaucoma	308	81.8
Pupillary block	39	79.5
Retinal detachment	56	33.9
Vitritis		63.3
Cyclitic membrane	22	50.0
Endophthalmitis		25.0
IOL[a] dislocation	64	82.8
		72.2
At 1 year (persisting)		
Corneal edema	69	43.5
Iritis	78	59.0
Macular edema	135	32.6
		42.5

[a]IOL = intraocular lens.

Source: From Ref. 12.

in Table 4. In Tables 5 and 6 are the number and percentages of patients with adverse reactions or sight-threatening complications and an overall final visual acuity of 20/40 or better.

These data attest to the success of the newest surgical modalities of treatment for senile cataract that is so popular with physicians and patients alike. However, what is popular and successful is not always affordable; to date, no survey of the economic impact of intraocular lens implantation on the overall costs of health care in the United States has been made. The impact of restored vision on the productivity of an individual and of society as a whole is undoubtedly positive; however, the cost at which this benefit is purchased will undoubtedly be assessed and factored into the equation governing the availability of federal and private funding to finance the delivery of this advanced care.

CATARACT AS A MEDICAL THERAPEUTIC PROBLEM

It is curious to note the abundance of allegedly efficacious anti-cataract drugs available throughout the world when cataract surgery is so successful. These drugs enjoy huge sales and yet none has been shown to be clinically effective in humans. A recent excellent comprehensive survey of the medical treatments of cataract has been made by Kador (28). He points out the widespread use of these drugs is based upon the patient's desire for a nonsurgical alternative treatment, the physician's wish to do something for the patient between the time a cataract is diagnosed and is ready for surgical extraction, even though this may be simply a placebo effect, and the aggressive promotion of their products by pharmaceutical houses required by their government to market safe but not necessarily effective drugs. It was Kador's conclusion that only potent aldose reductase inhibitors, drugs that block the conversion of glucose to sorbitol or galactose to galactitol, offer the promise of anti-cataract efficacy in the diabetics.

There are a number of issues surrounding, and to some degree complicating, the development of anti-cataract medications.

Anyone involved in cataract research as well as the care of patients with cataract is aware of the conflict between scientific curiosity about the mechanisms causing and preventing cataract formation and one's clinical commitment to patients to do them no harm. An ophthalmologist must treat cataract-related visual disability as safely and as effectively as possible. As lens science progresses, one senses an increasing optimism about the availability soon of a "nonsurgical treatment for cataract." Is a nonsurgical treatment of cataract a desirable goal when surgical treatment is so safe and successful? It is of the utmost importance that the answer to this question be conceived carefully if scientists and clinicians are to

1. Serve patients effectively and ethically
2. Satisfy scientific curiosity with substantial rather than trivial experimentation
3. Resist commercial pressures that may be applied to condone, if not to actually generate, pseudoscience and scientific fraud.

After all, one is not treating a cataract, but a patient with cataract-related visual disability. The debate about medical versus surgical therapy for cataract must be waged with the patient's welfare uppermost in mind. Risks versus benefits versus costs, and so on, for each treatment modality must be carefully defined. The decision for or against any treatment must be based on evidence gathered as objectively as possible. It is not sufficient to demonstrate that an alleged anti-cataract drug is safe; it must be effective as well. With safety as the sole criterion of marketability, pharmaceutical houses may respond primarily to the commercial appeal of anti-cataract elixirs. Governments throughout the world have not insisted on proof of efficacy, and we have witnessed the large-scale exploitation of the public by pharmaceutical houses marketing every conceivable concoction as a means of delaying cataract formation. The recent association of pharmaceutical companies with the American Cooperative Cataract Research Group (CCRG) gives them an opportunity to demonstrate their commitment to sound scientific methods, and it gives the CCRG an opportunity to assist them in pursuing common goals. In view of these recent developments, it is timely and reasonable to consider the advisability and practicality of seeking a medical means of modifying the natural evolutionary pattern of senile cataract. These comments do not apply to the use of surgeries for the prevention of senile cataracts.

Most research in the field of lens and cataract is problem oriented. The formation of the CCRG in 1976 was a response to the widespread belief that lens research should be directed more closely to solving problems of human cataractogenesis. The impact of senile cataract on our society has been described in terms of incidence and prevalence of legal blindness and significant visual disability. It is the third leading cause behind glaucoma and macular degeneration. In the United States alone, 400,000 cataract extractions are done each year. This figure has been used to imply, not that cataracts are epidemic, but that so many surgical procedures were by themselves an economic and public health problem. The goal of medically curing cataracts was adopted quickly by the lens research community as a desirable alternative to this presumed problem of cataract surgery. Supposedly by avoiding or delaying surgery, one maintains personal productivity (i.e., ability to work and live productively, to support rather than depend on society and ability to pay taxes, and so on), preserves normal visual qualities of life, and reduces or delays medical costs associated with cataract surgery. Although this analysis is simplistic, it is so appealing that one hesitates to reason further and precisely articulate its faults. One fault is the failure to see the different results of delaying versus preventing cataract development. If a medical treatment for cataract will allow a patient to avoid cataract surgery forever, it would be an acceptable alternative to surgery if

1. The rate of significant drug-related adverse reactions was less than 8% (the rate of surgical complications)
2. The cost was significantly less than those of hospitalization for cataract extractions
3. Patient compliance were 100%.

However, if a medical treatment for cataract only postpones the time of surgery, the only beneficiaries of this treatment would be patients who died before surgery was needed and their third-party insurers. For the survivors, the costs of medical therapy would merely be added to those of cataract surgery. Surgical costs would be paid for with inflated dollars; therefore, it is possible to argue that merely delaying the date of cataract extraction would be fiscally counterproductive. Since surgical risks would not be avoided, but added to the risks of medical therapy, there would be an actual increase in the total risk to the patient who undergoes treatment for cataract. The delay of surgery by a few years would have little effect on the net productivity, wages, and taxes paid by an individual, since employment frequently continues to the age of 70 years and surgical convalescence can occur within the usual allowances of sick time leave. There is also reason to doubt that patients would desire a postponement of surgery—no longer is it a dreaded, dreadful experience.

It is possible that medical and surgical therapies of cataract will not prove to be mutually exclusive. If the cost of the medical treatment is very low and the risks of its use negligible, then there would be no reason not to offer this treatment to the patients with cataract. Those patients who do not have access to surgical facilities would benefit from continued use, and those patients not wishing or able to have surgery at a particular time could use medical therapy until surgery was desired. The critical issues are cost, risk, and benefit.

Research to Prevent Blindness has documented the surprising fact that senile cataract, this surgically curable disease, is the third leading cause of legal blindness and significant visual disability in the United States. This may mean that there is a large population of patients with cataracts who may be reluctant to undergo surgery and, therefore, may be enthusiastic about a medical treatment for cataract, or it may mean that these patients do not have access to surgical care. Since the data needed to answer this question are not yet available, let us assume for the moment that these patients are awaiting a medical means of dealing with their cataracts.

Before such a medical treatment can be used on patients in the United States, it must be proven safe and effective to the satisfaction of scientists and the Food and Drug Administration. A large number of problems face the individual charged with the responsibility of designing a protocol to test the efficacy of an alleged anticataract drug.

The natural histories of different cataracts are not known, and if clinical impressions are at all useful, they suggest that only vague estimates of maturation rates can be made for an individual's cataract. Not knowing the natural history of a particular cataract, a clinical investigator does not know when to medically intervene. We know that slit-lamp evaluation of the lens with a dilated pupil will reveal nuclear and cortical opacification before the patient has any symptoms. For this patient without symptoms, a cataract does not exist and for the ophthalmologist, there is no problem commanding therapeutic intervention.

When symptoms appear, they may be uniocular and, therefore, not affecting binocular vision or the ability to cope with the demands of job-related and leisure activities. Still, a problem requiring treatment does not exist; physicians in general are loathe to interfere with a patient's health unless there is a manifest problem. When symptoms worsen to the point that a handicap exists, it may now be too late to introduce medical treatment for cataract. Such treatment should be able to reverse, not merely delay further change, if it is to help such a patient. Many patients with cataract-related symptoms are impatient and seek relief promptly. These patients may not be motivated to use a prolonged medical treatment knowing that surgery is available. Also, since most ophthalmologists are convinced that surgery is the treatment of choice, how are they to be convinced of the merits of testing anti-cataract drugs?

Let us assume that a time for initiating a medical treatment for cataract exists. What type of study should be done? Basically the design is simple: a double-masked, randomized clinical trial involving cataract and control populations with random assignment of drug and placebo. Although conceptually quite simple, the study will be extremely complex. Contributing to this complexity are the following questions.

Which cataracts are to be studied—pure or compound types? If only patients with nuclear, subcapsular, or other single forms of cataract are eligible, approximately 8 of 10 patients may be eliminated. In a study of extracted, classified cataracts (29–32), complex cataracts (combinations of cortical, subcapsular, supranuclear, and nuclear change) predominated. Prevalence data for very early classified cataracts do not yet exist, so the above references may underestimate the prevalence of single-region cataracts.

If complex cataracts are to be studied, a means of classifying and objectively measuring cataractous change in vivo will be needed so that each cataract can be used as its own control (i.e., baseline measurements will be used to assess the effect of treatment or placebo in individual patients). Since presumably there would be comparable numbers of simple and complex cataracts in each treatment group, the classification of the cataract would not be a variable.

How is the drug effect to be measured?

1. Visual acuity
2. Glare sensitivity or contrast sensitivity
3. Lens thickness
4. Red-reflex photos (with Neitz or Zeiss-Kawara camera)
5. Slit-lamp photos (with TOPCON SL-45 camera)
6. Laser light scattering

There are limitations to each of these techniques, but it is fair to say that some objective handle on cataract maturation can now be obtained with current commercially available instrumentation.

How many patients are needed?

1. For cortical and supranuclear cataracts that increase in size by as little as 1.6% every 4 years (33), an infinite number of patients would be needed to demonstrate a statistically significant drug effect.
2. For nuclear and subcapsular cataracts, less than an infinite number of patients would be needed.

Who will do these studies? Are ophthalmologists who believe that surgery is the treatment of choice likely to test a medical treatment for cataract?

Who will finance these studies? Government funding may be given only after a cost-effectiveness analysis is done. If the cost of evaluating and implementing a medical treatment of cataract equals the cost of removing the cataract, why study it, particularly if one is expecting only to delay, not avoid cataract surgery?

Pharmaceutical funding might be provided if the dividend/investment ratio were favorable. Certainly, many drug houses have profited from the worldwide distribution and sale of alleged anti-cataract drugs. Now that more and more countries are requiring proof of efficacy and safety, rather than safety alone, the dividend/investment ratio may drop to unacceptable levels. Simplifying study protocol design would favorably affect this ratio; complicating the protocol would have the opposite effect, raising study (i.e., investment) costs.

How will the risks of a drug be defined without doing a long-term study? Will ophthalmologists test drugs whose risks are unknown, likely to increase with continued use, and involve more than 5% of the treated population?

If a case for the applicability of a medical treatment for cataracts can be made forcefully and convincingly, it must be based on sociologic data documenting the acceptability of such therapy, epidemiologic data documenting the availability of a target population, biostatistical data documenting the feasibility of a study, as well as biochemical data on mechanisms of cataract formation. The CCRG may help to gather these data by establishing broader ties with sociologists, epidemiologists, biostatisticians, and, most importantly, other clinical ophthalmologists. It is this collaborative effort that may carry the advances in the laboratory to the patient in the community.

CATARACT AS A RESEARCH PROBLEM

Many of the issues raised in the preceding section are fundamentally research problems; to that already impressive list must be added some others. Foremost has been the animal orientation of American lens research. In 1976, the Cooperative Cataract Research Group was formed with one of its major objectives to reorient lens research in the United States to place more emphasis on the human than the animal cataract (34). The CCRG received funding from the National Eye Institute as a consortium in 1980 and has achieved considerable success in emphasizing human lens research (35-66). It has reversed the universal trend to discard extracted human cataracts after a cursory pathologic examination. Such cataracts are now photographed, classified in a standardized manner (29-31), and immediately made available to lens scientists throughout the country for research. A major objective of the consortium has been the coordinated and cooperative application of the limited resources of individual lens research laboratories to the unlimited complexity of the problem of human cataract. Through cooperation in protocol development and execution, it is more likely that an understanding of the key steps in the mechanisms of human cataract formation will be gained than if individual laboratories proceeded independently. The four goals of the CCRG—(1) increasing the researcher's supply of human cataracts, (2) developing a means of classifying cataractous changes (29-31), (3) implementing the PROPHET computer system to CCRG data (67,68), and (4) developing a means of pooling and sharing data—have to a large degree been realized. That such progress has been possible since 1978 is in large part due to the ready availability of cataracts extracted intracapsularly. It is ironic that the aforementioned shift from intracapsular to extracapsular cataract extraction, while of such benefit to the patient, is threatening the stability of the CCRG. If the supply of intracapsularly extracted cataracts decreases further, as planned extracapsular (either Phaco or PECCE) surgery increases, the CCRG will have to drastically reduce its research or develop alternative supplies of human lenses. One such alternative is the National Diabetes Research Interchange (NDRI), a very successful organization devoted to the distribution of tissues from human diabetics to investigators throughout the country. Many lenses have been made available to CCRG scientists from the NDRI. Unfortunately, eye banks have only exceptionally proven to be a ready and reliable source of human lenses.

The collaborative efforts of lens scientists have extended beyond the United States. There have been two meetings of American and Japanese scientists working on human cataract to consider ways of furthering international collaboration.

In the first meeting in 1980 (San Francisco), most impressive was the different emphasis in the two countries: in Japan, clinical and epidemiologic research predominated, but in the United States basic cataract research efforts far outnumbered clinical efforts. In the 1983 meeting in Hawaii, each group had well-developed basic and clinical research programs. In the United States, however, there is still a scarcity of ophthalmologists involved in basic or applied clinical research on the mechanisms of cataract formation. Recently (October 1983), a meeting between American and European lens scientists was held at Oakland University in Rochester, Michigan. It is likely that further international cooperation will evolve in the next decade.

That the switch to extracapsular surgery in the United States has hurt the CCRG research effort is understandable; however, this setback has spurred the development of in vivo techniques of human lens study. Such techniques are needed to measure the impact of anti-cataract drugs and have benefited from the thrusts of the international pharmaceutical industries in this area. The development of these techniques could not have occurred at a more opportune time in the history of human cataract research; just as the supply of human lenses is dwindling, in vivo alternatives appear.

A promising invasive technique for providing researchers with samples of lens capsule or epithelium, anterior, equatorial, and posterior cortex has been developed by Horwitz and Strattsma (69). They divert measured samples of aspirate taken at the time of surgery from a hand-operated irrigation-aspiration apparatus. These samples are then analyzed biochemically for soluble and insoluble proteins, as well as other static biochemical constituents. A similar technique was applied to microdissected extracted lenses (70). Nuclei are still available to lens scientists in large numbers, since most extracapsular surgeons do not sonicate (and thereby destroy) the nucleus. In vivo microdissection and sampling offer what may prove to be the best way of studying the biochemical abnormalities associated with regional opacification of the lens. A drawback, however, is the lengthening of operating time required by the refinements of the sampling procedure and the need to have a surgeon in collaboration with a biochemist interested in this type of sample.

There are several noninvasive photographic techniques now in use or about to be used for the study of human cataract. In the 1960s, Niesel (71,72) and Brown (73–75) designed a camera with a tilted film plane (Scheimpflug principle) to allow in-focus imaging from the anterior cornea to the posterior lens. This camera was used to study accommodation and to measure lens growth (74). Photographs of the lens obtained with Brown's camera were far superior to standard slit-lamp lens photographs, because the entire lens was in focus—not just a 1.0–1.5 mm portion in focus in the standard slit-lamp lens photo.

Hockwin and colleagues and TOPCON Instruments worked to improve the clinical applicability of Brown's camera and the fruit of their efforts is the

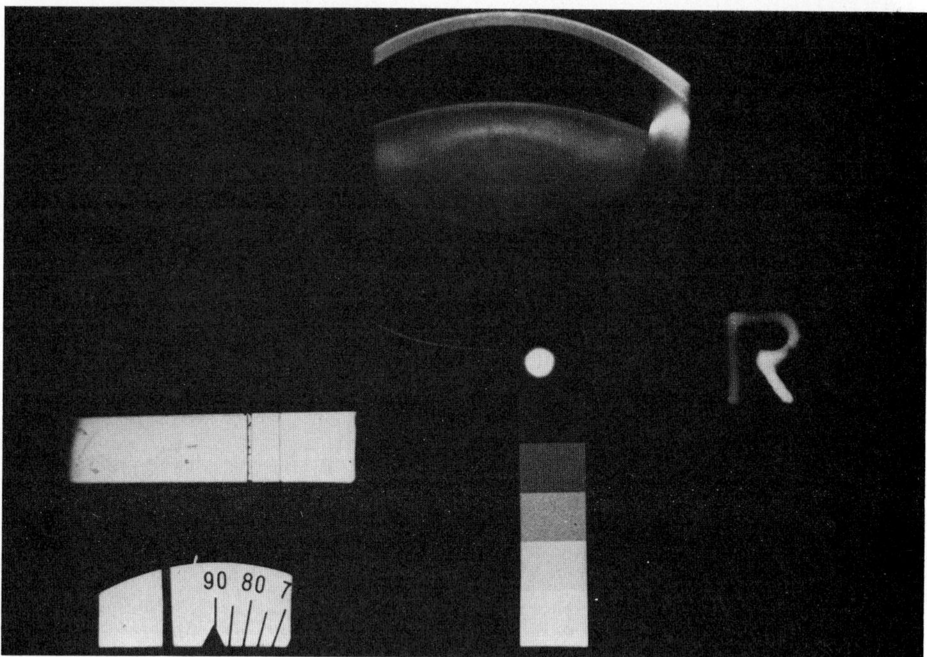

Figure 2 Optical section of the cornea and the lens, taken with the Topcon SL-45 Lens Densitograph. The lens appears clear.

SL-45, a slit-lamp camera that captures on a single 35 mm frame, an in-focus anterior segment slit photo, a step-wedge gray scale, a scale indicating the axis of the slit, and patient-identifying information (76). This camera is designed to allow quantitative microdensitometry of cataract images and has been applied in clinical trials of potential or alleged anti-cataract drugs (77). It is an excellent method for documenting early and moderately advanced nuclear opacification, but does not work as well with advanced nuclear or mixed cortical and nuclear cataracts (Figs. 2 to 4).

The challenge of measuring objectively the type and extent of cortical cataractous change has fascinated several scientists. In contrast to human nuclear cataracts, which possess polar symmetry (slit images of the nucleus rotated about the anterior-posterior pole of the lens are identical), cortical cataracts are devoid of symmetry (Fig. 5).

It is impossible to represent a cortical cataract with a single, slit image. To deal with this, ophthalmic photographers have resorted to retroillumination

Figure 3   Topcon photo showing a dense nuclear cataract.

photography of all forms of cortical cataracts (photography in which the
cataract is illuminated with light reflected from the retina). With conventional
clinical photoslit lamps and fundus cameras, there are a number of artifacts in
such photos: (1) shadowing of half of the photo, (2) visible flash lamp
filaments, and (3) Maltese cross, shadow patterns if crossed polarizers are used to
eliminate reflex from the lamp filament. During the past decade, considerable
progress has been made in eliminating these artifacts; Kawara and Obazawa (78)
developed an optical device that places crossed polarizing and orange filters in
the path of the illumination beam of a conventional, Zeiss photoslit lamp.
Photographs taken with this device are free of shadow and filament artifacts and
are amenable to quantitative analysis. Sasaki et al. (79) and Kawara et al. (80)
have used image analysis of digitized Zeiss-Kawara photos to measure the
maturation of a cataract with time. Maclean and Taylor (33) achieved the same
type of photograph by combining the unshadowed portions of each member of a
stereo pair taken with a conventional photoslit lamp. He documented the surprisingly slow growth of human cortical cataracts: 1.6% mean increase in area
over 4½ years. The latest development in retroillumination photography is the

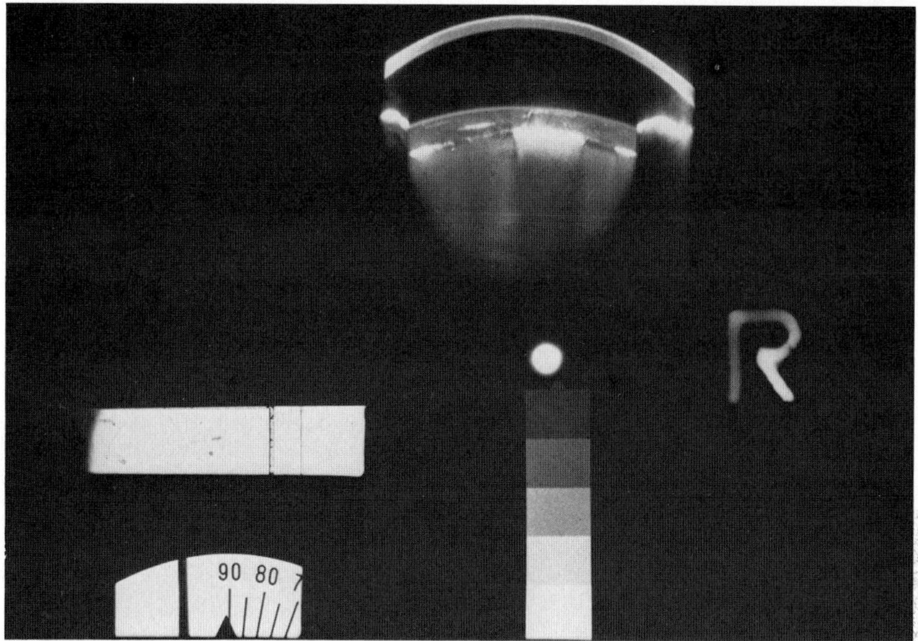

Figure 4   Topcon photo showing anterior cortical cataracts.

Neitz Cataract Camera, which combines Kawara optics with other advances to produce an affordable, easy-to-use camera that delivers artifact-free retro-illumination photos of the human cataractous lens (see Fig. 6).

Considerable progress has been made in solving the problems associated with computerized analysis of photographic images of human cataracts (81).

Other noninvasive techniques for measuring the extent and/or density of human cataractous changes are listed below:

1. Light scattering (37,38,82,83)
2. Quantitative biomicroscopy of light back-scattered from the lens (84)
3. Tyndall photometry (85,86)
4. Specular microscopy of the human lens (87)
5. Quasi-elastic laser light scattering (88–90)
6. Arden grating measurement of contrast sensitivity (91,92)
7. Fluorescence, UV, Raman, and visible spectroscopy (93–96)

It is reasonable to conclude that one or a combination of the above-mentioned techniques may be applied successfully to studies of (1) the natural history of

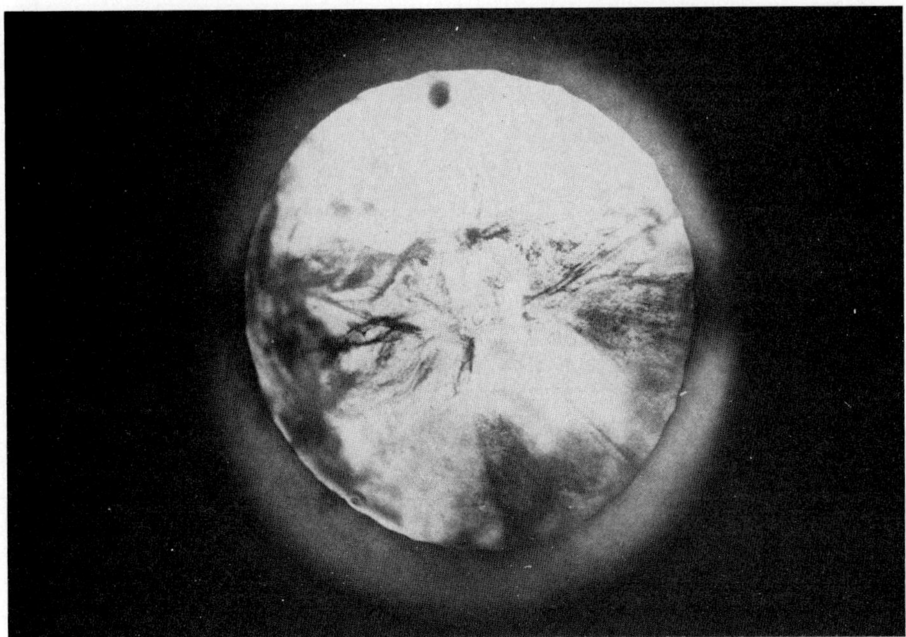

Figure 5   Cortical cataracts photographed with the Zeiss-Kawara camera.

certain forms of human cataract, (2) the efficacy of anticataract drugs, and
(3) the cataractogenic potential of drugs, foods or other environmental risk
factors.

Of great interest is the recent application of biophysical techniques to the
study of human lens metabolism in situ. Raman spectroscopy has recently been
applied successfully to the study of the in situ lens (96,97). This technique may
provide information about the secondary and tertiary structure of lens proteins,
as well as the concentration of sulfhydryl groups in different regions of the lens.
The application of nuclear magnetic resonance (NMR) spectroscopy to living
systems has recently been reviewed (98), and several publications attesting to the
utility of this technique in the study of the lens have recently appeared
(99–101). NMR spectroscopy in vivo may be most useful in detecting characteristic precataractous metabolic changes. For example, it may be possible to do
$[^{31}P]$NMR imaging of the lens and know precisely what metabolic status the
lens is in, or do proton imaging of the lens and know the state and distribution
of water and lipid in the lens, or do electron spin resonance (ESR) scans to see if
free radicals are formed and removed at acceptable rates. Any deviation from the

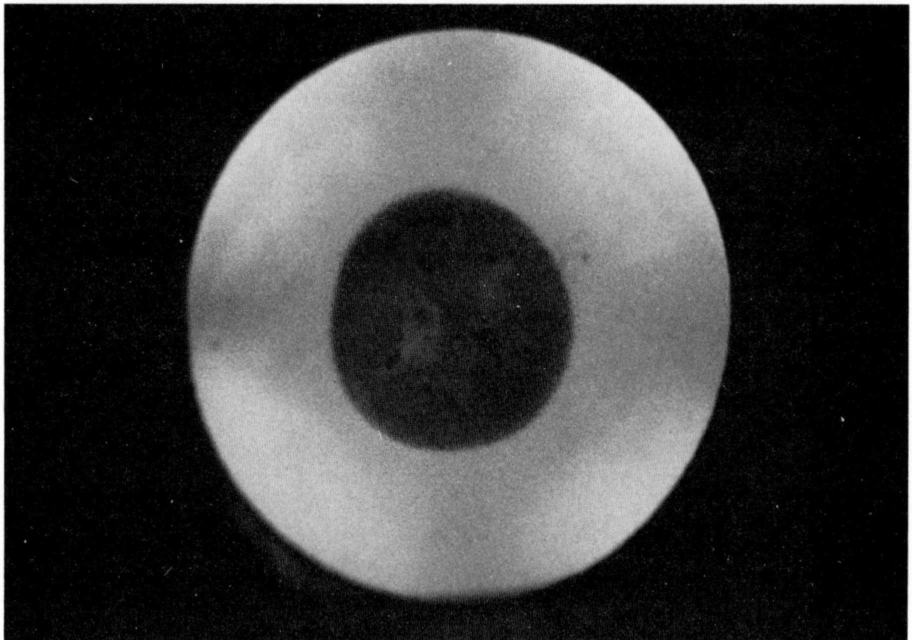

Figure 6 Congenital nuclear cataract photographed with the Neitz Cataract Camera.

norm should alert the clinician and the patient to the possibility of eventual cataract formation and cause them to consider initiating anti-cataract medical therapy.

CONCLUSIONS

Just as the past decade has witnessed a revolution in the surgical treatment of cataract, we believe that the next decade will see a similar revolution in our understanding of the mechanisms of cataractogenesis. Whether this understanding extends our ability to treat cataract medically remains to be seen. It is certain that the technological innovations in the areas of objective documentation of cataractous change and the renewed dedication of lens scientists throughout the world to the study of human cataract will enable promising anti-cataract drugs to be tested in a rigorous, scientifically sound manner. The dedication of the laboratory scientist and the clinical researcher to the goals of understanding and treating human cataract appears to have paid off. It remains to be seen whether additional research by epidemiologists, sociologists, biostatisticans, and clinical ophthalmologists will receive the support needed to answer the question of whether the delivery of medical treatment for cataract is feasible and affordable.

REFERENCES

1. G. R. Fraser and A. I. Friedman. *The Causes of Blindness in Childhood*, Baltimore, Johns Hopkins Press, 1967.
2. B. Penzani. Osserwazioni cliniche e studio statistico sula recita infantile. Am. Oh. Clin. Ocul., *80*: 189 (1954).
3. M. M. Parks. Posterior lens capsulectomy during primary cataract surgery in children. Ophthalmology, *90*: 344 (1983).
4. T. D. Vargan. Critical period deprivation for amblyopia in children. Trans. Ophthalmol. Soc. UK, *99*: 432 (1979).
5. G. L. Rogers, C. L. Tishler, B. H. Tsou, R. W. Hertle, and R. R. Fellows. Visual acuities in infants with congenital cataracts operated on prior to 6 months of age. Arch. Ophthalmol., *99*: 999 (1981).
6. R. Beller, C. S. Hayt, E. Mang, and J. V. Odorn. Good visual function after neonatal surgery for congenital monocular cataracts. Am. J. Ophthalmol., *91*: 559 (1981).
7. S. S. Gelbart, C. S. Hayt, G. Jastebski, and E. Mang. Long-term visual results in bilateral congenital cataracts. Am. J. Ophthalmol., *93*: 615 (1982).
8. M. M. Parks. Visual results in aphakic children. Am. J. Ophthalmol., *94*: 441 (1982).
9. C. D. Binkhorst and M. H. Gobin. Treatment of congenital and juvenile cataract with intraocular lens implants (pseudophakos). Br. J. Ophthalmol., *54*: 959 (1970).
10. National Society to Prevent Blindness. *Vision Problems in the U.S.*, New York, 1980.
11. Report of the Cataract Panel. Volume Two/Part Three, *Vision Research*, A National Plan 1983–1987, NIH Publ. No. 83-2473, U.S. DHHS. Public Health Service, NIH, 1983, p. 1.
12. W. J. Stark, D. M. Worthen, J. T. Holladay, et al. FDA Report on intra-ocular lenses. Ophthalmology, *90*: 311 (1983).
13. W. J. Stark, M. C. Leske, D. M. Worthen, and G. C. Murray. Trends in cataract surgery and intraocular lenses in the United States. Am. J. Ophthalmol., *96*: 304 (1983).
14. *Congenital Malformations Surveillance Report*, July 1978–June 1979. U.S. Department of Health and Human Services, Centers for Disease Control, 1980.
15. H. A. Kahn, H. M. Leibowitz, J. P. Ganley, et al. The Framingham Eye Study. I. Outline and major prevalence findings. Am. J. Epidemiol., *106*: 17 (1977).

16. C. D. Kelman. Phaco-emulsification and aspiration. A new technique of cataract removal. A preliminary report. Am. J. Ophthalmol., *64*: 23 (1967).
17. The Hospital Record Study, 1979–1982. Ann Arbor. The Commission on Professional and Hospital activities, pp. 10 and 11.
18. Detailed Diagnosis and Surgical Procedures for Patients Discharged from Short-Stay Hospitals, United States, 1979. Hyattsville, Maryland, U.S. Department of Health and Human Services, Publication Number PHS 82-1274, 1982, p. 268.
19. C. D. Kelman. Symposium: Phacoemulsification. History of emulsification and aspiration of senile cataracts. Trans. Am. Acad. Ophthalmol. Otolaryngol., *78*: OP5 (1974).
20. J. M. and J. H. Little (eds.). *Phacoemulsification and Aspiration of Cataracts*, C. V. Mosby, St. Louis, 1979.
21. D. J. McIntyre. The coaxial cannula connector system. Cont. Intraoc. Lens Med. J., *2*: 50 (1976).
22. J. M. Emery and D. J. McIntyre (eds.), *Extracapsular Cataract Surgery*, C. V. Mosby, St. Louis, 1983, p. 285.
23. N. S. Jaffe, S. M. Luscombe, H. M. Clayman, and J. D. Gass. A fluorescein angiographic study of cystoid macular edema. Am. J. Ophthalmol., *92*: 775 (1981).
24. D. Aron-Rosa, J. C. Grieseman, and J. J. Aron. Use of a pulsed neodymium Yag laser (picosecond) to open the posterior lens capsule in traumatic cataract. A preliminary report. Ophthalmic Surg., *12*: 496 (1981).
25. H. Ridley. Intra-ocular acrylic lenses. A recent development in surgery of cataract. Br. J. Ophthalmol., *36*: 113 (1952).
26. C. D. Binkhorst. De implantatie van kunststoflenzen in het oog. Een nieuwe fixatiemethode: De pupillens of iris-cliplens. Ned. Tijdschr. Geneeskd., *103*: 1289 (1959).
27. W. J. Stark. Unpublished estimates (1984).
28. P. F. Kador. Overview of the current attempts toward the medical treatment of cataract. Ophthalmology, *90*: 352 (1983) (141 references).
29. L. T. Chylack, Jr. Classification of human cataracts. Arch. Ophthalmol., *96*: 888 (1978).
30. L. T. Chylack, Jr., M. R. Lee, W. H. Tung, and H. M. Cheng. Classification of human senile cataractous change by the American Cooperative Cataract Research Group (CCRG) Method. I. Instrumentation and technique. Invest. Ophthalmol. Vis. Sci., *24*: 424 (1983).
31. L. T. Chylack, Jr., O. White, R. K. Neff, and W. H. Tung. Classification of human senile cataractous change by the American Cooperative Cataract Research Group (CCRG) method. II. Stages of simplification of cataract classification. Invest. Ophthalmol. Vis. Sci., *25*: 166 (1984).
32. L. T. Chylack, Jr., B. J. Ransil, and O. White. Classification of human senile cataractous change by the American Cooperative Cataract Research Group (CCRG) method. III. The association of nuclear color (sclerosis) with extent of cataract formation, age and visual acuity. Invest. Ophthalmol. Vis. Sci., *25*: 174 (1984).

33. H. Maclean and C. J. Taylor. An objective staging for cortical cataract in vivo aided by pattern-analyzing computer. Exp. Eye Res., *33*: 597 (1981).

34. L. T. Chylack, Jr., and J. H. Kinoshita. The Cooperative Cataract Research Group. Invest. Ophthalmol. Vis. Sci., *17*: 1131 (1978).

35. L. T. Chylack, Jr., H. F. Henriques, and W. H. Tung. Inhibition of sorbitol production in human lenses by an aldose reductase inhibitor. Doc. Ophthalmol. Proc. Ser., *18*: 65 (1979).

36. J. A. Jedziniak, L. T. Chylack, Jr., H. M. Cheng, M. K. Gillis, A. A. Kalustian, and W. H. Tung. The sorbitol pathway in the human lens: Aldose reductase and polyol dehydrogenase. Invest. Ophthalmol. Vis. Sci., *20*: 314 (1981).

37. L. T. Chylack, Jr., F. A. Bettelheim, and W. H. Tung. Studies on human cataracts. I. Evaluation of techniques of human cataract preservation after extraction. Invest. Ophthalmol. Vis. Sci., *20*: 327 (1981).

38. E. L. Siew, F. A. Bettelheim, L. T. Chylack, Jr., and W. H. Tung. Studies on human cataracts. II. Correlation between the clinical description and the light scattering parameters of human cataracts. Invest. Ophthalmol. Vis. Sci., *20*: 334 (1981).

39. F. A. Bettelheim, E. L. Siew, and L. T. Chylack, Jr. Studies on human cataracts. III. Structural elements in nuclear cataracts and their contribution to the turbidity. Invest. Ophthalmol. Vis. Sci., *20*: 348 (1981).

40. L. T. Chylack, Jr., H. F. Henriques, H. M. Cheng, and W. H. Tung. Efficacy of Alrestatin, an aldose reductase inhibitor, in human diabetic and non-diabetic lenses. Ophthalmology, *86*: 1579 (1979).

41. P. F. Kador, J. H. Kinoshita, W. H. Tung, and L. T. Chylack, Jr. Differences in the susceptibility of various aldose reductases in inhibition. II. Invest. Ophthalmol. Vis. Sci., *19*: 980 (1980).

42. C. V. Harding, L. T. Chylack, Jr., A. R. Susan, J. G. Decker, and W. K. Lo. Morphological changes in the cataract: The ultrastructure of human lens opacities, localized by Cooperative Cataract Research Group procedures. In *Red Blood Cells and Lens Metabolism* (S. K. Srivastava, ed.), Elsevier-North Holland, Amsterdam, 1980, p. 27.

43. H. M. Cheng, L. T. Chylack, Jr., and I. von Saltza. Supplementing glucose metabolism in human senile cataracts. Invest. Ophthalmol. Vis. Sci., *21*: 812 (1981).

44. H. M. Cheng and L. T. Chylack, Jr. Thiol oxidation in the crystalline lens. I. The rate-limiting role of hexokinase in aging and human lenses. Invest. Ophthalmol. Vis. Sci., *19*: 522 (1980).

45. H. M. Cheng, L. T. Chylack, Jr., C. N. Sang, N. Orzalesi, and G. Corongiu. GSSG-reducing activity in lenses deficient in glucose-6-phosphate dehydrogenase. Metab. Pediatr. Syst. Ophthalmol., *7*: 53 (1983).

46. J. A. Jedziniak and J. Rokita. Aldehyde metabolism in the human lens. Exp. Eye Res., *37*: 119 (1983).

47. C. V. Harding, L. T. Chylack, Jr., S. R. Susan, W. K. Lo, and W. F. Bobrowski. Elemental and ultrastructural analysis of specific human lens opacities. Invest. Ophthalmol. Vis. Sci., *23*: 1 (1982).

48.　F. A. Bettelheim, E. L. Siew, L. T. Chylack, Jr., and J. H. Seland. The effect of freezing on human cortical cataracts. Invest. Ophthalmol. Vis. Sci., *24*: 403 (1983).

49.　G. L. Feldman, L. S. Feldman, and G. Rouser. The isolation and partial characterization of gangliosides and ceramide polyhexosides from the lens of the human eye. Lipids, *1*: 21 (1966).

50.　A. Spector, M. H. Garner, W. H. Garner, D. Roy, P. N. Farnsworth, and S. Shyne. An extrinsic membrane polypeptide associated with high molecular weight aggregates in human cataract. Science, *204*: 1323 (1979).

51.　M. H. Garner, D. Roy, and A. Spector. Immunochemical characterization of the main intrinsic proteins on the human lens membrane. Exp. Eye Res., *34*: 781 (1982).

52.　L. J. Takemoto and J. S. Hansen. Intermolecular disulfide bonding of lens membrane-proteins during human cataractogenesis. Invest. Ophthalmol. Vis. Sci., *22*: 336 (1982).

53.　P. N. Farnsworth, S. Shyne, P. A. Burke, and A. V. Fasano. Cellular maturation of the human lens fibers. In *Red Blood Cell and Lens Metabolism* (S. K. Srivastava, ed.), Elsevier-North Holland, Amsterdam, 1980, p. 45.

54.　A. Spector, M. W. Garner, D. Roy, W. H. Garner, P. N. Farnsworth, and S. Shyne. Oxidation of lens proteins. In *Red Blood Cell and Lens Metabolism* (S. K. Srivastava, ed), Elsevier-North Holland, Amsterdam, 1980, p. 81.

55.　P. N. Farnsworth, S. Shyne, S. J. Caputo, A. V. Fasano, and A. Spector. The microtubule: A major cytoskeletal component of the human lens. Exp. Eye Res., *30*: 611 (1980).

56.　P. N. Farnsworth, S. Shyne, and P. Burke. Lens fiber differentiation. In *Fourth International Symposium on the Structure of the Eye*, Guadalajara, Mexico, 1980.

57.　F. A. Bettelheim, E. L. Siew, S. Shyne, P. Farnsworth, and P. Burke. A comparative study of human lens by light scattering and scanning electron microscopy. Exp. Eye Res., *32*: 125 (1981).

58.　W. K. Lo, C. V. Harding, and H. Maisel. Gap junctions in the human cataractous lens. J. Cell Biol., *87*: 56A (1980).

59.　S. Lerman, J. Megaw, and K. Gardner. P-UVA therapy and human cataractogenesis. Invest. Ophthalmol. Vis. Sci., *23*: 801 (1982).

60.　S. Lerman, J. Megaw, K. Gardner, R. Long, D. Ashley, and J. H. Goldstein. An NMR analysis of aging changes in human lens proteins. Invest. Ophthalmol. Vis. Sci., *23*: 218 (1982).

61.　V. N. Reddy, F. J. Giblin, and H. Matsuda. Defense system of the lens against oxidative damage. In *Red Blood Cell and Lens Metabolism* (S. K. Srivastava, ed.), Elsevier-North Holland, Amsterdam, 1980, p. 138.

62.　K. R. Hightower and V. N. Reddy. Calcium content and distribution in human cataract. Exp. Eye Res., *34*: 413 (1982).

63.　W. H. Garner, S. R. Leff, and S. Spector. Tritium incorporation into the major intrinsic membrane polypeptides of normal human lenses: 26,000 and 22,000 dalton species. In *Red Blood Cell and Lens Metabolism* (S. K. Srivastava, ed.), Elsevier-North Holland, Amsterdam, 1980, p. 367.

64. A. Spector and M. H. Garner. Interaction of human cataract fiber cell membrane polypeptides with cytoplasmic components. In *Red Blood Cell and Lens Metabolism* (S. K. Srivastava, ed.), Elsevier-North Holland, Amsterdam, 1980, p. 233.

65. M. Garner and A. Spector. Sulfur oxidation in selected human cortical cataracts and nuclear cataracts. Exp. Eye Res., *31*: 361 (1980).

66. S. Zigman, M. Datiles, and E. Torczynski. Sunlight and human cataracts. Invest. Ophthalmol. Vis. Sci., *18*: 462 (1979).

67. *PROPHET User's Manual*: Prepared by Bolt Beranek and Newman, Inc.: Cambridge, Mass. April (1982), Section 2–9.

68. *PROPHET Statistics Manual*: Prepared by Bolt Beranek and Newman, Inc.: Cambridge, Mass. April (1982).

69. J. Horwitz and B. Straatsma. Unpublished material presented at U.S. CCRG-EURAGE meeting, Oakland University, Rochester, Michigan, October 1983.

70. L. J. Takemoto, J. S. Hansen, and J. Horwitz. Biochemical analysis of microdissected sections from the normal and cataractous human lens. Curr. Eye Res., *2*: 443 (1982).

71. P. Niesel. Spaltlampenphotographic mit der Haag-Streit Spaltlampe 900. Ophthalmologica, *151*: 489 (1966).

72. P. Niesel. Spaltlampenphotographic der linse fur MeBzwecks. Schmeiz Ophthal. Ges. 58. Vero., Solothurn 1965. Ophthalmologica, *152*: 387 (1966).

73. N. Brown. Slit image photography. Trans. Ophthalmol. Soc. UK, *89*: 397 (1969).

74. N. Brown. Slit image photography and measurements of the eye. Med. Biol. Illust., *23*: 192 (1973).

75. N. Brown. Lens changes with age and cataract; slit image photography. In *The Human Lens in Relation to Cataract*, Ciba Foundation Symposium 19, Elsevier, Exerpta Medica/North Holland, Amsterdam, 1973, p. 65.

76. O. Hockwin, V. Dragomirescu, H. R. Koch, and K. Sasaki. Followup method for the documentation of lens opacities with a new photographic equipment (TOPCON lens densitography). XXIII. Intern. Congr. Ophthalm., Kyoto/Japan, Abstr. Exerpta Medica ICS No. 442 (1982) (exhibit).

77. E. Weigelin and O. Hockwin. Rapport sur l'etude clinique controlee randomisee de Phakan/Phakolen. Symposium International sur le Cristallin, Strasbourg, 1982. Diffusion: Laboratories Chauvin-Blanche-B.P. 1174-34009. Montpellier, Cedex-France.

78. T. Kawara and H. Obazawa. A new method for retroillumination photography of cataractous lens opacities. Am. J. Ophthalmol., *90*: 186 (1980).

79. K. Sasaki, T. Shibata, M. Hiiragi, and Y. Sakamoto. Documentation of coloration of crystalline lens in vivo. Jpn. J. Clin. Ophthalmol. (Rinsho Ganko), *37*: 832 (1983).

80. T. Kawara, H. Obazawa, R. Nakano, M. Sasaki, and T. Sakata. Quantitative evaluation of cataractous lens opacities with retroillumination photography. Jpn. J. Clin. Ophthalmol. (Rinsho Ganka), *33*: 21 (1979).

81. L. T. Chylack, Jr., L. Sher, and M. Adler. Objective documentation of human cataractous change in vivo with photography and computerized image analysis (manuscript in preparation).

82. F. A. Bettelheim. Syneresis and its possible role in cataractogenesis. Exp. Eye Res., *28*: 189 (1979).

83. L. T. Chylack, Jr., F. A. Bettelheim, and W. H. Tung. Studies on human cataracts. I. Evaluation of techniques of human cataract preservation after extraction. Invest. Ophthalmol. Vis. Sci., *20*: 327 (1981).

84. J. Sigelman, S. L. Trokel, and A. Spector. Quantitative biomicroscopy of lens light back scatter, changes with aging and opacification. Arch. Ophthalmol., *92*: 437 (1974).

85. M. Blumenthal, J. Bodenheimer, A. Kushelensky, and L. Rothkoff. New Tyndall photometer. Arch. Ophthalmol., *95*: 323 (1977).

86. I. Ben-Sira, D. Weinberger, J. Bodenheimer, and Y. Yasseu. Clinical method for measurement of light backscattering from the in vivo human lens. Invest. Ophthalmol. Vis. Sci., *19*: 435 (1980).

87. A. J. Bron and K. Matsuda. Specular microscopy of the human lens. Trans. Ophthalmol. Soc. UK, *101*: 163 (1981).

88. T. Tanaka and G. B. Benedek. Observation of protein diffusivity in intact human bovine lenses with application to cataract. Invest. Ophthalmol. Vis. Sci., *14*: 449 (1975).

89. T. Tanaka and C. Ishimoto. In-vivo observation of protein diffusivity in rabbit lenses. Invest. Ophthalmol. Vis. Sci., *16*: 135 (1977).

90. J. N. Weiss, I. Nishio, J. I. Clark, T. Tanaka, G. B. Benedek, F. J. Giblin, and V. N. Reddy. Early detection of cataractogenesis by laser light scattering. Invest. Ophthalmol. Vis. Sci. (ARVO abstract) (1982).

91. H. Singh, R. L. Cooper, V. A. Alder, G. J. Crawford, A. Tevell, and I. Constable. The Arden grating acuity: Effect of age and optical factors in the normal patient, with prediction of the false negative rate in screening for glaucoma. Br. J. Ophthalmol., *65*: 518 (1981).

92. H. W. Skalka. Arden grating test in evaluating "early" posterior subcapsular cataracts. South. Med. J., *74*: 1368 (1981).

93. S. Lerman and R. Borkman. Spectroscopic evaluation and classification of the normal, aging and cataractous lens. Ophthalmol. Res., *8*: 335 (1976).

94. S. Lerman, O. Hockwin, and V. Dragomirescu. In vivo lens fluorescence photography. Ophthalmic Res., *13*: 224 (1981).

95. S. Lerman and O. Hockwin. UV-visible slit lamp densitography of the human eye. Exp. Eye Res., *33*: 587 (1981).

96. N. T. Yu, J. F. R. Kuck, Jr., and C. C. Askreu. Laser Raman spectroscopy of the lens in situ, measured in an anesthetized rabbit. Curr. Eye Res., *1*: 615 (1982).

97. A. Mizuno, Y. Ozaki, Y. Kamada, H. Myazaki, K. Itoh, and K. Iriyama. Direct measurement of Raman spectra of intact lens in a whole eyeball. Curr. Eye Res., *1*: 609 (1981).

98. C. T. Burt. Minireview. NMR of live systems. Life Sci., *31*: 2793 (1982).

99. P. Racz, K. Tompa, and I. Pocsic. The state of water in normal and senile cataractous lenses studied by nuclear magnetic resonance. Exp. Eye Res., *28*: 129 (1979).

100. J. V. Greiner, S. J. Kopp, D. R. Sanders, and T. Glonek. Dynamic changes in the organophosphate profile of the experimental galactose-induced cataract. Invest. Ophthalmol. Vis. Sci., *22*: 613 (1982).

101. R. G. Gonzalez, J. Willis, J. Aguayo, P. Campbell, L. T. Chylack, Jr., and T. Schleich. ^{13}C-Nuclear magnetic resonance studies of sugar cataractogenesis in the single rabbit lens. Invest. Ophthalmol. Vis. Sci., *22*: 808 (1982).

Index